Lecture Notes in Computer Science   10575

Commenced Publication in 1973
Founding and Former Series Editors:
Gerhard Goos, Juris Hartmanis, and Jan van Leeuwen

Stefan Rass · Bo An
Christopher Kiekintveld · Fei Fang
Stefan Schauer (Eds.)

Decision and Game Theory for Security

8th International Conference, GameSec 2017
Vienna, Austria, October 23–25, 2017
Proceedings

 Springer

Editors
Stefan Rass 🆔
Universität Klagenfurt
Klagenfurt
Austria

Bo An 🆔
Nanyang Technological University
Singapore
Singapore

Christopher Kiekintveld 🆔
University of Texas at El Paso
El Paso, TX
USA

Fei Fang 🆔
Carnegie Mellon University
Pittsburgh, PA
USA

Stefan Schauer 🆔
AIT Austrian Institute of Technology GmbH
Klagenfurt
Austria

ISSN 0302-9743 ISSN 1611-3349 (electronic)
Lecture Notes in Computer Science
ISBN 978-3-319-68710-0 ISBN 978-3-319-68711-7 (eBook)
DOI 10.1007/978-3-319-68711-7

Library of Congress Control Number: 2017955778

LNCS Sublibrary: SL4 – Security and Cryptology

Printed on acid-free paper

This Springer imprint is published by Springer Nature
The registered company is Springer International Publishing AG
The registered company address is: Gewerbestrasse 11, 6330 Cham, Switzerland

Preface

Contemporary information and communication technology evolves fast not only in terms of sophistication, but also in diversity. The increasing complexity, pervasiveness, and connectivity of today's information systems raises new challenges to security, and cyberspace has become a playground for people with all levels of skills and all kinds of intention (positive and negative). With 24/7 connectivity having become an integral part of people's daily life, protecting information, identities, and assets has gained more importance than ever. While oil and coal have been the most important commodities in past centuries, information is the commodity of the twenty-first century, and cyber-warfare is widely about gaining the most of the resource "information," as much as past decades have seen wars for land or wealth.

Traditional security has successfully accomplished a long way toward protecting well-defined goals like confidentiality, integrity, availability, and authenticity (CIA+). However, the term "security" has evolved into meaning much more than CIA+ these days. The Internet is surely an indispensable supporting infrastructure, but also an equally rich source of threats. Around the beginning of the new millennium, a paradigm extension in the field can be observed, with the first scientific considerations on how game theory can be used for security. Although the situation between an attacker and a defender being the most natural incarnation of non-cooperative competition, it comes somewhat as a surprise that it took until the new millennium for the first scientific work on game theory applied to security. Ever since then, interest in the field has grown rapidly, and game theory and decision theory have become a systematic and well-proven powerful fundament of today's security research. Indeed, while conventional security aims at preventing an anticipated set of forbidden actions that make up the respective security model, game theory and decision theory take a different and more economic viewpoint: Security is not the absence of threats, but the point where attacking a system has become more expensive than not attacking. Starting from a game and decision theoretic root thus achieves the most elegant form of security, by analyzing and creating incentives to actively encourage honest behavior rather than preventing maliciousness. At the same time, the economic approach to security is essential as it parallels the evolution of today's attackers. Cybercrime has grown into a full-featured economy, maintaining black markets, supply chains, and widely resembling an illegal counterpart of the official software market. Traditional security remains an important fundament for tackling the issue from below, but game- and decision theory offer the top–down view by adopting the economic and strategic view of the attackers too, and as such complements purely technological security means.

The optimum is, of course, achieved when both routes are taken toward meeting in the middle, and this is what the GameSec conference series initiated in 2010 in Berlin, Germany. It brings together internationally recognized researchers from the security field, optimization, economics, and statistics, to discuss challenges and advance solutions to contemporary security issues. Following the success of this first scientific event of its kind, subsequent conferences were organized in College Park Maryland (USA, 2011), Budapest (Hungary, 2012), Fort Worth Texas (USA, 2013), Los Angeles (USA, 2014), London (UK, 2015), New York (USA, 2016), and this year in Vienna, Austria, during October 23–25.

In all these years, GameSec has showcased a continuously increasing number of novel, high-quality theoretical and practical contributions to address issues like privacy, trust, infrastructure security, green security, and many more, and densely connected a scientific community of experts all over the globe and from various fields of computer science, economics, and mathematics, under the common goal of security. This year continued this tradition, and we are proud to present a new set of high-quality scientific contributions to advance security. The program of GameSec 2017 featured 28 full papers, selected from a total of 71 submissions, based on three reviews per paper. Submissions were received from all over the world, which underpins the global relevance of security and the methods pursued by the community. In addition, a special track on "Data-Centric Models and Approaches" was introduced in recognition of the problem of gathering and analyzing data about security incidents. Companies and security agencies may be reluctant in releasing such information to protect their reputation or the targets of attack. The special track's focus was thus on gathering data and building models from it, and as such contributed to closing this gap between theory and practice.

We would like to thank the Austrian Institute of Technology for hosting this year's event, and we also thank Springer for its continuous support of the conference series, by publishing this book as part of the *Lecture Notes in Computer Science* (LNCS) series. We hope that you enjoy reading as much as we enjoyed compiling this volume. Let us together take this step toward the next level of security!

October 2017   Stefan Rass
                                                                    Bo An
                                                    Christopher Kiekintveld
                                                           Stefan Schauer
                                                                 Fei Fang

Organization

Steering Committee

Tansu Alpcan   University of Melbourne, Australia
Nick Bambos   Stanford University, USA
John S. Baras   University of Maryland, USA
Tamer Başar   University of Illinois at Urbana-Champaign, USA
Anthony Ephremides   University of Maryland, USA
Jean-Pierre Hubaux   EPFL, Switzerland
Milind Tambe   University of Southern California, USA

Organizers

General Chair

Stefan Rass   Universität Klagenfurt, Austria

Technical Program Committee Chairs

Bo An   Nanyang Technological University, Singapore
Christopher Kiekintveld   University of Texas at El Paso, USA

Special Track Chair

Fei Fang   Carnegie Mellon University, USA

Publication Chair

Stefan Schauer   AIT Austrian Institute of Technology, Austria

Local Arrangements and Registration

Birgit Merl   Universität Klagenfurt, Austria

Publicity Chairs

Daniel Xiapu Luo   The Hong Kong Polytechnic University, SAR China
Antonios Gouglidis   Lancaster University, UK
Jun Zhuang   University at Buffalo, USA

Web Chairs

Markus Blauensteiner   Universität Klagenfurt, Austria
Philipp Pobaschnig   Universität Klagenfurt, Austria

Technical Program Committee

Bo An (Chair)   Nanyang Technological University, Singapore
Christopher Kiekintveld   University of Texas at El Paso, USA
  (Chair)
Alvaro Cardenas   University of Texas at Dallas, USA
Andrew Odlyzko   University of Minnesota, USA
Anil Kumar Chorppath   TU Dresden, Germany
Arman Mhr Khouzani   Queen Mary University of London, UK
Aron Laszka   Vanderbilt University, USA
Arunesh Sinha   University of Michigan, USA
Bo An   Nanyang Technological University, Singapore
Carlos Cid   Royal Holloway, University of London, UK
Christopher Kiekintveld   University of Texas at El Paso, USA
David Pym   University College London, UK
Eduard Jorswieck   Technical University Dresden, Germany
Fei Fang   Carnegie Mellon University, USA
Fernando Ordonez   Universidad de Chile, Chile
George Theodorakopoulos   Cardiff University, UK
Habtamu Abie   Norsk Regnesentral, Norwegian Computing Center,
                                Norway
Jean Leneutre   Ecole Nationale Supérieure des télécommunications,
                                France
Jens Grossklags   Technical University of Munich, Germany
John S. Baras   University of Maryland, USA
Jun Zhuang   SUNY Buffalo, USA
Konstantin Avrachenkov   Inria Sophia Antipolis, France
Long Tran-Thanh   University of Southampton, UK
Emmanouil Panaousis   University of Brighton, UK
Mehrdad Nojoumian   Florida Atlantic University, USA
Mohammad Hossein   Isfahan University of Technology, Iran
  Manshaei
Murat Kantarcioglu   University of Texas at Dallas, USA
Pasquale Malacaria   Queen Mary University of London, UK
Pradeep Varakantham   Singapore Management University, Singapore
Quanyan Zhu   New York University Brooklyn, USA
Sandra König   AIT Austrian Institute of Technology, Austria
Stefan Rass   Universität Klagenfurt, Austria
Stefan Schauer   AIT Austrian Institute of Technology, Austria
Yee Wei Law   University of South Australia, Australia
Yevgeniy Vorobeychik   Vanderbilt University, USA
Yezekael Hayel   LIA/University of Avignon, France

Contents

Short Papers

Full Papers

Optimizing Traffic Enforcement:
From the Lab to the Roads

Ariel Rosenfeld[1]($\boxtimes$), Oleg Maksimov[2], and Sarit Kraus[2]

[1] Department of Computer Science and Applied Mathematics,
Weizmann Institute of Science, Rehovot, Israel
arielros1@gmail.com

[2] Department of Computer Science, Bar-Ilan University, Ramat-Gan, Israel

Abstract. Road accidents are the leading causes of death of youths and young adults worldwide. Efficient traffic enforcement has been conclusively shown to reduce high-risk driving behaviors and thus reduce accidents. Today, traffic police departments use simplified methods for their resource allocation (heuristics, accident hotspots, etc.). To address this potential shortcoming, in [23], we introduced a novel algorithmic solution, based on efficient optimization of the allocation of police resources, which relies on the prediction of accidents. This prediction can also be used for raising public awareness regarding road accidents. However, significant challenges arise when instantiating the proposed solution in real-world security settings. This paper reports on three main challenges: (1) Data-centric challenges; (2) Police-deployment challenges; and (3) Challenges in raising public awareness. We mainly focus on the data-centric challenge, highlighting the data collection and analysis, and provide a detailed description of how we tackled the challenge of predicting the likelihood of road accidents. We further outline the other two challenges, providing appropriate technical and methodological solutions including an open-access application for making our prediction model accessible to the public.

1 Introduction

Every year the lives of approximately 1.25 million people are cut short and between 20 and 50 million people suffer disability or other severe injuries as a result of severe road accidents (accidents that cause death or injury) [29]. Efficient traffic enforcement can reduce the number and severity of severe road accidents by giving drivers the feeling that they are likely to be caught and sanctioned when breaking the law [11]. Road safety agencies have already identified the need for improvement in traffic enforcement and it is now an integral part of many countries' road safety policies [12]. Traffic police resources cannot cover the entire road network given the limited number of police cars and officers [7], and therefore some allocation mechanism is needed.

To address this challenge, we introduced the Traffic Enforcement Allocation Problem (TEAP) in a previous work [23]. TEAP is represented as an optimization problem which is shown to be NPH for approximation within any constant

© Springer International Publishing AG 2017
S. Rass et al. (Eds.): GameSec 2017, LNCS 10575, pp. 3–20, 2017.
DOI: 10.1007/978-3-319-68711-7_1

factor. TEAP relies on a pre-defined road network which is associated with two functions: (1) a function for measuring the likelihood that a severe traffic accident will occur on any road segment at any time; (2) a function for measuring the effect that the police allocation (both past and present) has on the risk of accidents occurring on any road at any time. Despite its computational complexity, realistically sized TEAPs can be solved efficiently using a newly proposed relaxed optimization technique named ROSE which leverages the TEAPs' characteristics. Nevertheless, despite its appealing theoretical properties and its practical promise, several challenges need to be addressed before the proposed model and solution can be applied in real-world traffic enforcement settings.

This paper presents the challenges that arise from our current effort to apply the TEAP model and its proposed optimization solution for the Israeli Traffic Police (ITP). We focus on three main challenges: First, the TEAP assumes that the road network and all required models which are associated with the road network are pre-defined and given. Namely, the TEAP assumes that the road network has already been divided into equally sized road segments, that the risk that a severe traffic accident will occur on any road at any time is known and that the effect of police enforcement on the latter is given. We refer to the challenge of defining and estimating the above as the **Data-centric challenge**. Second, several logistical/technical issues need to be addressed. For example, during the morning shift, the traffic allocation needs to accommodate a lunch break for the police officers. We refer to these challenges as the **Police-deployment challenges**. Lastly, data was gathered during the design and development process. However, most of the gathered data is confidential and thus cannot be released to the public. The public may benefit from making some of this data available in some format which will not jeopardize confidential material. We refer to these challenges as **Challenges in raising public awareness**.

In addition to our discussion on the above challenges and the provided technical and methodological solutions, this paper also provides a "behind the scenes" view of the process of moving from a theoretical model, tested in lab-settings, to an actual deployed system in field trials on the roads.

2   Related Work and Background

Recent studies suggest that drivers respect traffic laws mainly due to enforcement concerns, rather than safety concerns (e.g., [24]). As a result, efficient traffic enforcement has been shown to reduce a wide range of high-risk, illegal driving behaviors, including driving while under the influence of drugs/alcohol, speeding, lack of seatbelt use and red-light running, and thus reduces road accidents (e.g., [1,3,25]).

Within the Security Games (SG) field, optimal security resource allocation mechanisms and applications for mitigating various types of crimes have been developed. The generic SG framework consists of a defender (traffic police) which has a limited number of resources (police cars) to protect a large set of targets (road segments) from an adversary (reckless drivers). This approach has led to a variety of successfully deployed applications for the security of infrastructure and wildlife [26].

Other than our previous work [23], Brown et al. [4] is the only work in the scope of SG which addresses traffic enforcement. The authors model the problem as a Stackelberg Security Game (SSG) where traffic police seek to apprehend reckless drivers who in turn seek to avoid apprehension. In a SSG, the traffic police commit to a mixed strategy that drivers can first observe and then use in order to best respond. However, several practical issues make existing SG-based solutions seem unsuitable for the task of mitigating severe car accidents. First, traffic enforcement seeks to reduce road accidents (and not necessarily to apprehend reckless drivers) [12]. Second, over 4 decades of traffic enforcement literature shows that drivers are acting in a less strategical manner, *reacting* to observed police presence in the past and the present as well as on the current road and other roads in the vicinity [9]. While non-strategical adversaries in SG settings have recently been addressed (e.g., modeled as opportunistic criminals [32]), existing solutions do not account for the above challenges when combined. If we translate these seemingly technical issues into theoretical ones, we can identify that traffic enforcement is in fact a *non-Markovian* and *coupled* allocation problem. Namely, past police actions and states influence future states (non-Markovian), and police cars should coordinate their actions (coupled).

The TEAP model, which we propose in [23], addresses the above issues. TEAP leverages on the fact that, according to police enforcement experts, if police cars are stationed at the same place and time, their effectiveness in reducing road accidents cannot be assumed to be greater than the effectiveness of a single police car at the same point and time. In a TEAP, the interaction between drivers and police is modeled as a repeated game over $T(< \infty)$ rounds, which takes place on a road network, represented as a graph $G = \langle V, E \rangle$ where $V = \{v\}$ is the set of intersections and $E = \{e = (u, v)\}$ is the set of road segments. The graph is then extended into a *transition graph* such that each vertex v (edge e) is replicated T times, one for each round, denoted v_t (e_t), assuming that it takes one unit of time to traverse each road (see [31] for an extended discussion of the use of transition graphs). Each v_t in the transition graph is associated with the number of police cars that start their trajectories in it minus the number of police cars that end their trajectory in it, denoted b_{v_t}. The b_{v_t} values are assumed to be known in advance and cannot be changed by the police. Every road segment and time, e_t, is associated with an indicator $H[e_t]$ which assumes the value of 1 if a police resource is allocated to e_t. The allocation history of the police resources until time t (including) is denoted as H_t. The *risk of accidents* occurring at e_t is denoted as $\text{risk}(e_t)$. The risk function measures the likelihood that a severe traffic accident will occur at e_t in the absence of police enforcement (in the $[0, 1]$ range). The *effectiveness of enforcement* is denoted as $\text{eff}(H_t, e_t)$. eff measures the effect that the police allocation history has on the risk of accidents occurring at e_t. Consequently, the TEAP is formulated as the following mathematical program:

$$\min_{H_T} \sum_t \sum_{e_t} \text{risk}(e_t) \cdot (1 - \text{eff}(e_t, H_t)) \tag{1}$$

$$\text{s.t} \sum_{v'_{t-1}} H_t[(v'_{t-1}, v_t)_{t-1}]$$

$$- \sum_{v'_{t+1}} H_{t+1}[(v_t, v'_{t+1})_{t+1}] = b_{v_t} \quad \forall v_t \in G_T \tag{2}$$

$$H_T[e_t] \in \{0, 1\} \quad \forall e, t \tag{3}$$

Complete details and source code, including a master-slave optimization technique for solving the TEAP, are presented and evaluated in [23]. We refer the reader to the original paper for a thorough discussion about the TEAP's benefits and limitations.

Generally speaking, security settings are often very dynamic and complex, which may make it impractical to capture all of the necessary characteristics of the designated domain in a general game model built in a lab [21]. As a result, when moving from a theoretical model, tested in lab-settings, to an actual deployed system in field trials, different challenges arise. Therefore, despite the evidence showing the benefits of the above TEAP formulation, several data and logistical challenges prevent it from being implemented in the field as an "off-the-shelf" product. Furthermore, some of the assumptions made in the original formulation do not hold in practice, which necessitates the modification of the proposed formulation.

It is common to consider ARMOR as the first deployed SG system [22]. The system was deployed at the Los Angeles International Airport (LAX) in 2007 in order to randomize checkpoints on the roadways entering the airport as well as canine patrol routes within the airport terminals. For its deployment, the authors faced different challenges, mainly in instantiating their model to the LAX environment and increasing organizational acceptance of the proposed solution [16]. For example, the defender's payoffs were hard-coded after a series of interviews with airport security personnel. In addition, in order to allow security personnel the needed flexibility to adjust and change the provided allocation, an "override" option was added to the system. These insights are integrated in this work as well. Note that ARMOR provides a static allocation of security resources, which does not account for the spatio-temporal aspects of traffic enforcement.

A more similar system to ours, which also requires transition-based scheduling, is the TRUSTS system which is designed for fare-evasion deterrence in urban transit systems [31]. Similar to traffic enforcement settings, the TRUSTS system also allows defender resources to move across a graph structure over time. In deploying TRUSTS, several issues had to be addressed, with the prominent one being execution uncertainty [8]. In real world trials, a significant fraction of the executions of the pre-generated schedules got interrupted for a variety of reasons (e.g., writing citations, felony arrests, etc.), causing the officers to miss the train that they were supposed to take. Despite the resemblances between the two domains, exact timing is of far lesser significance in traffic enforcement as temporal constraints are more flexible than in transit system enforcement. Specifically, traffic enforcement officers are not bound to a fixed train schedule and thus delays have less far-reaching consequences on a planned schedule. According to

ITP experts, traffic police schedules are usually macro-managed, for example, enforcement of a road segment is scheduled in hours and not minutes, thus officers can better plan their actions and adapt to changes in real time. Therefore, delays do not pose a big concern in our setting. Several parameters of the TRUSTS formulation are estimated from available data such as ridership of different trains. Other parameters, which are more complex to estimate given available data, are estimated using non-data-driven methods (e.g., the uncertainty parameter). Similarly, in this work, most parameters are learned directly from Israeli-based data such as the risk of accidents occurring at each road segment and time which are estimated based on 11 years of collected data. Other parameters, such as the effectiveness of enforcement efforts which are much harder to learn, are estimated using other sources (in our case, past literature).

To the best of our knowledge, the most recently deployed application is PAWS [30]. PAWS is designed to combat illegal poaching through the optimization of human patrol resources. However, initial field tests of PAWS have revealed several data-centric and deployment-centric challenges [13]. For example, it turns out that the PAWS grid-based model does not capture important factors which may hinder the quality of the provided allocation and even prevents patrollers from completing their tasks, such as terrain elevation and accessibility. Furthermore, in the original model, PAWS assumed that many parameters are fixed and known, however, due to animal movement and seasonal changes, this assumption does not hold. Similarly, in this work, in order to deploy the proposed traffic enforcement solution, important factors are integrated within our modified model and certain assumptions are relaxed.

Data gathering always poses a great challenge in the development and deployment of security-based applications. Unfortunately, this data is usually withheld from the public. In traffic enforcement, one usually encounters aggregated statistics on the number of accidents and their severity on a monthly or even yearly basis. Releasing some of the gathered data and its analysis to the public in an efficient and natural way could potentially help raise public awareness to road dangers, help drivers trade off travel time, distance and safety, and help in achieving our main goal – reducing the number of road accidents. In this paper, we provide such a publicly available system without jeopardizing the police's interests.

3 Data-Centric Challenges

Obtaining and leveraging the "right" data is a challenge for most SG-based systems (e.g., [17,19]). To instantiate the TEAP formulation in Israel, we had to address four cardinal data-related challenges: (1) building a road network (Sect. 3.1); (2) extending the road network into a transition graph (Sect. 3.2); (3 + 4) deriving risk and eff models which are associated with the constructed graph (Sects. 3.3 and 3.4). The solution for each of the above four challenges constitutes a component in our solution architecture as illustrated in Fig. 1.

3.1 Building a Road Network

The TEAP formulation, similar to many other transition-based formulations, builds on a graph G. In traffic enforcement settings, G represents a road network where the vertices indicate intersections and edges indicate road segments for enforcement. While it may seem trivial to obtain a network graph from open source systems such as OpenStreetMap[1], the translation of this *raw map* into a useful graph raises a number of practical issues and fundamental questions. First, the ITP does not *enforce* traffic laws on local roads (e.g., inner-city roads) which are in the jurisdiction of the local police departments. A naïve solution would be to omit the roads and intersections in question from the graph. However, this omission may disconnect the graph, which is undesired. Second, some of a road's features such as the speed limit and number of lanes may change over different parts of the road. In order to construct a suitable prediction model for risk, a challenge we address in Sect. 3.3, one should be able to identify these features for each road segment e. Unfortunately, these features are hard to obtain and are usually aggregated according to some road segmentation by the data collector. In Israel, this data is collected by both the ITP and the Central Bureau of Statistics (denoted CBS, www.cbs.org.il). Finally, it is well known that achieving organizational acceptance of SG-based solutions is a highly complex task (see Sect. 4). Therefore, using a police-defined road segmentation, when available, is preferable. Police-based segmentation can also encapsulate other domain specific knowledge and constraints, for example, on a narrow road segment without shoulders it may be impossible or highly undesirable to perform enforcement from the police's perspective. Therefore, using police-planned enforcement segments has significant benefits.

Based on the discussion above, we sought to use the police-based segmentation. However, in the initial stage of the research, the ITP did not allow us access to their segmentation (see Sect. 4). Therefore, we used the CBS's road network. In a much later phase of the research, after the publication of [23] which relied on the CBS's road network, more than a year and a half after the project was initiated, the police segmentation was made available to us. Unfortunately, the police segmentation is far from perfect. Despite its resemblance to the CBS's segmentation, ample effort had to be invested to transform the provided segmentation into a complete, connected and valid road network. Namely, in many cases, adjunct roads were not marked as such; some road segments were overlapping by mistake while others were disjointed despite being physically connected, etc. Therefore, we manually processed the police segmentation on the basis of the CBS's network (about 40 human hours). We denote the resulting road network graph as G from this point onwards. G consists of several hundred road segments.

The above is illustrated as **Component A** in our solution architecture, Fig. 1.

---

[1] https://www.openstreetmap.org.

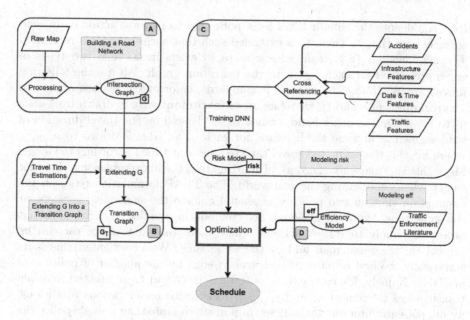

Fig. 1. High-level solution architecture

3.2 Extending G into a Transition Graph

According to the TEAP's original formulation, G is extended into a transition graph by replicating each vertex and edge T times. Thus, the formulation assumes that at each time step a police car is assigned to enforce a different road segment. Namely, each police car can perform a single action at each road e given that it is present at e. The formulation also assumes that the enforcement action takes exactly 1 time unit. While this is reasonable in theory, in practice a police car may either engage in various types of enforcement (e.g., using speed-guns to catch speeding cars or randomly choosing drivers on whom to perform breathalyzer tests in order to catch drunk drivers) or in transit (moving across the road network without enforcement so as to reach the intended enforcement site). As a result, the solution to the original TEAP does not prescribe which enforcement should be conducted but only where and when. Furthermore, the solution is limited to only enforcement actions, which according to ITP experts take significantly longer than simple transit actions.

In order to adequately extend G into a practical transition graph, we amend the TEAP formulation in the following way: We denote $A = \{\alpha_1, \alpha_2, \ldots, \alpha_n\}$ as the set of actions that a police car can take at every road segment and time. Jain et al. [16] found that providing a very detailed allocation and micro-managing resources does not get as positive a reception from users. Instead, the authors suggest using more abstract instructions, which they found to be better received. Therefore, we simplify the model by assuming two actions, *Enforce* and *Transit*, and leave the investigation of different enforcement options for future work. Let

$l(e, t, \alpha_i)$ denote the time it takes for a police car to perform action α_i in road segment e at time t. Then, G is extended such that each vertex v is replicated T times, denoted $\{v_t\}$. If an edge $e = (u, v)$ exists in G, then two types of edges are added for each $t < T$ to the transition graph. We use the following procedure: With accordance to ITP standards, enforcement actions are set to 1 h (regardless of e and t), therefore an **enforcement** edge is drawn from each u_t to v_{t+60}. On the other hand, transit edges depend on the travel duration of road segment e at time t. Therefore, for each u_t an edge is drawn to $v_{t+l(e,t)}$ where $l(e, t)$ is the estimated travel time to cross e at time t according to Google Maps (https://maps.google.com).[2] Unfortunately, the above procedure does not suffice when representing the real world. The TEAP formulation relies on the assumption that no two police cars should enforce the same road segment at the same time. However, although not common in practice, this rule does not necessarily apply to *transit* actions, where more than one police car can be present on the same road and at the same time. We investigated this issue empirically, we first duplicated each transit edge by the number of police cars available. Namely, for each edge $e = <u, v>$ in G and t, we created multiple transit edges to connect u_t and $v_{t+l(e,t)}$. Practically, under various conditions, we did not encounter any realistic settings in which more than a single police car was present on the same road segment at the same time in Israel. Therefore, while theoretically justified, we avoid replicating transit edges for our deployment. We denote the resulting transition graph as G_T from this point onwards. Note that the notation v_t is still used in its usual meaning. However, notation e_t is not well defined since G_T has multiple edges (e_t may denote an enforcement edge or a transit edge), and therefore will not be used from this point onwards.

Per the ITP's request, we allow all police cars to start and end their routes at any intersection by introducing a dummy source vertex which is connected to all intersections at time 0 and a dummy sink vertex, accessible from all intersections at time $T + 1$. Thus, we assign each $b_{v_t} = 0$. Note that in practical deployment (Sect. 4) b_{v_t} may assume a value other than zero in cases where a police car is scheduled to visit a specific intersection at a specific time (e.g., due to road work) or plans to start its route at a certain intersection (e.g., once the road work is completed). In such cases b_{v_t} is set appropriately.

The above is illustrated as **Component B** in our solution architecture, Fig. 1.

3.3   Modeling risk

The risk function captures the risk of accidents occurring at each road segment and time. Recently, traffic police forces began implementing the predictive policing paradigm [20] through which police officers can identify people and locations which are at increased risk. From a methodological standpoint, the effort of predicting road accidents has mainly focused on aggregative analysis, most commonly the prediction of the *annual* number of severe accidents per road

---

[2] Time was discretized in 10 min time-frames.

segment, using statistical methods such as empirical Bayes, Poisson or negative binomial regression models [5,15]. Such aggregation is limited in its use to police forces as the allocation of traffic police enforcement requires a prediction on a much more finely grained level. According to experts in traffic enforcement, the state-of-the-art prediction model is the one used by the Indiana traffic police https://www.in.gov/isp/ispCrashApp/main.html. The Indiana system does not consider each road segment separately but instead covers the Indiana state map (including residential areas) with a grid of 1 km by 1 km squares and provides a prediction as to the risk of a severe accident occurring at each square in three-hour time-frames. According to our discussion with the Indiana traffic police, the prediction model uses approximately 90 features and achieves an Area Under the Curve (AUC) of approximately 0.8. In this paper, we were able to construct a prediction model that provides beneficial predictions for each road segment on *one hour time-frames* by using a unique set of 122 features and 11 years of collected data that achieves an AUC of 0.89. Our model is available at http://www.biu-ai.com/trafficPolice/ in order to encourage other researchers to tackle the important and challenging task of preventing severe road accidents.

We obtained the records of 11 years of accident reports from the Israeli CBS (2005–2015). By cross-referencing these reports with additional sources such as the Israeli GIS database and the Israeli Meteorological Service (IMS, www.ims.gov.il) weather reports, we were able to characterize each accident using a unique set of 122 features. The features are divided into 3 categories: (1) infrastructure features; (2) date and time features; and (3) traffic features. To the best of our knowledge, this is the largest set of features ever to be used to predict severe car accidents.

Infrastructure Features. The geography of Israel is very diverse, with desert conditions in the south and snow-capped mountains in the north. It is customary to divide Israeli into 3 regions: North, South and Center. These three regions differ significantly in their population and land use. For example, the central region is a metropolitan area (e.g., the Tel-Aviv metropolis) characterized by dense urban building and high-tech land use whereas the southern region is mostly a desert which consists of rural low-density residential areas for the most part [3 features]. The ITP further divides Israeli into 15 districts according to geographic criteria [15 features].

Each road segment is characterized according to its type (e.g., highway) [7 features], its length in km [1 feature], the number of lanes [7 features], the posted speed limit [5 features], road signals [2 features], road width [5 features], whether a traffic light is present on the road segment [2 feature], road surface conditions (e.g., gravel/paved) [6 features] and whether the road is lit up at night [5 features]. Unfortunately, to date, we were unable to obtain additional features that have been shown to affect the prevalence of road accidents in past literature. These features include the existence of road shoulders, the road segment's curvature, incline/decline etc.

Date and Time Characteristics. We characterize the date using the month of the year [12 features], day of the week [7 features] and an indicator whether it is a weekday, weekend, holiday, holiday evening or another type of special day [5 features]. Time is characterized on an hourly scale [24 features] and by an indicator of whether it is daytime or nighttime [2 features].

In addition, we characterize the weather in the vicinity of the road segment at the given time using the publicly available IMS reports and forecasts [4 features].

Traffic Characteristics. While the infrastructure characteristics do not change frequently, the traffic that goes through the road segments changes rapidly over time. We characterize the traffic by its volume [1 feature] and average speed [5 features]. Traffic volume is provided by the CBS and average speeds are provided by the ITP. We further identify the number of severe accidents which have occurred on that road segment in the prior 30, 90, 180 and 365 days [8 features].

Training a Deep Neural Network. Using more than 30,000 accident records (accidents that took place between 2005 and 2015 in Israel) and undersampling the "non-accident" class (see [6]), we trained a deep neural network model. The model receives, as input, vectors of 122 features (as described above) representing a road segment and time. The model returns a value in the [0, 1] range, *acting as a proxy* to the likelihood of an accident occurring on that road at that time. Note that severe accidents are sporadic events in both time and space. Therefore, directly estimating the probability of accidents occurring on any road segment at any time is extremely challenging.

Our network consists of 3 layers, $1024 \times 512 \times 1$, where the hidden layer uses the common RelU activation function. Several other architectures were tested and found to be of lower quality in terms of AUC.

We compared our prediction model to several baseline prediction models, such as logistic regression, SVM and XGBoost. The latter is currently in use by the Indiana traffic police for the same task. Our model achieves an AUC of 0.89, outperforming logistic regression, SVM and XGBoost, which recorded 0.78, 0.77 and 0.82, respectively, using 10-cross validation.

Using entropy-based ranking feature selection [14], we identify 5 groups of high ranking features in the following order of importance: (1) number of past accidents; (2) traffic volume; (3) road type; (4) speed limit; and (5) weather. Contrary to what the authors initially expected, among the lowest ranking features one would find: (1) day of the week; (2) month of the year; and (3) enforcement district. We plan to further analyze these results in order to provide additional practical suggestions for the ITP.

The above is illustrated as **Component C** in our solution architecture, Fig. 1.

3.4   Modeling eff

The eff function measures the effect that the police allocation history has on the risk of accidents occurring on any road segment at any time. Unfortunately, as discussed in Sect. 4, getting the ITP's allocation history was very difficult. We only gained access to a single month's allocation in 2017. Unfortunately, the allocation was recorded in the format of GPS coordinates which police cars report once every few minutes while on duty. Translating these GPS coordinates into a usable format which will allow us to understand when and where a police car was enforcing the law, driving through a road segment or handling other events (e.g., a traffic police car may be temporarily assigned to assist police patrol cars) is extremely complex. For example, a police car may stay for some time at a gas station. The police car may be refueling, there may be a technical problem with the police car that needs fixing, or the police may be having a lunch break or enforcing the law on a nearby road. We are currently working with the ITP on a methodology to address the above issues in the future.

Isolating the effect that the police allocation has on the likelihood of road accidents is extremely complex regardless of the above mentioned issues. First, endogeneity is a big concern. Naturally, police cars are likely to be stationed on dangerous roads. Naïve statistical inference may conclude that the presence of a police car *increases* the likelihood of accidents. The endogeneity problem is particularly relevant in the context of time series analysis of causal processes, which is the case in traffic enforcement. Even if we assume that `eff` is independent of all other factors within a given period, but is influenced by the average speed of traffic in the preceding period, then `eff` is exogenous within the period but endogenous over time, which poses an additional statistical challenge.

As a result of the above discussion, we base our estimation of `eff` on [28]. The author used a unique database to track the exact location of the Dallas Police Department's patrol cars throughout 2009 and cross-referenced it with the car accidents of that year. To the best of our knowledge, this is the most recent investigation of the topic. The author found that on a given road at a given time, if enforced, `eff` should assume a value of 36%. However, enforcement effects are not restricted to the specific time and space in which the enforcement is performed. For example, *time halo* is the time and the intensity to which the effects of enforcement on drivers' behavior continue after the enforcement operations have been concluded. It has been recorded that longer enforcement efforts cause more intense time halo effects that can last for hours and influence the next day(s) or even week(s) during the same time of day as the enforcement. *Distance halo* is defined as the distance over which the effects of an enforcement operation last after a driver passes the enforcement site. The most frequent distance halo effects are in the range of 1.5–3.5 km from the enforcement site (see [9] for a review). In accordance with the ITP's expert estimations, we define time halo effects in the exponential diminishing form $\frac{36}{2^k}\%$ where $k \geq 0$ is the hours that have passed since the enforcement effort. To avoid negligible effects, we prune the effect at $k = 3$. The distance halo effect is defined to be 5%, given that the two road segments are adjutant. Given the police allocation, `eff` assumes a simple submodular form where `eff` takes the largest applicable effect and adds half of each of the smaller appropriate effects to it. For example, if a road segment is enforced for two hours straight (and no other time or distance halo effects are appropriate), `eff` assumes 45% ($= 36\% + \frac{18}{2}\%$).

As discussed before, we are currently working towards a more data-driven approach for modeling `eff` in Israel.

The above is illustrated as **Component D** in our solution architecture, Fig. 1.

4   Police-Deployment Challenges

In order to deploy our model and solution in Israel, several logistical/technical issues had to be addressed. In this section we discuss challenges which arise from our interaction with the ITP and do not focus on data.

4.1   Security Clearance

Before any meaningful intersection with the ITP could take place (e.g., allowing us access to their data), the authors had to obtain security clearance, including a 2-hour background check and interview at the ITP headquarters in the Israeli capital (Jerusalem). The clearance came through about 6 months into the process.

The clearance that we obtained allowed us full access to ITP information. However, due to bureaucratic reasons, obtaining each piece of information required a long approval process. As a result, at the moment, we do not have access in *real time* to important pieces of information such as average speeds. Note that the ITP does have accurate estimations thereof using anonymized cellular reports [2].

4.2   Logistics

In order to deploy our model and solution in the real world, logistical constraints need to be addressed. There are three main constraints: First, some police resources may have specific schedule constraints of the form "Officer X must arrive at traffic court at 2pm and stay there until 3pm to testify in a trial". Such constraints are easily integrated within our model by setting the b values of the intersections in the transition graph appropriately. Second, according to the ITP, during an 8-hour shift, each police car should have a break of about 1 h to eat and reach its next destination. The rationale is that the ITP has arranged various different places for police officers to eat and therefore no special requirements should be implemented as to where a police car should have its break. This break is scheduled for different times, for example, interleaving during the 4th hour of work so as to avoid having all officers on break at the same time. Specifically, officers are interleaved as to when they would go on a break during the 4th hour of work such that at least k police cars are not on break at any given moment (k is a police defined constant). We amend our model by adding designated "break" vertices during the 4th hour. These vertices are accessible from any vertex during the 4th hour and are connected to all vertices which are one hour later. For example, a police car can go on a break from any location at 12:00, and continue its schedule from any vertex at 13:00. This formulation was specifically tailored at the request of the ITP. To make sure each police car goes on a single break, nodes during the 4th hour were duplicated such that every node had two copies – "pre-break" and "post-break". Then, pre-break nodes were disconnected from 5th hour nodes and post-break nodes were disconnected such that they are only accessible from break nodes or other post-break nodes. Simply put, a police car can only reach the 5th hour of the shift if it goes though a post-break node. Naturally, the post-break nodes do not allow re-access to a break node, ensuring that each police car visits only a single break node on its path. Third, an unexpected event may cause a police car to deviate from its schedule. For example, a police car may be sent to clear an unexpected road block, making its schedule infeasible. The ITP claims that there is no easy way to determine the likelihood of these unexpected events in the real world, making the MDP-based approach used by TRUSTS inapplicable. We resolve this issue as proposed in [8], by *assuming* perfect execution and only after a non-default transition occurs does the central command resolve the TEAP starting from the current state. Yet this requires a quick solution, as the ITP would not accept a long wait time. Given that the original TEAP formulation (as presented in [23]) was modified in this work, we reevaluated the runtime results of the proposed solution given the new formulation. Similar to our previous findings, the runtime of the amended solution does not impose a significant concern. Specifically, given the modified formulation for allocating 10 police cars for 96 h (4 days, 12 shifts) in a designated district, the proposed solution runs in approximately 1 min, compared to more than 30 min by a naïve solver as described in the original paper. Namely, the modifications proposed in this paper do not jeopardize the solution method's superiority over baseline methods.

Note that, given a non-default transition, we recalculate the allocation for all police cars, as local adjustments may produce suboptimal allocations. We plan to investigate local methods for adjusting infeasible or undesired allocations in future work.

4.3 Deployment and Evaluation

As a first step, the ITP wanted to know how different our provided allocations are from their current hand-crafted, time-consuming allocations. Unfortunately, as mentioned above, usable records of past police allocations are currently unavailable. Instead, we were given a list of road segments which the ITP considers to be of "special enforcement interest". The list, which consists of approximately 5% of the road segments, was constructed by the ITP's researchers and acts as a guideline as to which road segments are the most important to enforce during a given month. Generally speaking, the ITP focuses on these road segments and their surroundings. We generated a schedule for a whole month using the modified TEAP formulation as described above, and identified the number of times a police car was assigned to one of the designated road segments. On average, a road segment on the list is enforced 25 times more often than other road segments. Note that the entire list was associated with very high risk scores, specifically, if road segments were to be sorted according to their average risk score on any day or at any time, the entire list would be at the top 15% of all road segments.

There are many challenges when attempting to evaluate deployed SG-based systems in the field [27]. Specifically, unlike conceptual ideas, where we can run thousands of careful simulations under controlled conditions, we cannot conduct such experiments in the real world [18]. We are currently working towards a head-to-head comparison of our proposed solution against an expert-generated allocation. Our controlled experiment would take place in two very similar enforcement districts. During a period of at least one month, one district will use our system while the other will use an expert-generated allocation. It is hard to believe that such a comparison will yield a statistically significant difference, due to the fact that road accidents are very sporadic. A similar problem was encountered in PAWS, where the authors faced a similar issue of how to quantify the number of saved wildlife due to their provided solution. The authors instead use human and animal signs as indicators that PAWS patrols prioritize areas with higher animal and poacher activity. In the same spirit, we would also record other statistics such as the average speeds and the number of citations issued by the police officers, as well as other statistics, as proxies to the allocation's quality. Speed has been identified as a key risk factor in road traffic safety, influencing both the risk of a road crash and the severity of the injury that results from crashes [10]. Furthermore, a higher number of issued citations can suggest that the provided solution can avoid the human-generated predictability. Note that the benefit of our solution, as well as other deployed solutions such as PAWS, should be expected in the long-term.

5   Raising Public Awareness

Recently, the World Health Organization (WHO) has released a report on road traffic injuries and how they can be reduced [29]. The WHO mentioned that governments should take a more holistic approach to mitigating road accidents, which includes not only better enforcement but also the modification of infrastructure and the raising of public awareness. The WHO further mentions the latter as one of its own tasks, "sharing information with the public on (road) risks...".

As discussed in Sect. 3.3, the Indiana traffic police has provided a visual tool that uses color shading to show a low, moderate or high probability of a crash occurring in each 1-square-kilometer area in the state. This interactive map predicts where crashes are likely to occur across the state of Indiana, so citizens and law enforcement can be more proactive in avoiding or preventing accidents.

Despite its publication in the general media, according to the developers, only a few citizens use the system. We speculate that there are two main reasons for this disappointing feedback: First, the use of a grid-cell instead of road segments and the relatively large time frames (3-hours long) make the system less practical for drivers. Consider a driver who tries to decide which route to take at 8am from one point to the other. The driver is interested in the risk associated with the different route options at 8am rather than grid-cells between 8am and 11am. Second, the system uses a newly designed interface. However, in creating solutions for people, we must be cognizant to how difficult it will be for a user to adopt our solution. Each deviation from existing methodology is a step away from the familiar that we must convince the user to accept [26].

To address these two issues we provide the *www.SafeRoad.today* open-access system, which is mounted on the popular Google maps interface. Users can access our website and review the risk in each road segment in Israel in 1-hour time frames discretized into 5 risk levels – "very low", "low", "average", "high" and "very high". However, unlike the ITP, drivers may not be interested in the risk of an accident on a road segment but rather **their own risk** of being involved in an accident. Therefore we provide 2 layers, one illustrating the `risk` function as described in Sect. 3.3, and the other illustrating the `risk` function after the normalization by the expected traffic volume. Users can query the system for risks associated with any road at any time. Another type of query is a route query. The user can query the system for routes to take her from one point to another using the regular Google maps interface. Once a query for a route is made, in addition to the travel time and distance for each possible route provided by Google maps, the system provides a `risk` estimation for each route, enabling the driver to consider the safety factor in her route selection. An additional layer allows the user to query the system for past road accidents. Given dates and locations, provided by the user, all severe accidents that occurred during the designated time and at the designated locations will appear on the map, each according to the place of the crash. Each accident appears alongside some basic information regarding the crash such as the date and time, type of crash (e.g., a car and a motorcycle), number of injuries, etc. Note that our system does not jeopardize the confidentiality of police data – it simply does not contain any restricted data. To the best of our knowledge, our system is the first of its kind.

The *SafeRoad.today* system joins existing publicly accessible systems designed for raising public awareness of other road dangers. For example, WAZE[3] provides road danger alerts for drivers such as road work, Sustrans[4] provides safe cycle routes, factoring in bike lanes and traffic free routes and Rudder[5] provides safe pedestrian routes, factoring in street lighting.

A snapshot of our system is provided in Fig. 2.

---

[3] https://www.waze.com/livemap.
[4] http://www.sustrans.org.uk/ncn/map.
[5] https://walkrudder.com/.

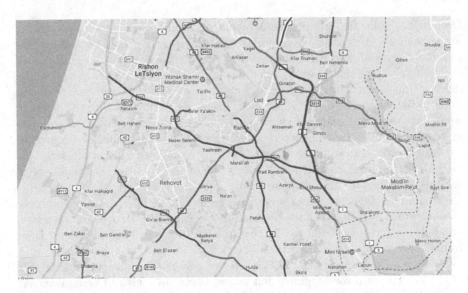

Fig. 2. A snapshot from *SafeRoad.today* system. Risk illustration is as follows: Red = "Very High", Orange = "High", Azure = "Low" and Dark Blue = "Very Low". Average risks are not depicted. (Color figure online)

6 Conclusions

In this paper we present key challenges and solutions in transforming a lab-based theoretical model for mitigating road accidents through efficient enforcement to an operational system in field trials. We focus on three main challenges: data, deployment and raising public awareness. These challenges, and specifically the data-centric challenges, are very common in security-based applications. Two important components of our provided solution include a novel traffic accident prediction model, available at http://www.biu-ai.com/trafficPolice/, and an open-access risk visualization system, *SafeRoad.today*, which is available for public use. Our prediction model provides a state-of-the-art prediction tool, based on 122 features and 11 years of collected data, that we hope will encourage other researchers and practitioners to tackle the important and challenging task of preventing severe road accidents. In addition, our *SafeRoad.today* system is designed for and targeted at raising public awareness and allowing drivers to make better decisions. These components, which amend and extend our police allocation mechanism from [23], combine to provide a viable tool for mitigating road accidents in Israel and can be adapted to other countries as well.

This "behind the scenes" paper also provides a unique look into the considerations and decisions that developers of deployed security-based applications have to face. Since the challenges discussed and addressed in this work are not unique to traffic enforcement, we hope that the provided discussion and insights will assist others in the process of deploying their security-based systems in the real world.

References

1. Adler, N., Hakkert, A.S., Kornbluth, J., Raviv, T., Sher, M.: Location-allocation models for traffic police patrol vehicles on an interurban network. Ann. Oper. Res. **221**, 9–31 (2014)
2. Bar-Gera, H.: Evaluation of a cellular phone-based system for measurements of traffic speeds and travel times: a case study from israel. Transp. Res. Part C Emerg. Technol. **15**(6), 380–391 (2007)
3. Bates, L., Soole, D., Watson, B.: The effectiveness of traffic policing in reducing traffic crashes. In: Prenzler, T. (ed.) Policing and Security in Practice, pp. 90–109. Springer, London (2012). doi:10.1057/9781137007780_6
4. Brown, M., Saisubramanian, S., Varakantham, P.R., Tambe, M.: Streets: game-theoretic traffic patrolling with exploration and exploitation. In: Proceedings of the 2014 AAAI Conference on Artificial Intelligence (2014)
5. Chang, L.Y.: Analysis of freeway accident frequencies: negative binomial regression versus artificial neural network. Saf. Sci. **43**(8), 541–557 (2005)
6. Chawla, N.V.: Data mining for imbalanced datasets: an overview. In: Maimon, O., Rokach, L. (eds.) Data Mining and Knowledge Discovery Handbook, pp. 853–867. Springer, Boston (2005). doi:10.1007/0-387-25465-X_40
7. DeAngelo, G., Hansen, B.: Life and death in the fast lane: police enforcement and traffic fatalities. Am. Econ. J. Econ. Policy **6**(2), 231–257 (2014)
8. Delle Fave, F.M., Jiang, A.X., Yin, Z., Zhang, C., Tambe, M., Kraus, S., Sullivan, J.P.: Game-theoretic patrolling with dynamic execution uncertainty and a case study on a real transit system. J. Artif. Intell. Res. **50**, 321–367 (2014)
9. Elliott, M., Broughton, J., et al.: How methods and levels of policing affect road casualty rates. Transport Research Laboratory (2005)
10. Elvik, R.: A re-parameterisation of the power model of the relationship between the speed of traffic and the number of accidents and accident victims. Accid. Anal. Prev. **50**, 854–860 (2013)
11. Elvik, R., Vaa, T., Erke, A., Sorensen, M.: The Handbook of Road Safety Measures. Emerald Group Publishing, Bingley (2009)
12. European Transport Safety Council: How traffic law enforcement can contribute to safer roads: PIN flash report 31, June 2016. http://etsc.eu/wp-content/uploads/NEW_PIN_FLASH31_final.pdf/
13. Fang, F., Nguyen, T.H., Pickles, R., Lam, W.Y., Clements, G.R., An, B., Singh, A., Tambe, M., Lemieux, A.: Deploying paws: field optimization of the protection assistant for wildlife security. In: Twenty-Eighth IAAI Conference (2016)
14. Guyon, I., Elisseeff, A.: An introduction to variable and feature selection. J. Mach. Learn. Res. **3**, 1157–1182 (2003)
15. Hauer, E.: Empirical bayes approach to the estimation of "unsafety": the multi-variate regression method. Accid. Anal. Prev. **24**(5), 457–477 (1992)
16. Jain, M., Tsai, J., Pita, J., Kiekintveld, C., Rathi, S., Tambe, M., Ordóñez, F.: Software assistants for randomized patrol planning for the LAX airport police and the federal air marshal service. Interfaces **40**(4), 267–290 (2010)
17. Kar, D., Ford, B., Gholami, S., Fang, F., Plumptre, A., Tambe, M., Driciru, M., Wanyama, F., Rwetsiba, A.: Cloudy with a chance of poaching: adversary behavior modeling and forecasting with real-world poaching data. In: Proceedings of the 16th Conference on Autonomous Agents and Multi-Agent Systems, pp. 159–167 (2017)

18. Nguyen, T.H., Kar, D., Brown, M., Sinha, A., Jiang, A.X., Tambe, M.: Towards a science of security games. In: Toni, B. (ed.) Mathematical Sciences with Multidisciplinary Applications, pp. 347–381. Springer, Cham (2016). doi:10.1007/978-3-319-31323-8_16

19. Park, N., Serra, E., Subrahmanian, V.: Saving rhinos with predictive analytics. IEEE Intell. Syst. **30**(4), 86–88 (2015)

20. Perry, W.L.: Predictive Policing: The Role of Crime Forecasting in Law Enforcement Operations. Rand Corporation, Santa Monica (2013)

21. Pita, J., Bellamane, H., Jain, M., Kiekintveld, C., Tsai, J., Ordóñez, F., Tambe, M.: Security applications: lessons of real-world deployment. ACM SIGecom Exch. **8**(2), 5 (2009)

22. Pita, J., Jain, M., Marecki, J., Ordóñez, F., Portway, C., Tambe, M., Western, C., Paruchuri, P., Kraus, S.: Deployed ARMOR protection: the application of a game theoretic model for security at the Los Angeles International Airport. In: Proceedings of the 2008 International Conference on Autonomous Agents and Multiagent Systems, pp. 125–132 (2008)

23. Rosenfeld, A., Kraus, S.: When security games hit traffic: optimal traffic enforcement under one sided uncertainty. In: Proceedings of the 26th International Conference on Artificial Intelligence, IJCAI (2017)

24. Schechtman, E., Bar-Gera, H., Musicant, O.: Driver views on speed and enforcement. Accid. Anal. Prev. **89**, 9–21 (2016)

25. Simandl, J.K., Graettinger, A.J., Smith, R.K., Jones, S., Barnett, T.E.: Making use of big data to evaluate the effectiveness of selective law enforcement in reducing crashes. Transp. Res. Rec. J. Transp. Res. Board **2584**, 8–15 (2016)

26. Tambe, M.: Security and Game Theory: Algorithms, Deployed Systems, Lessons Learned. Cambridge University Press, New York (2011)

27. Taylor, M.E., Kiekintveld, C., Western, C., Tambe, M.: A framework for evaluating deployed security systems: is there a chink in your ARMOR? Informatica **34**(2), 129–139 (2010)

28. Weisburd, S.: Does police presence reduce car accidents? Technical report, August 2016

29. World Health Organization: Road traffic injuries fact sheet, May 2017. http://www.who.int/mediacentre/factsheets/fs358/en/

30. Yang, R., Ford, B., Tambe, M., Lemieux, A.: Adaptive resource allocation for wildlife protection against illegal poachers. In: Proceedings of the 2014 International Conference on Autonomous Agents and Multi-Agent Systems, pp. 453–460 (2014)

31. Yin, Z., Jiang, A.X., Johnson, M.P., Kiekintveld, C., Leyton-Brown, K., Sandholm, T., Tambe, M., Sullivan, J.P.: TRUSTS: scheduling randomized patrols for fare inspection in transit systems. In: Proceedings of the 2012 Innovative Applications of Artificial Intelligence (2012)

32. Zhang, C., Gholami, S., Kar, D., Sinha, A., Jain, M., Goyal, R., Tambe, M.: Keeping pace with criminals: an extended study of designing patrol allocation against adaptive opportunistic criminals. Games **7**(3), 15 (2016)

Incentive Compatibility of Pay Per Last N Shares in Bitcoin Mining Pools

Yevhen Zolotavkin, Julian García[✉], and Carsten Rudolph

Faculty of IT, Monash University, Clayton, Australia
{yevhen.zolotavkin,julian.garcia,carsten.rudolph}@monash.edu

Abstract. Pay per last N shares (PPLNS) is a popular pool mining reward mechanism on a number of cryptocurrencies, including Bitcoin. In PPLNS pools, miners may stand to benefit by delaying reports of found shares. This attack may entail unfair or inefficient outcomes. We propose a simple but general game theoretical model of delays in PPLNS. We derive conditions for incentive compatible rewards, showing that the power of the most powerful miner determines whether incentives are compatible or not. An efficient algorithm to find Nash equilibria is put forward, and used to show how fairness and efficiency deteriorate with inside-pool inequality. In pools where all players have comparable computational power incentives to deviate from protocol are minor, but gains may be considerable in pools where miner's resources are unequal. We explore how our findings can be applied to ameliorate delay attacks by fitting real-world parameters to our model.

1 Introduction

Blockchain is a distributed ledger technology with demonstrated potential to revolutionize industry and commerce [10]. A number of popular cryptocurrencies based on blockchains have been launched in recent years to unprecedented adoption. These include Bitcoin (BTC) [11], Litecoin (LTC) and Zcash (ZEC) [13], among others [4]. The main technological innovation behind this drive is the proof-of-work consensus mechanism [7], which allows for the ledger integrity to be maintained in a distributed fashion. To achieve this level of decentralization, the system relies on miners who are incentivized to verify transactions. When incentives are compatible, rational players will find it in their best interest to stick to protocol. This paper uses game theory to derive conditions under which a popular mining reward mechanism, Pay per last N shares (PPLNS), is incentive compatible.

The Blockchain is a public ledger that keeps transaction information in a sequence of transaction blocks. Each block contains a hash of the previous block, and the chain grows as new transactions are verified and added to the chain. Any agent can add a block to the chain, so the approach relies on cryptographic puzzles, known as proofs of work, in order to reach consensus. The longest chain, as measured by computational effort exerted, is assumed to be the consensus chain. The agents solving the cryptographic puzzles are known as miners, and

© Springer International Publishing AG 2017
S. Rass et al. (Eds.): GameSec 2017, LNCS 10575, pp. 21–39, 2017.
DOI: 10.1007/978-3-319-68711-7_2

they exchange their computational power for new currency and transaction fees. The puzzle is randomized in such a way that each miner has a probability of discovering the next block proportional to their share of computational power in the network [7].

The mining market is very competitive. Individual miners face large variances in income. Consequently, most miners pool their computational resources, sharing the rewards of the pool amongst all members in proportion to the computational effort invested in mining [9]. Through pooling, miners ensure a more stable income flow. Mining pools are managed by an administrator who will often collect fees from miners, distributing the rewards when blocks are discovered in the pool. Miners prove their work on behalf of the pool by discovering "shares", which are partial proofs of work. It is assumed that every share requires equal computational effort. In addition to satisfying the requirement for partial proof of work, every computed share may in addition qualify as full proof of work. In the latter case, the pool is rewarded by the Bitcoin network, which issues new coins and transfers them to the pool's account. The reward obtained by the pool is then distributed to the members of the pool, according to its reward scheme and the submission behaviour of all the pool members. Reward mechanisms serve to aggregate shares reported in the pool, so as to perform a fair distribution according to work.

Early reward mechanisms often rewarded miners in proportion to the amount of shares submitted by a miner in each round [14]. However, since the distribution of rewards is exponential, under this scheme, miners may increase their reward expectation by changing pools frequently. This attack is known as pool hopping, and discourages honest mining to unsustainable rates [3]. Pay per last N shares (PPLNS) addresses this issue.

In PPLNS, each miner gets a reward that is proportional to the effort exerted during the last shares preceding a submitted solution. Since solutions are not predictable, this reward scheme discourages hoppers who risk losing shares outside the range given by N. A simplified scheme of PPLNS is shown in Fig. 1.

In Fig. 1, time flows from left to right, so that the right-most share is the most recent. A discovered block is marked with a $ sign, and not counted as a share in PPLNS. In this simple example, we consider only two miners forming the pool with power $\alpha_1 = 0.6$ and $\alpha_2 = 0.4$ for *Miner 1* and *Miner 2*, respectively. The length of the window N is 8 shares.

PPLNS is used by many Bitcoin pools, such as Kano [8], P2Pool [12], AntPool, BCMonster [2], among others. While this reward scheme is resilient to pool hopping attacks, other vulnerabilities are hypothesized to encourage dishonest mining [5]. In other words, the incentive compatibility of PPLNS is questionable [15].

We investigate a new type of attack for PPLNS pools. The idea is that miners can dishonestly increase their revenue by delaying reports of some of the shares that were obtained during a round. Instead of submitting share(s) to the pool manager when these are discovered, an attacker submits them at the end of the mining round, which will happen only if she finds a full solution. The

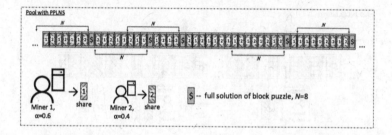

Fig. 1. Schematic explanation of mining in PPLNS pool with 2 miners.

purpose of this paper is to model the strategic incentives behind this kind of attack, as well as to estimate how damaging it can be to the pool. To do so, we formulate a simple game capturing the incentives of pool mining, and solve for Nash equilibria. A PPLNS scheme is incentive compatible if there are Nash equilibria in which miners do not delay their reports.

The rest of this paper is organized as follows: Sect. 2 contains detailed description of the attack and model that can be used to find equilibria. Conditions for incentive compatibility are discussed in Sect. 3, followed by Sect. 4, which addresses how severe attacks may be in pools that are not incentive compatible. We discuss our results and their implications in Sect. 5.

2   Model

Our model starts by computing the expected revenue of a pool member, given the pool composition, pool parameters as well as the rest of Bitcoin network. We consider the puzzle difficulty, D, to be pre-set at the network level. The PPLNS window size, N, is set by the pool manager. We also assume a given distribution of mining power τ_i for i in $1, \ldots, m$, where m is the size of the pool.

Each miner has two actions upon mining every single share: delay or report. For every miner i, we compute how the expected monetary reward changes given these options. The marginal profit for every share depends on the previous decisions made by the miner as well as the strategies of other miners in the pool.

For an attacking miner, there are two separate phases during every round. During the first phase, a miner collects shares for delay (does not report any single share). During the second phase, she reports every newly mined share immediately. For every miner, there is an individual turning point between these phases, which depends on the marginal profit of the two actions (delay or not). The turning point corresponds to the condition when the marginal profit for both actions is equal, or, when the strategy of the miner reaches its natural limit. The rationale behind these limits dictates that the number of delayed shares cannot be less than 0 and cannot exceed N. As soon as the individual turning condition is satisfied, the miner is in the second phase. In terms of time flow, equilibrium arises when every miner is beyond their turning point. Throughout the paper, we

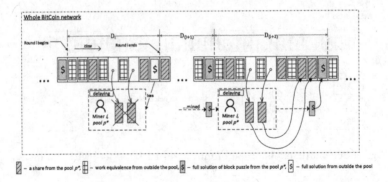

Fig. 2. Whole network schematic picture of a successful and an unsuccessful delay attack.

assume that rounds end some time after this turning point (the validity of this assumption is addressed in Sect. 4). For simplicity, we define an honest miner as one who is always in the second phase (delays 0). Likewise, attacking miners are those who delay at least 1 share in the first phase.

We also assume the following order for the submission of the delayed shares: if an attacker discovers a full solution of the Bitcoin puzzle, she reports all her delayed shares first, and reveals the full solution immediately after that. In our model, reporting shares collected during the first phase happens without time delays in revealing the full solution.

For an honest miner, the expected reward depends on N and D. Parameter D is the complexity of finding a full solution and can be expressed as the average number of shares that need to be mined to discover a full block. Every miner submits a share that he/she has mined and expects that a number of payments will be received for that share during the period in which the next N shares are sent by the pool members (a share will carry no value after this period). The expectation for that number of payments is $\frac{N}{D}$ and the value of a single payment is $Rew * \frac{1}{N}$, where Rew is a standard monetary reward for discovering a block. For simplicity, we omit the constant Rew. Therefore, every miner expects that every submitted share is worth $\frac{N}{ND}$.

These honest expectations for share payments change under delay attacks. A player j can delay an amount of $x_j \in \mathbb{N}$ shares. The effective window size is then $\hat{N}$ instead of N; and the effective expected number of shares submitted between two full solutions, found by the pool, is $\hat{D}$ instead of D. The reasons causing this are illustrated on Figs. 2 and 3. There are several immediate observations: (1) if an attacker is successful in finding a full solution she will report her delayed shares first; (2) due to delaying, the majority of the attackers will lose all the shares collected during the first phase.

Every reported share will be rewarded in a form of monetary payoff from the pool manager within the next $\hat{N}$ subsequent steps. Observation (1) above, implies

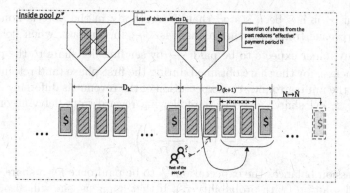

Fig. 3. Inside-pool schematic picture showing how D and N are affected by the delay attack.

that the expected number of steps when a potential reward can be received will be reduced (Fig. 3). This quantity can be computed as follows:

$$\hat{N} = \sum_{j=1}^{m}(N - x_j)\tau_j + (1 - p^*)N \leq N, \tag{1}$$

here, p^* is the probability that the solution is discovered by someone inside the miner's pool, i.e., $p^* = \sum_{j=1}^{m} x_j$. Expression (1) can be explained as follows: The first term, $\sum_{j=1}^{m}(N - x_j)\tau_j$, accounts for the probability τ_j, that miner j finds the full solution and will reduce the effective period for payment to $N - x_j$. The second term, $(1 - p^*)N$, accounts for the probability of finding the full solution outside the pool, $(1 - p^*)$. In this case, all the attacking miners lose their delayed shares and the effective period for payment is N.

Because the majority of the attackers will lose all the shares collected during the first phase, we can conclude that the amount of shares submitted between the nearest two full solutions is less than D. This is reflected in expression (2), which specifies the effective expected number $\hat{D}$ of shares submitted in the pool between the full solutions.

$$\hat{D} = \sum_{j=1}^{m}\left(x_j - \sum_{k=1}^{m} x_k + D\right)\tau_j + (1 - p^*)\left(D - \sum_{j=1}^{m} x_j\right) = D - \sum_{j=1}^{m} x_j + \sum_{j=1}^{m} x_j\tau_j. \tag{2}$$

Expression (2) can be explained as follows. Miner j will be able to publish her delayed shares with probability τ_j. In this case, all shares delayed by other attackers will be lost, and, expected number of shares (submitted in the pool since the last full solution was reported) is $(x_j - \sum_{k=1}^{m} x_k + D)$. Summing up such expectation for all the miners in the pool, we obtain $\sum_{j=1}^{m}(x_j - \sum_{k=1}^{m} x_k + D)\tau_j$. In addition, with probability $1 - p^*$ all delayed shares in the pool will be lost (because the full solution is found by miners outside the pool). This is expressed via term $(1 - p^*)(D - \sum_{j=1}^{m} x_j)$. From (1) and (2) it can be noted that when $x_j = 0, \forall j$, then $\hat{D} = D$ and $\hat{N} = N$.

Previously, it has been stated that if everybody in the pool is honest, the expected revenue from reporting a share is $\frac{N}{ND}$. In contrast, when delaying is possible any miner expects to be paid $\frac{\hat{N}}{N\hat{D}}$ by sending her share to the pool.

Nonetheless, for the share obtained during the first phase (and retained until the end of the mining round) the expectation of the revenue is different. A player j delaying $x_j - 1$ shares, expects the following reward from delaying one more share:

$$\frac{\tau_j}{N}\left(1 + \frac{N - x_j}{\hat{D}}\right).$$

This expression balances the expectation $\frac{\tau_j}{N}$ to be paid once for a share, when j finds a full solution (with probability τ_j). If that happens, she will also be paid $\frac{N - x_j}{\hat{D}}$ times in the subsequent rounds.

Now, we can sum up: some of the miners may never delay because it is not profitable for them to delay a single share; some can delay every mined share until they collect N; and, some will collect a number between 0 and N. Thus, a situation in which miners have no incentive to deviate is found by solving:

$$\frac{\hat{N}}{N\hat{D}} = \begin{cases} \frac{\tau_i}{N}\left(1 + \frac{N-x_i}{\hat{D}}\right), & \text{if } 0 \leq x_i < N, \\ \frac{\tau_i}{N}\left(1 + \frac{N-x_i}{\hat{D}}\right) + C_i, & \text{if } x_i = 0, \quad \forall i\,(C_i \geq 0). \\ \frac{\tau_i}{N}\left(1 + \frac{N-x_i}{\hat{D}}\right) - C_i, & \text{if } x_i = N, \end{cases} \tag{3}$$

This equation can be explained by the following constraints: *(i)* x_i cannot be negative – it is impossible to delay a negative number of shares; *(ii)* x_i cannot exceed N because under PPLNS, only the most recent N shares preceding the full solution (found by that pool) can be paid. The parameter C_i here compensates unequal profitability of delaying versus honest reporting. One can see that at $x_i = 0$, reporting may be more profitable for the i-th miner. On the other hand, at $x_i = N$, delaying can be more profitable than reporting.

The symbols listed in Table 1 will be used to define incentive compatibility and to estimate changes in parameters of PPLNS in case the pool is not incentive compatible (Sects. 3 and 4, respectively).

3   Incentive Compatibility

In this section, we will investigate a condition that guarantees honest mining. From Eq. (3), the only kind of incentive compatible equilibrium is described as $\frac{\hat{N}}{N\hat{D}} = \frac{\tau_i}{N}\left(1 + \frac{N-x_i}{\hat{D}}\right) + C_i, \forall i\,(x_i = 0, C_i \geq 0)$ which is equivalent to the following inequality:

$$\frac{\hat{N}}{N\hat{D}} \geq \frac{\tau_i}{N}\left(1 + \frac{N - x_i}{\hat{D}}\right), \forall i, x_i = 0. \tag{4}$$

Table 1. Notation and parameters

Notation	Meaning
p^*	Total mining power of the pool
α_i	Power of miner i relative to the mining power of the pool
τ_i	Absolute power of miner i, e.g., $\tau = \alpha_i p^*$
N	Window size, parameter of PPLNS
D	Complexity Bitcoin network expressed in (average) number of shares
$\hat{N}$	Expected number of steps when a reported share can be rewarded by the pool (case of more than 2 miners that may delay more than 1 share)
$\hat{D}$	Expected number of shares submitted into the pool during the period between reporting two consequent full solutions in the same pool (case of more than 2 miners that may delay more than 1 share)
x_j	Number of shares delayed by miner j
m	Number of miners in the pool
m'	Number of miners who delay shares in the pool

Inability to satisfy expression (4) for a single i, would mean that the pool will not mine honestly. For a pool of size m, there are $2^m - 1$ possible types of deviations from the mining protocol (each miner can either delay or always report). This yields a brute force search unfeasible for large values of m. Nonetheless, we will show that in order to verify incentive compatibility, we do not require exhaustive search. Instead, we derive a condition that can be checked in a linear time.

To derive conditions for incentive compatibility, it is useful to observe the following:

1. The set of all deviations needs to be reduced to a set $\mathcal{F}$, $|\mathcal{F}| \leq m$, of the deviations which (and only which) may produce an equilibrium (based on *Lemma 1*)
2. We show that if there is an incentive compatible equilibrium as described by (4), this equilibrium is unique (*Lemma 2*).
3. A single condition is sufficient and necessary to guarantee (4) (*Lemma 3*).

We start discussing cases that differ from (4). It will be demonstrated that there are only m other profiles that can be equilibria. We point to the fact that a delay attack requires that at least one miner delays a positive number of shares. Further, we show that an equilibrium where for a miner with power τ_i delays are only possible when all other miners with $\tau_k \geq \tau_i$ delay too.

Lemma 1. *If there is an equilibrium and a set $\mathcal{M}$ of delaying miners with power τ_i, $i \in \mathcal{M}$, delaying positive number of shares, then a miner with power τ_k is also delaying if $\exists k \notin \mathcal{M}, \tau_k \geq \tau_i$.*

(see Appendix for the proof).

As result, a miner with power τ_k should also be added to the set $\mathcal{M}$ of delaying miners. In the rest of the paper, we assume that miners are assigned indices according to their mining power sorted in descending order, e.g. $\tau_i \geq \tau_{i+1}$. This allows us to label an equilibrium compactly – specifying the index of the least powerful miner who can delay profitably. Since there are only m miners, we have at most m types of equilibria that differ from (4). The result from *Lemma* 1, showing that $x_i \geq x_{i+1}$ will be used in *Lemmas* 2 and 3.

Lemma 2. *The conditions that support incentive compatibility are inconsistent with any other kind of deviation represented by $\mathcal{F}$.*

For delaying miners included in set $\mathcal{M}$ information about other delaying miners may be incomplete. *Lemma* 2 implies that: under certain conditions, a miner with power τ_i will delay a non-negative number of shares irrespectively of its inclusion in the set of delaying miners $\mathcal{M}$; expressions (8) and (9) (Appendix) can be used to calculate directly the number of shares delayed by miner i.

For incentive compatibility, it is necessary that for the most powerful miner (with power τ_1) the delay is not profitable. Using *Lemmas* 1 and 2, we will show that a sufficient and necessary condition for incentive compatibility can be expressed in terms of τ_1.

Lemma 3. *For incentive compatible mining under PPLNS it is sufficient and necessary that $\tau_1 \leq \frac{N}{N+D}$.*

In other words, an incentive compatible pool requires a bound on the computational power of the most powerful miner. This condition for honest mining is important, but even if pools are not incentive compatible the incentives to deviate may be small. The next section explores how these incentives change when we instantiate our model with realistic parameters.

4   Severity of Delay Attacks in the Real World

We propose an algorithm for equilibrium search, and this allows us to show how the parameters of the pool affect the likelihood of delaying attacks. The precondition for our algorithm is existence of equilibria.

To quantify the effect of incentive compatibility it is important to find equilibrium in the form of (3). The main obstacle here is that (3) represents a system of piece-wise expressions. For every single expression with index i, the choice of one out of three different domains affects all expressions in the system.

We use an iterative approach. Consider the schematic illustration on Fig. 4. Here, pool miners are classified into 3 classes ($x = \{0, (0, N), N\}$) according to the power they have. As it has been discussed previously in *Lemma* 1, miners with more power can profitably delay a greater number of shares, which cannot exceed the size of reward window N. Also, the number of shares cannot be negative. According to (3), to make C_i non-negative, for separate cases $x_i = 0$

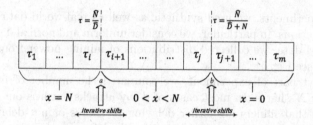

Fig. 4. Illustration of iterative algorithm for equilibrium search.

and $x_i = N$ the mining power should be $\tau_i \leq \frac{\hat{N}}{\hat{D}+N}$ and $\tau_i \geq \frac{\hat{N}}{\hat{D}}$, respectively. However, both $\hat{N}$ and, $\hat{D}$ depend on the selection of points a and b (see Fig. 4).

As soon as a, b are known, values of x for the domain $(0, N)$ can be calculated by solving a system of linear equations:

$$N - \sum_{j=1}^{m} x_j \tau_j = \tau_i \left(D - \sum_{j=1}^{m} x_j + \sum_{j=1}^{m} x_j \tau_j + N - x_i \right), \forall i, \tau_i \in (a, b),$$

where one should first substitute $x = 0$ and $x = N$ for corresponding indices.

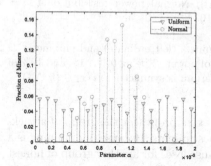

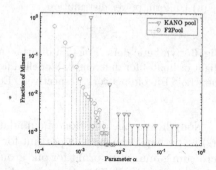

(a) Normalized distribution for uniform and normal ($\mu = 10^{-3}$, $\sigma = 10^{-3}$) cases.

(b) Normalized distribution for Kano pool and F2Pool.

Fig. 5. Distribution of mining power.

The size of the window (a, b) can potentially change from 0 to m. Therefore, the left endpoint a can be placed in any position between 1 and $m - l$, $l = $ length$[(a, b)]$. This requires $\sum_{l=0}^{m}(m - l)$ iterations with each requiring at most 2 computations (at the endpoints) to check validity of the assumption about a and b for that iteration. If the assumption is correct, the other $l - 2$ roots inside the window should be calculated. In terms of computation complexity, the whole procedure requires $O(m^2)$.

In our experiments, we used synthetic as well as real-world data for mining power distributions. In particular, we consider uniform and normal distributions. For real-world data, we collected distributions of mining power from Kanopool and F2pool (see Fig. 5).

In the first part of experiment, we compared the number of miners, who delay exactly N shares. In most cases of delay attacks it turns out that $a = b$ which means that miners are either delaying N shares or not delaying at all. The number of delaying miners is plotted for the left ordinate versus parameter k, where $N = kD$. In addition, the right ordinate scale was used to represent dependency of parameter $\frac{\hat{D}}{D}$ from k (Figs. 6 and 7).

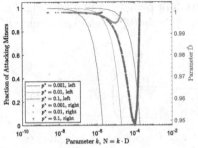
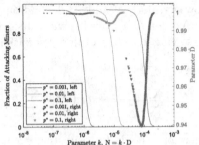

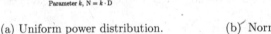

(a) Uniform power distribution.

(b) Normal power distribution ($\mu = 10^{-3}$, $\sigma = 10^{-3}$).

Fig. 6. Synthetic data. Fraction of attacking miners (left ordinate) and parameter $\hat{D}$ (right ordinate) for different k. Modelled for pool power p^* being 0.1%, 1% and 10% of the whole Bitcoin network, respectively. Equilibrium is symmetric, $\forall i \, (x_i \in \{0, N\})$.

Nonetheless, the question of cumulative extra profit (for the group of attackers) is, perhaps, the most important for honest miners. Because pool mining is a zero-sum game, extra profit for one group cause loses for another group of honest miners in that pool. There are several important differences with the concept of marginal profit for a share that has been used to find equilibrium [6]. In order to calculate cumulative extra profit one should consider: (a) extra profit is collected from those rounds where the full solution is submitted by honest miners of that pool; (b) an assumption about the duration of mining round is important and its validity is expressed with certain level of confidence (Fig. 8).

Extra profit is examined for the case when every attacker delays exactly N shares to the end of a round. Since extra profit is discussed in the context of successful solving of a puzzle by the pool, for each miner i we will refer to the power α_i in relation to the pool (not the whole Bitcoin network).

If one considers only the circumstances when attackers win a round, their expected profit is proportional to their power and is equal to what they can earn in fair mining. This is due to the fact that every miner submits N shares before

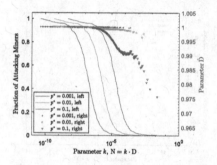

(a) Power distribution from Kano pool.

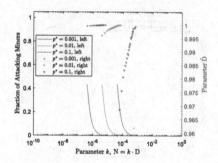

(b) Power distribution from F2pool.

Fig. 7. Real-world mining pools. Fraction of attacking miners (left ordinate) and parameter $\hat{D}$ (right ordinate) for different k. Modelled for pool power p^* being 0.1%, 1% and 10% of the whole Bitcoin network, respectively. Equilibrium is symmetric, $\forall i\,(x_i \in \{0, N\})$.

releasing a full solution. Such reward distribution is equivalent to solo mining when a miner collects all the revenue in the case of success.

However, if one considers circumstances when honest miners win, it is clear that each attacker collects a fraction of the reward which is proportional to her power in that pool. This can be seen as an additional profit (because they have already collected their fair portion). Such model of extra profit has one limitation: we assume that every attacker manages to collect her N shares (for the delay attack), and, after that, submits no less than $\alpha_i N$ shares to the pool. Therefore, a round should last the time which exceeds that estimation. For a subgroup of attackers, this happens with a probability determined by the least powerful miner in that subgroup (because collecting N shares for the attack takes her the most time). Hence, collective extra payment of any subgroup of attackers can be obtained with certain level of confidence.

It is assumed that a subgroup of attackers of size l includes all miners with power greater or equal than α_l (see *Lemma* 1 for support of this assumption). For every integer $l \in [1, m']$ (m' is the number of attackers in the pool) we will calculate: *(a)* collective extra profit E_l; *(b)* the conditional probability for a round to last longer than it takes for the l-th miner (time t_l) to mine $N + \alpha_l N$ shares, given that the round is won by that pool (i.e., probability $p(t_l|p^*)$). In Fig. 8, for every value of N we calculated maximum extra profit E_l where conditional probability $p(t_l|p^*)$ is greater than or equal to the corresponding confidence level C.

The subgroup of attackers exploits honest miners, who earn $Rew \sum_{m'+1}^{m} \tau_i$, where Rew is the current reward for discovering a full solution in the network (consisting of 12.5 BTC and transaction fees of up to 13.9 BTC on average). For the subgroup (size l) of attackers whose total power is $\sum_1^l \alpha_i$, the expected

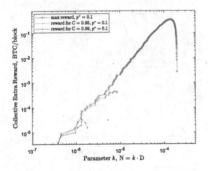

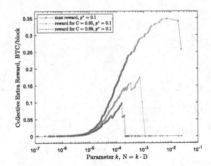

(a) Uniform distribution of mining power.

(b) Power distribution from Kano pool.

Fig. 8. Cumulative extra profit versus parameter k. Pool power is 10% of total network. Different colors represent profit for infinite length of mining round (max), for an average round with confidence levels 0.95 and 0.99, respectively. (Color figure online)

collective extra profit E_l is

$$E_l = Rew\left(\sum_{m'+1}^{m} \tau_i \times \sum_{1}^{l} \alpha_i\right).$$

The value of $p(t_l|p^*)$ is calculated as follows:

$$p(t_l|p^*) = 1 - \int_0^{t_l} f(t|p^*)dt,$$

where $f(t|p^*) = \frac{1}{Dp^*}e^{-\frac{t}{Dp^*}}$ is the conditional *pdf* for finding a full solution. The time t_l, necessary for l-th miner to collect $N + \alpha_l N$ shares is specified as $t_l = \frac{N + \alpha_l N}{\alpha_l}$. Hence, $p(t_l|p^*) = e^{-N\frac{1+\alpha_l}{\alpha_l Dp^*}}$, and, requiring that $p(t_l|p^*) \geq C$ we arrive to $\alpha_l(p^* \ln C + k) \leq -k$, $N = kD$. Considering that α_l is positive, there is an additional requirement $k < -p^* \ln C$ (it can be seen from Fig. 8 that blue and green plots are rising from zero level only for $k < 5 \times 10^{-3}$). If the latter is satisfied, we further require that $\alpha_l \geq -\frac{k}{p^* \ln C + k}$, $N = kD$.

For every k and corresponding C, we find l, such that $\max_{(\alpha_l \geq \frac{-k}{p^* \ln C + k})}[l] \leq m'$, and compute E_l (other attackers with indices $\leq l$ also pass the test and form the subgroup that has C-confident cumulative extra profit).

As one can see from the graphs, the extra profit of attackers can be quite substantial in terms of BTC. Remarkably, real-world power distributions (e.g., from Kano pool) lead to sufficiently higher levels of vulnerability to the attack, when compared with a benchmark uniform distribution of power.

5 Discussion

Incentive compatibility is an easily verifiable condition. It only requires information about the computational power of the most powerful miner in the pool. This verification can thus happen in linear time $O(m)$.

It should be stressed that known PPLNS pools comply with the requirement of incentive compatibility. For the existing majority of the pools, k varies between 1 and 5. Nonetheless, this parameter is under the sole control of the pool administrator who may decide to reduce it in order to satisfy requests from the majority of the miners.

Looking at pool miner forums, one can easily observe that a substantial number of miners would like to collect their payments faster. That aspect is especially important for pools that are not very large and infrequently discover complete solutions. Miners who join such pools during the winning round often find themselves in unfair and underpaid situations. The only way to satisfy their expectations fairly is reducing N, which increases the odds for delay attacks.

In pools that are not incentive compatible, our experimental results show that the fraction of delaying miners decreases with k, regardless of power distribution. Also, the shapes of the plots for the pools of different size (but same power distribution), e.g. 0.1%, 1%, 10% of total network power, are similar. However, a comparison between different pools reveals that for the same value of k, known real-world pools may have a higher proportion of attackers compared to artificially simulated data. This is due to the greater inequality in mining power distribution in real-world pools such as Kano. For instance, the most powerful member of a pool can sometimes account for up to a quarter of the pools total power. This may also be a significant obstacle in satisfying the condition for incentive compatibility, $\tau_1 \leq \frac{k}{k+1}$, in large pools with relatively small k.

Interestingly, $\hat{D}$ is non-monotonic on k. Obviously, $\hat{D}$ cannot be greater than D, however, the position of its minimum reflects differences in distribution of mining power in different pools. In addition, greater pool size (e.g. 10% vs 0.1% of network power) allows for attacks with greater k and that causes a greater decline in $\hat{D}$. The non-monotonic behaviour is due to the following property. For very small k, the changes in $\hat{D}$ (compared to D) are insignificant because the amount of shares that are delayed by every miner is negligibly small. For k close to the maximum, changes in $\hat{D}$ are also small due to the fact the number of attackers is small. Interestingly, the position of minimum in $\hat{D}$ for Kano pool (modelling 10% of network power) corresponds to the attack when only two most powerful miners delay. In contrast to that, for simulated data the same effect is achieved only when a majority of pool miners attack. Drops in the number of submitted shares (around 5% for large pools) can serve as a flag feature for pool administrators, who might detect the anomaly even before the attackers collect their first extra revenue.

Our plots for cumulative extra profit for a subgroup of attackers are also non-monotonic. That is because attackers exploit honest miners: when honest miners earn most the fraction of attackers is small; when fraction of attackers is large, honest miners earn little. It should be noted that the red plot (for the both types of power distribution) stands for maximum collective revenue of attackers when the whole group of attackers can exploit honest miners. That may happen only if a round is unlimited in time. Comparing extra profit in real pools with synthetic data one can notice that for high confidence of estimation, uniform distribution produces insignificant incentives for dishonest miners (even though the pool is large, 10%). On the other hand, incentives for dishonest miners may be quite substantial (up to 0.17 BTC) for a pool with power a distribution that is like that of the Kano pool.

A PPLNS variant that is adopted in several large pools uses the concept of sharechain [12]. This assumes that every share is included in a simplified version of the main Blockchain, making delay attacks impossible by protocol. On the other hand, it may also cause a negative effect on honest miners. If for some reason (e.g., network latency) a share is out of sync, it is lost. Dead on Arrival rates can reach up to 15% of all submitted shares with this scheme. This is a disadvantage for miners whose network connection is unreliable. In that sense, traditional PPLNS has an advantage and is unlikely to be replaced in the near future. Hence, aspects of traditional PPLNS scheme should be analysed with greater attention. Our model shows, in summary, that equitable pools and smaller pools are more resilient. This in sharp contrast to the state of the Bitcoin network.

The analysis of incentive compatibility and related strategic models provide an opportunity to better understand reward functions in the Blockchain. The mechanism design of reward functions is a nascent and promising application of non-cooperative game theory. These models are also useful to evaluate implementation trade-offs. For example, the so-called Block Withholding Attack [1], may become less attractive for an attacker who can benefit from delaying. An adversary delaying shares until the end of the round would be unwilling to discard complete solutions. Also note, for example, that the average number of shares submitted per discovered block, $\hat{D}$, decreases with positive delays. This reduction may be significant from the perspective of computational and network load on pool administrators.

A   Appendix

A.1   Remarks

In the proofs, several aspects related to the concept of incentive compatibility are discussed. For that purpose, it is important to show that:

(1) for the current proofs, we will distinguish only two cases (instead of 3 in Eq. 3) $0 < x \leq N$ and $x = 0$. That can be explained by the fact that pool mining is either entirely honest or not (incentive compatibility questions only that aspect). The state of incentive compatibility when nobody delays can be derived from Eq. 3, $x_i = 0, \forall i$:

$$\frac{\hat{N}}{N\hat{D}} = \frac{\tau_i}{N}\left(1 + \frac{N - x_i}{\hat{D}}\right) + C_i, \ C_i \geq 0.$$

That is equivalent to

$$\hat{N} \geq \tau_i \left(\hat{D} + N - x_i \right),$$

or, this is equivalent to the requirement

$$\hat{N} = \tau_i \left(\hat{D} + N - x_i \right), \ x_i \leq 0, \ \forall i. \tag{5}$$

The latter notation will be used as it allows to analyse conditions for incentive compatibility using the roots of a system of linear equations.

(2) One should distinguish between two different situations: a miner i may have incentives to delay a positive number of shares even if $i \notin \mathcal{M}$; or, a miner i is included in $\mathcal{M}$ and definitely has an incentive to delay. It is assumed that miners in $\mathcal{M}$ do not have information about other delaying miners from outside $\mathcal{M}$. As a result of inclusion (or not inclusion) in the group of delaying miners $\mathcal{M}$, the incentive may be different. That is easy to see on the following example: the amounts of the shares delayed by miners in $\mathcal{M}$ depend on their information about $\mathcal{M}$, but, for i-th miner who is not in $\mathcal{M}$ the amount of delayed shares depends on the information about himself (τ_i) and the information about the number of shares that are delayed by miners in $\mathcal{M}$. However, in case i is the only miner in $\mathcal{M}$, e.g. $\mathcal{M} = \{i\}$ the incentive of the miner i is the same as if $\mathcal{M} = \varnothing$.

According to the definition, incentive compatibility is an equilibrium when $\mathcal{M} = \varnothing$ and nobody has an incentive to delay. Nonetheless, it is not clear if a pool with incentive compatible conditions can be in a state of another equilibrium when $\mathcal{M} \neq \varnothing, |M| > 1$. Information about $\mathcal{M}$ may be incomplete, and, answer to the question about other (delaying) equilibrium may require certain assumption about $\mathcal{M}$. In order to resolve that obstacle, we will produce some intermediate results in *Lemmas* 1 and 2.

A.2   Lemmas

Lemma 1. If there is an equilibrium and a set $\mathcal{M}$ of delaying miners τ_i, $i \in \mathcal{M}$, delaying positive number of shares, then miner with power τ_k is also delaying if $\exists k \notin \mathcal{M}, \ \tau_k \geq \tau_i$.

Proof. Let's assume that $l = \arg \min_{\mathcal{M}} \tau_i$. Considering ONLY delaying by miners in the system described by set $\mathcal{M}$, we rewrite (5) and express x_l as

$$x_l = \frac{\sum_{j \in \mathcal{M}} x_j \tau_j - N}{\tau_l} + D + N - \sum_{j \in \mathcal{M}} x_j + \sum_{j \in \mathcal{M}} x_j \tau_j. \tag{6}$$

Now, we investigate incentive of a miner with τ_k, $k \notin \mathcal{M}$, who has information about delaying miners from $\mathcal{M}$. As previously, we use (5), however, in that case additional components with index k is included:

$$x_k \tau_k = \sum_{j \in \mathcal{M}} x_j \tau_j + x_k \tau_k - N + \tau_k \left(D + N - \sum_{j \in \mathcal{M}} x_j + \sum_{j \in \mathcal{M}} x_j \tau_j - x_k + x_k \tau_k \right),$$

$$x_k(1-\tau_k) = \frac{\sum_{j\in M} x_j\tau_j - N}{\tau_k} + D + N - \sum_{j\in M} x_j + \sum_{j\in M} x_j\tau_j. \qquad (7)$$

Right hand sides of (6) and (7) are identical except of the difference in denominators of terms $\frac{\sum_{j\in M} x_j\tau_j - N}{\tau_l}$ and $\frac{\sum_{j\in M} x_j\tau_j - N}{\tau_k}$, respectively. Nominator $\sum_{j\in M} x_j\tau_j - N$ is definitely negative. In the opposite case it would mean that at least one miner $g\in M$, has incentive to delay $x_g > N$ shares. One can conclude this from the fact that $\sum_{j\in M}\tau_j < p^* \leq 1$. Delaying $x_g > N$ is clearly irrational because PPLNS reward scheme considers only the last N submitted shares.

Therefore, $\frac{\sum_{j\in M} x_j\tau_j - N}{\tau_k} \geq \frac{\sum_{j\in M} x_j\tau_j - N}{\tau_l}$ as long as $\tau_k \geq \tau_l$. Finally, we arrive to $x_k(1-\tau_k) \geq x_l$, and because $x_l, (1-\tau_k)$ are non-negative, x_k is non-negative. $\qquad\square$

Lemma 2: Conditions that support incentive compatibility are inconsistent with any other kind of deviation represented by $\mathcal{F}$.

Proof. We organize our proof in the following order. First, some $\mathcal{M}$, $|\mathcal{M}| = l$, is considered. That can be expanded by adding index $l+1$ which represents a miner who can delay profitably. As a result, $\mathcal{M} \to \mathcal{M}'$, $|\mathcal{M}'| = l+1$. Two cases of delay attack will be accounted for a miner with τ_{l+1}: attack with $\mathcal{M}$, attack with $\mathcal{M}'$. Expressions for the number of delayed shares ($x_{l+1}^{\mathcal{M}}$ and $x_{l+1}^{\mathcal{M}'}$, respectively) will be elaborated for the both cases. It will be demonstrated that if $x_{l+1}^{\mathcal{M}}$ is positive, then $x_{l+1}^{\mathcal{M}'}$ is positive too, and, vice versa.

Second, we are going show that by reducing $\mathcal{M}$ we will arrive to $\mathcal{M}^1$, $|\mathcal{M}^1| = 1$, containing only the most powerful miner of that pool with power τ_1. That would mean that a single deviation from incentive compatibility is profitable, which contradicts with the requirement for equilibrium. This conflicts with our assumption about incentive compatibility.

(1) Recalling (5) and (6) we can write

$$x_j\tau_j = \sum_{j\in M} x_j\tau_j - N + \tau_j\left(D + N - \sum_{j\in M} x_j + \sum_{j\in M} x_j\tau_j\right),$$

$$x_j = \frac{\sum_{j\in M} x_j\tau_j - N}{\tau_j} + D + N - \sum_{j\in M} x_j + \sum_{j\in M} x_j\tau_j.$$

There are l possible variants for the first and the second equation, respectively, where $j = 1, 2, \ldots, l$. Summing up all the l variations for each of the equations, one will obtain:

$$\sum_{j\in M} x_j\tau_j = l\left(\sum_{j\in M} x_j\tau_j - N\right) + \left(D + N - \sum_{j\in M} x_j + \sum_{j\in M} x_j\tau_j\right)\sum_{j\in M}\tau_j,$$

$$\sum_{j\in M} x_j = \left(\sum_{j\in M} x_j\tau_j - N\right)\sum_{j\in M}\frac{1}{\tau_j} + l\left(D + N - \sum_{j\in M} x_j + \sum_{j\in M} x_j\tau_j\right),$$

respectively. For simplicity, we use the following substitutions: $X = \sum_{j \in \mathcal{M}} x_j \tau_j$, $Y = \sum_{j \in \mathcal{M}} x_j$, $\dot{p} = \sum_{j \in \mathcal{M}} \tau_j$, $S = \sum_{j \in \mathcal{M}} \frac{1}{\tau_j}$. Solving system

$$
\begin{cases}
X = l\,(X - N) + \dot{p}\,(D + N + X - Y) \\
Y = S\,(X - N) + l\,(D + N + X - Y)
\end{cases},
$$

in respect to X and Y we will arrive to the answers $X = N + \frac{N(l+1-2\dot{p})-D\dot{p}}{l^2-1-\dot{p}(S-1)}$, $Y = 2N + D + \frac{N(2-l+S-2\dot{p})+D(1-l-\dot{p})}{l^2-1-\dot{p}(S-1)}$. The obtained results are for the system of configuration $\mathcal{M}$ and dimensionality l. In order to re-calculate X, Y for configuration $\mathcal{M}'$ (dimensionality $l+1$) one would need to replace l with $l+1$, $\dot{p}$ with $\dot{p} + \tau_{l+1}$, S with $S + \frac{1}{\tau_{l+1}}$. For configuration $\mathcal{M}$ we express variable $x_{l+1}^{\mathcal{M}}$ (which is not yet included in the system) in terms of $X^{\mathcal{M}}, Y^{\mathcal{M}}$ using (7):

$$
\begin{aligned}
x_{l+1}^{\mathcal{M}}(1 - \tau_{l+1}) &= \frac{X^{\mathcal{M}} - N}{\tau_{l+1}} + D + N + X^{\mathcal{M}} - Y^{\mathcal{M}} \\
&= \frac{1}{\tau_{l+1}} \frac{N\left(l + 1 - 2\dot{p} + \tau_{l+1}(2l - S - 1)\right) - D\left(\dot{p} + \tau_{l+1}(1 - l)\right)}{l^2 - 1 - \dot{p}(S - 1)}.
\end{aligned}
\tag{8}
$$

For configuration $\mathcal{M}'$ we express $x_{l+1}^{\mathcal{M}'}$ as an in terms of $X^{\mathcal{M}'}, Y^{\mathcal{M}'}$ using (6):

$$
\begin{aligned}
x_{l+1}^{\mathcal{M}'} &= \frac{X^{\mathcal{M}'} - N}{\tau_{l+1}} + D + N + X^{\mathcal{M}'} - Y^{\mathcal{M}'} \\
&= \frac{1}{\tau_{l+1}} \frac{N\left(l + 1 - 2\dot{p} + \tau_{l+1}(2l - S - 1)\right) - D\left(\dot{p} + \tau_{l+1}(1 - l)\right)}{(l+1)^2 - 1 - (\dot{p} + \tau_{l+1})\left(S + \frac{1}{\tau_{l+1}} - 1\right)}.
\end{aligned}
\tag{9}
$$

Now, we are going to compare right-hand sides of Eq. (8) and (9). In the both cases nominators $N(l + 1 - 2\dot{p} + \tau_{l+1}(2l - S - 1)) - D(\dot{p} + \tau_{l+1}(1 - l))$ are identical. Our task is to prove that denominators in (8) and (9) $l^2 - 1 - \dot{p}(S - 1)$ and $(l+1)^2 - 1 - (\dot{p} + \tau_{l+1})\left(S + \frac{1}{\tau_{l+1}} - 1\right)$, respectively, are of the same sign.

We show that expression $l^2 - 1 - \dot{p}(S - 1) = l^2 - \dot{p}S - (1 - \dot{p})$ is negative. Clearly, $-(1 - p)$ is negative. Further, it will be proven that $l^2 - \dot{p}S \leq 0$. That expression can be represented as $l^2 - \sum_{j=1}^{l} \tau_j \times \sum_{j=1}^{l} \frac{1}{\tau_j}$. Component $\sum_{i=1}^{l} \sum_{j=1}^{l} \frac{\tau_i}{\tau_j}$ has l^2 terms. Exactly l out of l^2 terms are $\frac{\tau_j}{\tau_j} = 1$. Among the rest $l^2 - l$ (this number is obviously even for any natural l) terms, there are $\frac{l^2 - l}{2}$ pairs $\left(\frac{\tau_i}{\tau_j}, \frac{\tau_j}{\tau_i}\right)$, $i \neq j$. We conclude that $\frac{\tau_i}{\tau_j} + \frac{\tau_j}{\tau_i} = \frac{\tau_i^2 + \tau_j^2}{\tau_i \tau_j} \geq 2$ because $(\tau_i - \tau_j)^2 \geq 0$.

Denominator $(l+1)^2 - 1 - (\dot{p} + \tau_{l+1})\left(S + \frac{1}{\tau_{l+1}} - 1\right)$ from (9) is obtained from $l^2 - 1 - \dot{p}(S - 1)$ by substituting l with $l + 1$, $\dot{p}$ with $\dot{p} + \tau_{l+1}$, S with $S + \frac{1}{\tau_{l+1}}$. Therefore, its sign is identical to $l^2 - 1 - \dot{p}(S - 1)$ from (8) because in the proof we generalized values for $l, \dot{p}, S$. Hence, the both of $x_{l+1}^{\mathcal{M}}$ and $x_{l+1}^{\mathcal{M}'}$ are the numbers of the same sign.

(2) Further, the following technique will be used. Posit that the same conditions that provide incentive compatibility may be exploited by a set of miners $\mathcal{M}$, $|\mathcal{M}| = l$, to delay profitably. Also, let us assume another case of a set $\mathcal{M}^{l-1}$, $|\mathcal{M}^{l-1}| = l - 1$, and a miner with power τ_l who has information about $\mathcal{M}^{l-1}$. In those two cases, miner with power τ_l delays profitably according to the proof provided above. For the latter case, the configuration for delaying equilibrium can be represented as $\{\mathcal{M}^{l-1},\ l\}$. According to the results from *Lemma* 1, miner $(l - 1) \in \mathcal{M}^{l-1}$ also delays profitably. Therefore, we may consider another possible configuration $\{\mathcal{M}^{l-2},\ l - 1\}$ for whom delaying is definitely profitable. Finally, we may arrive to the configuration $\{\mathcal{M}^1,\ 2\}$ where $\mathcal{M}^1$ contains only 1-st miner with power τ_1, who can delay profitably. In such case he has an incentive to deviate from honest mining even though the information about actions of others is not taken into account. That clearly contradicts with the assumption that incentive compatibility is an equilibrium.   □

Lemma 3: For incentive compatible mining under PPLNS it is sufficient and necessary that $\tau_1 \leq \frac{N}{N+D}$.

Proof. Condition $\tau_1 \leq \frac{N}{N+D}$ can be derived from the requirement $\hat{N} \geq \tau_1 \left(\hat{D} + N - x_1 \right)$, $x_1 = 0$, for special case when $\mathcal{M} = \varnothing$ meaning that for the most powerful miner it is not profitable to delay. From the second part of *Lemma* 2 it is easy to see why such condition is necessary for incentive compatibility. In addition, it will be illustrated that it is sufficient. We consider $\mathcal{M}^1$ which includes only the 1-st miner. According to *Lemma* 1, the number of delayed shares for the second powerful miner with power of τ_2 (who is not yet included in $\mathcal{M}^1$) is not positive either, $x_2^{\mathcal{M}^1} (1 - \tau_2) \leq x_1^{\mathcal{M}^1} \leq 0$. If we consider $\mathcal{M}^2$ that includes the 1-st and 2- miners, according to *Lemma* 2, sign of x_2 does not change. Hence, neither further expansion of $\mathcal{M}$ nor considering delay from miners that are not included in $\mathcal{M}$ can produce roots that are entirely positive. This means that no delaying configuration can be in a state of equilibrium.   □

References

1. Bag, S., Ruj, S., Sakurai, K.: Bitcoin block withholding attack: analysis and mitigation. IEEE Trans. Inf. Forensics Secur. **12**(8), 1967–1978 (2017)
2. BCmonster: Mining statistics (2017). http://www.bcmonster.com/. Accessed 22 Mar 2017
3. Chávez, J.J.G., da Silva Rodrigues, C.K.: Automatic hopping among pools and distributed applications in the bitcoin network. In: 2016 XXI Symposium on Signal Processing, Images and Artificial Vision (STSIVA), pp. 1–7, August 2016
4. Dziembowski, S.: Introduction to cryptocurrencies. In: Proceedings of the 22nd ACM SIGSAC Conference on Computer and Communications Security, CCS 2015, pp. 1700–1701, NY, USA (2015). http://doi.acm.org/10.1145/2810103.2812704
5. Fisch, B.A., Pass, R., Shelat, A.: Socially optimal mining pools. ArXiv e-prints March 2017

6. Fudenberg, D., Tirole, J.: Game Theory, 11th edn. The MIT Press, Cambridge (1991)
7. Gervais, A., Karame, G.O., Wüst, K., Glykantzis, V., Ritzdorf, H., Capkun, S.: On the security and performance of proof of work blockchains. In: Proceedings of the 2016 ACM SIGSAC Conference on Computer and Communications Security, CCS 2016, pp. 3–16, NY, USA (2016). http://doi.acm.org/10.1145/2976749.2978341
8. Kano pool: Pool payout (2017). https://kano.is/index.php?k=payout. Accessed Mar 23 2017
9. Lewenberg, Y., Bachrach, Y., Sompolinsky, Y., Zohar, A., Rosenschein, J.S.: Bitcoin mining pools: A cooperative game theoretic analysis. In: Proceedings of the 2015 International Conference on Autonomous Agents and Multiagent Systems, AAMAS 2015, pp. 919–927, International Foundation for Autonomous Agents and Multiagent Systems, Richland, SC (2015). http://dl.acm.org/citation.cfm?id=2772879.2773270
10. Morabito, V.: Business Innovation Through Blockchain, vol. 1. Springer International Publishing AG, Heidelberg (2017)
11. Nakamoto, S.: Bitcoin: A peer-to-peer electronic cash system (2008). https://bitcoin.org/bitcoin.pdf. Accessed 29 Jan 2016
12. P2Pool: P2Pool bitcoin mining pool global statistics (2017). http://p2pool.org/stats/index.php. Accessed 19 Mar 2017
13. Peck, M.: A blockchain currency that beat s bitcoin on privacy [news]. IEEE Spectr. **53**(12), 11–13 (2016)
14. Rosenfeld, M.: Analysis of bitcoin pooled mining reward systems. arXiv preprint (2011). arXiv:1112.4980
15. Schrijvers, O., Bonneau, J., Boneh, D., Roughgarden, T.: Incentive compatibility of bitcoin mining pool reward functions. In: Grossklags, J., Preneel, B. (eds.) FC 2016. LNCS, vol. 9603, pp. 477–498. Springer, Heidelberg (2017). doi:10.1007/978-3-662-54970-4_28

Adaptivity in Network Interdiction

Bastián Bahamondes[1], José Correa[1], Jannik Matuschke[2(✉)],
and Gianpaolo Oriolo[3]

[1] Department of Industrial Engineering, Universidad de Chile, Santiago, Chile
{bastian.bahamondes,correa}@uchile.cl
[2] TUM School of Management and Department of Mathematics,
Technische Universität München, Munich, Germany
jannik.matuschke@tum.de
[3] Department of Civil Engineering and Computer Science,
Università di Roma Tor Vergata, Rome, Italy
oriolo@disp.uniroma2.it

Abstract. We study a network security game arising in the interdiction of fare evasion or smuggling. A defender places a security checkpoint in the network according to a chosen probability distribution over the links of the network. An intruder, knowing this distribution, wants to travel from her initial location to a target node. For every traversed link she incurs a cost equal to the transit time of that link. Furthermore, if she encounters the checkpoint, she has to pay a fine.

The intruder may adapt her path online, exploiting additional knowledge gained along the way. We investigate the complexity of computing optimal strategies for intruder and defender. We give a concise encoding of the intruders optimal strategy and present an approximation scheme to compute it. For the defender, we consider two different objectives: (i) maximizing the intruder's cost, for which we give an approximation scheme, and (ii) maximizing the collected fine, which we show to be strongly NP-hard. We also give a paramterized bound on the worst-case ratio of the intruders best adaptive strategy to the best non-adaptive strategy, i.e., when she fixes the complete route at the start.

1 Introduction

Network interdiction problems model the control or halting of an adversary's activity on a network. Typically, this is modelled as the interaction between two adversaries—an *intruder* and a *defender*—in the context of a Stackelberg game. The defender allocates (or removes) scarce resources on the network in order to thwart the objective of the intruder, who—knowing the defender's strategy—reacts by choosing the response strategy optimizing his own objective. Such models are used to great effect in applications such as disease containment [11,13], drug traffic interdiction [17], airport security [16], or fare inspection [5].

In order to mitigate the intruder's advantage of observing the defender's actions first, the defender may opt to employ a randomized strategy. The intruder can only observe the probability distribution of the defender's actions, but she

S. Rass et al. (Eds.): GameSec 2017, LNCS 10575, pp. 40–52, 2017.
DOI: 10.1007/978-3-319-68711-7_3

does not know the exact realization. In this work, we study a variant of a network interdiction problem in which the defender employs such randomization, but the intruder gains additional information about the realization while she is acting, and may use this information to adapt her strategy.

Our game is played on a network. The defender randomly establishes a security checkpoint on one of the arcs. The intruder wants to move from her initial location to a designated target node, preferably without being detected by the defender. Her objective is to minimize her expected cost, which consists of movement costs for traversing arcs and a fine, which has to be paid if she traverses the arc with the checkpoint. Knowing the probability distribution specified by the defender and that only one arc is subjected to inspection, the intruder gains additional information while traveling through the network, observing whether or not the inspected arc was among those she traversed so far. She may use this information in order to decide which arc to take next. This type of path-finding strategy is called *adaptive*, as opposed to a *non-adaptive strategy*, in which she commits to an origin-destination-path at the start and does not deviate from it.

In this paper, we investigate the computational complexity of finding optimal adaptive and non-adaptive strategies for the intruder as well as optimal randomized strategies for the defender, considering two objectives: (i) the *zero-sum* objective of maximizing the intruder's cost and (ii) the *profit maximization* objective of maximizing the expected collected fine. We also provide bounds on the cost ratio between optimal adaptive and non-adaptive strategies and the impact of adaptivity on the defender's objective.

1.1 Related Work

Stackelberg games, and in particular network interdiction models, are widely used in the context of security applications; see the textbook by Tambe for an overview of applications in airport security [16].

A very basic version of a Stackelberg game is the security game studied by Washburn and Wood [17]. In this zero-sum game, an inspector strives to maximize the probability of catching an evader, who chooses a path minimizing that probability. The authors show that optimal strategies for both players can be computed by a network flow approach. The optimization problem of maximizing the defender's profit has been extensively studied in the context of *Stackelberg pricing games* [3,9,14]. Here, the defender sets tolls for a subset of the edges of the network, trying to collect as much tolls as possible from the intruder, who chooses a path minimizing the sum of the travel costs plus the tolls. As opposed to the zero-sum game mentioned above, these pricing games are usually computationally hard to solve.

The particular game we study in this article arises from a variation of two toll/fare inspection models introduced by Borndörfer et al. [2] and Correa et al. [5], respectively. In these models, the defender, who represents the network operator, decides an inspection probability for each arc, subject to budget limiting the total sum of inspection probabilities. The intruder (toll evading truck drivers/fare evading passengers) tries to get to her destination minimizing a combined objective of travel time and expected cost for the fine when being discovered. Correa et al. [5] also study an adaptive version of the problem, in which the intruders adapt their behavior as they traverse the network. They propose an efficient algorithm based on a generalized flow decomposition, and give a tight bound on the adaptivity gap of 4/3; see Sect. 5 for details. In both the above models, the event of an inspection occurring on a given arc is independent to that on all other arcs. In contrast, in our model, the checkpoint can only be located on a single arc, leading to a different optimization problem for the intruder.

A different notion of adaptive path-finding was previously studied by Adjiashvili et al. [1] in the so-called *Online Replacement Path* problem. Here, a routing mechanism must send a package between two nodes in a network trying to minimize transit cost. An adversary, knowing the intended route, may make one of the arcs fail. Upon encountering the failed arc, the package may be rerouted to its destination along a different path. Note that in this setting the failing arc is chosen by the adversary *after* the routing has started, whereas in our settings the inspection probabilities are determined *before* the intruder chooses her path. Computationally, adaptive path-finding is related to shortest path problems in which there is a trade-off between two cost functions. The restricted shortest path problem [6,8,10] and the parametric shortest path problem [4,12] are representative examples of such problems.

1.2 Contribution

We study both adaptive and the non-adaptive path-finding strategies for the intruder. After observing that the non-adaptive intruder's problem reduces to the standard shortest path problem, we turn into the adaptive version, which turns out to be much more intricate. We show that an optimal adaptive strategy of the intruder can always be represented by a simple, i.e., cycle-free, path. We then devise fully polynomial time approximation scheme (FPTAS) for computing the a near-optimal adaptive strategy with adjustable precision.

By using an approximate version of the equivalence of separation and optimization [15], we also obtain an FPTAS for maximizing the defender's zero-sum objective. For the profit objective, on the other hand, we show that the defender's optimization problem is strongly NP-hard, ruling out the existence of an FPTAS (unless $P = NP$).

We further study the impact of adaptivity on the intruder's and defender's objective. Extending a result by Correa et al. [5], we show that the intruder's best non-adaptive strategy is within a factor of 4/3 of the optimal adaptive strategy and that this ratio decreases for instances where the intruder does not deviate significantly from her shortest path (which is a natural assumption, e.g., in the context of transit networks). We also mention that our bound on the adaptivity gap for the intruder directly translates to several guarantees for the defender's zero sum game, e.g., bounding his loss in pay-off when he wrongly believes the intruder is non-adaptive.

2   The Model

Before we can describe our model in detail, we establish some notation. Throughout this article, we are given a directed graph $G = (V, E)$ with $n := |V|$ nodes and $m := |E|$ arcs. For two nodes $u, v \in V$ an u-v-walk in G is a sequence of edges $(e_1, \ldots, e_k)$ with $e_i = (v_{i-1}, v_i) \in E$ and $v_0 = u$ and $v_k = v$. A u-v-path is a u-v-walk in which no arc or node is repeated, i.e., $e_i \neq e_j$ and $v_i \neq v_j$ for $i \neq j$. For a u-v-path P, we let $V(P)$ be the set of nodes visited by P and for $u', v' \in V(P)$ such that P visits u' before v', we let $P[u', v']$ denote the u'-v'-path contained in P. We denote the set of all u-v-walks in G by $\mathcal{W}_{uv}$ and the set of all u-v-paths in G by $\mathcal{P}_{uv}$.

In our model, the intruder starts at a designated node s and wants to reach a node t (both nodes are also known to the defender). Each arc $e \in E$ is equipped with a cost $c_e \in \mathbb{Z}_+$ that is incurred to the intruder when she traverses e. Furthermore, there is a fine F, which the defender charges to the intruder, if she runs into the defender's security checkpoint. In the first level of our interdiction game, the defender specifies the random distribution of the checkpoint, i.e., he specifies for every arc $e \in E$ the probability π_e of placing the checkpoint at e. In the second level, the intruder takes her way from s to t, having full knowledge of the probability distribution chosen by the defender. We distinguish two variants of the intruder's path-finding strategy:

non-adaptive: At the start, the intruder selects an s-t-path $P \in \mathcal{P}_{st}$ and follows this path to t. For every arc $e \in P$ she pays the transit cost c_e of that arc. In addition, if the security checkpoint is located on one of the arcs of P, she has to pay the fine F.

adaptive: From her current location, the intruder moves along one of the outgoing arcs e to a neighboring vertex, paying the transit cost c_e. She observes whether the security checkpoint is located at the arc she traverses (in which case she additionally has to pay the fine F). Knowing this information, she decides which arc to take next. This procedure continues until she reaches her destination (after a finite number of steps).

The intruder's objective is to minimize her expected cost. For a set of arcs S, we use $c(S) := \sum_{e \in S} c_e$ to denote the sum of the transit times and $\pi(S) := \sum_{e \in S} \pi_e$ to denote the probability that the security checkpoint is located within the set of arcs S (note that we can sum up these probabilities since there is a single checkpoint, so these are disjoint events). Therefore, in the non-adaptive case, the expected cost of following a path P is

$$f_{N,\pi}(P) := c(P) + \pi(P)F = \sum_{e \in P} (c_e + \pi_e F).$$

We denote the optimization problem of finding an optimal non-adaptive strategy for the intruder by

$$\min_{P \in \mathcal{P}_{st}} f_{N,\pi}(P). \tag{INT$_N$}$$

Thus, it is straightforward to note that an optimal non-adaptive strategy for the intruder is to follow a shortest path with respect to arc weights $c_e + \pi_e F$. Such a path can be computed efficiently, e.g., using Dijkstra's Algorithm. Therefore we conclude the following result.

Proposition 1. INT$_N$ *reduces to the Shortest Path Problem and can be solved in polynomial time.*

The optimal adaptive strategy is less obvious. In principle, the intruder's choice of where to go next from her current location can depend on the set of arcs she has visited so far and the information whether the security checkpoint is located at one of these arcs. Let us consider any such adaptive strategy. Note that, because the intruder has to reach t after a finite number of steps, for each fixed realization of the checkpoint location, the strategy determines an s-t-walk. We distinguish two cases.

First, assume the intruder encounters the checkpoint in every realization. Then for the given strategy, she pays the fine with probability 1. Obviously, the non-adaptive strategy of simply following the shortest path with respect to c has at most the same cost than the considered strategy.

Now assume that there is a realization in which the intruder reaches t without being inspected. Let $W = (e_1, \dots, e_k)$ be the walk she takes in this realization, with $e_i = (v_{i-1}, v_i)$, $v_0 = s$, and $v_k = t$. Observe that W is the same for all realizations where the intruder is not inspected, as her decisions are based only on whether or not she encountered the checkpoint so far. We now define a new adaptive strategy, in which the intruder follows W starting at s until she either reaches t or encounters the security checkpoint at some arc e_i of W. In the latter case, after traversing e_i she simply follows a shortest path with respect to c from her current location v_i to t. It is easy to check that the cost of the new strategy is at most the cost of the strategy considered originally.

We have thus shown that for every adaptive strategy there is a strategy of at most the same cost which is completely defined by an s-t-walk W that the intruder follows while not being inspected. Note that W can contain cycles and arcs can appear multiple times along W. Define $\tilde{\pi}_i := \pi_{e_i}$ if $e_i \neq e_j$ for all $j < i$, i.e., the ith position is the first appearance of the arc e_i on W, and $\tilde{\pi}_i := 0$ otherwise, i.e., if arc e_i occurred on W before the ith position. Furthermore, let $\text{SP}_c(v, w) := \min_{P \in \mathcal{P}_{vw}} c(P)$ be the length of a shortest path w.r.t. c from v to w. Then the intruder's expected cost for following W can be expressed as follows:

$$f_{A,\pi}(W) := \sum_{i=1}^{k} \tilde{\pi}_i \left(\sum_{j=1}^{i} c_{e_j} + F + \text{SP}_c(v_i, t) \right) + \left(1 - \sum_{i=1}^{k} \tilde{\pi}_i \right) \sum_{i=1}^{k} c_{e_i}$$

Here, each summand of the first sum corresponds to the event that the checkpoint is encountered at arc e_i (which can only happen if it is the first occurrence of this arc along the walk). In this case, the intruder traverses the walk W until e_i, pays the fine, and then follows the shortest path from v_i to t. The second sum

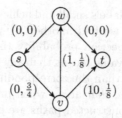

Fig. 1. Example network for the intruder's best response problem. Labels (c_e, π_e) at the arcs denote transit times and inspection probabilities. A possible adaptive strategy for the intruder is to follow s-t-walk s-v-w-s-v-t and deviating to a shortest path when encountering the security checkpoint. For a fine $F = 7$, the expected cost of this strategy is 9.25, whereas following the underlying simple path s-v-t deviating to a shortest path after inspection has a higher expected cost of 9.375.

represents the event that none of the arcs in W contains the checkpoint, in which case the intruder simply traverses W from start to end.

In the above discussion, we assumed that the intruder may walk along cycles and even traverse arcs multiple times. Although all transit costs are non-negative, such detours cycles could—in principle—help the intruder, because along the way she gains additional information. In fact, Fig. 1 depicts an example of an s-t-walk containing a cycle where the intruder's expected cost increases when omitting the cycle. However, one can show that there always exists an optimal adaptive solution without a cycle, i.e., defined by an s-t-path.

Lemma 1. *Let P be a shortest s-t-path w.r.t. c. Then $f_{A,\pi}(P) \leq \mathrm{SP}_c(s,t) + F$.*

Lemma 2. *There is an s-t-path P such that $f_{A,\pi}(P) \leq f_{A,\pi}(W)$ for all s-t-walks W.*

The problem of finding an optimal adaptive strategy thus reduces to finding an s-t-path minimizing $f_{A,\pi}$. We denote this optimization problem by

$$\min_{P \in \mathcal{P}_{st}} f_{A,\pi}(P). \tag{$\mathrm{INT_A}$}$$

3   Approximating the Intruder's Optimal Strategy

A *fully polynomial time approximation scheme* (FPTAS) for a minimization problem is an algorithm that takes as input an instance of the problem as well as a precision parameter $\varepsilon > 0$, and computes in polynomial time in the size of the input and $1/\varepsilon$ a solution to that instance with cost at most $(1 + \varepsilon)\mathrm{OPT}$, where OPT denotes the cost of the optimal solution.

In this section, we design such an FPTAS for $\mathrm{INT_A}$. The algorithm is based on a label propagating approach, where each label at node v represents an s-v-path, that is extended by propagating the label along the outgoing edges of v. In order to keep the number of distinct labels small and achieve polynomial running time, we discard an s-v-path when we find another s-v-path with similar objective function value but higher inspection probability (intuitively higher inspection probability at equal objective value means that any completion of the new path to an s-t-path will be cheaper than the corresponding completion of the former path). An additional challenge, that arises when propagating the labels in the graph, is to ensure that the constructed paths are cycle free. To deal with this issue we argue that there is a way to avoid cycles without overlooking potentially good paths.

The Algorithm. Given $\varepsilon > 0$, let $\alpha := 1 + \frac{\varepsilon}{2n}$. From Lemma 1, we know that the cost of an optimal strategy is in the interval $[0, \mathrm{SP}_c(s,t) + F]$. We divide this interval geometrically by powers of α. Let $K := \lceil \log_\alpha (\mathrm{SP}_c(s,t) + F) \rceil$ and define $I_0 := [0,1)$ as well as $I_k := [\alpha^{k-1}, \alpha^k)$ for $k \in \{1, \dots, K\}$. At every node v we maintain an array $L_v^0, \dots, L_v^K$, where L_v^i is either empty or contains a label (f, q, P) such that P is an s-v-path with $f = f_{A,\pi}(P) \in I_k$ and $q = \pi(P)$.

Initially, only the label $L_s^0 = (0, 0, \emptyset)$ is present. In each iteration, the algorithm propagates all labels at each vertex v along all outgoing arcs (v, w). When propagating label (f, q, P) at node v along arc $e = (v, w)$, we get a label (f', q', P') at node w with $f' = f + (1 - q)c_e + \pi_e(F + \mathrm{SP}_c(w,t))$, $q' = q + \pi_e$, and $P' = P \cup \{e\}$. In order to avoid cycles, the propagation of (f, q, P) along $e = (v, w)$ only takes place if $w \notin V(P)$. Moreover, if the propagation of a label along an arc gives rise to two different labels (f', q', P') and (f'', q'', P'') for a node such that $f', f'' \in I_k$ for some k, we discard the label with the lower inspection probability (breaking ties arbitrarily). The full description is given in Algorithm 1.

From the previous discussion, the following lemma is straightforward:

Lemma 3. *If Algorithm 1 creates a label (f, q, P) in a node $v \in V$, then P is a (s, v)-path with $f_{A,\pi}(P) = f$ and $\pi(P) = q$.*

Now let P^* be an s-t-path minimizing $f_{A,\pi}(P^*)$. Let $(e_1, \dots, e_k)$ be the arcs of P^*, with $e_i = (v_{i-1}, v_i)$, $v_0 = s$ and $v_k = t$. Define $f_i^* := f_{A,\pi}(P^*[s, v_i])$ and $q_i^* := \pi(P^*[s, v_i])$. For $x \in \mathbb{R}$, let $(x)_+$ denote the positive part of x, i.e. $(x)_+ := \max\{x, 0\}$. We call an iteration of the outer for loop of Algorithm 1 a *round*. The following lemma can be proved by induction on the rounds of the algorithm, using a sequence of careful estimates on the cost of paths and subpaths.

Lemma 4. *After round i of Algorithm 1, there is a label (f_i, q_i, P_i) at node v_i with $f_i \leq \alpha^i f_i^* - (q_i^* - q_i)_+ \cdot c(P^*[v_i, t])$.*

Lemma 4 in particular implies that, at the end of round n, the algorithm has found an s-t-path P with $f_{A,\pi}(P) \leq \alpha^n f_{A,\pi}(P^*)$. Note that $\alpha^n = (1 + \frac{\varepsilon}{2n})^n \leq (1 + \varepsilon)$ for all $\varepsilon < 1$. It is also easy to verify that the algorithm runs in time polynomial in $1/\varepsilon$ and the input size.

Theorem 1. *Algorithm 1 is an FPTAS for* INT_A.

Algorithm 1. FPTAS for INT_A

1: Compute $\text{SP}_c(v,t)$ for all $v \in V$.
2: Let $\alpha \leftarrow 1 + \frac{\varepsilon}{2n}$ and $K \leftarrow \lceil \log_\alpha (\text{SP}_c(s,t) + F) \rceil$
3: Let $L_s^0 \leftarrow (0, 0, \emptyset)$ and $L_v^k \leftarrow \emptyset$ for all $(v,k) \in V \times \{0, \ldots, K\} \setminus \{(s,0)\}$
4: **for** $i = 1, \ldots, (n-1)$ **do**
5:     **for all** $e = (v,w) \in E$ **and** $k = 0, \ldots, K$ **do**
6:         **if** $L_v^k \neq \emptyset$ **then**
7:             $\text{PUSH}(L_v^k, e)$
8: Let $(f^*, q^*, P^*) \in \text{argmin} \{ f \; : \; (f,q,P) \in L_t^k \text{ for some } k \}$
9: Return P^*

10: **procedure** $\text{PUSH}(L = (f,q,P), e = (v,w))$
11:     **if** $w \notin V(P)$ **then**
12:         Let $f' \leftarrow f + (1-q)c_e + \pi_e (\text{SP}_c(w,t) + F)$
13:         Let $q' \leftarrow q + \pi_e$
14:         Let $P' \leftarrow P \cup \{e\}$
15:         Let $k \leftarrow \min \{\ell \in \mathbb{Z}_+ \; : \; f' < \alpha^\ell\}$
16:         **if** $L_w^k = \emptyset$ **then**
17:             $L_w^k \leftarrow (f', q', P')$
18:         **else**
19:             Let $(f'', q'', P'') \leftarrow L_w^k$
20:             **if** $q' > q''$ **then**
21:                 $L_w^k \leftarrow (f', q', P')$

4   Complexity of the Defender's Problem

We study the defender's optimization problem for deciding the inspection probabilities on every edge of the network, for both the adaptive and non-adaptive intruder. We analyze two different objectives: maximizing the minimum expected intruder's cost and collecting the highest possible fine from inspections.

4.1   The Zero-Sum Objective

We first consider the defender's problem of maximizing the intruder's expected cost. This problem can be stated as

$$\max_{\substack{\sum_{e \in E} \pi_e = 1 \\ \pi \geq 0}} \min_{P \in \mathcal{P}_{st}} f_{X,\pi}(P), \qquad (\text{DEF}_X^{\text{cost}})$$

where $X \in \{A, N\}$, depending on whether the intruder is adaptive or non-adaptive. Note that for a fixed path $P \in \mathcal{P}_{st}$, the function $f_{X,\pi}(P)$ is affine linear in π, both for $X = $ A and $X = $ N. Therefore, we can reformulate the defender's problem as a linear program:

$$\max_{\lambda \in \mathbb{R}, \pi \in \mathbb{R}^E} \quad \lambda$$

$$\text{s.t.} \quad \lambda \leq f_{X,\pi}(P) \quad \forall P \in \mathcal{P}_{st}$$

$$\sum_{e \in E} \pi_e = 1 \qquad\qquad\qquad (\text{LP}_X^{\text{cost}})$$

$$\pi_e \geq 0 \qquad\qquad \forall e \in E.$$

Note that the number of constraints in the above LP can be exponential in the size of the network, as it contains one constraint for every path. A standard way to solve such non-compact LPs is to devise a *separation routine*: A famous result by Grötschel, Lovasz, and Schrijver [7] shows that in order to solve a linear program with the ellipsoid method, it is sufficient to determine for a given setting of the variables, whether it is a feasible solution, and if not, find a violated inequality.

Indeed checking whether a given solution (π, λ) is feasible for $\text{LP}_X^{\text{cost}}$ boils down to determining whether there is a path P with $f_{X,\pi}(P) < \lambda$. For this, it is sufficient to determine the intruder's optimal path. As discussed in Sect. 2, this can be done efficiently for the non-adaptive setting. We thus obtain the following theorem.

Theorem 2. $\text{DEF}_N^{\text{cost}}$ *can be solved in polynomial time.*

For the adaptive intruder problem, we do not know an exact polynomial time algorithm. However, we can use the FPTAS presented in Sect. 3 as an *approximate separation routine*. This enables us to employ an approximation version of the equivalence of separation and optimization [15], obtaining an FPTAS for $\text{DEF}_A^{\text{cost}}$.

Theorem 3. *There is an FPTAS for* $\text{DEF}_A^{\text{cost}}$.

4.2 The Profit Maximization Objective

Next we address the problem of maximizing the expected fine collected by the defender through inspections, that is

$$\max \sum_{e \in P} \pi_e F \qquad\qquad\qquad (\text{DEF}_X^{\text{fine}})$$

$$\text{s.t.} \sum_{e \in E} \pi_e = 1, \quad \pi \geq 0$$

$$P \in \text{argmin}\{f_{X,\pi}(P') : P' \in P \in \mathcal{P}_{st}\},$$

where again $X \in \{\text{A}, \text{N}\}$ specifies whether the intruder employs an adaptive or non-adaptive path-finding strategy, respectively.

This problem shares many features with the Stackelberg network pricing problem, which is defined as follows: in the first stage, the defender sets tolls on a given subset of "tollable" edges. In the second stage the intruder chooses a path between two fixed nodes minimizing the sum of travel times plus the tolls of the traversed arcs. The defender's objective is to maximize the collected revenue from the tolls. Roch et al. [14] showed that this problem is NP-hard.

We show that also $\mathrm{DEF}_N^{\mathrm{fine}}$ is NP-hard, even when all arc costs are in $\{0, 1, 2\}$. Such a hardness for instances with small input numbers is referred to as *strong* NP-hardness. Our reduction resembles that of Roch et al., but we have to introduce some modifications to accommodate for non-tollable arcs, which exist in the Stackelberg network pricing problem but not in $\mathrm{DEF}_N^{\mathrm{fine}}$.

Theorem 4. $\mathrm{DEF}_N^{\mathrm{fine}}$ *is strongly NP-hard.*

Although we do not provide a hardness result for $\mathrm{DEF}_A^{\mathrm{fine}}$, we expect it to be NP-hard as well, as the adaptive intruder's first stage problem becomes as least as hard than it is in the $\mathrm{DEF}_N^{\mathrm{fine}}$ setting.

5 The Impact of Adaptivity

5.1 Adaptivity Gap for the Follower

Let OPT_A and OPT_N the optimal values for INT_A and INT_N respectively. Correa et al. [5] showed that for their model (in which inspections are independent events) the ratio of the best non-adaptive strategy to the best adaptive strategy is bounded by 4/3. Indeed, their proof does not use the fact that arc inspections are independent events and thus translates to our setting.

Theorem 5 (Correa et al. [5]). $\mathrm{OPT}_N \le \frac{4}{3}\mathrm{OPT}_A$.

In many real-life scenarios, it is reasonable to assume that the ratio of the length of the path chosen by the intruder to the shortest path (w.r.t. c) is not too large. E.g., most passengers in transit systems would pay a ticket rather than choosing a path with twice the transit time just in order to avoid inspection. We extend the proof by Correa et al. [5] to give a parameterized bound that takes this ratio into account and gives stronger guarantees for realistic values; also see Fig. 2.

Theorem 6. *If* $\mathrm{SP}_c(s, t) > 0$, *then* $\mathrm{OPT}_N \le \frac{\Delta^2}{2(1-\Delta)^{3/2}+3\Delta-2}\mathrm{OPT}_A$, *where* $\Delta :=$ $\mathrm{SP}_c(s, t)/c(P^*)$ *and* P^* *is an optimal solution to* INT_A.

Proof. We first observe that $\mathrm{OPT}_N \le \min\{\mathrm{SP}_c(s, t) + F, c(P^*) + \pi(P^*)F\}$ as both following the shortest path or following P^* are feasible non-adaptive strategies. On the other hand, observe that $\mathrm{OPT}_A = f_{A,\pi}(P^*) \ge (1 - \pi(P^*))c(P^*) + \pi(P^*)(\mathrm{SP}_c(s, t) + F)$, as the total amount of transit cost will always be at least

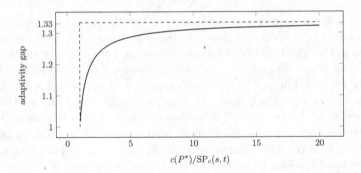

Fig. 2. Upper bound on the adaptivity gap $\text{OPT}_N/\text{OPT}_A$ given in Theorem 6 parameterized by $\varDelta^{-1} = c(P^*)/\text{SP}_c(s,t)$, where P^* is an optimal solution to INT_A.

as much as the length of a shortest s-t-path. Defining $S := \text{SP}_c(s,t)$, $C := c(P^*)$, and $Q := \pi(P^*)$, we obtain

$$\frac{\text{OPT}_A}{\text{OPT}_N} \geq \frac{(1-Q)C + Q(S+F)}{\min\{S+F, \, C+QF\}} = \frac{(1-Q)C + Q(\varDelta C + F)}{\min\{\varDelta C + F, \, C+QF\}}.$$

In order to prove the bound, we fix $\varDelta$ and treat C, F, Q as variables of an optimization problem subject to $Q \in [0,1]$ and $F, C \geq 0$.

$$\frac{\text{OPT}_A}{\text{OPT}_N} \geq \min_{F,C\geq0,Q\in[0,1]} \frac{(1-Q)C + Q(\varDelta C + F)}{\min\{\varDelta C + F, \, C+QF\}}.$$

It is easy to see that in an optimal solution, the minimum in the denominator is attained by both terms, i.e., $\varDelta C + F = C + QF$. Substituting $F = \frac{1-\varDelta}{1-Q}C$ we get

$$\frac{\text{OPT}_A}{\text{OPT}_N} \geq \min_{C\geq0,Q\in[0,1]} \frac{(1-Q)C}{\left(1+Q\frac{1-\varDelta}{1-Q}\right)C} + Q = \min_{Q\in[0,1]} \frac{(1-Q)^2}{1+\varDelta Q} + Q.$$

By computing the derivative of the righthand side term, we observe that the minimum is attained at $Q = \frac{1-\sqrt{1-\varDelta}}{\varDelta}$, which gives the desired bound. $\square$

5.2   Defender Gaps

We consider three gaps concerning the defender in the context of the zero-sum objective. Let π_A and π_N be the inspection probabilities that maximize the intruder's costs against an adaptive and non-adaptive intruder respectively, and let $f_X(\pi_Y) := \min_{P \in \mathcal{P}_{st}} f_{X,\pi_Y}(P)$ denote the defender's pay-off, where $X, Y \in \{\text{A}, \text{N}\}$.

Adaptivity Gap (η_A): This measures the defender's pay-off loss when the intruder is adaptive, as opposed to when she is non-adaptive.

Pay-off Gap (η_P)**:** When the intruder is adaptive, this gap measures the deviation of the defender's pay-off from his own estimation if he wrongly assumes she is non-adaptive.

Approximation Gap (η_{App})**:** This is the approximation factor achieved by the defender against an adaptive intruder when playing the optimal strategy for non-adaptive intruders π_N as an approximation for π_A.

$$\eta_A = \frac{f_N(\pi_N)}{f_A(\pi_A)}, \quad \eta_P = \frac{f_N(\pi_N)}{f_A(\pi_N)}, \quad \eta_{App} = \frac{f_A(\pi_A)}{f_A(\pi_N)}.$$

As a straightforward consequence of Theorem 5, all of these gaps are upper bounded by 4/3.

6   Conclusion

In this paper, we investigated different variants of a Stackelberg network game in which the follower can gain and exploit information about the realization of the leader's random strategy while traversing the network. In the present work, we confined ourselves to the model in which a single arc is subjected to inspections. Future work will focus on the natural generalization in which several checkpoints are placed simultaneously and possibly in a correlated fashion, getting closer to real-world security scenarios.

Acknowledgements. This work was supported by the Alexander von Humboldt Foundation with funds of the German Federal Ministry of Education and Research (BMBF), by the Millennium Nucleus Information and Coordination in Networks Grant ICM/FIC RC130003, and by a CONICYT grant (CONICYT-PCHA/MagísterNacional/2014 - 22141563).

References

1. Adjiashvili, D., Oriolo, G., Senatore, M.: The online replacement path problem. In: Bodlaender, H.L., Italiano, G.F. (eds.) ESA 2013. LNCS, vol. 8125, pp. 1–12. Springer, Heidelberg (2013). doi:10.1007/978-3-642-40450-4_1
2. Borndörfer, R., Omont, B., Sagnol, G., Swarat, E.: A Stackelberg game to optimize the distribution of controls in transportation networks. In: Krishnamurthy, V., Zhao, Q., Huang, M., Wen, Y. (eds.) GameNets 2012. LNICSSITE, vol. 105, pp. 224–235. Springer, Heidelberg (2012). doi:10.1007/978-3-642-35582-0_17
3. Brotcorne, L., Labbé, M., Marcotte, P., Savard, G.: A bilevel model for toll optimization on a multicommodity transportation network. Transp. Sci. **35**(4), 345–358 (2001)
4. Carstensen, P.J.: The complexity of some problems in parametric linear and combinatorial programming (1983)
5. Correa, J.R., Harks, T., Kreuzen, V.J.C., Matuschke, J.: Fare evasion in transit networks. Oper. Res. **65**(1), 165–183 (2017)
6. Garey, M.R., Johnson, D.S.: Computers and Intractability: A Guide to the Theory of NP-Completeness. W. H. Freeman, New York (2002)

7. Grötschel, M., Lovász, L., Schrijver, A.: Geometric Algorithms and Combinatorial Optimization. Algorithms and Combinatorics, vol. 2. Springer, Heidelberg (1988)

8. Hassin, R.: Approximation schemes for the restricted shortest path problem. Math. Oper. Res. **17**(1), 36–42 (1992)

9. Joret, G.: Stackelberg network pricing is hard to approximate. Networks **57**(2), 117–120 (2011)

10. Lorenz, D.H., Raz, D.: A simple efficient approximation scheme for the restricted shortest path problem. Oper. Res. Lett. **28**(5), 213–219 (2001)

11. Manfredi, P., Posta, P.D., d'Onofrio, A., Salinelli, E., Centrone, F., Meo, C., Poletti, P.: Optimal vaccination choice, vaccination games, and rational exemption: an appraisal. Vaccine **28**(1), 98–109 (2009)

12. Nikolova, E.V.: Strategic algorithms. Ph.D. thesis, Massachusetts Institute of Technology (2009)

13. Panda, S., Vorobeychik, Y.: Stackelberg games for vaccine design. In: Proceedings of the 2015 International Conference on Autonomous Agents and Multiagent Systems, pp. 1391–1399. International Foundation for Autonomous Agents and Multiagent Systems (2015)

14. Roch, S., Savard, G., Marcotte, P.: An approximation algorithm for Stackelberg network pricing. Networks **46**(1), 57–67 (2005)

15. Schulz, A.S., Uhan, N.A.: Approximating the least core value and least core of cooperative games with supermodular costs. Discrete Optim. **10**(2), 163–180 (2013)

16. Tambe, M.: Security and Game Theory: Algorithms, Deployed Systems, Lessons Learned. Cambridge University Press, Cambridge (2012)

17. Washburn, A., Wood, K.: Two-person zero-sum games for network interdiction. Oper. Res. **43**(2), 243–251 (1995)

Efficient Rational Proofs for Space Bounded Computations

Matteo Campanelli[✉] and Rosario Gennaro

The City University of New York, New York, USA
mcampanelli@gradcenter.cuny.edu, rosario@ccny.cuny.edu

Abstract. We present new protocols for the verification of *space bounded polytime computations* against a rational adversary. For such computations requiring sublinear space our protocol requires only a verifier running in sublinear-time. We extend our main result in several directions: (i) we present protocols for randomized complexity classes, using a new *composition theorem* for rational proofs which is of independent interest; (ii) we present lower bounds (i.e. conditional impossibility results) for Rational Proofs for various complexity classes.

Our new protocol is the first rational proof not based on the circuit model of computation, and the first *sequentially composable* protocols for a well-defined language class.

1 Introduction

Consider the problem of Outsourced Computation where a computationally "weak" client hires a more "powerful" server to store data and perform computations on its behalf. This paper is concerned with the problem of designing outsourced computation schemes that incentivize the server to perform correctly the tasks assigned by the client.

The rise of the *cloud computing* paradigm where business do not maintain their own IT infrastructure, but rather hire "providers" to run it, has brought this problem to the forefront of the research community. The goal is to find solutions that are efficient and feasible in practice for problems such as: How do we check the integrity of data that is stored remotely? How do we check computations performed on this remotely stored data? How can a client do this in the most efficient way possible?

For all the scenarios above, what mechanisms can be designed to incentivize parties to perform correctly no matter what the cost of the correct behavior might be?

1.1 Complexity Theory and Cryptography

The problem of efficiently checking the correctness of a computation performed by an untrusted party has been central in Complexity Theory for the last 30 years since the introduction of Interactive Proofs by Babai and Goldwasser, Micali and Rackoff [5,14].

© Springer International Publishing AG 2017
S. Rass et al. (Eds.): GameSec 2017, LNCS 10575, pp. 53–73, 2017.
DOI: 10.1007/978-3-319-68711-7_4

Verifiable Outsourced Computation is now a very active research area in Cryptography and Network Security (see [27] for a survey) with the aim to design protocols where it is impossible (under suitable cryptographic assumptions) for a provider to "cheat" in the above scenarios. While much progress has been done in this area, we are still far from solutions that can be deployed in practice. Part of the reason is that Cryptographers consider a very strong adversarial model that prevents any adversary from cheating. A different approach is to restrict ourselves to *rational adversaries*, whose motivation is not just to disrupt the protocol or computation, but simply to maximize a well defined utility function (e.g. profit).

1.2 Rational Proofs

In our work we use the concept of Rational Proofs introduced by Azar and Micali in [3] and refined in a subsequent paper [4].

In a Rational Proof, given a function f and an input x, the server returns the value $y = f(x)$, and (possibly) some auxiliary information, to the client. The client will in turn pay the server for its work with a reward which is a function of the messages sent by the server and some randomness chosen by the client. The crucial property is that this reward is maximized in expectation when the server returns the correct value y. Clearly a rational prover who is only interested in maximizing his reward, will always answer correctly.

The most striking feature of Rational Proofs is their simplicity. For example in [3], Azar and Micali show single-message Rational Proofs for any problem in $\#P$, where an (exponential-time) prover convinces a (poly-time) verifier of the number of satisfying assignment of a Boolean formula.

For the case of "real-life" computations, Azar and Micali in [4] consider the case of efficient provers (i.e. poly-time) and "super-efficient" (log-time) verifiers and present d-round Rational Proofs for functions computed by (uniform) Boolean circuits of depth d, for $d = O(\log n)$.

Recent work [16] shows how to obtain Rational Proofs with sublinear verifiers for languages in NC. Recalling that $L \subseteq NL \subseteq NC_2$, one can use the protocol in [16] to verify a logspace polytime computation (deterministic or nondeterministic) in $O(\log^2 n)$ rounds and $O(\log^2 n)$ verification.

The work by Chen et al. [9] focuses on rational proofs with multiple provers and the related class MRIP of languages decidable by a polynomial verifier interacting with an arbitrary number of provers. Under standard complexity assumptions, MRIP includes languages not decidable by a verifier interacting only with one prover. The class MRIP is equivalent to $EXP^{\|NP}$.

1.3 Repeated Executions with a Budget

In [8] we present a critique of the rational proof model in the case of "repeated executions with a budget". This model arises in the context of "volunteer computations" [1,22] where many computational tasks are outsourced and provers compete in solving as many as possible to obtain rewards. In this scenario assume that a prover has a certain budget B of "computational effort": how can one guarantee that the rational strategy is to provide the correct answer in *all* the proof he provides? The notion of rational proof guarantees that if the prover engages in a single rational proof then it is in his best interest to provide the correct output. But in [8] we show that in the presence of many computations, it might be more profitable for the prover to use his budget B to provide many incorrect answers than to provide a single correct answer. That's because incorrect (e.g. random) answers are "cheaper" to compute than the correct one and with the same budget B the prover can provide many of them while the entire budget might be necessary to solve a single problem correctly. If the difference in reward between correct and incorrect answers is not high enough then many incorrect answers may be more profitable and a rational prover will choose that strategy, and indeed this is the case for many of the protocols in [3,4,15,16].

In [8] we put forward a stronger notion of *sequentially composable rational proofs* which avoids the above problem and guarantees that the rational strategy is always the one to provide correct answers. We also presented sequentially composable rational proofs, but only for some ad-hoc cases, and were not able to generalize them to well-defined complexity classes.

1.4   Our Contribution

This paper presents new protocols for the verification of *space-bounded polytime computations* against a rational adversary. More specifically, let L be a language in the class $\mathsf{DTISP}(T(n), S(n))$, i.e. L is recognized by a deterministic Turing Machine M_L which runs in time $T(n)$ and space $S(n)$. We construct a protocol where a rational prover can convince the verifier that $x \in L$ or $x \notin L$ with the following properties:

- The verifier runs in time $O(S(n) \log n)$
- The protocol has $O(\log n)$ rounds and communication complexity $O(S(n) \log n)$
- The prover simply runs $M_L(x)$

Under suitable assumptions, our protocol can be proven to correctly incentivize a prover in **both** the stand-alone model of [3] and the sequentially composable definition of [8]. This is the first protocol which is sequentially composable for a well-defined complexity class.

For the case of "real-life" computations (i.e. poly-time computations verified by a "super-efficient" verifier) we note that for computations in sublinear space our general results yields a protocol in which the verifier is sublinear-time. More specifically, we introduce the first rational proof for SC (also known as $\mathsf{DTISP}(\mathrm{poly}(n), \mathrm{polylog}(n))$) with polylogarithmic verification and logarithmic rounds.

To compare this with the results in [16], we note that it is believed that $\mathsf{NC} \neq \mathsf{SC}$ and that the two classes are actually incomparable (see [10] for a discussion). For these computations our results compare favorably to the one in [16] in at least one aspect: our protocol requires $O(\log n)$ rounds and has the same verification complexity[1].

We present several extensions of our main result:

- Our main protocol can be extended to the case of space-bounded randomized computations using Nisan's pseudo-random generator [24] to derandomize the computation.
- We also present a different protocol that works for BPNC (bounded error randomized NC) where the Verifier runs in polylog time (note that this class is not covered by our main result since we do not know how to express NC with a polylog-space computation). This protocol uses in a crucial way a new *composition theorem* for rational proofs which we present in this paper and can be of independent interest.
- Finally we present lower bounds (i.e. conditional impossibility results) for Rational Proofs for various complexity classes.

1.5   The Landscape of Rational Proof Systems

Rational Proof systems can be divided in roughly two categories, both of them presented in the original work [3].

SCORING RULES. The work in [3] uses *scoring rules* to compute the reward paid by the verifier to the prover. A scoring rule is used to asses the "quality" of a prediction of a randomized process. Assume that the prover declares that a certain random variable X follows a particular probability distribution D. The verifier runs an "experiment" (i.e. samples the random variable in question) and computes a "reward" based on the distribution D announced by the prover and the result of the experiment. A scoring rule is maximized if the prover announced the real distribution followed by X. The novel aspect of many of the protocols in [3] was how to cast the computation of $y = f(x)$ as the announcement of a certain distribution D that could be tested efficiently by the verifier and rewarded by a scoring rule.

A simple example is the protocol for $\#P$ in [3] (or its "scaled-down" version for Hamming weight described more in detail in Sect. 2.1). Given a Boolean formula $\Phi(x_1, \ldots, x_n)$ the prover announces the number m of satisfying assignments. This can be interpreted as the prover announcing that if one chooses an assignment at random it will be a satisfying one with probability $m \cdot 2^{-n}$. The verifier then chooses a random assignment and checks if it satisfies Φ or not and uses m and the result of the test to compute the reward via a scoring rule. Since the scoring rule is maximized by the announcement of the correct m, a rational prover will announce the correct value.

---

[1] We also point out that in [16] a rational protocol for P, polytime computations, is presented, but for the case of a computationally bounded prover, i.e. a *rational argument*.

As pointed out in [8] the problem with the scoring rule approach is that the reward declines slowly as the distribution announced by the Prover becomes more and more distant from the real one. The consequence is that incorrect results still get a substantial reward, even if not a maximal one. Since those incorrect results can be computed faster than the correct one, a Prover with "budget" B might be incentivized to produce many incorrect answers instead of a single correct one. All of the scoring rule based protocols in [3,4,15,16] suffer from this problem.

WEAK INTERACTIVE PROOFS. In the definition of rational proofs we require that the expected reward is maximized for the honest prover. This definition can be made stronger (as done explicitly in [15]) requiring that every *systematically dishonest* prover would incur a polynomial loss (this property is usually described in terms of a *noticeable reward gap*). Obviously we can use classical interactive proofs to trivially obtain this property. In fact, recall standard interactive proofs: at the end of the interaction with a prover, the verifier applies a "decision function" D to a transcript in order to *accept* or *reject* the input x. A verifier may then pay the prover a reward $R = \mathsf{poly}(|x|)$ iff D accepts. The honest prover will clearly maximize its reward since, by definition of interactive proof, the probability of a wrong acceptance/rejection is negligible. Notice hoverer that we can obtain rational proofs with noticeble reward gap even if the protocol has a much higher error probability. In fact, for an appropriate choice of a (polynomial) reward R, the error probability can be as high as $1 - n^{-k}$ for some $k \in \mathbb{N}$. We call an interactive proof with such a high error probability a *weak interactive proof*[2].

Weak interactive proofs can be turned into strong (i.e. with negligible error) classical ones by repetition, which however increases the computational cost of the verifier. But since to obtain a rational proof it is not necessary to repeat them, we can use them to obtain rational proofs which are very efficient for the verifier. Indeed, some of the protocols in [3,8] are rational proofs based on weak interactive proofs. This approach is also the main focus in the present work.

DISCUSSION. There are two intriguing questions when we compare the "scoring rules" approach to build rational proofs, to the one based on "weak interactive proofs".

- Is one approach more powerful than the other?
- All the known sequentially composable proofs are weak interactive proofs. Does sequential composition requires a weak interactive proof?

We do not know the answers to the above questions. For a more detailed discussion we refer the reader to the end of Sect. 7.

---

[2] This is basically the *covert adversary* model for multiparty computation introduced in [2].

1.6 Other Related Work

INTERACTIVE PROOFS. As already discussed, a "traditional" interactive proof (where security holds against any adversary, even a computationally unbounded one) would work in our model. In this case the most relevant result is the recent independent work in [26] that presents breakthrough protocols for the deterministic (and randomized) restriction of the class of language we consider. If L is a language which is recognized by a deterministic (or randomized) Turing Machine M_L which runs in time $T(n)$ and space $S(n)$, then their protocol has the following properties:

- The verifier runs in $O(\text{poly}(S(n)) + n \cdot \text{polylog}(n))$ time;
- The prover runs in polynomial time;
- The protocol runs in *constant* rounds, with communication complexity $O(\text{poly}(S(n)n^\delta)$ for a constant δ.

Apart from round complexity (which is the impressive breakthrough of the result in [26]) our protocols fares better in all other categories. Note in particular that a sublinear space computation does not necessarily yield a sublinear-time verifier in [26]. On the other hand, we stress that our protocol only considers weaker rational adversaries.

COMPUTATIONAL ARGUMENTS. There is a large class of protocols for *arguments* of correctness (e.g. [12,13,19]) even in the rational model [15,16]. Recall that in an argument, security is achieved only against computationally bounded prover. In this case even single round solutions can be achieved. We do not consider this model in this paper, except in Sect. 5.2 as one possible option to obtain sequential composability.

COMPUTATIONAL DECISION THEORY. Other works in theoretical computer science have studied the connections between cost of computation and utility in decision problems. The work in [17] proposes a framework for *computational decision problems*, where the Decision Maker's (DM) utility depends on the algorithm chosen for computing its strategy. The Decision Maker runs the algorithm, assumed to be a Turing Machine, on the input to the computational decision problem. The output of the algorithm determines the DM's strategy. Thus the choice of the DM reduces to the choice of a Turing Machine from a certain space. The DM will have beliefs on the running time (cost) of each Turing Machine. The actual cost of running the chosen TM will affect the DM's reward. Rational proofs with costly computation could be formalized in the language of *computational decision problems* in [17]. There are similarities between the approach in this work and that in [17], as both take into account the cost of computation in a decision problem.

2 Rational Proofs

The following is the definition of Rational Proof from [3]. As usual with $\text{neg}(\cdot)$ we denote a *negligible* function, i.e. one that is asymptotically smaller than the inverse of any polynomial. Conversely a *noticeable* function is the inverse of a polynomial.

Definition 1 (Rational Proof). *A function* $f: \{0,1\}^n \to \{0,1\}^*$ *admits a rational proof if there exists an interactive proof* (P,V) *and a randomized reward function* rew $: \{0,1\}^* \to \mathbb{R}_{\geq 0}$ *such that*

1. *For any input* $x \in \{0,1\}^n$, $\Pr[\text{out}((P,V)(x)) = f(x)] \geq 1 - \text{neg}(n)$.
2. *For every prover* $\widetilde{P}$, *and for any input* $x \in \{0,1\}^n$ *there exists a* $\delta_{\widetilde{P}}(x) \geq 0$ *such that* $\mathbb{E}[\text{rew}((\widetilde{P},V)(x))] + \delta_{\widetilde{P}}(x) \leq \mathbb{E}[\text{rew}((P,V)(x))]$.

The expectations and the probabilities are taken over the random coins of the prover and verifier.

We note that differently than [3] we allow for non-perfect completeness: a negligible probability that even the correct prover will prove the wrong result. This will be necessary for our protocols for randomized computations.

Let $\epsilon_{\widetilde{P}} = \Pr[\text{out}((P,V)(x)) \neq f(x)]$. Following [15] we define the reward gap as $\Delta(x) = \min_{P^*: \epsilon_{P^*}=1}[\delta_{P^*}(x)]$, i.e. the minimum reward gap over the provers that always report the incorrect value. It is easy to see that for arbitrary prover $\widetilde{P}$ we have $\delta_{\widetilde{P}}(x) \geq \epsilon_{\widetilde{P}} \cdot \Delta(x)$. Therefore it suffices to prove that a protocol has a strictly positive reward gap $\Delta(x)$ for all x.

Definition 2 [3,4,15]. *The class* DRMA[r,c,T] *(Decisional Rational Merlin Arthur) is the class of boolean functions* $f : \{0,1\}^* \to \{0,1\}$ *admitting a rational proof* $\Pi = (P,V,\text{rew})$ *s.t. on input* x:

- Π *terminates in* $r(|x|)$ *rounds;*
- *The communication complexity of* P *is* $c(|x|)$;
- *The running time of* V *is* $T(|x|)$;
- *The function* rew *is bounded by a polynomial;*
- Π *has noticeable reward gap.*

Remark 1. The requirement that the reward gap must be noticeable was introduced in [4,15] and is explained in Sect. 5.

2.1 A Warmup Example

Consider the function $f : \{0,1\}^n \to [0 \ldots n]$ which on input x outputs the Hamming weight of x (i.e. $\sum_i x_i$ where x_i are the bits of x).

In [4] the prover announces a number $\tilde{m}$ which he claims to be equal to $m = f(x)$. This can be interpreted as the prover announcing that if one chooses an input bit x_i at random it will be equal to 1 with probability $\tilde{p} = \tilde{m}/n$. The verifier then chooses a random input bit x_i and uses $\tilde{m}, x_i$ to compute the reward via a scoring rule. Since the scoring rule is maximized by the announcement of the correct m, a rational prover will announce the correct value. The scoring rule used in [4] (and in all other rational proofs based on scoring rules) is Brier's rule where the reward is computed as $BSR(\tilde{p}, x_i)$ where $BSR(\tilde{p}, 1) = 2\tilde{p}(2 - \tilde{p})$ and $BSR(\tilde{p}, 0) = 2(1 - \tilde{p}^2)$. Notice that $p = m/n$ is the actual probability to get 1 when selecting an input bit at random, so the expected reward of the prover is $pBSR(\tilde{p}, 1) + (1 - p)BSR(\tilde{p}, 0)$ which is easily seen to be maximized for $\tilde{p} = p$, i.e. $\tilde{m} = m$.

In [8] we propose an alternative protocol for f (motivated by the issues we discuss in Sect. 5). In our protocol we compute f via an "addition circuit", organized as a complete binary tree with n leaves which are the input, and where each internal node is a (fan-in 2) addition gate – note that this circuit has depth $d = \log n$. The protocol has d rounds: at the first round the prover announces $\tilde{m}$ (the claimed value of $f(x)$) and its two "children" y_L, y_R in the output gate, i.e. the two input values of the last output gate G. The Verifier checks that $y_L + y_R = \tilde{m}$, and then asks the Prover to verify that y_L or y_R (chosen a random) is correct, by recursing on the above test. At the end the verifier has to check the last addition gate on two input bits: she performs this test on her own by reading just those two bits. If any of the tests fails, the verifier pays a reward of 0, otherwise she will pay R. The intuition is that a cheating prover will be caught with probability 2^{-d} which is exactly the reward gap (and for log-depth circuits like this one is noticeable). Note that the first protocol is a scoring-rule based one, while the second one is a weak-interactive proof.

3   Rational Proofs for Space-Bounded Computations

We are now ready to present our protocol. It uses the notion of a Turing Machine *configuration*, i.e. the complete description of the current state of the computation: for a machine M, its state, the position of its heads, the non-blank values on its tapes.

Let $L \in \mathsf{DTISP}(T(n), S(n))$ and M be the deterministic TM that recognizes L. On input x, let $\gamma_1, \ldots, \gamma_N$ (where $N = T(|x|)$) be the configurations that M goes through during the computation on input x, where γ_{i+1} is reached from γ_i according to the transition function of M. Note, first of all, that each configuration has size $O(S(n))$. Also if $x \in L$ (resp. $x \notin L$) then γ_N is an accepting (resp. rejecting) configuration.

The protocol presented below is a more general version of the one used in [8] and described above. The prover shows the claimed final configuration $\hat{\gamma}_N$ and then prover and verifier engage in a "chasing game", where the prover "commits" at each step to an intermediate configuration. If the prover is cheating (i.e. $\hat{\gamma}_N$ is wrong) then the intermediate configuration either does not follow from the initial configuration or does not lead to the final claimed configuration. At each step and after P communicates the intermediate configuration γ', the verifier then randomly chooses whether to continue invoking the protocol on the left or the right of γ'. The protocol terminates when V ends up on two previously declared adjacent configurations that he can check. Intuitively, the protocol works since, if $\hat{\gamma}_N$ is wrong, for any possible sequence of the prover's messages, there is at least one choice of random coins that allows V to detect it; the space of such choices is polynomial in size.

We assume that V has oracle access to the input x. What follows is a formal description of the protocol.

---

1. P sends to V:
    - γ_N, the final accepting configuration (the starting configuration, γ_1, is known to the verifier);
    - N, the number of steps between the two configurations.
2. Then V invokes the procedure $\mathsf{PathCheck}(N, \gamma_1, \gamma_N)$.

The procedure $\mathsf{PathCheck}(m, \gamma_l, \gamma_r)$ is defined for $1 \le m \le N$ as follows:

- If $m > 1$, then:
    1. P sends intermediate configurations γ_p and γ_q (which may coincide) where $p = \lfloor \frac{l+m-1}{2} \rfloor$ and $q = \lceil \frac{l+m-1}{2} \rceil$.
    2. If $p \neq q$, V checks whether there is a transition leading from configuration γ_p to configuration γ_q. If yes, V accepts; otherwise V halts and rejects.
    3. V generates a random bit $b \in_R \{0, 1\}$
    4. If $b = 0$ then the protocol continues invoking $\mathsf{PathCheck}(\lfloor \frac{m}{2} \rfloor, \gamma_l, \gamma_p)$;
       If $b = 1$ the protocol continues invoking $\mathsf{PathCheck}(\lfloor \frac{m}{2} \rfloor, \gamma_q, \gamma_r)$
- If $m = 1$, then V checks whether there is a transition leading from configuration γ_l to configuration γ_r. If $l = 1$, V checks that γ_l is indeed the initial configuration γ_1. If $r = N$, V checks that γ_r is indeed the final configuration sent by P at the beginning. If yes, V accepts; otherwise V rejects.

Theorem 1. $\mathsf{DTISP}[\mathsf{poly}(n), S(n)] \subseteq \mathsf{DRMA}[O(\log n), O(S(n) \log n), O(S(n) \log n)]$

Proof. Let us consider the efficiency of the protocol above. It requires $O(\log n)$ rounds. Since the computation is in $\mathsf{DTISP}[\mathsf{poly}(n), S(n)]$, the configurations P sends to V at each round have size $O(S(n))$. The verifier only needs to read the configurations and, at the last round, check the existence of a transition leading from γ_l to γ_r. Therefore the total running time for V is $O(S(n) \log n)$.

Let us now prove that this is a rational proof with noticeable reward gap. Observe that the protocol has perfect completeness. Let us now prove that the soundness is at most $1 - 2^{-\log N} = 1 - \frac{1}{O(\mathsf{poly}(n))}$. We aim at proving that, if there is no path between the configurations γ_1 and γ_N then V rejects with probability at least $2^{-\log N}$. Assume, for sake of simplicity, that $N = 2^k$ for some k. We will proceed by induction on k. If $k = 1$, P provides the only intermediate configuration γ' between γ_1 and γ_N. At this point V flips a coin and the protocol will terminate after testing whether there exists a transition between γ_1 and γ' or between γ' and γ_N. Since we assume the input is not in the language, there exists at most one of such transitions and V will detect this with probability $1/2$.

Now assume $k > 1$. At the first step of the protocol P provides an intermediate configuration γ'. Either there is no path between γ_1 and γ' or there is no path between γ' and γ_N. Say it is the former: the protocol will proceed on the left with probability $1/2$ and then V will detect P cheating with probability 2^{-k+1} by induction hypothesis, which concludes the proof.

The theorem above implies the results below.

Corollary 1. $\mathsf{L} \subseteq \mathsf{DRMA}[O(\log n), O(\log^2 n), O(\log^2 n)]$

This improves over the construction of rational proofs for L in [16] due to the better round complexity.

Corollary 2. $\mathsf{SC} \subseteq \mathsf{DRMA}[O(\log n), O(\mathrm{polylog}(n)), O(\mathrm{polylog}(n))]$

No known result was known for SC before.

3.1 Rational Proofs for Randomized Bounded Space Computation

We now describe a variation of the above protocol, for the case of randomized bounded space computations. Let $\mathsf{BPTISP}[t, s]$ denote the class of languages recognized by randomized machines using time t and space s with error bounded by $1/3$ on both sides. In other words, $L \in \mathsf{BPTISP}[\mathrm{poly}(n), S(n)]$ if there exists a (deterministic) Turing Machine M such that for any $x \in \{0,1\}^*$ $\Pr_{r \in_R \{0,1\}^{\rho(|x|)}}[M(x,r) = L(x)] \geq \frac{2}{3}$ and that runs in $S(|x|)$ space and polynomial time. Let $\rho(n)$ be the maximum number of random bits used by M for input $x \in \{0,1\}^n$; $\rho(\cdot)$ is clearly polynomial.

We can bring down the $2/3$ probability error to $\mathsf{neg}(n)$ by constructing a machine M'. M' would simulate the M on x iterating the simulation $m = \mathrm{poly}(|x|)$ times using fresh random bits at each execution and taking the majority output of $M(x; \cdot)$. The machine M' uses $m\rho(|x|)$ random bits and runs in polynomial time and $S(|x|) + O(\log(n))$ space.

The work in [24] introduces pseudo-random generators (PRG) resistant against space bounded adversaries. An implication of this result is that any randomized Turing Machine M_1 running in time T and space S can be simulated by a randomized Turing Machine M_2 running in time $O(T)$, space $O(S \log(T))$ and using only $O(S \log(T))$ random bits[3] (see in particular Theorem 3 in [24]). Let $L \in \mathsf{BPTISP}[(\mathrm{poly}(n), S(n)]$ and M' defined as above. We denote by $\hat{M}$ the simulation of M' that uses Nisan's result described above.

---

[3] We point out that the new machine M_2 introduces a small error. For our specific case this error keeps the overall error probability negligible and we can ignore it.

By using the properties of the new machine $\hat{M}$, we can directly construct rational proofs for $\mathsf{BPTISP}(\mathsf{poly}(n), S(n))$. We let the verifier picks a random string r (of length $O(S \log(T))$) and sends it to the prover. They then invoke a rational proof for the computation $\hat{M}(x; r)$.

By the observations above and Theorem 1 we have the following result:

Corollary 3. $\mathsf{BPTISP}[\mathsf{poly}(n), S(n)] \subseteq \mathsf{DRMA}[\log(n), S(n) \log^2(n), S(n) \log^2(n)]$

We note that for this protocol, we need to allow for non-perfect completeness in the definition of DRMA in order to allow for the probability that the verifier chooses a bad random string r.

4   A Composition Theorem for Rational Proofs

In this Section we prove an intuitively simple, but technically non-trivial, *composition theorem* that states that we while proving the value of a function f, we can replace oracle access to a function g, with a rational proof for g. The technically interesting part of the proof is to make sure that the *total* reward of the prover is maximized when the result of the computation of f is correct. In other words, while we know that lying in the computation of g will not be a rational strategy for just that computation, it may turn out to be the best strategy as it might increase the reward of an incorrect computation of f. A similar issue (arising in a particular rational proof for depth d circuits) was discussed in [4]: our proof generalizes their technique.

Definition 3. *We say that a rational proof (P, V, rew) for f is a g-oracle rational proof if V has oracle access to the function g and carries out at most one oracle query. We allow the function g to depend on the specific input x.*

Theorem 2. *Assume there exists a g-oracle rational proof $(P_f^o, V_f^o, \mathsf{rew}_f^o)$ for f with noticeable reward gap and with round, communication and verification complexity respectively r_f, c_f and T_f. Let t_I the time necessary to invoke the oracle for g and to read its output. Assume there exists a rational proof $(P_g, V_g, \mathsf{rew}_g)$ with noticeable reward gap for g with round, communication and verification complexity respectively r_g, c_g and T_g. Then there exists a (non g-oracle) rational proof with noticeable reward gap for f with round, communication and verification complexity respectively $r_f + 1 + r_g, c_f + t_I + c_g$ and $T_f - t_i + T_g$.*

Before we embark on the proof of Theorem 2 we state a technical Lemma whose simple proof is omitted for lack of space. The definition of rational proof requires that the expected reward of the honest prover is not lower than the expected reward of any other prover. The following intuitive lemma states we necessarily obtain this property if an honest prover has a polynomial expected gain in comparison to provers that *always* provide a wrong output.

Lemma 1. *Let (P, V) be a protocol and rew a reward function as in Definition 1. Let f be a function s.t. $\forall x \Pr[\mathsf{out}(P, V)(x)] = 1$. Let Δ be the corresponding reward gap w.r.t. the honest prover P and f. If $\Delta > \frac{1}{\mathsf{poly}(n)}$ then (P, V, rew) is a rational proof for f and admits noticeable reward gap.*

Now we can start the proof of Theorem 2.

Proof. Let rew_f^o and rew_g be the reward functions of the g-oracle rational proof for f and the rational proof for g respectively. We now construct a new verifier V for f. This verifier runs exactly like the g-oracle verifier for f except that every oracle query to g is now replaced with an invocation of the rational proof for g. The new reward function rew is defined as: $\mathsf{rew}(\mathcal{T}) = \delta\mathsf{rew}_f^o(\mathcal{T}_f^o \circ y_g) + \mathsf{rew}_g(\mathcal{T}_g)$. Here $\mathcal{T}$ is the complete transcript of the new rational proof, $\mathcal{T}_f^o$ is the transcript of the oracle rational proof for f, $\mathcal{T}_g$ and y_g are respectively the transcript and the output of the rational proof for g. Finally δ is multiplicative factor in $(0, 1]$). The intuition behind this formula is to "discount" the part of the reward from f so that the prover is incentivized to provide the true answer for g. In turn, since rew_f^o rewards the honest prover more when the verifier has the right answer for a query to g (by hypothesis), this entails that the whole protocol is rational proof for f.

To prove the theorem we will use Lemma 1 and it will suffice to prove that the new protocol has a noticeable reward gap.

Consider a prover $\tilde{P}$ that always answer incorrectly on the output of f. Let p_g be the probability that the prover outputs a correct y_g. Then the difference between the expected reward of the honest prover and $\tilde{P}$ is:

$$\delta(R_f^o - \tilde{R}_f^o) + (R_g - \tilde{R}_g) = \tag{1}$$

$$\delta(R_f^o - p_g\tilde{R}_f^{o,\text{good}(g)} - (1 - p_g)\tilde{R}_f^{o,\text{wrong}(g)})$$
$$+ (R_g - p_g\tilde{R}_g^{\text{good}(g)} - (1 - p_g)\tilde{R}_g^{\text{wrong}(g)}) = \tag{2}$$

$$\delta(p_g(R_f^o - \tilde{R}_f^{o,\text{good}(g)}) + (1 - p_g)(R_f - \tilde{R}_f^{o,\text{wrong}(g)}))$$
$$+ p_g(R_g - \tilde{R}_g^{\text{good}(g)}) + (1 - p_g)(R_g - \tilde{R}_g^{\text{wrong}(g)}) > \tag{3}$$

$$p_g\delta\Delta_f^o + (1 - p_g)(\Delta_g - \delta b_f^o(n)) \geq \tag{4}$$

$$\min\{\delta\Delta_f^o, \Delta_g - \delta b_f^o(n)\} > \tag{5}$$

$$\frac{1}{\mathsf{poly}(n)} \tag{6}$$

where the last inequality holds for $\delta = \frac{\Delta_g}{2b_f^o(n)}$.

The round, communication and verification complexity of the construction is given by the sum of the respective complexities from the two rational proofs modulo minor adjustments. These adjustments account for the additional round by which the verifier communicates to the prover the requested instance for g. □

We can use this result as a design tool of rational proofs for a function f: First build a rational proof for a function g and then one for f where we assume the verifier has oracle access to g. This automatically provides a complete rational proof for f.

Remark 2. Theorem 2 assumes that verifier in the oracle rational proof for f carries out a single oracle query. Notice however that the proof of the theorem can be generalized to any verifier carrying out a constant number of adaptive oracle queries, possibly all for distinct functions. This can be done by iteratively applying the theorem to a sequence of $m = O(1)$ oracle rational proofs for functions $f_1, ..., f_m$ where the i-th rational proof is f_{i+1}-oracle for $1 \le i < m$.

4.1 Rational Proofs for Randomized Circuits

As an application of the composition theorem described above we present an alternative approach to rational proofs for randomized computations. We show that by assuming the existence of a *common reference string (CRS)*[4] we obtain rational proofs for randomized circuits of polylogarithmic depth and polynomial size, i.e. BPNC the class of uniform polylog-depth poly-size randomized circuits with error bounded by $1/3$ on both sides.

If we insist on a "super-efficient" verifier (i.e. with sublinear running time) we cannot use the same approach as in Sect. 3.1 since we do not know how to bound the space $S(n)$ used by a computation in NC (and the verifier's complexity in our protocol for bounded space computations, depends on the space complexity of the underlying language). We get around this problem by assuming a CRS, to which the verifier has oracle access.

We start by describing a rational proof with oracle access for BPP and then we show how to remove the oracle access (via our composition theorem) for the case of BPNC.

Let $L \in$ BPP and let M a PTM that decides L in polynomial time and $\rho(\cdot)$ the randomness complexity of M. For $x \in \{0,1\}^*$ we denote by L_x the (deterministically decidable) language $\{(x,r) : r \in \{0,1\}^{\rho(|x|)} \wedge M(x,r) = L(x)\}$.

Lemma 2. *Let L be a language in BPP. Then there exists a L_x-oracle rational proof with CRS σ for L where $|\sigma| = \mathsf{poly}(n)\rho(n)$.*

Proof. Our construction is as follows. W.l.o.g. we will assume σ to be divided in $\ell = \mathsf{poly}(n)$ blocks $r_1, ..., r_\ell$, each of size $\rho(n)$.

1. The honest prover P runs $M(x, r_i)$ for $1 \le i \le \ell$ and announces m the number of strings r_i s.t. $M(x, r_i)$ accepts, i.e. $\sum_i M(x, r_i)$;
2. P sends m to x.
3. The Verifier accepts if $m > \ell/2$

We note that if we set $y_i = M(x, r_i)$ then the prover is announcing the Hamming weight of the string $y_1, ..., y_\ell$. At this point we can use the Hamming weight verification protocol in Sect. 2.1 where the Verifier use the oracle for L_x to verify on her own the value of y_i.

---

[4] A common reference string is a string generated by a trusted party to which both the prover and the verifier have access; it is a common assumption in cryptographic literature, e.g. Non-Interactive Zero Knowledge [7].

We note that no matter which protocol is used, round complexity, communication complexity and verifier running time (not counting the oracle calls) are all polylog(n).

To obtain our result for BPNC we invoke the following result from [16]:

Theorem 3. NC $\subseteq$ DRMA[polylog(n), polylog(n), polylog(n)]

The theorem above, together with Theorem 2 and Lemma 2 yields:

Corollary 4. *Let* $x \in \{0,1\}^n$ *and* $L \in$ BPNC. *Assuming the existence of a (polynomially long) CRS then there exists a rational proof for* L *with polylogarithmically many rounds, polylogarithmic communication and verification complexity.*

Notice that some problems (e.g. perfect matching) are not known to be in NC but are known to be in RNC $\subseteq$ BPNC [20].

5   Sequential Composability

Until now we have only considered agents who want to maximize their reward. But the reward alone, might not capture the complete utility function that the Prover is trying to maximize in his interaction with the Verifier. In particular we have not considered the *cost* incurred by the Prover to compute f and engage in the protocol. It makes sense then to define the *profit* of the Prover as the difference between the reward paid by the verifier and such cost.

As already pointed out in [4, 15] the definition of Rational Proof is sufficiently robust to also maximize the profit of the honest prover and not just the reward. Indeed consider the case of a "lazy" prover $\tilde{P}$ that does not evaluate the function: let $\tilde{R}(x), \tilde{C}(x)$ be the reward and cost associated with $\tilde{P}$ on input x (while $R(x), C(x)$ are the values associated with the honest prover).

Obviously we want $R(x) - C(x) \geq \tilde{R}(x) - \tilde{C}(x)$ or equivalently $R(x) - \tilde{R}(x) \geq C(x) - \tilde{C}(x)$. Recall the notion of *reward gap*: the minimum difference between the reward of the honest prover and any other prover $\Delta(x) \leq R(x) - \tilde{R}(x)$. To maximize the profit it is then sufficient to change the reward by a a multiplier $M = C(x)/\Delta(x)$. Thus we have that $M(R(x) - \tilde{R}(x)) \geq C(x) \geq C(x) - \tilde{C}(x)$ as desired. This explains why we require the reward gap to be at least the inverse of a polynomial: this will maintain the total reward paid by the Verifier bounded by a polynomial.

5.1   Profit in Repeated Executions

In [8] we showed how if Prover and Verifier engage in repeated execution of a Rational Proof, where the Prover has a "budget" of computation cost that he is willing to invest, then there is no guarantee anymore that the profit is maximized by the honest prover. The reason is that it might be more profitable for the prover to use his budget to provide many incorrect answers than to provide a single correct answer. That's because incorrect (e.g. random) answers are "cheaper" to compute than the correct one and with the same budget B the prover can provide many of them while the entire budget might be necessary to solve a single problem correctly. If incorrect answers still receive a substantial reward then many incorrect answers may be more profitable and a rational prover will choose that strategy.

We refer the reader to [8] for concrete examples of situations where this happens in many of the protocols in [3, 4, 15, 16].

This motivated us to consider a stronger definition which requires the reward to be somehow connected to the "effort" paid by the prover. The definition (stated below) basically says that if a (possibly dishonest) prover invests less computation than the honest prover then he must collect a smaller reward.

Definition 4 (Sequential Rational Proof). *A rational proof (P, V) for a function $f : \{0,1\}^n \rightarrow \{0,1\}^n$ is (ϵ, K)-sequentially composable for an input distribution $\mathcal{D}$, if for every prover $\widetilde{P}$, and every sequence of inputs $x, x_1, \ldots, x_k$ drawn according to $\mathcal{D}$ such that $C(x) \geq \sum_{i=1}^{k} \tilde{C}(x_i)$ and $k \leq K$ we have that $\sum_i \tilde{R}(x_i) - R \leq \epsilon$.*

The following Lemma is from [8].

Lemma 3. *Let (P, V) and rew be respectively an interactive proof and a reward function as in Definition 1; if rew can only assume the values 0 and R for some constant R, let $\tilde{p}_x = \Pr[\text{rew}((\widetilde{P}, V)(x)) = R]$. If for $x \in \mathcal{D}$, $\tilde{p}_x \leq \frac{\tilde{C}(x)}{C} + \epsilon$ then (P, V) is $(KR\epsilon, K)$-sequentially composable for $\mathcal{D}$.*

The intuition behind our definition and Lemma 3 is that to produce the correct result, the prover must run the computation and incur its full cost; moreover for a dishonest prover his probability of "success" has to be no bigger than the fraction of the total cost incurred.

This intuition is impossible to formalize if we do not introduce a probability distribution over the input space. Indeed, for a specific input x a "dishonest" prover $\widetilde{P}$ could have the correct $y = f(x)$ value "hardwired" and could answer correctly without having to perform any computation at all. Similarly, for certain inputs x, x' and a certain function f, a prover $\widetilde{P}$ after computing $y = f(x)$ might be able to "recycle" some of the computation effort (by saving some state) and compute $y' = f(x')$ incurring a much smaller cost than computing it from scratch. This is the reason our definition is parametrized over an input distribution $\mathcal{D}$ (and all the expectations, including the computation of the reward, are taken over the probability of selecting a given input x).

A way to address this problem was suggested in [6] under the name of *Unique Inner State Assumption (UISA)*: when inputs x are chosen according to $\mathcal{D}$, then we assume that computing f requires cost T from any party: this can be formalized by saying that if a party invests $t = \gamma T$ effort (for $\gamma \leq 1$), then it computes the correct value only with probability negligibly close to γ (since a party can always have a "mixed" strategy in which with probability γ it runs the correct computation and with probability $1 - \gamma$ does something else, like guessing at random).

Using this assumption [6] solve the problem of the "repeated executions with budget" by requiring the verifier to check the correctness of a random subset of the the prover's answer by running the computation herself on that subset. This makes the verifier "efficient" only in an amortized sense.

In [8] we formalized the notion of Sequential Composability in Definition 4 and, using a variation of the UISA, we showed protocols that are sequentially composable where the verifier is efficient (i.e. polylog verification time) on *each execution*. Unfortunately that proof of sequential composability works only for a limited subclass of log-depth circuits.

5.2   Sequential Composability of Our New Protocol

To prove our protocol to be sequentially composable we need two main assumptions which we discuss now.

HARDNESS OF GUESSING STATES. Our protocol imposes very weak requirements on the prover: the verifier just checks a single computation step in the entire process, albeit a step chosen at random among the entire sequence. We need an equivalent of the UISA which states that for every correct transition that the prover is able to produce he must pay "one" computation step. More formally for any Turing Machine M we say that pair of configuration γ, γ' is M-correct if γ' can be obtained from γ via a single computation step of M.

Definition 5 (Hardness of State Guessing Assumption). *Let M be a Turing Machine and let L_M be the language recognized by M. We say that the Hardness of State Guessing Assumption holds for M, for distribution $\mathcal{D}$ and security parameter ϵ if for any machine A running in time t the probability that A on input x outputs more than t, M-correct pairs of configurations is at most ϵ (where the probability is taken over the choice of x according to the distribution $\mathcal{D}$ and the internal coin tosses of A).*

ADAPTIVE VS. NON-ADAPTIVE PROVERS. Assumption 5 guarantees that to come up with t correct transitions, the prover must invest at least t amount of work. We now move to the ultimate goal which is to link the amount of work invested by the prover, to his probability of success. As discussed in [8] it is useful to distinguish between *adaptive and non-adaptive provers*.

When running a rational proof on the computation of M over an input x, an *adaptive* prover allocates its computational budget *on the fly* during the execution of the rational proof. Conversely a *non-adaptive* prover $\tilde{P}$ uses his computational budget to compute as much as possible about $M(x)$ before starting the protocol with the verifier. Clearly an adaptive prover strategy is more powerful than a non-adaptive one (since the adaptive prover can direct its computation effort where it matters most, i.e. where the Verifier "checks" the computation).

As an example, it is not hard to see that in our protocol an adaptive prover can succesfully cheat without investing much computational effort at all. The prover will answer at random until the very last step when he will compute and answer with a correct transition. Even if we invoke Assumption 5 a prover that invests only one computational step has a probability of success of $1 - \frac{1}{\mathsf{poly}(n)}$ (indeed the prover fails only if we end up checking against the initial configuration – this is the attack that makes Theorem 1 tight.).

Is it possible to limit the Prover to a non-adaptive strategy? As pointed out in [8] this could be achieved by imposing some "timing" constraints to the execution of the protocol: to prevent the prover from performing large computations while interacting with the Verifier, the latter could request that prover's responses be delivered "immediately", and if a delay happens then the Verifier will not pay the reward. Similar timing constraints have been used before in the cryptographic literature, e.g. see the notion of *timing assumptions* in the concurrent zero-knowedge protocols in [11]. Note that in order to require an "immediate" answer from the prover it is necessary that the latter stores all the intermediate configurations, which is why we require the prover to run in space $O(T(n)S(n))$ – this condition is not needed for the protocol to be rational in the stand-alone case, since even the honest prover could just compute the correct transition on the fly. Still this could be a problematic approach if the protocol is conducted over a network since the network delay will most likely be larger than the computation effort required by the above "cheating" strategy.

Another option is to assume that the Prover is computationally bounded (e.g. the rational argument model introduced in [15]) and ask the prover to commit to all the configurations in the computation before starting the interaction with the verifier. Then instead of sending the configuration, the prover will decommit it (if the decommitment fails, the verifier stops and pays 0 as a reward). If we use a Merkle-tree commitment, these steps can be performed and verified in $O(\log n)$ time.

In any case, for the proof we assume that non-adaptive strategies are the only rational ones and proceed in analyzing our protocol under the assumption that the prover is adopting a non-adaptive strategy.

THE PROOF. Under the above two assumptions, the proof of sequential composability is almost immediate.

Theorem 4. *Let $L \in \mathsf{NTISP}[\mathsf{poly}(n), S(n)]$ and M be a TM recognizing L. Assume that Assumption 5 holds for M, under input distribution $\mathcal{D}$ and parameter ϵ. Moreover assume the prover follows a non-adaptive strategy. Then the protocol of Sect. 3 is a $(KR\epsilon, K)$-sequentially composable rational proof under $\mathcal{D}$ for any $K \in \mathbb{N}, R \in \mathbb{R}_{\geq 0}$.*

Proof. Let $\widetilde{P}$ be a prover with a running time of t on input x. Let T be the total number of transitions required by M on input x, i.e. the computational cost of the honest prover.

Observe that $\tilde{p}_x$ is the probability that V makes the final check on one of the transitions correctly computed by $\widetilde{P}$. Because of Assumption 5 we know that the probability that $\widetilde{P}$ can compute more than t correct transitions is ϵ, therefore an upper bound on $\tilde{p}_x$ is $\frac{t}{T} + \epsilon$ and the Theorem follows from Corollary 3. $\square$

6   Lower Bounds for Rational Proofs

In this section we discuss how likely it is will be able to find very efficient non-cryptographic rational protocols for the classes P and NP.

We denote by BPQP the class of languages decidable by a randomized algorithm running in quasi-polynomial time, i.e. $\mathsf{BPQP} = \bigcup_{k>0} \mathsf{BPTIME}[2^{O(\log^k(n))}]$. Our theorem follows the same approach of Theorem 16 in [15][5].

Theorem 5. $\mathsf{NP} \not\subseteq \mathsf{DRMA}[\mathrm{polylog}(n), \mathrm{polylog}(n), \mathrm{poly}(n)]$ *unless* $\mathsf{NP} \subseteq \mathsf{BPQP}$.

Proof Sketch. Assume there exists a rational proof π_L for a language $L \in \mathsf{NP}$ with parameters as the ones above. We can build a PTM M to decide L as follows: *(i)* M generates all possible transcripts $\mathcal{T}$ for π_L; *(ii)* for each $\mathcal{T}$, M estimates the expected reward $R_\mathcal{T}$ associated to that transcript by sampling $\mathrm{rew}(\mathcal{T})$ t times (recall the reward function is probabilistic); *(iii)* M returns the output associated to transcript $\mathcal{T}^* = \arg\max_\mathcal{T} R_\mathcal{T}$.

Consider the space of the transcripts with a polylogarithmic number of rounds and bits exchanged. The number of possible transcripts in such protocol is bounded by $(2^{\mathrm{polylog}(n)})^{\mathrm{polylog}(n)} = 2^{\mathrm{polylog}(n)}$. Let Δ be the (noticeable) reward gap of the protocol. By using Hoeffding's inequality we can prove M can approximate each $R_\mathcal{T}$ within $\Delta/3$ with probability $2/3$ after $t = \mathrm{poly}(n)$ samples. Recalling the definition of reward gap (see Remark 1), we conclude M can decide L in randomized time $2^{\mathrm{polylog}(n)}$. □

It is not known whether $\mathsf{NP} \not\subseteq \mathsf{BPQP}$ is true, although this assumption has been used to show hardness of approximation results [21,23]. Notice that this assumption implies $\mathsf{NP} \not\subseteq \mathsf{BPP}$ [18].

Let us now consider rational proofs for P. By the following theorem they might require $\omega(\log(n))$ total communication complexity (since we believe $\mathsf{P} \subseteq \mathsf{BPNC}$ to be unlikely [25]).

Theorem 6. $\mathsf{P} \not\subseteq \mathsf{DRMA}[O(1), O(\log(n)), \mathrm{polylog}(n)]$ *unless* $\mathsf{P} \subseteq \mathsf{BPNC}$.

Proof Sketch. Given a language $L \in \mathsf{P}$ we build a machine M to decide L as in the proof of Theorem 5. The only difference is that M can be simulated by a randomized circuit of $\mathrm{polylog}(n)$ depth and polynomial size. In fact, all the possible $2^{O(\log(n))} = \mathrm{poly}(n)$ transcripts can be simulated in parallel in $O(\log(n))$ sequential time. The same holds computing the $t = \mathrm{poly}(n)$ sample rewards for each of these transcripts. By assumption on the verifier's running time, each reward can be computed in polylogarithmic sequential time. Finally, the estimate of each transcript's expected reward and the maximum among them can be computed in $O(\log(n))$ depth. □

Remark 3. Theorem 6 can be generalized to rational proofs with round and communication complexities r and c such that $r \cdot c = O(\log(n))$.

---

[5] Since we only sketch our proof the reader is invited to see details of the proof [15].

7   Conclusions and Open Problems

We presented a rational proof for languages recognized by (deterministic) space-bounded computations. Our protocol is the first rational proof for a general class of languages that does not use circuit representations. Our protocol is secure both in the standard stand-alone notion of rational proof [3] and in the stronger composable version in [8].

Our work leaves open a series of questions:

- Can we build efficient rational proofs for arbitrary poly-time computations, where the verifier runs in sub-linear (or even linear) time?
- Our proof of sequential composability considers only non-adaptive adversaries, and enforces this condition by the use of timing assumptions or computationally bounded provers. Is it possible to construct protocols that are secure against adaptive adversaries? Or is it possible to relax the timing assumption to something less stringent than what is required in our protocol?
- It would be interesting to investigate the connection between the model of Rational Proofs and the work on Computational Decision Theory in [17]. In particular looking at realistic cost models that could affect the choice of strategy by the prover particularly in the sequentially composable model.

In Sect. 1.5 we described the two main approaches to Rational Proofs design: scoring rules and weak interactive proofs. Trying and compare the power of these approaches, two natural questions arise:

- Does one approach systematically lead to more efficient rational proofs (in terms of rounds, communication and verifying complexity) than the other?
- Is one approach more suitable for sequential composability than the other?

We believe these two open questions are worth pursuing. Some discussion follows.

Regarding the first question: in the context of "stand-alone" (non sequential) rational proofs it is not clear which approach is more powerful. We know that for every language class known to admit a scoring rule based protocol we also have a weak interactive proof with similar performance metrics (i.e. number of rounds, verifier efficiency, etc.). Our result is the first example of a language class for which we have rational proofs based on weak interactive proofs but no example of a scoring rule based protocol exist[6]. This suggests that the weak interactive proof approach might be the more powerful technique. It is open if all rational proofs are indeed weak interactive proofs: i.e. that given a rational proof with certain efficiency parameters, one can construct a weak interactive proof with "approximately" the same parameters.

On the issue of sequential composability, we have already proven in [8] that some rational proofs based on scoring rules (such as Brier's scoring rule) are not

---

[6] We stress that in this comparison we are interested in protocols with similar efficiency parameters. For example, the work in [3] presents several large complexity classes for which we have rational proofs. However, these protocols require a polynomial verifier and do not obtain a noticeable reward gap.

sequentially composable. This problem might be inherent at least for scoring rules that pay a substantial reward to incorrect computations. What we can say is that all known sequentially composable proofs are based on weak interactive proofs ([4,8][7] and this work). Again it is open if this is required, i.e. that all sequentially composable rational proofs are weak interactive proofs.

Acknowledgments. The authors would like to thank Jesper Buus Nielsen for suggesting the approach of the construction in Theorem 1.

References

1. Anderson, D.P., Cobb, J., Korpela, E., Lebofsky, M., Werthimer, D.: SETI@ home: an experiment in public-resource computing. Commun. ACM **45**(11), 56–61 (2002)
2. Aumann, Y., Lindell, Y.: Security against covert adversaries: efficient protocols for realistic adversaries. J. Cryptology **23**(2), 281–343 (2010)
3. Azar, P.D., Micali, S.: Rational proofs. In: Proceedings of the Forty-fourth Annual ACM Symposium on Theory of Computing, pp. 1017–1028. ACM (2012)
4. Azar, P.D., Micali, S.: Super-efficient rational proofs. In: Proceedings of the Fourteenth ACM Conference on Electronic Commerce, pp. 29–30. ACM (2013)
5. Babai, L.: Trading group theory for randomness. In: Proceedings of the Seventeenth Annual ACM Symposium on Theory of Computing, pp. 421–429. ACM (1985)
6. Belenkiy, M., Chase, M., Christopher Erway, C., Jannotti, J., Küpçü, A., Lysyanskaya, A.: Incentivizing outsourced computation. In: Proceedings of the ACM SIGCOMM 2008 Workshop on Economics of Networked Systems, NetEcon 2008, Seattle, WA, USA, 22 August 2008, pp. 85–90 (2008)
7. Blum, M., De Santis, A., Micali, S., Persiano, G.: Noninteractive zero-knowledge. SIAM J. Comput. **20**(6), 1084–1118 (1991)
8. Campanelli, M., Gennaro, R.: Sequentially composable rational proofs. In: Khouzani, M.H.R., Panaousis, E., Theodorakopoulos, G. (eds.) GameSec 2015. LNCS, vol. 9406, pp. 270–288. Springer, Cham (2015). doi:10.1007/978-3-319-25594-1_15
9. Chen, J., McCauley, S., Singh, S.: Rational proofs with multiple provers. In: Proceedings of the 2016 ACM Conference on Innovations in Theoretical Computer Science, pp. 237–248. ACM (2016)
10. De Agostino, S., Silvestri, R.: Bounded size dictionary compression: SC_k-completeness and NC algorithms. Inf. Comput. **180**(2), 101–112 (2003)
11. Dwork, C., Naor, M., Sahai, A.: Concurrent zero-knowledge. J. ACM **51**(6), 851–898 (2004)
12. Gennaro, R., Gentry, C., Parno, B.: Non-interactive verifiable computing: outsourcing computation to untrusted workers. In: Rabin, T. (ed.) CRYPTO 2010. LNCS, vol. 6223, pp. 465–482. Springer, Heidelberg (2010). doi:10.1007/978-3-642-14623-7_25
13. Gennaro, R., Gentry, C., Parno, B., Raykova, M.: Quadratic span programs and succinct NIZKs without PCPs. In: Johansson, T., Nguyen, P.Q. (eds.) EUROCRYPT 2013. LNCS, vol. 7881, pp. 626–645. Springer, Heidelberg (2013). doi:10.1007/978-3-642-38348-9_37

---

[7] The construction in Theorem 5.1 in [4] is shown to be sequentially composable in [8].

14. Goldwasser, S., Micali, S., Rackoff, C.: The knowledge complexity of interactive proof systems. SIAM J. Comput. **18**(1), 186–208 (1989)
15. Guo, S., Hubáček, P., Rosen, A., Vald, M.: Rational arguments: single round delegation with sublinear verification. In: Proceedings of the 5th Conference on Innovations in Theoretical Computer Science, pp. 523–540. ACM (2014)
16. Guo, S., Hubáček, P., Rosen, A., Vald, M.: Rational sumchecks. In: Kushilevitz, E., Malkin, T. (eds.) TCC 2016. LNCS, vol. 9563, pp. 319–351. Springer, Heidelberg (2016). doi:10.1007/978-3-662-49099-0_12
17. Halpern, J.Y., Pass, R.: I don't want to think about it now: Decision theory with costly computation. arXiv preprint arXiv:1106.2657 (2011)
18. Johnson, D.S.: The np-completeness column: the many limits on approximation. ACM Trans. Algorithms (TALG) **2**(3), 473–489 (2006)
19. Kalai, Y.T., Raz, R., Rothblum, R.D.: How to delegate computations: the power of no-signaling proofs. In: Proceedings of the 46th Annual ACM Symposium on Theory of Computing, pp. 485–494. ACM (2014)
20. Karp, R.M., Upfal, E., Wigderson, A.: Constructing a perfect matching is in random nc. In: Proceedings of the Seventeenth Annual ACM Symposium on Theory of Computing, pp. 22–32. ACM (1985)
21. Khot, S., Ponnuswami, A.K.: Better inapproximability results for maxclique, chromatic number and min-3lin-deletion. In: Bugliesi, M., Preneel, B., Sassone, V., Wegener, I. (eds.) ICALP 2006. LNCS, vol. 4051, pp. 226–237. Springer, Heidelberg (2006). doi:10.1007/11786986_21
22. Larson, S.M., Snow, C.D., Shirts, M., Pande, V.S.: Folding@ home and genome@ home: Using distributed computing to tackle previously intractable problems in computational biology. arXiv preprint arXiv:0901.0866 (2009)
23. Makarychev, K., Manokaran, R., Sviridenko, M.: Maximum quadratic assignment problem: reduction from maximum label cover and LP-based approximation algorithm. In: Abramsky, S., Gavoille, C., Kirchner, C., Meyer auf der Heide, F., Spirakis, P.G. (eds.) ICALP 2010. LNCS, vol. 6198, pp. 594–604. Springer, Heidelberg (2010). doi:10.1007/978-3-642-14165-2_50
24. Nisan, N.: Pseudorandom generators for space-bounded computation. Combinatorica **12**(4), 449–461 (1992)
25. Papakonstantinou, P.A.: Constructions, lower bounds, and new directions in Cryptography and Computational Complexity. Ph.D. Thesis, University of Toronto (2010)
26. Reingold, O., Rothblum, R., Rothblum, G.: Constant-round interactive proofs for delegating computation. In: Proceedings of the Forty-Eighth Annual ACM on Symposium on Theory of Computing. ACM (2016). (page to appear)
27. Walfish, M., Blumberg, A.J.: Verifying computations without reexecuting them. Commun. ACM **58**(2), 74–84 (2015)

Game-Theoretical Analysis of PLC System Performance in the Presence of Jamming Attacks

Yun Ai[1,5], Manav R. Bhatnagar[2], Michael Cheffena[1], Aashish Mathur[3], and Artem Sedakov[4(✉)]

[1] Norwegian University of Science and Technology (NTNU), 2815 Gjøvik, Norway
{yun.ai,michael.cheffena}@ntnu.no
[2] Indian Institute of Technology Delhi, Hauz Khas, Delhi 110016, India
manav@ee.iiid.ac.in
[3] Indian Institute of Technology (BHU) Varanasi,
Varanasi 221005, Uttar Pradesh, India
amathur.ece@iitbhu.ac.in
[4] Saint Petersburg State University, Saint Petersburg 199034, Russia
a.sedakov@spbu.ru
[5] University of Oslo, 0316 Oslo, Norway

Abstract. In this paper, we investigate the performance of power line communication (PLC) network in the presence of jamming attacks. The legitimate nodes of the PLC network try to communicate with the anchor node of the network while the jamming node attempts to degrade the system performance. The fading, attenuation and colored noise of the PLC channel with dependence on the frequency and transmission distance are taken into account. To investigate the jamming problem, we frame the adversarial interaction into a Bayesian game, where the PLC network tries to maximize the overall expected network capacity and the jammer node has the opposite goal. In the Bayesian game, both players have imperfect knowledge of their opponents. We study effects of total power available to the players on the equilibrium of the game by formulating it into zero-sum and non-zero-sum games, respectively. It is found that under some network setup, there exists a threshold power for which the actual gameplay of the legitimate nodes does not depend upon the actions of the jamming node, and vice versa. This allows us to choose the appropriate power allocation schemes given the total power and the action of the jamming node in some cases.

Keywords: Security · Jamming attack · Game theory · Zero-sum game · Non-zero-sum game · Bayesian nash equilibrium · Power line communication

1 Introduction

In recent years, power line communication (PLC) has gained increasing interests from both the industry and academia due to the vision of widespread information

© Springer International Publishing AG 2017
S. Rass et al. (Eds.): GameSec 2017, LNCS 10575, pp. 74–90, 2017.
DOI: 10.1007/978-3-319-68711-7_5

transmission through power lines. With the advantages of omnipresence of power line and no need to invest in new infrastructure, PLC is set to be a promising technology with wide applications in smart grid, home automation and networking, etc. [1–3].

As in the case of wireless communications, PLC system is inherently based on broadcast transmission. This open and shared nature of the PLC transmission medium poses significant challenges for the communication secrecy and privacy in the presence of potential malicious attacks [4]. The nature of the malicious attacks generally indicates conflict and cooperation between the participants in the communication system. These kind of problems can be often addressed with the game theory approach, which has been widely used by the communication and networking research community to tackle various problems [5–7]. The anti-eavesdropping problem in the presence of selfish jamming is studied as a Bertrand game by assuming the single-channel multi-jammer and multi-channel single-jammer models in [5]. In [6], the authors consider a scenario where a jammer attacks one sub-band of a multi-channel wireless communication system. The strategies for both players are about the sub-channels to transmit or attack. The dependence of the equilibria of the formulated game on the relative position of the jammer is investigated. A reactive jamming scenario where the jammer may not always be able to accurately detect the legitimate transmissions is considered in [7]. Overall, depending on the specific scenario and the proposed strategy, different games and solutions can be formulated.

In this paper, we consider the jamming problem of PLC network. The PLC channel tremendously differs from the wireless channel in terms of the attenuation characteristics, fading distributions, and noise characteristics; the nature of wire transmission also makes the scenario of jamming different from the wireless case [8–10]. All these differences make the vast number of analysis and solutions for the wireless communication systems under malicious attacks inapplicable for the PLC systems. More specifically, we investigate the PLC system in the presence of jamming attack, where a malicious node attempts to degrade the network performance by contrasting the transmission at the physical layer. We interpret the legitimate nodes of the PLC network as one player (denoted as player L) with the aim of maximizing the system performance in terms of capacity while the malicious jamming node is considered as another player (denoted as player J) with the goal of minimizing the overall system performance. Therefore, the considered jamming problem can be well framed as a zero-sum game and analyzed with the game theory approach. Additionally, we consider a setup where the jammer has a goal of minimizing its losses assuming it can be tracked and then fined, thus the game becomes non-zero-sum.

The overall capacity of the PLC network, depends on the received signal-to-noise ratio (SNR) or signal-to-noise-plus-interference ratio (SINR) in case of jamming attack of each subchannel. The SNR or SINR highly depends upon the transmission power and the used frequency since the distances from the legitimate nodes to the anchor node in the PLC network are generally fixed. We assume that the legitimate nodes can allocate their spectrum depending on its power situation and their distances to the anchor node. Any feasible allocation of the spectrum by the legitimate nodes, we call type of player L. Meanwhile, the position of the jamming node is determined by its distance to the anchor node, which is supposed to be the type of the player J. We additionally assume that (i) the jammer has an imperfect knowledge on a particular spectrum allocation of the legitimate nodes but it knows all feasible allocations, and (ii) the legitimate nodes have imperfect information on a particular distance of the jammer to the anchor node but they have a knowledge on all feasible distances. Under these assumptions, the investigated game becomes a Bayesian game [11]. In our analysis, our objective is to investigate the role of the power allocation for both players and understand the corresponding effects on the resulting Bayesian equilibrium.

The remainder of the paper is organized as follows. In Sect. 2, we describe the considered system and PLC channel models. In Sect. 3, the investigated problem is formulated as Bayesian games (zero-sum and non-zero-sum); the Bayesian Nash equilibria (BNE) and the equilibrium payoffs to the formulated games are presented. The numerical results are presented in Sect. 4; and the impact of the number of sub-bands on the system performance is discussed. Section 5 concludes the paper.

2   System and Channel Model

PLC channel is tremendously different from the wireless channel. Attenuation in PLC systems depends on the characteristics of the power cables, length of transmission, and the operating frequency. The wireless channel noise stems from the thermal noise, which is modeled as additive white Gaussian noise (AWGN) [12]. However, the background noise in the PLC channel is not white but colored. The amplitude fading statistics in PLC environments are not well established compared to wireless communications. A vast number of measurement results show that distributions such as Rayleigh, Rician, and lognormal are recommended for defining the path amplitudes in PLC channels [13]. In our analysis, we will assume the amplitude following Rayleigh distribution, which was found to be the best fit for a wealth of PLC field measurements [14–18].

The input/output model of a PLC system over Rayleigh fading channel can be expressed as

$$y = h \cdot x + w, \tag{1}$$

where x is the channel input with unit energy, i.e., $E[|x|^2] = 1$, w represents the PLC background noise modeled as colored Gaussian distributed additive noise, and y is the channel output. The envelope of the channel gain, i.e., $|h|$, is Rayleigh distributed with PDF given by

$$f_{|h|}(z) = \frac{z}{\sigma^2} \cdot \exp\left(-\frac{z^2}{2\sigma^2}\right), \quad z \geqslant 0. \tag{2}$$

where $\sigma > 0$ is the scale parameter of the distribution, which determines the statistical average and the variance of the random variable as $E[|h|] = \sigma\sqrt{\pi/2}$

and $\mathrm{Var}[|h|] = (2 - 0.5\pi)\sigma^2$, respectively. In model (1), the average power of $h \cdot x$ depends on the transmit power P_L and the power attenuation $a(D_L, f)$ over transmission distance D_L at operating frequency f^1, i.e.,

$$\mathrm{E}[|h|^2 \cdot |x|^2] = \mathrm{E}[|h|^2] = P_L \cdot a(D_L, f). \tag{3}$$

Due to the nature of the cable propagation environment, the PLC attenuation model is significantly different from that of wireless channel and the attenuation $a(D, f)$ can be modeled by [19]

$$a(D_L, f) = e^{-2(\alpha_1 + \alpha_2 \cdot f^k) \cdot D_L}, \tag{4}$$

where α_1 and α_2 are constants with dependence on the system configurations; the exponent k is the attenuation factor with typical values between 0.5 and 1. It is obvious from (4) that the attenuation increases dramatically with higher frequency and larger transmission distance.

The widely used assumption of white noise for wireless channel does not hold for PLC channel. Instead, the background noise is colored and the average power per unit bandwidth, namely, the power spectral density (PSD), can be written as [19]

$$N(f) = \mathrm{E}[|w|^2] = 10^{0.1 \cdot (\beta_1 + \beta_2 \cdot e^{-f/\beta_3})} \quad [\mathrm{mW/Hz}], \tag{5}$$

where β_1, β_2, and β_3 are some constants.

With the aforementioned system, in case of no jamming, the received average SNR $\bar{\gamma}$ at the transmission distance D_L and frequency f can be expressed as

$$\bar{\gamma}(D_L, f) = \frac{P_L \cdot a(D_L, f)}{N(f)}. \tag{6}$$

A jammer J is located D_J away from the receiver and is transmitting noise-like power P_J over the concerned channel. To simplify our analysis, we can approximate the average SINR expression by using the variance of the Jammer's channel. This practice leads to an approximation, which is found reasonable in practice. Then, the corresponding average SINR can be simply expressed as

$$\bar{\gamma}(D_L, D_J, f) \approx \frac{P_L \cdot a(D_L, f)}{P_J \cdot a(D_J, f) + N(f)}. \tag{7}$$

It is well-known that under the Rayleigh fading channel, the instantaneous SNR or SINR γ is distributed according to an exponential distribution given by

$$f_\gamma(z) = \frac{1}{\bar{\gamma}} \cdot \exp\left(-\frac{z}{\bar{\gamma}}\right), \quad z \geqslant 0, \tag{8}$$

where the parameter $\bar{\gamma}$ is expressed in (6) or (7).

---

[1] The frequency f is in MHz throughout the paper.

The ergodic capacity (a.k.a. Shannon capacity) is defined as the expectation of the information rate over all states of the fading channel. The ergodic capacity of the PLC channel pertaining to the frequency f is expressed as [20]

$$C_f(P_L, D_L; P_J, D_J) = \int_0^\infty \log_2(1 + z) f_\gamma(z)\, dz = \log_2(e) \cdot e^{\frac{1}{\bar{\gamma}}} \cdot E_1\left(\frac{1}{\bar{\gamma}}\right),$$

where the function $E_1(\cdot)$ is the exponential integral of first order given by $E_1(x) = \int_1^\infty \frac{e^{-xt}}{t}\, dt$ [21].

For transmission over a frequency band B, the corresponding ergodic capacity per bandwidth becomes

$$C_B(P_L, D_L; P_J, D_J) = \frac{1}{|B|} \cdot \int_B \log_2(e) \cdot e^{\frac{1}{\bar{\gamma}}} \cdot E_1\left(\frac{1}{\bar{\gamma}}\right) df, \tag{9}$$

integrating over all frequencies f within frequency band B, where the average SNR or SINR $\bar{\gamma}$ is expressed in (6) or (7) depending on the presence of the jammer [22]. It is not possible to obtain closed-form expressions for the integral in (9), but it is simple and straightforward to evaluate it numerically using mathematical softwares such as Matlab and Mathematica.

3   Game-Theoretical Approach of the Jamming Attacks in PLC Network

3.1   Case of a Zero-Sum Game

We assume that the considered PLC network is represented by a finite set of legitimate nodes $\{1, \ldots, m\}$, all of which transmit information to an anchor node. The PLC system operates in the frequency division multiple access (FDMA) mode by using $n \geqslant m$ equal subchannels $B_1, \ldots, B_n$ within the available frequency band B. Denote by $B(\ell) \in \{B_1, \ldots, B_n\}$ the subchannel assigned to legitimate node ℓ. The legitimate nodes may transmit at different power levels depending upon its available power and distance. One jammer node exists in the PLC network which launches "brute-force" hostile attacks at the physical layer by raising the interference level on the transmitting frequency band. The jamming node is also intelligent enough to attack different subchannels with different powers. This considered scenario is quite practical as the PLC network can be readily extended into a core network in future smart grid network (e.g., in Fig. 1, the anchor node is the router connecting all devices over power line and a jamming node can potentially attack the PLC network). As a more practical illustration of usage, the legitimate node might be a wifi access point within a room where there exists no fiber or a sensor node which collects data on the surrounding environment, etc. The anchor node might be a router, which transfers the accumulated information within the PLC network to the data center or the Internet [23].

For player L, let D_ℓ, $\ell = 1, \ldots, m$, represent the distance between the anchor node and the ℓth legitimate node. The total available power for player L is

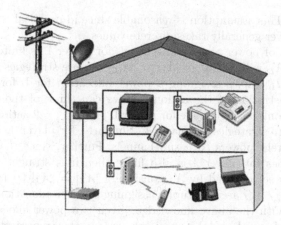

Fig. 1. A typical PLC network where jamming security can be an issue. [24]

denoted as P_L. As explained in Sect. 1, the uncertainty of the allocation scheme of the legitimate nodes on B is modeled by a set $\mathcal{T}_L$ of different types of allocation schemes of the spectrum, where $|\mathcal{T}_L|$ is finite. A type $t_L \in \mathcal{T}_L$ is an assignment profile $(B(1),\ldots,B(m))$ whose components are subchannels assigned to legitimate nodes. As the available frequency band B is equally divided into n subchannels, it is straightforward to see that $|\mathcal{T}_L| = n(n-1)\cdots(n-m+2)$. The corresponding prior probability for the type t_L is denoted as $p_L(t_L)$, and $\sum_{t_L \in \mathcal{T}_L} p_L(t_L) = 1$. For player J, the uncertainty of the distance from the jamming node to the anchor node is also simulated by a set $\mathcal{T}_J$ of different types of the distance, where $|\mathcal{T}_J|$ is also assumed to be finite. A type $t_J \in \mathcal{T}_J$ describes the distance D_J between the jamming node and the anchor node. Similarly, the corresponding prior probability for the type t_J in the finite set $\mathcal{T}_J$ is written as $p_J(t_J)$, and $\sum_{t_J \in \mathcal{T}_J} p_J(t_J) = 1$. The total available power for player J is denoted as P_J. The set of types of allocation schemes of the spectrum $\mathcal{T}_L$ (for player L), the set of types of distance $\mathcal{T}_J$ (for player J), available powers P_L and P_J, distances between legitimate nodes and the anchor node D_ℓ, $\ell = 1,\ldots,m$, and probabilities $p_L(t_L)$, $t_L \in \mathcal{T}_L$, and $p_J(t_J)$, $t_J \in \mathcal{T}_J$, are common knowledge. We suppose that the types of players are selected by Nature, the terminology in game theory standing for a fictitious player which introduces randomness to the game, according to the commonly known prior probability distributions [25].

Once the type of the allocation scheme for player L has been assigned by Nature according to the probability distribution, player L chooses its action $A_L = (P_{L,1},\ldots,P_{L,m})$ from a finite set of actions $\mathcal{A}_L$ to its advantage subject to the constraint $\sum_{\ell=1}^m P_{L,\ell} = P_L$. Similarly for player J, the type of the distance from the jamming node to the anchor node is first assigned by Nature according to the probability distribution, player J takes its action $A_J = (P_{J,1},\ldots,P_{L,n})$ from a finite set of actions $\mathcal{A}_J$ to its advantage subject to the constraint $\sum_{j=1}^n P_{J,j} = P_J$. It should be noted that the strategies, i.e., different power levels allocated to different nodes for player L or subchannels for player J, should take discrete values. This assumption is reasonable since in practical communication systems, the power generally takes discrete values.

Since the set of power allocation schemes for player L is finite, its strategy set denoted by $\mathcal{X}_L$ will consist of $|\mathcal{X}_L| = |\mathcal{A}_L|^{|\mathcal{T}_L|}$ pure strategies whose entries $X_L \in \mathcal{X}_L$ are $|\mathcal{T}_L|$-tuples assigning a power allocation for L for any possible type. As an illustration of the action set, we can think of the simplest case where there are only a single node for player L and $|\mathcal{A}_L| = 2$ actions, then there are only two pure strategies for player L. Similarly, for player J, its strategy is to allocate different powers to n subchannels. Further, since the set of power allocation schemes for player J is also finite then it is straightforward to see that its strategy set denoted by $\mathcal{X}_J$ consists of $|\mathcal{X}_J| = |\mathcal{A}_J|^{|\mathcal{T}_J|}$ pure strategies whose entries $X_J \in \mathcal{X}_J$ are $|\mathcal{T}_J|$-tuples assigning a power allocation for J for any possible type. With the above knowledge, given two power allocation schemes A_L and A_J for players L and J respectively, we can then represent the expected ex ante payoff of player L (with its goal to maximize the overall expected network capacity) as follows

$$E[U_L(X_L, X_J)] = \sum_{t_L \in \mathcal{T}_L} \sum_{t_J \in \mathcal{T}_J} p_L(t_L) \cdot p_J(t_J) \tag{10}$$
$$\times \left(\sum_{\ell=1}^{m} C_{B(\ell)}(P_{L,\ell}, D_\ell; P_{J,B(\ell)}, D_J) \right).$$

The capacity for each subchannel used by each legitimate node can be readily obtained by substituting (6) or (7) into (9). Since player J has the opposite goal (it aims at minimizing the overall network capacity), its expected payoff can be expressed as $E[U_J(X_L, X_J)] = -E[U_L(X_L, X_J)]$ for all $X_L \in \mathcal{X}_L$ and $X_J \in \mathcal{X}_J$. Thus the problem under consideration can be modeled by means of a zero-sum game.

In summary, the investigated Bayesian zero-sum game $\mathcal{G}$ can be characterized as

$$\mathcal{G} = \{\mathcal{P}, \mathcal{T}, \theta, \mathcal{A}, \mathcal{U}\}, \tag{11}$$

where the parameters are elaborated as follows:

- Player set $\mathcal{P} = \{L, J\}$ consists of two players, namely player L: the all legitimate nodes and player J: the jamming node;
- Type sets $\mathcal{T} = \{\mathcal{T}_L, \mathcal{T}_J\}$, where the type of player L is determined by the frequency band allocation scheme, and the type of player J is determined by the distance from the jamming node to the anchor node;
- Probability set $\theta = \{\theta_L, \theta_J\}$, where θ_L and θ_J are the prior probability distributions of the types on $\mathcal{T}_L$ and $\mathcal{T}_J$ assigning probabilities $p_L(t_L)$, $t_L \in \mathcal{T}_L$, and $p_J(t_J)$, $t_J \in \mathcal{T}_J$ for players L and J, respectively;
- Action sets $\mathcal{A} = \{\mathcal{A}_L, \mathcal{A}_J\}$, where $\mathcal{A}_L$ and $\mathcal{A}_J$ being the transmitting power allocations of the available frequency band B of players L and J, respectively;
- Utility functions $\mathcal{U} = \{E[U_L], E[U_J]\}$ where $E[U_L]$ is determined by (10) and $E[U_J] = -E[U_L]$.

3.2 Bayesian Nash Equilibrium

With a goal to find a BNE, which is a saddle point in a zero-sum game, we note that the equilibrium may not exist in pure strategies. For this reason, we introduce a mixed strategy ξ_L of player L as a probability distribution over set $\mathcal{X}_L$ of its pure strategies, where $\xi_L(X_L)$ denotes the probability of choosing pure strategy $X_L \in \mathcal{X}_L$ with $\sum_{X_L \in \mathcal{X}_L} \xi_L(X_L) = 1$. Similarly, a mixed strategy ξ_J of player J is a probability distribution over set $\mathcal{X}_J$, where $\xi_J(X_J)$ stands for the probability of choosing pure strategy $X_J \in \mathcal{X}_J$ with $\sum_{X_J \in \mathcal{X}_J} \xi_J(X_J) = 1$. Let Ξ_L and Ξ_J denote the sets of mixed strategies of players L and J, respectively. Given two mixed strategies ξ_L and ξ_J of players L and J, the expected payoff of player L (with its goal to maximize the overall expected network capacity) is given by

$$E[U_L(\xi_L, \xi_J)] = \sum_{X_L \in \mathcal{X}_L} \sum_{X_J \in \mathcal{X}_J} \xi_L(X_L) \xi_J(X_J) E[U_L(X_L, X_J)], \qquad (12)$$

and $E[U_J(\xi_L, \xi_J)] = -E[U_L(\xi_L, \xi_J)]$ for all $\xi_L \in \Xi_L$ and $\xi_J \in \Xi_J$. We call a pair (ξ_L^*, ξ_J^*) BNE, if $E[U_L(\xi_L, \xi_J^*)] \leqslant E[U_L(\xi_L^*, \xi_J^*)] \leqslant E[U_L(\xi_L^*, \xi_J)]$ for any $\xi_L \in \Xi_L$ and $\xi_J \in \Xi_J$. The expected payoff $E[U_L(\xi_L^*, \xi_J^*)]$ for BNE (ξ_L^*, ξ_J^*) is called the value of the game, which we denote by v.

BNE of the zero-sum game can be found with Minmax Theorem, which is closely related to the linear programming. According to Minmax Theorem, there exists at least one Nash equilibrium and all equilibria yield the same payoff for each player [25]. The mixed strategy under BNE ensures that the value v is maximized in the worst case due to the strategy played by the opponent [26]. This is mathematically expressed as

$$v = \max_{\xi_L \in \Xi_L} \underbrace{\min_{X_J \in \mathcal{X}_J} \sum_{X_L \in \mathcal{X}_L} \xi_L(X_L) E[U_L(X_L, X_J)]}_{v_L}$$

$$\doteq \underbrace{\min_{\xi_J \in \Xi_J} \max_{X_L \in \mathcal{X}_L} \sum_{X_J \in \mathcal{X}_J} \xi_J(X_J) E[U_L(X_L, X_J)]}_{v_J}. \qquad (13)$$

The above optimization can be further reformulated as the following dual linear programs:

$$\max_{\xi_L \in \Xi_L,\, v_L} v_L \qquad (14)$$

$$\text{subject to} \begin{cases} v_L \leqslant \sum_{X_L \in \mathcal{X}_L} \xi_L(X_L) E[U_L(X_L, X_J)], \ \ \forall X_J \in \mathcal{X}_J, \\ \sum_{X_L \in \mathcal{X}_L} \xi_L(X_L) = 1, \\ \xi_L(X_L) \geqslant 0, \ \ \forall X_L \in \mathcal{X}_L, \end{cases}$$

and

$$\min_{\xi_J \in \Xi_J, \, v_J} \quad v_J \tag{15}$$

$$\text{subject to} \begin{cases} v_J \geqslant \sum_{X_J \in \mathcal{X}_J} \xi_J(X_J) \mathrm{E}[U_L(X_L, X_J)], & \forall X_L \in \mathcal{X}_L, \\ \sum_{X_J \in \mathcal{X}_J} \xi_J(X_J) = 1, \\ \xi_J(X_J) \geqslant 0, & \forall X_J \in \mathcal{X}_J. \end{cases}$$

Let (ξ_L^*, v_L^*) and (ξ_J^*, v_J^*) represent the optimal solutions to the above linear programs (14)–(15). From Minmax Theorem it follows that pair (ξ_L^*, ξ_J^*) is a BNE in mixed strategies and $v = v_L^* = v_J^*$ is the value of the game. The optimal solutions of the linear programs can be readily obtained using Matlab command 'linprog()' [27].

3.3   Case of a Non-Zero-Sum Game

In the previous subsection we formulated and examined the problem of jamming attacks in the PLC network when legitimate nodes and the jamming node have opposite goals. However in some cases this approach seems less practical: for example, players do not necessarily aim at maximizing (minimizing) the overall network capacity. Below we propose an extension of the formulated problem to a case of a non-zero-sum game. Let as previously player L transmit the signal over the distance D_L at the frequency band B with the transmit power P_L whereas player J being at the distance D_J away from the receiver transmit noise with the transmit power P_J over the concerned frequency band. To be as close to the previous model as possible and at the same time extending it in line with [28], we define the payoffs of players L and J as $C_B(P_L, D_L; P_J, D_J)$ and $-(1 - \varrho) \cdot C_B(P_L, D_L; P_J, D_J) - \varrho \cdot F$, respectively, where the newly introduced parameters will be described followingly.

From the definitions of players' payoffs, we observe that player L still aims at maximizing its network capacity when transmitting the signal over the distance D_L at the frequency band B with power level P_L under the presence of the jammer (alternatively, player L maximizes its profit from the transmission of a signal receiving one unit of utility for providing one unit of capacity). On the other hand, player J minimizes his expected losses assuming he can be tracked when transmitting the noise signal over the frequency band B with a given constant probability ϱ and then fined a constant penalty $F > 0$. In practice, the penalty might be a fine to the jammer by the utility company after finding the jamming actions (with probability ϱ). Thus the goal of player J is to minimize the expected losses when transmitting the noise at power level P_J being at the distance D_J away from player L. Note that players L and J have completely opposite goals when the jammer can never be tracked, i.e., when the probability $\varrho = 0$. In this case the game becomes zero sum.

It is worth mentioning that players' behavior patterns remain unchanged: a strategy of player L, X_L, is a power allocation A_L among selected subchannels based on an assignment profile t_L realized with probability $p_L(t_L)$, while a strategy of player J, X_J, is a power allocation A_J among all subchannels within the available frequency band based on a distance to the anchor node t_J selected with probability $p_J(t_J)$. Given a strategy profile (X_L, X_J), we represent the expected payoffs of players. Since the goal of player L is still in the maximization of the expected network capacity, the expected payoff of L will have the form of (10), but the expected payoff of player J is given by

$$\mathrm{E}[U_J(X_L, X_J)] = -\sum_{t_L \in T_L} \sum_{t_J \in T_J} p_L(t_L) \cdot p_J(t_J) \qquad (16)$$

$$\times \left((1 - \varrho) \sum_{\ell=1}^{m} C_{B(\ell)}(P_{L,\ell}, D_\ell; P_{J,B(\ell)}, D_J) + \varrho m F \right).$$

The formulated game is not zero sum and it can be characterized by the same components as in (11) with the only difference that players' utility functions $\mathcal{U} = \{\mathrm{E}[U_L], \mathrm{E}[U_J]\}$ represented by their expected payoffs in the PLC network are determined by (10) and (16), respectively. Similarly, introducing mixed strategies ξ_L for player L and ξ_J for player J, we can write the expected payoffs of players as follows

$$\mathrm{E}[U_L(\xi_L, \xi_J)] = \sum_{X_L \in \mathcal{X}_L} \sum_{X_J \in \mathcal{X}_J} \xi_L(X_L)\xi_J(X_J)\mathrm{E}[U_L(X_L, X_J)], \qquad (17)$$

$$\mathrm{E}[U_J(\xi_L, \xi_J)] = \sum_{X_L \in \mathcal{X}_L} \sum_{X_J \in \mathcal{X}_J} \xi_L(X_L)\xi_J(X_J)\mathrm{E}[U_J(X_L, X_J)], \qquad (18)$$

where $\xi_L(X_L)$ and $\xi_J(X_J)$ stand for the probabilities of choosing pure strategies $X_L \in \mathcal{X}_L$ and $X_J \in \mathcal{X}_J$, respectively, with $\sum_{X_L \in \mathcal{X}_L} \xi_L(X_L) = 1$ and $\sum_{X_J \in \mathcal{X}_J} \xi_J(X_J) = 1$, whereas $\mathrm{E}[U_L(X_L, X_J)]$ and $\mathrm{E}[U_J(X_L, X_J)]$ are defined by (10) and (16). We call a pair (ξ_L^*, ξ_J^*) BNE in the non-zero-sum game if $\mathrm{E}[U_L(\xi_L, \xi_J^*)] \leqslant \mathrm{E}[U_L(\xi_L^*, \xi_J^*)]$ for any $\xi_L \in \Xi_L$ and at the same time the relationship $\mathrm{E}[U_J(\xi_L^*, \xi_J)] \leqslant \mathrm{E}[U_J(\xi_L^*, \xi_J^*)]$ holds for any $\xi_J \in \Xi_J$. We denote the expected equilibrium payoffs $\mathrm{E}[U_L(\xi_L^*, \xi_J^*)]$ and $\mathrm{E}[U_J(\xi_L^*, \xi_J^*)]$ for BNE (ξ_L^*, ξ_J^*) by v_L^* and v_J^*, respectively.

It is well-known that the Nash theorem guarantees the existence of at least one BNE in the game [25]. Moreover from the theory of non-zero-sum games we conclude that BNE satisfies the conditions:

$$v_L^* = \max_{X_L \in \mathcal{X}_L} \sum_{X_J \in \mathcal{X}_J} \xi_J^*(X_J)\mathrm{E}[U_L(X_L, X_J)],$$

$$v_J^* = \max_{X_J \in \mathcal{X}_J} \sum_{X_L \in \mathcal{X}_L} \xi_L^*(X_L)\mathrm{E}[U_J(X_L, X_J)].$$

For a two-person games with finite sets of strategies, there has been developed a combinatorial algorithm for finding an equilibrium (so-called the Lemke–Howson algorithm [29]). The mixed BNE can be obtained using Matlab function 'LemkeHowson()' [30].

4   Numerical Results

In this section, the analytical results derived in the previous sections are evaluated numerically with the use of Matlab. We adopt the PLC channel parameter values shown in Table 1, which are the experimental data from field measurements conducted in the industrial environments [31,32]. For simulation purpose, we investigate the simplest case where there are two nodes for player L and one jamming node for player J. Unless stated otherwise, the distances of the two nodes to the anchor node are $D_1 = 20$ m, $D_2 = 28$ m. The two frequencies used by PLC network are $B_1 = [10, 20]$ MHz and $B_2 = [20, 30]$ MHz. The frequency bands are the types of player L, i.e., $\mathcal{T}_L = \{t_{L1}, t_{L2}\}$ where $t_{L1} = B_1$, $t_{L2} = B_2$, which are assigned with probabilities $p_L(t_{L1}) = 1/3$ and $p_L(t_{L2}) = 2/3$. There is no complete information on the position of the jammer except that it is located either 21 or 26 m away from the anchor node, thus these two distances are the types of player J and $\mathcal{T}_J = \{t_{J1}, t_{J2}\}$ where $t_{J1} = 21$ m, $t_{J2} = 26$ m. The probability distribution is $p_J(t_{J1}) = 3/7$ and $p_J(t_{J2}) = 4/7$.

It is known that $P_L = 16$ dBm/Hz and $P_J = 12$ dBm/Hz. The action spaces for players are as follows. For player L, $\mathcal{A}_L = \{A_{L1}, A_{L2}\}$ where $A_{L1} = (0.75P_L, 0.25P_L)$, $A_{L2} = (0.5P_L, 0.5P_L)$, and $\mathcal{A}_J = \{A_{J1}, A_{J1}\}$ where $A_{J1} = (0.25P_J, 0.75P_J)$, $A_{J2} = (0.75P_J, 0.25P_J)$, thus both players have two power allocation schemes. This implies both players' strategy sets consist of four pure strategies: $\mathcal{X}_L = \{X_{L1}, X_{L2}, X_{L3}, X_{L4}\}$ and $\mathcal{X}_J = \{X_{J1}, X_{J2}, X_{J3}, X_{J4}\}$. Players' strategies should be read as follows. The strategy X_{L1} dictates player L to choose A_{L1} if he is of type t_{L1} and A_{L1} if he is of type t_{L2}. The strategy X_{L2} prescribes him to choose A_{L1} if he is of type t_{L1} and A_{L2} if he is of type t_{L2}. When selecting X_{L3}, player L chooses A_{L2} if he is of type t_{L1} and A_{L1} if he is of type t_{L2}. And finally, when selecting X_{L3}, player L chooses A_{L2} if he is of type t_{L1} and A_{L2} if he is of type t_{L2}. Similarly for player J.

4.1   Results for the Case of Zero-Sum Game

By solving the linear programs (14) and (15), the zero-sum game admits the mixed Bayesian Nash equilibrium which is given by $\xi_L^* = (0.0038, 0.6224, 0, 0.3738)$, $\xi_J^* = (0.0424, 0.9576, 0, 0)$, that is, player L with probability 0.0038 plays X_{L1}, with probability 0.6224 plays X_{L2}, and with probability 0.3738 plays X_{L4} whereas player J with probability 0.0424 chooses X_{J1} and with probability 0.9576 chooses X_{J2}. The value of the game $v^* = 1.21912$. Figures 2 and 3 show the equilibrium

Table 1. PLC Channel Parameters

Attenuation model parameters		
$\alpha_1 = 9.33 \times 10^{-3}$ m^{-1}	$\alpha_2 = 5.1 \times 10^{-3}$ s/m	$k = 0.7$
Noise model parameters (industrial environment)		
$\beta_1 = -123$	$\beta_2 = 40$	$\beta_3 = 8.6$

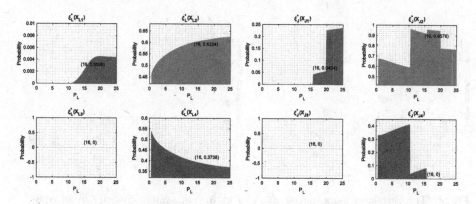

Fig. 2. Components of the equilibrium strategy ξ_L^* as a function of P_L with fixed value of $P_J = 12$ dBm/Hz.

Fig. 3. Components of the equilibrium strategy ξ_J^* as a function of P_L with fixed value of $P_J = 12$ dBm/Hz.

mixed strategies ξ_L^* and ξ_J^* for both players as function of P_L with the fixed value of $P_J = 12$ dBm/Hz. We note that under the considered range of values of P_L, player L (J) never uses his strategy $X_{L3}(X_{J3})$ in any equilibrium, while he starts using strategy $X_{L1}(X_{J1})$ when P_L exceeds some threshold. At the same time, when P_L exceeds this threshold, player J stops using strategy X_{J4} in any equilibrium. Similar figures can be provided for ξ_L^* and ξ_J^* as functions of P_J with fixed value of P_L.

Figure 4 shows the PLC system capacity at the Bayesian Nash equilibrium (the value of the zero-sum game v^*) as a function of the PSDs P_L and P_J. It is clear that in the equilibrium, the system capacity is proportional to the power of the legitimate nodes and inversely proportional to the power from the jamming node. In order to compare the system capacity resulting from different strategies for both players, we investigate the special case of two legitimate nodes for player L and one jamming node for player J. In this scenario, the three-dimensional Fig. 5 suffices to illustrate all the strategies of both players as well as the corresponding payoff. The PSDs for the legitimate nodes and the jamming node are set as $P_L = 16$ dBm/Hz and $P_J = 12$ dBm/Hz, respectively. It can be seen from Fig. 5 that the system capacity for a fixed strategy of player L is a convex function of $P_{J,1}$ while the capacity becomes a concave function of $P_{L,1}$ with fixed $P_{J,1}$. We can see that different strategies from both players lead to quite different system performances. However, there is a saddle point, which the legitimate nodes and the jammer both have no incentive to deviate. This saddle point or the Nash equilibrium is achieved while $P_{L,1} = 6$ dBm/Hz and $P_{J,1} = 12$ dBm/Hz .

Figure 6 demonstrates the relationship between the value of the game (Nash equilibrium payoff), the maxmin and minmax payoffs and the available power to player L, P_L. The maxmin payoff is simply the best payoff for player L when player J plays the most hostile strategy while the minmax payoff is player L's worst payoff when player J plays the least harmful strategy for player L. Clearly,

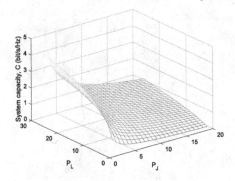

Fig. 4. The zero-sum game payoff at Bayesian Nash equilibrium (the value of the zero-sum game) as a function of the PSDs P_L and P_J.

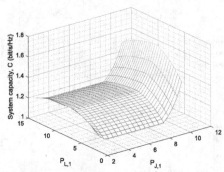

Fig. 5. The system capacity in an equilibrium in the zero-sum game for different strategies of both players as a function of $P_{L,1}$ and $P_{J,1}$ with fixed $P_L = 16$ dBm/Hz and $P_J = 12$ dBm/Hz.

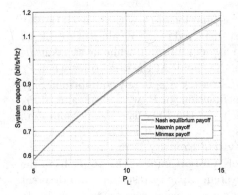

Fig. 6. The Bayesian Nash equilibrium payoff for player L as a function of P_L with the fixed value of $P_J = 12$ dBm/Hz.

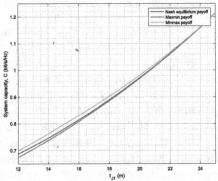

Fig. 7. The Bayesian Nash equilibrium payoff for player L as a function of the type t_{J1} with the other type t_{J2} fixed.

the Nash equilibrium payoff is bounded by the maxmin and minmax payoffs. However, the three payoffs converge while the power available to L is larger than the threshold, which indicates that player L behaves, in this scenario, almost independently of player J's strategy, and vice versa. The similar pattern is presented in Fig. 7, where the relationship is shown between the value of the game (Nash equilibrium payoff), the maxmin and minmax payoffs and the one selected distance (type) of player J, t_{J1}, with the other distance t_{J2} being fixed.

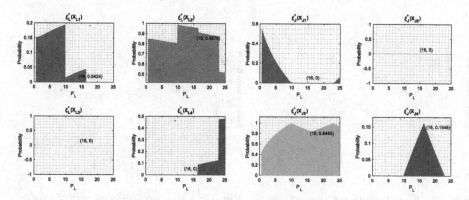

Fig. 8. Components of the equilibrium strategy ξ_L^* as a function of P_L with fixed value of $P_J = 12$ dBm/Hz.

Fig. 9. Components of the equilibrium strategy ξ_J^* as a function of P_L with fixed value of $P_J = 12$ dBm/Hz.

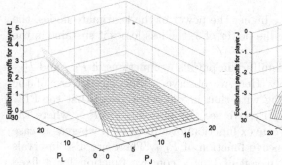

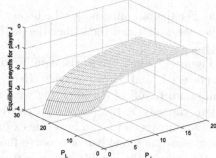

Fig. 10. The equilibrium payoff v_L^* as a function of the PSDs P_L and P_J.

Fig. 11. The equilibrium payoff v_J^* as a function of the PSDs P_L and P_J.

4.2   Results for the Case of Non-Zero-Sum Game

In the case of the non-zero-sum game we additionally assume that the probability of tracking the jammer equals $\varrho = 0.2$ and the fine $F = 50$. In the non-zero-sum game the mixed Bayesian Nash equilibrium is given using the Lemke-Howson algorithm as follows: $\xi_L^* = (0.0424, 0.9576, 0, 0)$, $\xi_J^* = (0, 0, 0.8455, 0.1545)$. Under this equilibrium profile, player L with probability 0.0424 plays X_{L1} and with probability 0.9576 plays X_{L2} whereas player J with probability 0.8455 chooses X_{J3} and with probability 0.1545 chooses X_{J4}. The equilibrium payoffs are: $v_L^* = 1.21141$ and $v_J^* = -1.46913$. Figures 8 and 9 show the equilibrium mixed strategies for both players as a function of P_L with fixed value of P_J.

Figures 10 and 11 show the PLC system capacity v_L^* and jammer's losses v_J^* at the Bayesian Nash equilibrium as a functions of the PSDs P_L and P_J. Again in the equilibrium, the system capacity is proportional to the power of the legitimate nodes and inversely proportional to the power from the jamming

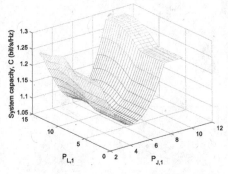

Fig. 12. The equilibrium payoff for player L as a function of $P_{L,1}$ and $P_{J,1}$ with fixed values of $P_L = 16$ dBm/Hz and $P_J = 12$ dBm/Hz.

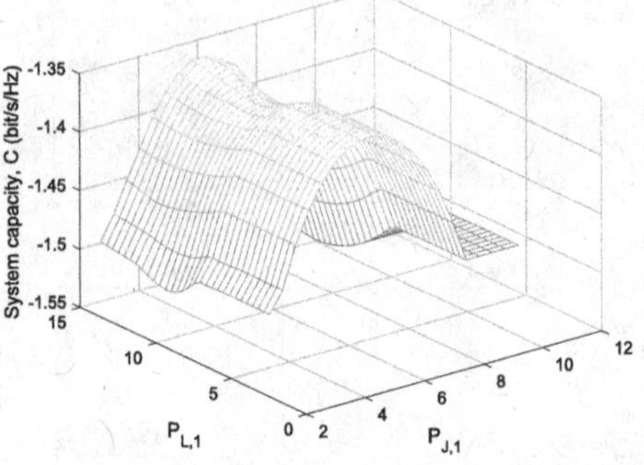

Fig. 13. The equilibrium payoff for player J as a function of $P_{L,1}$ and $P_{J,1}$ with fixed values of $P_L = 16$ dBm/Hz and $P_J = 12$ dBm/Hz.

node. For the jamming node, the higher the power of the legitimate nodes, the higher losses of J and the higher the power of J, the less losses it sustains in the system.

Figures 12 and 13 show the equilibrium payoffs as functions of $P_{L,1}$ and $P_{J,1}$ for players L and J, respectively. Here we observe a different pattern of the equilibrium payoff of L comparing with that in Fig. 5 (we recall that the PLC system capacity is the payoff of L). We see the intervals for $P_{L,1}$ where the equilibrium payoff of L can be a convex function for a fixed $P_{J,1}$, whereas in case of the zero-sum game it is a concave function of $P_{L,1}$. There are also intervals for $P_{J,1}$ where the equilibrium payoff of L is a concave function for a fixed $P_{L,1}$, whereas in Fig. 5 it is a convex function of $P_{J,1}$. Similar conclusion can be made also from Fig. 13 where the equilibrium payoff of player J is demonstrated, recalling that in the zero-sum game the payoff of J differs of L only in sign.

5   Conclusion

In this paper, we formulate the performance of a PLC network with the presence of jamming attacks into a Bayesian game. It was assumed that both players of the game have imperfect knowledge of the opponents, namely the spectrum allocation scheme for the legitimate nodes and the distance of the jamming node to the anchor node. Under some assumptions, we derived the Bayesian Nash equilibrium of the game. We further studied the effects of total power available to both players on the equilibrium. It is found that the equilibrium is unique in many setups, where the jamming node adopts a strategy following which it does not attack the subchannels used by legitimate nodes with specific power allocation. This allows the PLC network to choose the allocation schemes to its advantages in some cases.

It should be noted that the present model can be extended to the case when players have asymmetric information about types: when one player knows his own type but does not observe the type of his opponent what seems to be more practical in most cases. We leave this for future research.

Acknowledgment. The authors gratefully acknowledge the Regional Research Fund of Norway (RFF) for supporting our research. Yun Ai would also like to thank the Norwegian University Center in St. Petersburg for funding his research visit to the Saint Petersburg State University, Russia. Artem Sedakov acknowledges the Russian Foundation for Basic Research (grant No. 17-51-53030).

References

1. Pavlidou, N., Han Vinck, A.J., Yazdani, J., Honary, B.: Power line communications: state of the art and future trends. IEEE Commun. Mag. **41**(4), 34–40 (2003)
2. Liu, W., Sigle, M., Dostert, K.: Channel characterization and system verification for narrowband power line communication in smart grid applications. IEEE Commun. Mag. **49**(12), 28–35 (2011)
3. Zhang, L., Ma, H., Shi, D., Wang, P., Cai, G., Liu, X.: Reliability oriented modeling and analysis of vehicular power line communication for vehicle to grid (V2G) information exchange system. IEEE Access **5**, 12449–12457 (2017)
4. Pittolo, A., Tonello, A.M.: Physical layer security in power line communication networks: an emerging scenario, other than wireless. IET Commun. **8**(8), 1239–1247 (2014)
5. Wang, K., Yuan, L., Miyazaki, T., Guo, S., Sun, Y.: Antieavesdropping with self-ish jamming in wireless networks: a Bertrand game approach. IEEE Trans. Veh. Technol. **66**(7), 6268–6279 (2016)
6. Scalabrin, M., Vadori, V., Guglielmi, A.V., Badia, L.: A zero-sum jamming game with incomplete position information in wireless scenarios. In: Proceedings of European Wireless Conference (EW), pp. 1–6. VDE (2015)
7. Tang, X., Ren, P., Wang, Y., Du, Q., Sun, L.: Securing wireless transmission against reactive jamming: a Stackelberg game framework. In: Proceedings of IEEE Global Communications Conference (GLOBECOM), pp. 1–6. IEEE (2015)
8. Mathur, A., Bhatnagar, M.R., Panigrahi, B.K.: Performance evaluation of PLC under the combined effect of background and impulsive noises. IEEE Commun. Lett. **19**(7), 1117–1120 (2015)
9. Mathur, A., Bhatnagar, M.R., Panigrahi, B.K.: On the performance of a PLC system assuming differential binary phase shift keying. IEEE Commun. Lett. **20**(4), 668–671 (2016)
10. Salem, A., Rabie, K.M., Hamdi, K.A., Alsusa, E., Tonello, A.M.: Physical layer security of cooperative relaying power-line communication systems. In: Proceedings of IEEE International Symposium on Power Line Communications and Its Applications (ISPLC), pp. 185–189. IEEE (2016)
11. Osborne, M.J., Rubinstein, A.: A Course in Game Theory. MIT Press, Cambridge (1994)
12. Goldsmith, A.: Wireless Communications. Cambridge University Press, New York (2005)
13. Güzelgöz, S., Arslan, H., Islam, A., Domijan, A.: A review of wireless and PLC propagation channel characteristics for smart grid environments. J. Electr. Comput. Eng. **2011**, 1–12 (2011)

14. Karols, P., Dostert, K., Griepentrog, G., Huettinger, S.: Mass transit power traction networks as communication channels. IEEE J. Sel. Areas Commun. **24**, 1339–1350 (2006)
15. Kim, S.-C., Lee, J.-H., Song, H.-H., Kim, Y.-H.: Wideband channel measurements and modeling for in-house power line communication. In: Proceedings of IEEE International Symposium on Power Line Communications and its Applications (ISPLC) (2002)
16. Tlich, M., Zeddam, A., Moulin, F., Gauthier, F.: Indoor power-line communications channel characterization up to 100 MHz-part I: one-parameter deterministic model. IEEE Trans. Power Delivery **23**(3), 1392–1401 (2008)
17. Hoque, K.R., Debiasi, L., De Natale, F.G.B.: Performance analysis of MC-CDMA power line communication system. In: Proceedings of International Conference on Wireless and Optical Communications Networks (WOCN), pp. 1–5. IEEE (2007)
18. Karols, P.: Nachrichtentechnische Modellierung von Fahrleitungsnetzen in der Bahntechnik. Mensch-und-Buch-Verlag (2004)
19. Cheng, X., Cao, R., Yang, L.: Relay-aided amplify-and-forward powerline communications. IEEE Trans. Smart Grid **4**(1), 265–272 (2013)
20. Tse, D., Viswanath, P.: Fundamentals of Wireless Communication. Cambridge University Press, New York (2005)
21. Simon, M.K., Alouini, M.-S.: Digital Communication over Fading Channels, vol. 95. Wiley, Hoboken (2005)
22. Ai, Y., Cheffena, M.: Capacity analysis of PLC over Rayleigh fading channels with colored Nakagami-m additive noise. In: Proceedings of IEEE Vehicular Technology Conference (VTC-Fall), pp. 1–5. IEEE (2016)
23. Mathur, A., Bhatnagar, M.R., Ai, Y., Cheffena, M.: Performance analysis of a dual-hop wireless-powerline mixed cooperative system. Submitted to IEEE Transactions on Industrial Informatics
24. (2012). https://seminarprojects.blogspot.com/2012/08/power-line-communication-plc.html. Accessed 28 June 2017
25. Myerson, R.B.: Game Theory. Harvard University Press, Cambridge (2013)
26. Han, Z.: Game Theory in Wireless and Communication Networks: Theory, Models, and Applications. Cambridge University Press, Cambridge (2012)
27. Dantzig, G.B.: Linear Programming and Extensions. Princeton University Press, Princeton (1998)
28. Garnaev, A., Baykal-Gursoy, M., Poor, H.V.: A game theoretic analysis of secret and reliable communication with active and passive adversarial modes. IEEE Trans. Wireless Commun. **15**(3), 2155–2163 (2016)
29. Lemke, C.E., Howson Jr., J.T.: Equilibrium points of bimatrix games. J. Soc. Ind. Appl. Math. **12**(2), 413–423 (1964)
30. https://www.mathworks.com/matlabcentral/fileexchange/44279-lemke-howson-algorithm-for-2-player-games. Accessed 28 June 2017
31. Zimmermann, M., Dostert, K.: A multipath model for the powerline channel. IEEE Trans. Commun. **50**(4), 553–559 (2002)
32. Philipps, H.: Development of a statistical model for powerline communication channels. In: Proceedings of IEEE International Symposium on Power Line Communications and its Applications (ISPLC), pp. 5–7 (2000)

Secure Sensor Design for Cyber-Physical Systems Against Advanced Persistent Threats

Muhammed O. Sayin[(✉)] and Tamer Başar

Coordinated Science Laboratory, University of Illinois at Urbana-Champaign,
Urbana, IL 61801, USA
{sayin2,basar1}@illinois.edu

Abstract. We introduce a new paradigm to the field of control theory: "secure sensor design". Particularly, we design sensor outputs cautiously against advanced persistent threats that can intervene in cyber-physical systems. Such threats are designed for the very specific target systems and seeking to achieve their malicious goals in the long term while avoiding intrusion detection. Since such attacks can avoid detection mechanisms, the controller of the system could have already been intervened in by an adversary. Disregarding such a possibility and disclosing information without caution can have severe consequences. Therefore, through secure sensor design, we seek to minimize the damage of such undetected attacks in cyber-physical systems while impacting the ordinary operations of the system at minimum. We, specifically, consider a controlled Markov-Gaussian process, where a sensor observes the state of the system and discloses information to a controller that can have friendly or adversarial intentions. We show that sensor outputs that are memoryless and linear in the state of the system can be optimal, in the sense of game-theoretic hierarchical equilibrium, within the general class of strategies. We also provide a semi-definite programming based algorithm to design the secure sensor outputs numerically.

Keywords: Stackelberg games · Stochastic control · Cyber-physical systems · Security · Advanced persistent threats · Sensor design · Semi-definite programming

1 Introduction

A cyber-physical system can be considered as a system equipped with sensing and actuation capabilities in the physical part, and monitoring or controlling capabilities using computer-based algorithms in the cyber part, e.g., process control systems, robotics, smart grid, and autonomous vehicles [9]. However, due to the cyber part, such systems are very prone to cyber-attacks. Reference [10] reveals such vulnerabilities of the inner vehicle networks to cyber attacks

This research was supported by the U.S. Office of Naval Research (ONR) MURI grant N00014-16-1-2710.

experimentally, e.g., an attacker has been able to control the brake system of a moving vehicle remotely. In 2010, StuxNet worm targeted very specifically certain supervisory control and data acquisition (SCADA) systems and managed to cause substantial damage, which was an eye-opener pointing to insufficiency of the existing, isolation based, security mechanisms for such systems [8]. Recently in 2014, Dragonfly Malware infiltrated into the cyber-physical systems across the energy and pharmaceutical industries and intervened in the systems over a long period of time stealthily [16]. In a nutshell, those experiences show that once an adversarial attacker infiltrates into the cyber part of the system, he/she can monitor and control the physical processes away from the system's desired target, which can lead to severe consequences. Therefore, developing novel formal security mechanisms plays a vital role in the security of these systems.

Existing studies mainly focus on characterizing the vulnerabilities of cyber-physical systems against various attack models. Reference [14] formulates necessary conditions for an undetected attack that can cause unbounded error in the state estimation. In [18], the authors characterize necessary and sufficient conditions for an undetected attack when the system does not have any sensor and process noises. In [5,6], the authors formulate the optimal cyber-attacks with control objectives, where the attacker both seeks to be undetected and drive the state of the systems according to his/her adversarial goals by manipulating sensor outputs and control inputs together. Recently, Reference [20] has analyzed the optimal attack strategies seeking to increase the quadratic cost of a system with linear Gaussian dynamics, while maintaining certain degree of stealthiness.

There are also studies that aim to provide formal security guarantees against false data injection attacks, where attackers infiltrate into a subset of multiple sensors and report false outputs into the system. In order to detect and recover from such attacks, Reference [7] provides a security mechanism for estimation and control based applications, and in [13], the authors propose a coding scheme for the outputs of multiple sensors. Apart from these two separate approaches, i.e., analyzing optimal attacks with control objectives and encoding outputs of multiple sensors against false data injection attacks, we aim to combine them together in the secure sensor design framework. Particularly, closed-loop control is essential in cyber-physical systems due to the uncertainty of the state noise, i.e., a controller needs the sensor outputs to be able to drive the state toward his/her desired path [11]. By designing sensor outputs in advance, we seek to provide security against the attacks with control objectives.

Economics also plays an essential role while developing defense strategies for cyber-security of systems [4]. As an example, investment on security measures should not exceed the value of the protected asset. Furthermore, adversarial attacks are also costly and an attack would be feasible, therefore expected, if the attack costs the attacker less than the damage at the target. Therefore reducing the damage that can be caused by such threats as much as possible is crucial to reduce the feasibility, therefore the likelihood, of such attacks. To this end, in the secure sensor design framework, we seek to minimize the damage by the attacks, with minimum impact on the ordinary operations of the system.

We propose a new approach for the security of cyber-physical systems by minimizing the damage of cyber-attacks on the system. We focus on undetectable, or difficult to detect, attacks, which we call "advanced persistent threats". These attacks are advanced by targeting very specific systems with knowledge about the underlying dynamics, and persistent by attacking stealthily, i.e., avoiding detection mechanisms. Since such attackers can intervene in the system for a long period of time without being detected, this rises the possibility of adversarial intervention in cyber part of the systems at any time. Therefore, the system designer should take such possibilities into consideration. However, the designer should also not take precautions as if the cyber part of the system is compromised due to such a possibility since that would impact the intended operations of the system substantially. In particular, there is a trade-off between securing the system and maintaining a certain performance in the system.

In this paper, to obtain explicit results, we specifically consider systems with linear quadratic Gaussian dynamics and control objectives, which have various applications in industry [20] from manufacturing processes to aerospace control. We consider the possibility for adversarial interventions in the controller by advanced persistent threats, and seek to design sensor outputs cautiously in advance. Therefore, there is a hierarchical structure between the sensor and the controller of the system. The controller constructs a closed-loop control input based on the sensor output, knowing the relationship between the sensor output and the state. Furthermore, if the controller is an adversary, then the objectives of the sensor and the controller mismatch. Therefore, we can analyze the interactions between the sensor and the controller through a game-theoretic hierarchical equilibrium, which implies that, as a sensor designer, we should anticipate the controller's reaction by also taking into account that the controller can have both friendly or adversarial objectives. We show that for controlled Markov-Gaussian processes, the equilibrium achieving sensor outputs are memoryless and linear in the underlying state of the system. Additionally, we provide a semi-definite programming (SDP) based algorithm to design secure sensor outputs numerically.

The main contributions of this paper are as follows:

- This appears to be the first work in the literature to study sensor design against advanced persistent threats that can infiltrate into the controller of a cyber-physical system.
- We provide a formal problem formulation from a game-theoretical perspective to design sensor outputs cautiously due to the possibility of undetected interventions in the controllers.
- Given any sensor strategies, we compute the optimal control strategies for both friendly and adversarial objectives. Note that the adversary seeks to construct control inputs that are close to the control inputs that would have been constructed if he/she had a friendly objective in order to avoid detection and accomplish his/her malicious goals in the long term over the time horizon by exploiting the uncertainties in the system.
- We show that the optimal sensor strategies in the sense of game-theoretic hierarchical equilibrium are memoryless and linear in the underlying state.

Correspondingly, friendly as well as adversarial control strategies are linear in the sensor outputs.

– We also provide a practical algorithm to design secure sensors numerically.

The paper is organized as follows: In Sect. 2, we provide the secure sensor design framework. In Sect. 3, we formulate the associated multi-stage static Bayesian Stackelberg game. In Sect. 4, we characterize the optimal controller response strategies for given sensor strategies. We compute the corresponding optimal sensor strategies in Sect. 5. We conclude the paper in Sect. 6 with several remarks and possible research directions. An Appendix A includes proof of a technical result.

Notations: For an index-ordered set of variables, e.g., $x_1, \cdots, x_n$, we define $x_{[k,l]} := x_k, \cdots, x_l$, where $1 \leq k \leq l \leq n$. $\mathbb{N}(0, .)$ denotes the multivariate Gaussian distribution with zero mean and designated covariance. We denote random variables by bold lower case letters, e.g., $\boldsymbol{x}$. For a vector x and a matrix A, x' and A' denote their transposes, respectively, and $\|x\|$ denotes the Euclidean (L^2) norm of the vector x. For a matrix A, $\text{tr}\{A\}$ denotes its trace. We denote the identity and zero matrices with the associated dimensions by I and O, respectively. For positive semi-definite matrices A and B, $A \succeq B$ means that $A - B$ is also a positive semi-definite matrix.

2   Problem Formulation

Consider a controlled stochastic system [11] described by the following state equation:

$$\boldsymbol{x}_{k+1} = A\boldsymbol{x}_k + B\boldsymbol{u}_k + \boldsymbol{v}_k, \; k = 1, 2, \ldots, n, \tag{1}$$

where[1] $A \in \mathbb{R}^{m \times m}$, $B \in \mathbb{R}^{m \times r}$, $\boldsymbol{x}_1 \sim \mathbb{N}(0, \Sigma_1)$. The additive noise sequence $\{\boldsymbol{v}_k\}$ is a white Gaussian vector process, i.e., $\boldsymbol{v}_k \sim \mathbb{N}(0, \Sigma_v)$, and is independent of the initial state $\boldsymbol{x}_1$. The closed loop control vector $\boldsymbol{u}_k \in \mathbb{R}^r$ is given by

$$\boldsymbol{u}_k = \gamma_k(\boldsymbol{s}_{[1,k]}), \tag{2}$$

where $\gamma_k(\cdot)$ can also be any Borel measurable function from $\mathbb{R}^{mk}$ to $\mathbb{R}^r$, and $\boldsymbol{s}_k \in \mathbb{R}^m$ is the sensor output, which is given by

$$\boldsymbol{s}_k = \eta_k(\boldsymbol{x}_{[1,k]}), \tag{3}$$

where $\eta_k(\cdot)$ can be any Borel measurable function from $\mathbb{R}^{mk}$ to $\mathbb{R}^m$.

As seen in Fig. 1, we have two non-cooperating agents: Sensor (S) and Controller (C). C can be a friend or an adversary while S does not know C's type. Only S has access to the state $\boldsymbol{x}_k$ and can construct sensor output $\boldsymbol{s}_k$. C observes $\boldsymbol{s}_k$, knows S's strategy $\eta_k(\cdot)$ due to a hierarchy between the agents, and, by using $\boldsymbol{s}_{[1,k]}$, can construct a closed loop control input $\boldsymbol{u}_k$, which cannot be monitored by the system.

---

[1] Even though we consider time invariant matrices A and B for notational simplicity, the provided results could also be extended to time-variant cases.

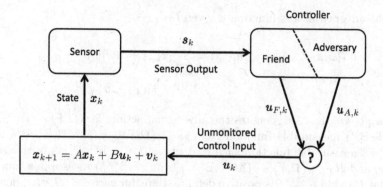

Fig. 1. Cyber physical system including a sensor and a controller.

Remark 1. A hierarchy between the agents is a reasonable assumption in control system design since sensors are designed and implemented in advance, and system engineers design the controllers knowing the relation between the sensor output and the underlying state.

The agents S and C construct s_k and u_k according to their own objectives. In particular, S chooses $\eta_k(\cdot)$ from the strategy space Υ_k, which, for each k, is the set of all Borel measurable functions from $\mathbb{R}^{mk}$ to $\mathbb{R}^m$, i.e., $\eta_k \in \Upsilon_k$ and $s_k = \eta_k(x_{[1,k]})$. C chooses $\gamma_k(\cdot)$ from the strategy space Γ_k, which is the set of all Borel measurable functions from $\mathbb{R}^{mk}$ to $\mathbb{R}^r$, i.e., $\gamma_k \in \Gamma_k$ and $u_k = \gamma_k(s_{[1,k]})$.

Normally, in a stochastic control scenario [11], S and C would have a common finite horizon[2] quadratic loss function

$$L(x_{[2,n+1]}, s_{[1,n]}, u_{[1,n]}) = \sum_{k=1}^{n} \|x_{k+1}\|_{Q_{k+1}}^2 + \|u_k\|_{R_k}^2, \tag{4}$$

where $Q_{k+1} \in \mathbb{R}^{m \times m}$ is positive semi-definite and $R_k \in \mathbb{R}^{r \times r}$ is positive definite. Then, S would disclose the state directly so that C could drive the state in their commonly desired path [11,12]. However, in a cyber physical system, the system is vulnerable against adversarial attacks that seek to drive the state of the system away from the system's desired target. We call such attacks "advanced persistent threats", which are advanced by being designed very specifically for the targeted system, i.e., the attacker knows, or can learn stealthily, the underlying state recursion, and persistent by avoiding intrusion detection. Therefore, S, i.e., the sensor designer, should anticipate the likelihood of adversarial intrusions into C, i.e., the possibility that C can be an adversary, and construct s_k accordingly.

We denote the set of all adversarial objectives by Ω, the appropriate σ-algebra on Ω by F, and the probability distribution over Ω by **P**. In particular, we have the probability space $(\Omega, \mathsf{F}, \mathbf{P})$. And for a point $\omega \in \Omega$ drawn from Ω according

---

[2] E.g., horizon length is n.

to $\mathbf{P}$, the adversarial loss function is given by

$$L_A(\omega, \boldsymbol{x}_{[2,n+1]}, \boldsymbol{s}_{[1,n]}, \boldsymbol{u}_{[1,n]}) = \sum_{k=1}^{n} \|\boldsymbol{x}_{k+1} - z(\omega)\|^2_{Q_{A,k+1}(\omega)}$$
$$+ \|\boldsymbol{u}_{A,k} - \boldsymbol{u}^*_{F,k}\|^2_{R_{A,k}(\omega)}, \tag{5}$$

where $\boldsymbol{u}_{A,k}$, $k = 1, \ldots, n$, denotes the adversarial action, $z : (\Omega, \mathsf{F}) \to (\mathbb{R}^m, \mathsf{B}^m)$ is an $(\mathsf{F}, \mathsf{B}^m)$ measurable function[3], $Q_{A,k+1} : (\Omega, \mathsf{F}) \to (\mathbb{R}^{m \times m}, \mathsf{B}^{m \times m})$ is an $(\mathsf{F}, \mathsf{B}^{m \times m})$ measurable function such that $Q_{A,k+1}(\omega) \in \mathbb{R}^{m \times m}$ is positive semi-definite, and $R_{A,k} : (\Omega, \mathsf{F}) \to (\mathbb{R}^{r \times r}, \mathsf{B}^{r \times r})$ is an $(\mathsf{F}, \mathsf{B}^{r \times r})$ measurable function such that $R_{A,k}(\omega) \in \mathbb{R}^{r \times r}$ is positive definite. Here, for each $\omega \in \Omega$, $z(\omega)$ denotes the desired state that the adversary seeks to drive the system to, and $\boldsymbol{u}^*_{F,k}$ is the optimal action that would have been taken if C was a friend so that the adversary can avoid intrusion detection by being close to $\boldsymbol{u}^*_{F,k}$. We further assume that $z(\omega)$ is a second-order random vector.

Remark 2. We note that if the control inputs could have been monitored, then any deviation of the control input from the optimal control input of a friend type C could have been detected instantly.

3   A Multi-stage Static Bayesian Stackelberg Game

In order to model undetected adversarial interventions, let $\boldsymbol{\theta}$ be a Bernoulli random variable, with a commonly known p, corresponding to the likelihood of C being an adversary, i.e., $\mathbb{P}\{\boldsymbol{\theta} = 1\} = p$, and $\boldsymbol{\theta} = 1$ if C is an adversary. Since the type of C is not known by S, we can consider this incomplete information scenario as an imperfect information scenario [15]; in which Nature moves first, draws a realization of $\boldsymbol{\theta}$, then if the realization $\theta = 1$, also draws $\omega \in \Omega$, and reveals these only to C.

Furthermore, the multiple interactions between non-cooperating S and C can be considered as a *multi-stage* game [1]. Since S's actions $\boldsymbol{s}_{[1,n]}$ do not depend on C's actions $\boldsymbol{u}_{[1,n]}$, i.e., S cannot update his/her strategies after observing $\boldsymbol{u}_{[1,n]}$, this is a multi-stage *static* game. The underlying state recursion is common knowledge of both S and C (even if C can be an adversary). The type of C and, if C is an adversary, his/her objective are not known by S. However, S knows the probability space $(\Omega, \mathsf{F}, \mathbf{P})$ and p, which implies that this is a multi-stage static *Bayesian* game. There is also a hierarchy [1,17] between the agents in the announcement of the strategies such that S leads the game by announcing and sticking to his/her strategies in advance, i.e., C knows $\eta_{[1,n]}$ in advance. Therefore, we can model such a scheme as a multi-stage static Bayesian *Stackelberg* game, in which S is the leader.

Remark 3. Once any adversarial intrusion has been detected due to C's anomalous behavior through external defense mechanisms, this multi-stage static

---

[3] B^m denotes the Borel σ-algebra on $\mathbb{R}^m$.

Bayesian Stackelberg game terminates since the uncertainty about C's type is removed. The reaction of the system after the detection is beyond this paper's scope. Therefore, we consider that the game continues over the horizon and continuation of the game implies that any adversarial intervention has not been detected while the possibility of undetected adversarial intervention still exists.

Remark 4. Even though the attacker can also inject false data into the sensor outputs in order to avoid detection as in integrity attacks, e.g., [5,6], the attacker still needs the actual sensor outputs, which are designed by the system designer in advance, in order to construct the optimal control input according to his/her objective. Therefore, secure sensor design framework also plays a crucial role for the security of the systems against integrity attacks.

S and C aim to minimize their expected loss functions through the actions $s_{[1,n]}$ and $u_{[1,n]}$ by choosing the strategies $\eta_{[1,n]}$ and $\gamma_{[1,n]}$ accordingly. Given the realizations of S's actions, i.e., $s_{[1,k]}$, C constructs the control input $u_{F,k}$ or $u_{A,k}$ depending on his/her type, which not only depends on $s_{[1,k]}$, but also the associated strategies $\eta_{[1,k]}$. In order to show this dependence explicitly, we denote C's strategies by $u_{F,k} = \gamma_{F,k}(s_{[1,k]}; \eta_{[1,k]})$ instead of $\gamma_{F,k}(s_{[1,k]})$ if C is a friend, or $u_{A,k} = \gamma_{A,k}(\omega, s_{[1,k]}; \eta_{[1,k]})$ instead of $\gamma_{A,k}(\omega, s_{[1,k]})$ if C is an adversary. Furthermore, given S's strategies $\eta_{[1,n]}$, we let $\Pi_F(\eta_{[1,n]}), \Pi_A(\omega, \eta_{[1,n]}) \subset \mathbb{R}^{r \times n}$ be C's reaction set. And these reaction sets are given by:

$$\Pi_F(\eta_{[1,n]}) := \underset{\substack{u_{F,k} \in \mathbb{R}^r \\ k=1,\dots,n}}{\operatorname{argmin}} \ \mathbb{E}\{L(x_{[2,n+1]}, s_{[1,n]}, u_{F,[1,n]})\},$$

$$\Pi_A(\omega, \eta_{[1,n]}) := \underset{\substack{u_{A,k} \in \mathbb{R}^r \\ k=1,\dots,n}}{\operatorname{argmin}} \ \mathbb{E}\{L_A(\omega, x_{[2,n+1]}, s_{[1,n]}, u_{A,[1,n]})\},$$

where $\mathbb{E}$ denotes the expectation taken over $\{x_1, v_{[1,n]}\}$. Due to the positive definiteness assumptions on R_k and $R_{A,k}(\omega)$, for all $\omega \in \Omega$, L and L_A are strictly convex in C's actions $u_{F,[1,n]}, u_{A,[1,n]}$. This implies that the corresponding reaction sets Π_F and Π_A are singletons and the best C actions $u^*_{F,k}, u^*_{A,k}$ are unique.

Corresponding to the loss functions L and L_A, depending on the agents' actions s_k and u_k, there exist certain cost functions depending on the agents' strategies: $J(\eta_{[1,n]}, \gamma_{[1,n]})$ and $J_A(\omega, \eta_{[1,n]}, \gamma_{[1,n]})$, while each strategy implicitly depends on the other. Therefore let $\tilde{\Pi}_F$ and $\tilde{\Pi}_A$ be the sets of best C strategies, as subsets of $\times_{k=1}^n \Gamma_k$:

$$\tilde{\Pi}_F(\eta_{[1,n]}) := \underset{\substack{\gamma_{F,k} \in \Gamma_k \\ k=1,\dots,n}}{\operatorname{argmin}} \ J(\eta_{[1,n]}, \gamma_{F,[1,n]}),$$

$$\tilde{\Pi}_A(\omega, \eta_{[1,n]}) := \underset{\substack{\gamma_{A,k}(\omega,\cdot) \in \Gamma_k \\ k=1,\dots,n}}{\operatorname{argmin}} \ J_A(\omega, \eta_{[1,n]}, \gamma_{A,[1,n]}),$$

which are equivalence classes such that $\forall\, \gamma^*_{F,[1,n]} \in \tilde{\Pi}_F$ (or $\forall\, \gamma^*_{A,[1,n]} \in \tilde{\Pi}_A$), we have $\boldsymbol{u}^*_{F,k} = \gamma^*_{F,k}(\boldsymbol{s}_{[1,k]}; \eta_{[1;k]})$ (or $\boldsymbol{u}^*_{A,k} = \gamma^*_{A,k}(\omega, \boldsymbol{s}_{[1,k]}; \eta_{[1;k]})$). Therefore, the pair of strategies $\left[\eta^*_{[1,n]}; (\gamma^*_{F,[1,n]}, \gamma^*_{A,[1,n]})\right]$ attains the Stackelberg equilibrium provided that

$$\eta^*_{[1,n]} = \operatorname*{argmin}_{\substack{\eta_k \in \Upsilon_k, \\ k=1,\dots,n}} (1-p) J\big(\eta_{[1,n]}, \gamma^*_{F,[1,n]}(\cdot\,; \eta_{[1,n]})\big)$$

$$+ \, p \int_\Omega J\big(\eta_{[1,n]}, \gamma^*_{A,[1,n]}(\omega, \cdot\,; \eta_{[1,n]})\big) \mathbf{P}(d\omega) \tag{6a}$$

$$\gamma^*_{F,[1,n]}(\cdot\,; \eta_{[1,n]}) = \operatorname*{argmin}_{\substack{\gamma_{F,k} \in \Gamma_k, \\ k=1,\dots,n}} J\big(\eta_{[1,n]}, \gamma_{F,[1,n]}(\cdot\,; \eta_{[1,n]})\big), \tag{6b}$$

$$\gamma^*_{A,[1,n]}(\omega, \cdot\,; \eta_{[1,n]}) = \operatorname*{argmin}_{\substack{\gamma_{A,k}(\omega,\cdot) \in \Gamma_k, \\ k=1,\dots,n}} J_A\big(\omega, \eta_{[1,n]}, \gamma_{A,[1,n]}(\omega, \cdot\,; \eta_{[1,n]})\big). \tag{6c}$$

In the following sections, we analyze these equilibrium achieving strategies, i.e., $\left[\eta^*_{[1,n]}; (\gamma^*_{F,[1,n]}, \gamma^*_{A,[1,n]})\right]$.

4   Optimal Follower (Controller) Reactions

By (4), for a given $\boldsymbol{s}_{[1,n]}$, the friendly C also seeks to minimize

$$\sum_{k=1}^{n} \mathbb{E}\left\{ \|\boldsymbol{x}_{k+1}\|^2_{Q_{k+1}} + \|\boldsymbol{u}_k\|^2_{R_k} \right\}, \tag{7}$$

over $\gamma_{F,k} \in \Gamma_k$, $k = 1,\dots,n$, such that $\boldsymbol{u}_{F,k} = \gamma_{F,k}(\boldsymbol{s}_{[1,k]})$ subject to (1)–(3). In order to facilitate the subsequent analysis, in the following, we rewrite the state equations (1)–(2) and the expected loss function (7) without altering the optimization problem.

Lemma 1. *The friendly objective* (7) *is equivalent to:*

$$\min_{\substack{\gamma_{F,k} \in \Gamma_k \\ k=1,\dots,n}} \sum_{k=1}^{n} \mathbb{E}\|\boldsymbol{u}_{F,k} + K_k \boldsymbol{x}_k\|^2_{\Delta_k} + G, \tag{8}$$

where

$$K_k = \Delta_k^{-1} B'_k \tilde{Q}_{k+1} A \tag{9a}$$

$$\Delta_k = B' \tilde{Q}_{k+1} B + R_k \tag{9b}$$

$$G = \operatorname{tr}\{\Sigma_1 \tilde{Q}_1\} + \sum_{k=1}^{n} \operatorname{tr}\{\Sigma_v \tilde{Q}_{k+1}\} \tag{9c}$$

and $\{\tilde{Q}_k\}$ *is a sequence defined through the following discrete-time Riccati equation:*

$$\tilde{Q}_{k+1} = Q_k + A'\left(\tilde{Q}_{k+1} - \tilde{Q}_{k+1}B\Delta_k^{-1}B'\tilde{Q}_{k+1}\right)A, \tag{10a}$$

$$\tilde{Q}_{n+1} = Q_{n+1} \text{ and } Q_1 = O. \tag{10b}$$

Proof. This follows from the extensively used "completing the squares" technique [2,11]. □

Note that in (8), $\boldsymbol{x}_k$ depends on the previous control inputs $\boldsymbol{u}_{[1,k-1]}$. Through a change of variables [2], the friendly C's objective (8) can be written as

$$\min_{\substack{\gamma_{F,k}\in\Gamma_k \\ k=1,\dots,n}} \sum_{k=1}^{n} \mathbb{E}\|\boldsymbol{u}_{F,k}^o + K_k\boldsymbol{x}_k^o\|_{\Delta_k}^2 + G \tag{11}$$

subject to (9)–(10) and

$$\boldsymbol{x}_{k+1}^o = A\boldsymbol{x}_k^o + \boldsymbol{v}_k, \; k = 1,\dots,n, \text{ and } \boldsymbol{x}_1^o = \boldsymbol{x}_1, \tag{12a}$$

$$\boldsymbol{u}_{F,k}^o = \boldsymbol{u}_{F,k} + K_kB\boldsymbol{u}_{F,k-1} + K_kAB\boldsymbol{u}_{F,k-2} + \cdots + K_kA^{k-2}B\boldsymbol{u}_{F,1}. \tag{12b}$$

Note also that, now, the process $\{\boldsymbol{x}_k^o\}$ is independent of the control inputs $\boldsymbol{u}_{F,k}$ (and $\boldsymbol{u}_{F,k}^o$). Therefore, by (11), given the sensor outputs $\boldsymbol{s}_{[1,k]} = s_{[1,k]}$, the optimal *transformed* control input $u_{F,k}^o$ (12b) is given by

$$u_{F,k}^{o*} = -K_k\mathbb{E}\{\boldsymbol{x}_k^o|\boldsymbol{s}_{[1,k]} = s_{[1,k]}\},$$

which implies

$$\boldsymbol{u}_{F,k}^{o*} = -K_k\mathbb{E}\{\boldsymbol{x}_k^o|\boldsymbol{s}_{[1,k]}\} \tag{13}$$

almost everywhere on $\mathbb{R}^r$. By (12b), we have

$$\begin{bmatrix} \boldsymbol{u}_{F,n}^o \\ \boldsymbol{u}_{F,n-1}^o \\ \vdots \\ \boldsymbol{u}_{F,1}^o \end{bmatrix} = \underbrace{\begin{bmatrix} I & K_nB & \cdots & K_nA^{n-2}B \\ & I & \cdots & K_{n-1}A^{n-3}B \\ & & \ddots & \vdots \\ & & & I \end{bmatrix}}_{=:\, \Phi} \underbrace{\begin{bmatrix} \boldsymbol{u}_{F,n} \\ \boldsymbol{u}_{F,n-1} \\ \vdots \\ \boldsymbol{u}_{F,1} \end{bmatrix}}_{=:\, \boldsymbol{u}},$$

$$\underbrace{\phantom{\begin{bmatrix} \boldsymbol{u}_{F,n}^o \\ \vdots \end{bmatrix}}}_{=:\, \boldsymbol{u}^o}$$

which can also be written as $\boldsymbol{u}_F^o = \Phi\boldsymbol{u}_F$. And (13) leads to

$$\boldsymbol{u}_F^{o*} = -\underbrace{\begin{bmatrix} K_n & & \\ & \ddots & \\ & & K_1 \end{bmatrix}}_{=:\, K} \underbrace{\begin{bmatrix} \mathbb{E}\{\boldsymbol{x}_n^o|\boldsymbol{s}_{[1,n]}\} \\ \vdots \\ \mathbb{E}\{\boldsymbol{x}_1^o|\boldsymbol{s}_1\} \end{bmatrix}}_{=:\, \hat{\boldsymbol{x}}^o}, \tag{14}$$

which yields that the actual optimal control inputs are given by

$$\boxed{\boldsymbol{u}_F^* = -\Phi^{-1}K\,\hat{\boldsymbol{x}}^o.} \tag{15}$$

While the friendly C has the same objective (4) with S, by (5), for each $\omega \in \Omega$, the adversarial C's objective is to minimize

$$\sum_{k=1}^{n} \mathbb{E}\left\{ \|\boldsymbol{x}_{k+1} - z(\omega)\|_{Q_{A,k+1}(\omega)}^2 + \|\boldsymbol{u}_{A,k} - \boldsymbol{u}_{F,k}^*\|_{R_{A,k}(\omega)}^2 \right\}, \qquad (16)$$

over $\gamma_{A,k}(\omega, \cdot) \in \Gamma_k$, $k = 1, \ldots, n$, such that $\boldsymbol{u}_{A,k} = \gamma_{A,k}(\omega, \boldsymbol{s}_{[1,k]})$ subject to (1)–(3). Next, we aim to rewrite the state equations and the expected loss functions as in Lemma 1 and (11) for the minimization of the adversarial objective.

Let $\delta \boldsymbol{u}_k := \boldsymbol{u}_{A,k} - \boldsymbol{u}_{F,k}^*$ and instead of (1), consider the following recursion:

$$\begin{bmatrix} \boldsymbol{x}_{k+1} \\ \boldsymbol{u}_F^* \\ z(\omega) \end{bmatrix} = \underbrace{\begin{bmatrix} A & \vdots & B & \cdots \\ \hline O & \vdots & I \end{bmatrix}}_{=: \bar{A}} \underbrace{\begin{bmatrix} \boldsymbol{x}_k \\ \boldsymbol{u}_F^* \\ z(\omega) \end{bmatrix}}_{= \bar{\boldsymbol{x}}_k} + \underbrace{\begin{bmatrix} B \\ O \end{bmatrix}}_{=: \bar{B}} \delta \boldsymbol{u}_k + \underbrace{\begin{bmatrix} I \\ O \end{bmatrix}}_{=: E} \boldsymbol{v}_k,$$

which can also be written as

$$\bar{\boldsymbol{x}}_{k+1} = \bar{A}\bar{\boldsymbol{x}}_k + \bar{B}\,\delta \boldsymbol{u}_k + E\boldsymbol{v}_k. \qquad (17)$$

Correspondingly, the objective can be rewritten as

$$\sum_{k=1}^{n} \mathbb{E}\left\{ \|\bar{\boldsymbol{x}}_{k+1}\|_{\tilde{Q}_{A,k+1}(\omega)}^2 + \|\delta \boldsymbol{u}_k\|_{R_{A,k}(\omega)}^2 \right\}, \qquad (18)$$

where

$$\begin{aligned} \tilde{Q}_{A,k+1}(\omega) &:= \begin{bmatrix} I \\ O \\ -I \end{bmatrix} Q_{A,k+1}(\omega) \begin{bmatrix} I & O & -I \end{bmatrix} \\ &= \begin{bmatrix} Q_{A,k+1}(\omega) & O & -Q_{A,k+1}(\omega) \\ O & O & O \\ -Q_{A,k+1}(\omega) & O & Q_{A,k+1}(\omega) \end{bmatrix}. \end{aligned}$$

We point out the resemblance between (7) and (18). Therefore, by Lemma 1 and (11), we have the following transformations:

Lemma 2. *The adversary's objective* (18) *is equivalent to:*

$$\min_{\substack{\gamma_{A,k}(\omega, \cdot) \in \Gamma_k \\ k=1,\ldots,n}} \sum_{k=1}^{n} \mathbb{E}\|\delta \boldsymbol{u}_k + K_{A,k}(\omega)\bar{\boldsymbol{x}}_k\|_{\Delta_{A,k}(\omega)}^2 + G_A(\omega), \qquad (19)$$

where

$$K_{A,k}(\omega) = \Delta_{A,k}(\omega)^{-1} \bar{B}' \tilde{Q}_{A,k+1}(\omega)\bar{A} \qquad (20a)$$

$$\Delta_{A,k}(\omega) = \bar{B}' \tilde{Q}_{A,k+1}(\omega)\bar{B} + R_{A,k}(\omega) \qquad (20b)$$

$$G_A(\omega) = \mathrm{tr}\{\bar{\Sigma}_1 \tilde{Q}_{A,1}(\omega)\} + \sum_{k=1}^{n} \mathrm{tr}\{\bar{\Sigma}_v \tilde{Q}_{A,k+1}(\omega)\}, \qquad (20c)$$

$$\bar{\Sigma}_1 := \begin{bmatrix} \Sigma_1 & \mathbb{E}\{\boldsymbol{x}_1^o(\boldsymbol{u}_F^*)'\} & O \\ \mathbb{E}\{\boldsymbol{u}_F^*(\boldsymbol{x}_1^o)'\} & \mathbb{E}\{\boldsymbol{u}_F^*(\boldsymbol{u}_F^*)'\} & O \\ O & O & z(\omega)z(\omega)' \end{bmatrix} \text{ and } \bar{\Sigma}_v := \begin{bmatrix} \Sigma_v & O \\ O & O \end{bmatrix},$$

and $\{\tilde{Q}_{A,k}(\omega)\}$ for each $\omega \in \Omega$ is a sequence defined through the following discrete-time Riccati equation:

$$\tilde{Q}_{A,k+1}(\omega) = Q_{A,k}(\omega) + \bar{A}'\left(\tilde{Q}_{A,k+1}(\omega) - \tilde{Q}_{A,k+1}(\omega)\bar{B}\Delta_{A,k}(\omega)^{-1}\bar{B}'\tilde{Q}_{A,k+1}(\omega)\right)\bar{A}, \quad (21a)$$

$$\tilde{Q}_{A,n+1}(\omega) = Q_{A,n+1}(\omega) \text{ and } Q_{A,1}(\omega) = O. \quad (21b)$$

And corresponding to (11), the adversarial objective (19) can be written as

$$\min_{\substack{\gamma_{A,k}(\omega,\cdot)\in\Gamma_k \\ k=1,\dots,n}} \sum_{k=1}^{n} \mathbb{E}\|\,\delta\boldsymbol{u}_k^o + K_{A,k}(\omega)\bar{\boldsymbol{x}}_k^o\|_{\Delta_{A,k}(\omega)}^2 + G_A(\omega) \quad (22)$$

subject to (20)–(21) and

$$\bar{\boldsymbol{x}}_{k+1}^o = \bar{A}\bar{\boldsymbol{x}}_k^o + E v_k, \ k = 1,\dots,n, \text{ and } \bar{\boldsymbol{x}}_1^o = \bar{\boldsymbol{x}}_1, \quad (23a)$$

$$\delta\boldsymbol{u}_k^o = \delta\boldsymbol{u}_k + K_{A,k}(\omega)\bar{B}\,\delta\boldsymbol{u}_{k-1} + K_{A,k}(\omega)\bar{A}\bar{B}\,\delta\boldsymbol{u}_{k-2} + \cdots + K_{A,k}(\omega)\bar{A}^{k-2}\bar{B}\,\delta\boldsymbol{u}_1. \quad (23b)$$

Note that in (22), $C_A(\omega)$ is independent from the adversary's optimization arguments even though it depends on $\boldsymbol{u}_F^*$ due to $\bar{\Sigma}_1$ in (20c). Furthermore, given the sensor outputs $\boldsymbol{s}_{[1,k]} = s_{[1,k]}$, the optimal *transformed* adversary action $\delta u_{A,k}^{o*}$ of (23b) is given by

$$\delta u_{A,k}^{o*} = -K_{A,k}(\omega)\mathbb{E}\{\bar{\boldsymbol{x}}_k^o | \boldsymbol{s}_{[1,k]} = s_{[1,k]}\},$$

which also implies

$$\delta\boldsymbol{u}_k^{o*} = -K_{A,k}(\omega)\mathbb{E}\{\bar{\boldsymbol{x}}_k^o | \boldsymbol{s}_{[1,k]}\} \quad (24)$$

almost everywhere on $\mathbb{R}^r$. By (23b), we have

$$\underbrace{\begin{bmatrix} \delta\boldsymbol{u}_n^o \\ \delta\boldsymbol{u}_{n-1}^o \\ \vdots \\ \delta\boldsymbol{u}_1^o \end{bmatrix}}_{=:\ \delta\boldsymbol{u}^o} = \underbrace{\begin{bmatrix} I & K_{A,n}(\omega)\bar{B}_{n-1} & \cdots & K_{A,n}(\omega)\bar{A}^{n-2}\bar{B} \\ & I & \cdots & K_{A,n-1}(\omega)\bar{A}^{n-3}\bar{B} \\ & & \ddots & \vdots \\ & & & I \end{bmatrix}}_{=:\ \Phi_A(\omega)} \underbrace{\begin{bmatrix} \delta\boldsymbol{u}_n \\ \delta\boldsymbol{u}_{n-1} \\ \vdots \\ \delta\boldsymbol{u}_1 \end{bmatrix}}_{=:\ \delta\boldsymbol{u}},$$

which can also be written as $\delta\boldsymbol{u}^o = \Phi_A(\omega)\,\delta\boldsymbol{u}$. And (24) leads to

$$\delta\boldsymbol{u}^{o*} = -\underbrace{\begin{bmatrix} K_{A,n}(\omega) & & \\ & \ddots & \\ & & K_{A,1}(\omega) \end{bmatrix}}_{=:\ K_A(\omega)} \begin{bmatrix} \mathbb{E}\{\bar{\boldsymbol{x}}_n^o | \boldsymbol{s}_{[1,n]}\} \\ \vdots \\ \mathbb{E}\{\bar{\boldsymbol{x}}_1^o | \boldsymbol{s}_1\} \end{bmatrix}. \quad (25)$$

Next, we seek to compute $\mathbb{E}\{\bar{\boldsymbol{x}}_k^o | \boldsymbol{s}_{[1,k]}\}$ in (24). To this end, let us take a closer look at (23a):

$$\begin{bmatrix} \check{\boldsymbol{x}}_{k+1} \\ \boldsymbol{u}_F^* \\ z(\omega) \end{bmatrix} = \begin{bmatrix} A & \cdots & B & \cdots \\ \hline O & & I & \end{bmatrix} \begin{bmatrix} \check{\boldsymbol{x}}_k \\ \boldsymbol{u}_F^* \\ z(\omega) \end{bmatrix} + \begin{bmatrix} I \\ O \end{bmatrix} v_k,$$

where we introduce $\check{\boldsymbol{x}}_k$, which is given by

$$
\begin{aligned}
\check{\boldsymbol{x}}_1 &= \boldsymbol{x}_1 = \boldsymbol{x}_1^o \\
\check{\boldsymbol{x}}_2 &= A\check{\boldsymbol{x}}_1 + B\boldsymbol{u}_{F,1}^* + \boldsymbol{v}_1 = \boldsymbol{x}_2^o + B\boldsymbol{u}_{F,1}^* \\
\check{\boldsymbol{x}}_3 &= A\check{\boldsymbol{x}}_2 + B\boldsymbol{u}_{F,2}^* + \boldsymbol{v}_2 = A(\boldsymbol{x}_2^o + B\boldsymbol{u}_{F,1}) + B\boldsymbol{u}_{F,2}^* + \boldsymbol{v}_2 = \boldsymbol{x}_3^o + AB\boldsymbol{u}_{F,1}^* + B\boldsymbol{u}_{F,2}^* \\
&\vdots
\end{aligned}
$$

$$
\check{\boldsymbol{x}}_k = \boldsymbol{x}_k^o + B\boldsymbol{u}_{F,k-1}^* + AB\boldsymbol{u}_{F,k-2}^* + \cdots + A^{k-2}B\boldsymbol{u}_{F,1}^*. \tag{26}
$$

Then, we have

$$
\begin{bmatrix} \check{\boldsymbol{x}}_n \\ \check{\boldsymbol{x}}_{n-1} \\ \vdots \\ \check{\boldsymbol{x}}_1 \end{bmatrix} = \begin{bmatrix} \boldsymbol{x}_n^o \\ \boldsymbol{x}_{n-1}^o \\ \vdots \\ \boldsymbol{x}_1^o \end{bmatrix} + \underbrace{\begin{bmatrix} O & B & AB & \cdots & A^{n-2}B \\ O & O & B & \cdots & A^{n-3}B \\ \vdots & & & & \vdots \\ O & O & \cdots & \cdots & O \end{bmatrix}}_{=:\, D} \begin{bmatrix} \boldsymbol{u}_{F,n}^* \\ \boldsymbol{u}_{F,n-1}^* \\ \vdots \\ \boldsymbol{u}_{F,1}^* \end{bmatrix}.
$$

Let D be partitioned as $D = [D_n' \cdots D_1']'$ such that

$$
\check{\boldsymbol{x}}_k = \boldsymbol{x}_k^o + D_k \boldsymbol{u}_F^*. \tag{27}
$$

Therefore, $\mathbb{E}\{\check{\boldsymbol{x}}_k^o | \boldsymbol{s}_{[1,k]}\}$ can be written as

$$
\mathbb{E}\{\check{\boldsymbol{x}}_k^o | \boldsymbol{s}_{[1,k]}\} = \begin{bmatrix} \mathbb{E}\{\boldsymbol{x}_k^o | \boldsymbol{s}_{[1,k]}\} + D_k \mathbb{E}\{\boldsymbol{u}_F^* | \boldsymbol{s}_{[1,k]}\} \\ \mathbb{E}\{\boldsymbol{u}_F^* | \boldsymbol{s}_{[1,k]}\} \\ z(\omega) \end{bmatrix}. \tag{28}
$$

Furthermore, (14) and (15) lead to

$$
\mathbb{E}\{\boldsymbol{u}_F^* | \boldsymbol{s}_{[1,k]}\} = -\Phi^{-1}K \begin{bmatrix} \mathbb{E}\{\mathbb{E}\{\boldsymbol{x}_n^o | \boldsymbol{s}_{[1,n]}\} | \boldsymbol{s}_{[1,k]}\} \\ \vdots \\ \mathbb{E}\{\mathbb{E}\{\boldsymbol{x}_1^o | \boldsymbol{s}_1\} | \boldsymbol{s}_{[1,k]}\} \end{bmatrix}. \tag{29}
$$

Note that we have

$$
\mathbb{E}\{\mathbb{E}\{\boldsymbol{x}_l^o | \boldsymbol{s}_{[1,l]}\} | \boldsymbol{s}_{[1,k]}\} = \begin{cases} \mathbb{E}\{\boldsymbol{x}_l^o | \boldsymbol{s}_{[1,k]}\} & \text{if } l \geq k \\ \mathbb{E}\{\boldsymbol{x}_l^o | \boldsymbol{s}_{[1,l]}\} & \text{if } l < k \end{cases},
$$

where the first case, i.e., $l \geq k$, follows due to the iterated expectations with nested conditioning sets, i.e., $\{\boldsymbol{s}_{[1,l]}\} \supseteq \{\boldsymbol{s}_{[1,k]}\}$ if $l \geq k$, and the second case, i.e., $l < k$, follows since $\mathbb{E}\{\boldsymbol{x}_l^o | \boldsymbol{s}_{[1,l]}\}$ is $\sigma\text{-}\boldsymbol{s}_{[1,k]}$ measurable if $l < k$. Therefore, (29) can be written as

$$
\mathbb{E}\{\boldsymbol{u}_F^* | \boldsymbol{s}_{[1,k]}\} = -\Phi^{-1}K \underbrace{\begin{bmatrix} & & A^{n-k} & \\ O & \vdots & & O \\ & A & \\ & I & \\ O & O & & I \end{bmatrix}}_{=:\, L_k} \hat{\underline{\boldsymbol{x}}}^o, \tag{30}
$$

where the middle block is the kth block column. Hence, we can rewrite (28) as

$$\mathbb{E}\{\bar{\boldsymbol{x}}_k^o|\boldsymbol{s}_{[1,k]}\} = \underbrace{\begin{bmatrix} E_k - D_k\Phi^{-1}KL_k \\ -\Phi^{-1}KL_k \\ O \end{bmatrix}}_{=:\ F_k} \hat{\boldsymbol{x}}^o + \underbrace{\begin{bmatrix} O \\ O \\ z(\omega) \end{bmatrix}}_{=:\ \underline{z}(\omega)}, \qquad (31)$$

where E_k is the indicator matrix such that $\mathbb{E}\{\boldsymbol{x}_k^o|\boldsymbol{s}_{[1,k]}\} = E_k\hat{\boldsymbol{x}}^o$, $k = 1,\ldots,n$. Then, by (24), (25), and (31), we have

$$\delta\boldsymbol{u}^{o*}(\omega) = -K_A(\omega)\underbrace{\begin{bmatrix} F_n \\ \vdots \\ F_1 \end{bmatrix}}_{=:\ F} \hat{\boldsymbol{x}}^o - K_A(\omega)\underline{z}(\omega).$$

Therefore, the actual optimal adversarial actions are given by

$$\boxed{\boldsymbol{u}_A^*(\omega) = \boldsymbol{u}_F^* - \Phi_A(\omega)^{-1}K_A(\omega)\big[F\hat{\boldsymbol{x}}^o + \underline{z}(\omega)\big].} \qquad (32)$$

In the following theorem, we recap the results.

Theorem 1. *Given S's strategies* $\boldsymbol{s}_k = \eta_k(\boldsymbol{x}_{[1,k]})$, $k = 1,\ldots,n$, *C's optimal reactions* $\boldsymbol{u}_{F,k}$ *and* $\boldsymbol{u}_{A,k}(\omega)$ *are given by* (15) *or* (32) *depending on whether C is a friend or an adversary, respectively.*

In the following section, we formulate S's optimal strategies.

5   Optimal Leader (Sensor) Actions

By Theorem 1, S's objective can be written as

$$\min_{\substack{\eta_k \in \Upsilon_k, \\ k=1,\ldots,n}} (1-p) \sum_{k=1}^{n} \mathbb{E}\left\{ \|\boldsymbol{x}_{k+1}\|_{Q_{k+1}}^2 + \|\boldsymbol{u}_{F,k}^*\|_{R_k}^2 \right\}$$

$$+ p \int_\Omega \sum_{k=1}^{n} \mathbb{E}\left\{ \|\boldsymbol{x}_{k+1}\|_{Q_{k+1}}^2 + \|\boldsymbol{u}_{A,k}^*(\omega)\|_{R_k}^2 \right\} \mathbf{P}(d\omega).$$

However, we should also take into account that $\boldsymbol{x}_k$ evolves according to (1), which implies that the state $\boldsymbol{x}_k$ depends on the control input, and therefore C's type. In order to show this explicit dependence, henceforth, we will denote the state by $\boldsymbol{x}_{F,k}$ when C is a friend or by $\boldsymbol{x}_{A,k}$ when C is an adversary. Correspondingly, the sensor outputs are denoted by $\boldsymbol{s}_{F,k}$ and $\boldsymbol{s}_{A,k}$, respectively. Therefore, an explicit representation for S's objective is given by

$$\min_{\substack{\eta_k \in \Upsilon_k, \\ k=1,\ldots,n}} (1-p) \sum_{k=1}^{n} \mathbb{E}\left\{ \|\boldsymbol{x}_{F,k+1}\|_{Q_{k+1}}^2 + \|\boldsymbol{u}_{F,k}^*\|_{R_k}^2 \right\}$$

$$+ p \int_\Omega \sum_{k=1}^{n} \mathbb{E}\left\{ \|\boldsymbol{x}_{A,k+1}(\omega)\|_{Q_{k+1}}^2 + \|\boldsymbol{u}_{A,k}^*(\omega)\|_{R_k}^2 \right\} \mathbf{P}(d\omega). \qquad (33)$$

Even though S constructs a single set of strategies $\{\eta_k \in \Upsilon_k\}$ without knowing C's type, the resulting sensor outputs $\{s_k = \eta_k(\boldsymbol{x}_{[1,k]})\}$ depend on the states, $\boldsymbol{x}_{[1,k]}$'s, hence C's type, i.e., $\boldsymbol{x}_k = \boldsymbol{x}_{F,k}$ if C is a friend or $\boldsymbol{x}_k = \boldsymbol{x}_{A,k}$ if C is an adversary.

Let $T := \Phi^{-1}K$,

$$T_A(\omega) := \Phi^{-1}K + \Phi_A(\omega)^{-1}K_A(\omega)F$$

$$\xi(\omega) := \Phi_A(\omega)^{-1}K_A(\omega)\underline{z}(\omega)$$

such that $\boldsymbol{u}_F^* = -T\hat{\underline{\boldsymbol{x}}}_F^o$ and $\boldsymbol{u}_A^*(\omega) = -T_A(\omega)\hat{\underline{\boldsymbol{x}}}_A - \xi(\omega)$, where $\hat{\underline{\boldsymbol{x}}}_\iota^o := [(\hat{\boldsymbol{x}}_{\iota,n}^o)' \cdots (\hat{\boldsymbol{x}}_{\iota,1}^o)']'$ and $\hat{\boldsymbol{x}}_{\iota,k}^o := \mathbb{E}\{\boldsymbol{x}_k^o | s_{\iota,[1,k]}\}$, for $\iota = \{F, A\}$. Note that the matrices T and $T_A(\omega)$, for each $\omega \in \Omega$, are block upper triangular. Furthermore, let $\hat{\underline{\boldsymbol{x}}}_{\iota,k}^o := \left[(\hat{\boldsymbol{x}}_{\iota,k}^o)' \cdots (\hat{\boldsymbol{x}}_{\iota,1}^o)'\right]'$, $\xi(\omega)$ be partitioned into $\xi(\omega) = [\xi_n(\omega)' \cdots \xi_1(\omega)']'$, and the block upper triangular matrices T and $T_A(\omega)$ be partitioned into the block matrices as

$$T = \begin{bmatrix} T_{n,n} & T_{n,n-1} & \cdots & T_{n,1} \\ & T_{n-1,n-1} & \cdots & T_{n-1,1} \\ & & \ddots & \\ & & & T_{1,1} \end{bmatrix}, T_A = \begin{bmatrix} T_{A,n,n} & T_{A,n,n-1} & \cdots & T_{A,n,1} \\ & T_{A,n-1,n-1} & \cdots & T_{A,n-1,1} \\ & & \ddots & \\ & & & T_{A,1,1} \end{bmatrix},$$

where we have dropped the argument ω for notational simplicity, and $\bar{T}_k := [T_{k,k} \cdots T_{k,1}]$, $\bar{T}_{A,k}(\omega) := [T_{A,k,k}(\omega) \cdots T_{A,k,1}(\omega)]$. Then, by Lemma 1 and (11), (33) is equivalent to

$$\min_{\substack{\eta_k \in \Upsilon_k, \\ k=1,\ldots,n}} (1-p) \sum_{k=1}^{n} \mathbb{E}\|K_k\boldsymbol{x}_k^o - \bar{T}_k\hat{\underline{\boldsymbol{x}}}_{F,k}^o\|_{\Delta_k}^2$$

$$+ p \int_\Omega \sum_{k=1}^{n} \mathbb{E}\|K_k\boldsymbol{x}_k^o - \bar{T}_{A,k}(\omega)\hat{\underline{\boldsymbol{x}}}_{A,k}^o(\omega) - \xi_k(\omega)\|_{\Delta_k}^2 \mathbf{P}(d\omega) + G. \quad (34)$$

The first summation in (34) can be written as

$$\sum_{k=1}^{n} \mathrm{tr}\{\mathbb{E}\{\boldsymbol{x}_k^o(\boldsymbol{x}_k^o)'\}K_k'\Delta_kK_k\} - 2\,\mathrm{tr}\{\mathbb{E}\{\hat{\underline{\boldsymbol{x}}}_{F,k}^o(\boldsymbol{x}_k^o)'\}K_k'\Delta_k\bar{T}_k\}$$

$$+ \mathrm{tr}\{\mathbb{E}\{\hat{\underline{\boldsymbol{x}}}_{F,k}^o(\hat{\underline{\boldsymbol{x}}}_{F,k}^o)'\}\bar{T}_k'\Delta_k\bar{T}_k\} \quad (35)$$

while the second summation can be written as

$$\sum_{k=1}^{n} \text{tr}\{\mathbb{E}\{\boldsymbol{x}_k^o(\boldsymbol{x}_k^o)'\}K_k'\Delta_k K_k\} + \int_{\Omega} \xi_k(\omega)'\Delta_k \xi_k(\omega)\mathbf{P}(d\omega)$$

$$+ \int_{\Omega} \text{tr}\{\mathbb{E}\{\hat{\underline{\boldsymbol{x}}}_{A,k}^o(\omega)\hat{\underline{\boldsymbol{x}}}_{A,k}^o(\omega)'\}\bar{T}_{A,k}(\omega)'\Delta_k \bar{T}_{A,k}(\omega)\}\mathbf{P}(d\omega)$$

$$- 2 \int_{\Omega} \text{tr}\{\mathbb{E}\{\hat{\underline{\boldsymbol{x}}}_{A,k}^o(\omega)(\boldsymbol{x}_k^o)'\}K_k'\Delta_k \bar{T}_{A,k}(\omega)\}\mathbf{P}(d\omega)$$

$$+ 2 \int_{\Omega} \text{tr}\{\mathbb{E}\{\hat{\underline{\boldsymbol{x}}}_{A,k}^o(\omega)\}\xi_k(\omega)'\Delta_k \bar{T}_{A,k}(\omega)\}\mathbf{P}(d\omega)$$

$$- 2 \int_{\Omega} \text{tr}\{\mathbb{E}\{\boldsymbol{x}_k^o\}\xi_k(\omega)'\Delta_k K_k\}\mathbf{P}(d\omega), \tag{36}$$

where the last term is zero since $\boldsymbol{x}_k^o$ is zero-mean. The following lemma says that the posterior covariances do not depend on ω.

Lemma 3. *The posterior $\hat{\boldsymbol{x}}_{A,k}^o(\omega)$ is independent of ω. Further, both posteriors $\hat{\boldsymbol{x}}_{F,k}^o$ and $\hat{\boldsymbol{x}}_{A,k}^o$ are equivalent and given by*

$$\boxed{\hat{\boldsymbol{x}}_k^o := \hat{\boldsymbol{x}}_{F,k}^o = \hat{\boldsymbol{x}}_{A,k}^o(\omega) = \mathbb{E}\left\{\boldsymbol{x}_k^o \mid \eta_1(\boldsymbol{x}_1^o), \ldots, \eta_k(\boldsymbol{x}_{[1,k]}^o)\right\}.} \tag{37}$$

Proof. Consider the state recursion when C is a friend:

$$\boldsymbol{x}_{F,k+1} = A\boldsymbol{x}_{F,k} + B\boldsymbol{u}_{F,k}^* + \boldsymbol{v}_k,$$

which can also be written as[4]

$$\boldsymbol{x}_{F,1} = \boldsymbol{x}_1^o$$
$$\boldsymbol{x}_{F,2} = A\boldsymbol{x}_{F,1} + B\boldsymbol{u}_{F,1}^* + \boldsymbol{v}_1 = \boldsymbol{x}_2^o + B\boldsymbol{u}_{F,1}^*$$
$$\boldsymbol{x}_{F,3} = A\boldsymbol{x}_{F,2} + B\boldsymbol{u}_{F,2}^* + \boldsymbol{v}_2 = A(\boldsymbol{x}_2^o + B\boldsymbol{u}_{F,1}^*) + B\boldsymbol{u}_{F,2}^* + \boldsymbol{v}_2$$
$$= \boldsymbol{x}_3^o + AB\boldsymbol{u}_{F,1}^* + B\boldsymbol{u}_{F,2}^*$$
$$\vdots$$
$$\boldsymbol{x}_{F,k} = \boldsymbol{x}_k^o + B\boldsymbol{u}_{F,k-1}^* + AB\boldsymbol{u}_{F,k-2}^* + \cdots + A^{k-2}B\boldsymbol{u}_{F,1}^*.$$

Let $M_k := [B \ AB \ \cdots \ A^{k-2}B]$ and $\underline{\boldsymbol{u}}_{F,k} := [\boldsymbol{u}_{F,k}' \ \cdots \ \boldsymbol{u}_{F,1}']'$. Then, for $k > 1$, we have

$$\boldsymbol{x}_{F,k} := \boldsymbol{x}_k^o + M_{k-1}\underline{\boldsymbol{u}}_{F,k-1}. \tag{38}$$

Furthermore, let

$$T_k := \begin{bmatrix} T_{k,k} & \cdots & T_{k,1} \\ & \ddots & \vdots \\ & & T_{1,1} \end{bmatrix}$$

---

[4] Note the resemblance to (26).

such that $\underline{u}_{F,k} = -T_k \hat{\underline{x}}^o_{F,k}$ and (38) can be written as

$$\boldsymbol{x}_{F,k} = \boldsymbol{x}^o_k - M_{k-1} T_{k-1} \hat{\underline{x}}^o_{F,k-1}. \tag{39}$$

Therefore, we have $\hat{\boldsymbol{x}}^o_{F,k} = \mathbb{E}\{\boldsymbol{x}^o_k | \eta_1(\boldsymbol{x}^o_1), \ldots, \eta_k(\boldsymbol{x}^o_1, \ldots, \boldsymbol{x}^o_k - c_{k,k})\}$, for certain deterministic $c_{i,j} \in \mathbb{R}^m$, $i,j = 1, \ldots, k$, since $\hat{\boldsymbol{x}}^o_{F,j}$ is $\sigma - \boldsymbol{x}^o_{[1,j]}$ measurable. Correspondingly, we have

$$\boldsymbol{x}_{A,k}(\omega) = \boldsymbol{x}^o_k - M_{k-1} T_{A,k-1}(\omega) \hat{\underline{x}}^o_{A,k-1} - M_{k-1} \underline{\xi}_{k-1}(\omega), \tag{40}$$

where

$$T_{A,k}(\omega) := \begin{bmatrix} T_{A,k,k}(\omega) & \cdots & T_{A,k,1}(\omega) \\ & \ddots & \vdots \\ & & T_{A,1,1}(\omega) \end{bmatrix} \text{ and } \underline{\xi}_k(\omega) := \begin{bmatrix} \xi_k(\omega) \\ \vdots \\ \xi_1(\omega) \end{bmatrix},$$

which leads to $\hat{\boldsymbol{x}}^o_{A,k}(\omega) = \mathbb{E}\{\boldsymbol{x}^o_k | \eta_1(\boldsymbol{x}^o_1), \ldots, \eta_k(\boldsymbol{x}^o_1, \ldots, \boldsymbol{x}^o_k - d_{k,k}(\omega))\}$, for certain other deterministic $d_{i,j}(\omega) \in \mathbb{R}^m$, $i,j = 1, \ldots, k$, since $\hat{\boldsymbol{x}}^o_{A,j}(\omega)$ is $\sigma - \boldsymbol{x}^o_{[1,j]}$ measurable.

Next, we employ the following lemma about shifting of random variables in order to compute $\hat{\boldsymbol{x}}^o_{F,k}$'s and $\hat{\boldsymbol{x}}^o_{A,k}(\omega)$'s.

Lemma 4. *Let $(\Omega, \mathsf{F}, \mathbf{P})$ be a probability space, where Ω is the outcome space with an appropriate σ-algebra F, and $\mathbf{P}$ is a distribution over Ω. Let also $\boldsymbol{x} : (\Omega, \mathsf{F}) \to (\mathbb{R}^m, \mathsf{B}^m)$ be a random variable, $h : (\mathbb{R}^m, \mathsf{B}^m) \to (\mathbb{R}^m, \mathsf{B}^m)$ be a Borel measurable function, and $c \in \mathbb{R}^m$ be a deterministic vector. Then, we have*

$$\mathbb{E}\{\boldsymbol{x}|h(\boldsymbol{x})\} = \mathbb{E}\{\boldsymbol{x}|h(\boldsymbol{x} + c)\}. \tag{41}$$

Proof. The proof is provided in the Appendix A. ∎

Therefore, Lemma 4 and (51) imply (37) and the proof is concluded. □

Next, by (35), (36), and Lemma 3, (34) can be written as

$$\min_{\substack{\eta_k \in \Upsilon_k, \\ k=1,\ldots,n}} \sum_{k=1}^n \mathrm{tr}\{\Sigma_k K'_k \Delta_k K_k\} + p\, \mathbb{E}_\Omega\{\xi_k(\omega)' \Delta_k \xi_k(\omega)\}$$

$$- 2\, \mathrm{tr}\Big\{\mathbb{E}\{\hat{\underline{x}}^o_k (\boldsymbol{x}^o_k)'\} K'_k \Delta_k \big((1-p)\bar{T}_k + p\, \mathbb{E}_\Omega\{\bar{T}_{A,k}(\omega)\}\big)\Big\}$$

$$+ p\, \mathrm{tr}\Big\{\mathbb{E}\{\hat{\underline{x}}^o_k (\hat{\underline{x}}^o_k)'\} \mathbb{E}_\Omega\{\bar{T}_{A,k}(\omega)' \Delta_k \bar{T}_{A,k}(\omega)\}\Big\}$$

$$+ (1-p)\, \mathrm{tr}\Big\{\mathbb{E}\{\hat{\underline{x}}^o_k (\hat{\underline{x}}^o_k)'\} \bar{T}'_k \Delta_k \bar{T}_k\Big\} + G, \tag{42}$$

where $\mathbb{E}_\Omega$ denotes the expectation taken over Ω with respect to the distribution $\mathbf{P}$ and $\Sigma_k := \mathbb{E}\{\boldsymbol{x}^o_k (\boldsymbol{x}^o_k)'\}$.

We note that for $l \le k$, $\mathbb{E}\{\hat{\boldsymbol{x}}^o_l (\boldsymbol{x}^o_k)'\} = \mathbb{E}\{\hat{\boldsymbol{x}}^o_l (\boldsymbol{x}^o_l)'\}(A')^{k-l}$ since $\boldsymbol{v}_j$, $j > l$, and $\hat{\boldsymbol{x}}^o_l$, which is $\sigma\text{-}\boldsymbol{s}_{[1,l]}$ measurable, are independent of each other and $\{\boldsymbol{v}_k\}$ is a zero-mean white noise process. Furthermore, we have

$$\mathbb{E}\{\hat{\boldsymbol{x}}^o_l (\boldsymbol{x}^o_l)'\} = \mathbb{E}\{\mathbb{E}\{\hat{\boldsymbol{x}}^o_l (\boldsymbol{x}^o_l)' | \boldsymbol{s}_{[1,l]}\}\}$$

$$= \mathbb{E}\{\hat{\boldsymbol{x}}^o_l (\hat{\boldsymbol{x}}^o_l)'\} \tag{43}$$

due to the law of iterated expectations. Let $H_k := \mathbb{E}\{\hat{x}_k^o(\hat{x}_k^o)'\}$. Then, we have

$$\mathbb{E}\{\underline{\hat{x}}_k^o(x_k^o)'\} = \begin{bmatrix} H_k \\ H_{k-1}A' \\ \vdots \\ H_1(A')^{k-1} \end{bmatrix}, \mathbb{E}\{\underline{\hat{x}}_{k-1}^o(x_k^o)'\} = \begin{bmatrix} H_{k-1}A' \\ \vdots \\ H_1(A')^{k-1} \end{bmatrix}$$

and

$$\mathbb{E}\{\underline{\hat{x}}_k^o(\underline{\hat{x}}_k^o)'\} = \begin{bmatrix} \mathbb{E}\{\hat{x}_k^o(\hat{x}_k^o)'\} & \cdots & \mathbb{E}\{\hat{x}_k^o(\hat{x}_1^o)'\} \\ \vdots & & \vdots \\ \mathbb{E}\{\hat{x}_1^o(\hat{x}_k^o)'\} & \cdots & \mathbb{E}\{\hat{x}_1^o(\hat{x}_1^o)'\} \end{bmatrix}$$

$$= \begin{bmatrix} H_k & AH_{k-1} & \cdots & A^{k-1}H_1 \\ H_{k-1}A' & H_{k-1} & \cdots & A^{k-2}H_1 \\ \vdots & \vdots & \ddots & \vdots \\ H_1(A')^{k-1} & H_1(A')^{k-2} & \cdots & H_1 \end{bmatrix} \tag{44}$$

since for $l < k$, we have

$$\mathbb{E}\{\hat{x}_l^o(\hat{x}_k^o)'\} = \mathbb{E}\{\mathbb{E}\{\hat{x}_l^o(\hat{x}_k^o)'|s_{[1,l]}\}\}$$
$$\overset{(a)}{=} \mathbb{E}\{\hat{x}_l^o\mathbb{E}\{\hat{x}_k^o|s_{[1,l]}\}'\}$$
$$\overset{(b)}{=} \mathbb{E}\{\hat{x}_l^o(\hat{x}_l^o)'\}(A')^{k-l},$$

where (a) holds since $\hat{x}_l^o$ is $\sigma\text{-}s_{[1,l]}$ measurable, and (b) follows due to the iterated expectations with nested conditioning sets, i.e., $\{s_{[1,l]}\} \subseteq \{s_{[1,k]}\}$.

Next, we can rewrite (42) as

$$\min_{\substack{\eta_k \in \Upsilon_k, \\ k=1,\ldots,n}} \sum_{k=1}^n \Xi_k^o + \text{tr}\left\{\begin{bmatrix} H_k \\ H_{k-1}A' \\ \vdots \\ H_1(A')^{k-1} \end{bmatrix} \Xi_k\right\} + \text{tr}\left\{\begin{bmatrix} H_k & AH_{k-1} & \cdots & A^{k-1}H_1 \\ H_{k-1}A' & H_{k-1} & \cdots & A^{k-2}H_1 \\ \vdots & \vdots & \ddots & \vdots \\ H_1(A')^{k-1} & H_1(A')^{k-2} & \cdots & H_1 \end{bmatrix} \bar{\Xi}_k\right\}, \tag{45}$$

where

$$\Xi_k^o := \text{tr}\{\Sigma_k K_k' \Delta_k K_k\} + p\,\mathbb{E}_\Omega\{\xi_k(\omega)'\Delta_k\xi_k(\omega)\} + \frac{1}{n}G$$
$$\Xi_k := -2K_k'\Delta_k\big((1-p)\bar{T}_k + p\,\mathbb{E}_\Omega\{\bar{T}_{A,k}(\omega)\}\big)$$
$$\bar{\Xi}_k := p\,\mathbb{E}_\Omega\{\bar{T}_{A,k}(\omega)'\Delta_k\bar{T}_{A,k}(\omega)\} + (1-p)\bar{T}_k'\Delta_k\bar{T}_k,$$

which are independent of the optimization arguments. Hence, the optimization problem (42) faced by S can be written as an affine function of H_k's as follows:

$$\min_{\substack{\eta_k \in \Upsilon_k, \\ k=1,\ldots,n}} \sum_{k=1}^n \text{tr}\{V_k H_k\} + \Xi^o, \tag{46}$$

for certain symmetric deterministic matrices $V_k \in \mathbb{R}^{m \times m}$, $k = 1, \ldots, n$, where $\Xi^o := \sum_{k=1}^n \Xi_k^o$. Note that as a sensor designer, we seek to solve this infinite-dimensional optimization problem (46) within the general class of strategies.

To this end, we employ the approach in [19], which considers a finite-dimensional optimization problem that bounds the original infinite dimensional one from below, and then, compute strategies for the original problem, which optimizes the lower bound. Based on this, the following theorem characterizes equilibrium achieving strategies of both agents S and C.

Theorem 2. *The multi-stage static Bayesian Stackelberg equilibrium between S and C, i.e., (6), can be attained through linear strategies, i.e., the secure sensor outputs $s_{[1,n]}$ are linear in the state $x_{[1,n]}$ and the corresponding, friendly or adversarial, control inputs, $u_{F,[1,n]}$ or $u_{A,[1,n]}$, are linear in the sensor outputs $s_{[1,n]}$.*

Proof. Based on Lemma 1 in [19], by characterizing necessary conditions on H_k's, we have

$$\min_{\substack{S_k \in \mathbb{S}^m \\ k=1,\ldots,n}} \sum_{k=1}^{n} \operatorname{tr}\{V_k S_k\} \leq \min_{\substack{\eta_k \in \varUpsilon_k \\ k=1,\ldots,n}} \sum_{k=1}^{n} \operatorname{tr}\{V_k H_k\}, \qquad (47)$$
$$\text{s.t. } \Sigma_j \succeq S_j \succeq A S_{j-1} A' \; \forall j$$

where $\Sigma_j := \mathbb{E}\{x_j^o(x_j^o)'\}$ and $\mathbb{S}^m$ denotes the set of $m \times m$ symmetric matrices. Note that the left hand side of (47) is a finite-dimensional optimization, indeed an SDP, problem. By invoking Theorem 3 in [19], we can characterize the solutions of this SDP problem, $S_1^*, \ldots, S_n^*$, as

$$S_k^* = A S_{k-1}^* A' + (\Sigma_k - A S_{k-1}^* A')^{1/2} P_k (\Sigma_k - A S_{k-1} A')^{1/2}, \qquad (48)$$

for $k = 1, \ldots, n$, where $S_0^* = O$ and P_k's are certain symmetric idempotent matrices. Note that by solving the SDP problem numerically, we can compute the corresponding P_k's.

Next, say that S employs memoryless linear policies $s_k = \eta_k(x_{F,k}) = C_k' x_{F,k}$ if C is friendly or $s_k = \eta_k(x_{A,k}(\omega)) = C_k' x_{A,k}(\omega)$. Then, by Lemma 3, we have

$$\hat{x}_k^o = \mathbb{E}\{x_k^o | C_1' x_1^o, \ldots, C_k' x_k^o\}.$$

which can also be written as

$$\hat{x}_k^o = A \hat{x}_{k-1}^o + (\Sigma_k - A H_{k-1} A') C_k (C_k'(\Sigma_k - A H_{k-1} A') C_k)^+ C_k'(x_k^o - A \hat{x}_{k-1}^o),$$

for $k = 1, \ldots, n$, $\hat{x}_{-1}^o := 0$ and $H_0 := O$. Therefore, $H_k = \mathbb{E}\{\hat{x}_k^o(\hat{x}_k^o)'\}$ is given by

$$H_k = A H_{k-1} A' + (\Sigma_k - A H_{k-1} A') C_k (C_k'(\Sigma_k - A H_{k-1} A') C_k)^+ C_k'(\Sigma_k - A H_{k-1} A'). \quad (49)$$

We emphasize the resemblance between (48) and (49). In particular, if we set $\bar{C}_k := (\Sigma_k - A H_{k-1} A')^{1/2} C_k$, $k = 1, \ldots, n$, (49) yields

$$H_k = A H_{k-1} A' + (\Sigma_k - A H_{k-1} A')^{1/2} \bar{C}_k (\bar{C}_k' \bar{C}_k)^+ \bar{C}_k'(\Sigma_k - A H_{k-1} A')^{1/2},$$

where $\bar{C}_k(\bar{C}_k'\bar{C}_k)^+\bar{C}_k'$ is also a symmetric idempotent matrix just like P_k in (48).

Therefore, given P_k's, let $P_k = U_k \Lambda_k U_k'$ be the eigen decomposition and set $\bar{C}_k = U_k \Lambda_k$, i.e., set

$$C_k = (\Sigma_k - A S_{k-1}^* A')^{-1/2} U_k \Lambda_k. \tag{50}$$

Then, we obtain $H_k = S_k^*$, which implies that S's optimal strategies are memoryless and linear in the underlying state. Correspondingly, the optimal control inputs for both friendly and adversarial C are linear in the sensor outputs by (15) or (32). □

In Table 1, we provide a numerical algorithm to design secure sensors in advance.

Table 1. Computation of equilibrium achieving sender policies.

Algorithm: Secure Sensor Design

SDP Problem:

  Compute V_k, for $k = 1, \ldots, n$, by (7)-(45).

  Solve the SDP problem on the left hand side of (47) through a numerical toolbox

   and obtain the solutions S_k^*, for $k = 1, \ldots, n$.

  Set $S_0^* = O$.

Equilibrium achieving sensor strategies:

  Compute the corresponding idempotent matrices $P_k, \forall k$, by using S_k^*, $\forall k$, and (48).

  Compute the eigen decompositions: $P_k = U_k \Lambda_k U_k'$.

  Compute C_k, $\forall k$, by using $S_{k-1}^*, U_k, \Lambda_k$, and (50).

---

6   Conclusion

In this paper, we have proposed and addressed secure sensor design problem for cyber-physical systems with linear quadratic Gaussian dynamics against the advanced persistent threats with control objectives. By designing sensor outputs cautiously in advance, we have sought to minimize the damage that can be caused by undetected target-specific threats. However, this is not an active defense strategy against a detected threat. Therefore, such a defense mechanism should also consider the maintenance of the ordinary operations of the system. To this end, we have modeled the problem formally in a game-theoretical setting. We have determined the optimal control inputs for both friendly and adversarial objectives. Then, we have characterized the secure sensor strategies, showing that the strategies that are memoryless and linear in the underlying state lead to the equilibrium. Finally, we have provided an algorithm to compute these strategies numerically.

Some future directions of research on this topic include secure sensor design when the sensor has access to the state only partially, e.g., noisy observation, or when the attackers infiltrate into the controller within the horizon. Note also that we have only considered the secure sensor design within optimal control framework. Formulations for, e.g., robust control or feedback stability of the systems, can also be interesting future research directions.

A   Appendix: Proof of Lemma 4

Let $y_1 = h(x)$ and $y_2 = h(x+c)$ be random variables, where c is a deterministic shift vector of the same dimension as x. Then, for any $B \in \mathsf{B}^p$, we have $y_1^{-1}(B) = \{\omega \in \Omega : y_1(\omega) \in B\} = \{\omega \in \Omega : h(x)(\omega) \in B\} = \{\omega \in \Omega : x(\omega) \in h^{-1}(B)\}$. Correspondingly, we also have $y_2^{-1}(B) = \{\omega \in \Omega : y_2(\omega) \in B\} = \{\omega \in \Omega : h(x+c)(\omega) \in B\} = \{\omega \in \Omega : x(\omega) \in h^{-1}(B) - c\}$. Note that the σ-algebras generated by the random variables y_1 and y_2 are given by $\sigma(y_i) = \{y_i^{-1}(B) : B \in \mathsf{B}^p\}$, for $i = 1, 2$ [3]. This implies that $\sigma(y_1) = \{\{\omega \in \Omega : x(\omega) \in h^{-1}(B)\} : B \in \mathsf{B}^p\}$ and $\sigma(y_2) = \{\{\omega \in \Omega : x(\omega) \in h^{-1}(B) - c\} : B \in \mathsf{B}^p\}$. Furthermore, for each $B \in \mathsf{B}^p$, there exists $B_2 \in \mathsf{B}^p$ such that

$$h^{-1}(B) = h^{-1}(B_2) - c \in \mathsf{B}^p$$

since Borel sets are shift invariant [3]. Therefore, we have

$$\sigma(y_1) = \sigma(y_2) \tag{51}$$

and correspondingly, we obtain (41).

References

1. Başar, T., Olsder, G.: Dynamic Noncoopertative Game Theory. Society for Industrial Mathematics (SIAM) Series in Classics in Applied Mathematics. SIAM, Philadelphia (1999)
2. Bansal, R., Başar, T.: Simultaneous design of measurement and control strategies for stochastic systems with feedback. Automatica **25**(5), 679–694 (1989)
3. Billingsley, P.: Probability and Measure. Wiley, New Jersey (2012)
4. Brangetto, P., Aubyn, M.K.-S.: Economic aspects of national cyber security strategies. Technical report, NATO Cooperative Cyber Defense Centre of Excellence Tallinn, Estonia (2015)
5. Chen, Y., Kar, S., Moura, J.M.F.: Cyber physical attacks constrained by control objectives. In: Proceedings of American Control Conference (ACC), pp. 1185–1190 (2016)
6. Chen, Y., Kar, S., Moura, J.M.F.: Cyber physical attacks with control objectives and detection constraints. In: Proceedings of the 55th IEEE Conference on Decision and Control (CDC), pp. 1125–1130 (2016)

7. Fawzi, H., Tauada, P., Diggavi, S.: Secure estimation and control for cyber physical systems under adversarial attacks. IEEE Trans. Autom. Control **59**(6), 1454–1467 (2014)
8. Karnouskos, S.: Stuxnet worm impact on industrial cyber-physical system security. In: Proceedings of IEEE Industrial Electronics Society (IECON) (2011)
9. Khaitan, S.K., McCalley, J.D.: Design techniques and applications of cyberphysical systems: a survey. IEEE Syst. J. **9**(2), 350–365 (2014)
10. Koscher, K., Czeskis, A., Roesner, F., Patel, S., Kohno, T., Checkoway, S., McCoy, D., Kantor, B., Anderson, D., Shacham, H., Savage, S.: Experimental security analysis of a modern automobile. In: Proceedings of IEEE Symposium on Security and Privacy, pp. 447–462, 2010
11. Kumar, P.R., Varaiya, P.: Stochastic Systems: Estimation, Identification and Adaptive Control. Prentice Hall, Englewood Cliffs (1986)
12. Liberzon, D.: Calculus of Variations and Optimal Control Theory: A Concise Introduction. Princeton University Press, Princeton (2011)
13. Miao, F., Zhu, Q., Pajic, M., Pappas, G.J.: Coding schemes for securing cyber-physical systems against stealthy data injection attacks. IEEE Trans. Autom. Control **4**, 106–117 (2017)
14. Mo, Y., Sinopoli, B.: Integrity attacks on cyber-physical systems. In: Proceedings of the 1st ACM International Conference on High Confidence Networked Systems, pp. 47–54, 2012
15. Myerson, R.B.: Game Theory: Analysis of Conflict. Harvard University Press, Cambridge (1997)
16. Nelson, N.: The impact of Dragonfly malware on industrial control systems. The SANS Institute (2016)
17. Paruchuri, P., Pearce, J.P., Marecki, J., Tambe, M., Ordonez, F., Karus, S.: Playing games for security: An efficient exact algorithm for solving Bayesian Stackelberg games. In: Proceedings of Autonomous Agents and Multiagent Systems (AAMAS) (2008)
18. Pasqualetti, F., Dorfler, F., Bullo, F.: Attack detection and identification in cyber-physical systems. IEEE Trans. Autom. Control **58**(11), 2715–2729 (2013)
19. Sayin, M.O., Akyol, E., Başar, T.: Hierarchical multi-stage Gaussian signaling games: strategic communication and control. Automatica, arXiv:1609.09448 (2017, submitted)
20. Zhang, R., Venkitasubramaniam, P.: Stealthy control signal attacks in linear quadratic Gaussian control systems: detectability reward tradeoff. IEEE Trans. Inf. Forensics Secur. **12**(7), 1555–1570 (2017)

An Ultimatum Game Model for the Evolution of Privacy in Jointly Managed Content

Sarah Rajtmajer[1], Anna Squicciarini[1(✉)], Jose M. Such[2], Justin Semonsen[3], and Andrew Belmonte[1]

[1] Pennsylvania State University, University Park, PA, USA
acs20@psu.edu
[2] King's College London, Strand, London, UK
[3] Rutgers University, Breswick, NJ, USA

Abstract. Content sharing in social networks is now one of the most common activities of internet users. In sharing content, users often have to make access control or privacy decisions that impact other stakeholders or co-owners. These decisions involve negotiation, either implicitly or explicitly. Over time, as users engage in these interactions, their own privacy attitudes evolve, influenced by and consequently influencing their peers. In this paper, we present a variation of the one-shot Ultimatum Game, wherein we model individual users interacting with their peers to make privacy decisions about shared content. We analyze the effects of sharing dynamics on individuals' privacy preferences over repeated interactions of the game. We theoretically demonstrate conditions under which users' access decisions eventually converge, and characterize this limit as a function of inherent individual preferences at the start of the game and willingness to concede these preferences over time. We provide simulations highlighting specific insights on global and local influence, short-term interactions and the effects of homophily on consensus.

1 Introduction

We aim to investigate the impact of multi-party decision sharing in a social network. In highly connected networks, content sharing is frequent and users make decisions about the amount and type of content they choose to share, as well as their preferred privacy preferences. Previous work has largely investigated how to reconcile users' (possibly conflicting) privacy preferences with respect to commonly owned (or jointly managed) content [16,34]. For instance, the typical example used in the literature is that of a photo in which multiple users are depicted, they have conflicting privacy preferences as to with whom the photo would be shared in a social network, and they use a (technology-aided) reconciliation method to resolve the conflicts. Despite the amount of work in this area, the impact of these interactions over time - both on users and on the content shared - regardless of the reconciliation method, is largely unexplored. In particular, we are yet to understand how individuals' sharing decisions change over time, who are the most influential users, how they benefit from it, and the privacy gains and losses from a collective perspective.

S. Rass et al. (Eds.): GameSec 2017, LNCS 10575, pp. 112–130, 2017.
DOI: 10.1007/978-3-319-68711-7_7

This "research gap" is possibly due to two (related) reasons. First, to our knowledge, proposed content sharing models to date have not been translated into practical features or applications: social networks provide minimal support for multi-party decision making tools. Hence, an exploration in the wild of the effects of multi-party sharing is fundamentally hard. Second, to date, work that focuses on multi-party sharing has adopted a micro-scale view of the interactions among users (i.e., one-on-one and one-shot interactions), in an attempt to minimize discomfort and other security properties from a one-interaction at a time standpoint.

In this paper, we aim to answer a broader and, we believe, more important set of questions about the potential longitudinal effects of repeated negotiations over jointly managed content among users in a social network. We assume, consistent with reality [3,20,39], that users wish to reach agreement and share content jointly. Over time, this will lead users to feel pressure to move away from their individual preferred settings and toward the preferences of their peers. In doing so, some users will experience sharing loss, while others will experience privacy loss. In this setting, our specific questions are:

- How does multi-party involvement in access control decisions affect the individual behaviors of social network users?
- What are the collective privacy gains and losses associated with multi-user sharing?
- Bearing in mind that users adopt individual strategies to respond to access decisions for shared content, which users are more likely to drive group decisions? Likewise, which users are most likely to benefit from repeated interactions?

We model user interactions through a repeated game. Specifically, evidence indicates that one-shot decisions for multi-party access control may be well-described using the language of the Ultimatum Game, specifically a natural tension between selfish preferences (i.e., maximizing a personal utility function) and a less-tangible desire to cooperate [3,20,34,39]. That is, empirical studies about multi-party access control showed that users are naturally selfish and seek to impose their preferences as much as they can even when they know other stakeholders may not be happy about it [34], but at the same time users do collaborate [39] as they do not want to cause any deliberate harm to other stakeholders and would normally consider their preferences and potential objections in a more cooperative way [3,20].

Accordingly, we present a variation of the one-shot Ultimatum Game, wherein individuals interact with peers to make a decision on a piece of shared content. The outcome of this game is either success or failure, wherein success implies that a satisfactory decision for all parties is made and failure instead implies that the parties could not reach an agreement. This approach was inspired by recent work of fairness in the Ultimatum Game [42].

Our proposed game is grounded on empirical data about individuals' behaviour in one-shot, multi-party access control decisions [34,35,39] mentioned above to structure repeated pairwise negotiations on jointly managed content in a social network. We theoretically demonstrate that over time, the system converges towards a "fair" state, wherein each individual's preferences are accounted for. In this state, users' preferred privacy values approach a constant value that is dependent on how stubborn individual users are, until all values are within a window of compromise (which in turn depends on the structure of the network). We also carry out a series of numerical experiments on simulated data, and provide insights on a number of interesting cases, e.g., when a number of perfectly stubborn users (i.e. users unwilling to compromise or adapt to other users' preferences) are at play, when highly connected users exist in the network, and when networks are homogeneous.

The paper is organized as follows. In the next section, we highlight our assumptions and the problem statement. In Sect. 3, we present our theoretical model. We discuss theoretical results in Sect. 4 and provide experimental insights in Sect. 5. We overview related work in Sect. 6. Finally, we conclude the paper with a discussion of limitations and future work in Sect. 7.

2   Problem Statement

We consider an online social network wherein linked users, i.e., two users connected by an "edge" in the social network graph, may jointly manage content. While one user is typically first to share a given piece of content, henceforth the "poster", other users, henceforth the "stakeholders", may also be affected by the content (e.g. a photo in which she is depicted). Users, both posters and stakeholders, likely differ in both structural and inherent qualities. Structurally, they have variable numbers of friends, i.e., degree ($deg(n)$), and variable (closeness, betweenness) centrality. Inherently, users may differ in propensity for sharing [22] and stubbornness [2,40].

As a piece of jointly managed content is considered, the stakeholder has the opportunity to accept or decline the privacy settings selected by the poster — a decision that is made based on a joint effect of inherent sharing preference, stubbornness, the personal relationship between the two users and the nature of the content itself. Access settings, then, are co-determined by posters and stakeholders using a one-round negotiation, which we model as a one-shot Ultimatum Game.

An important assumption underlying this game is that the proposer and responder *would like* to reach agreement. First, the underlying social network structure implies that the proposer and responder are friends, acquaintances or members of a social cohort. Reaching agreement represents social harmony that is preferable, and empirical evidence tells us that both posters and stakeholders listen to and consider each others' preferences and objections [39]. In some cases, agreement may be required for content to successfully be posted. In other cases, the proposer may have authority to post content at his desired privacy level without consent of other stakeholders, but she hesitates to do so understanding that her cohort may take the same liberty with future content, or because they put themselves in the position of stakeholders and understand they may not be happy with the content shared [34]. In order to reach agreement, both proposer and stakeholders understand they must concede (part of) their preferences and move toward some compromise privacy setting [3,20,39]. However, the amount each party shifts (or concedes) may not be the same, and its likely influenced by their individual propensity for sharing [22] and stubbornness [2,40] as stated above.

We study the impact of this variant of one-shot ultimatum games over time, and specifically, the extent to which these one-shot interactions, wherein users must compromise (as much as they feel comfortable) in order for content to be shared, is conducive of a "fair" system. Here, by fair system, we refer to a system wherein each user is given an equal opportunity to participate in an interaction, based on his/her current degree in the network graph. Furthermore, each user is free to respond based on his own preferences and inclinations, and each user's response for each game equally influences system dynamics. Given these equitable rules of the game, answers to the three research questions posed above may shed light on the ways in which outcomes are and are not as equitable.

Following, we discuss the model and its outcomes with focus on the case of one poster and one stakeholder, for simplicity of presentation. Note however that this is not a loss of generality, as k asynchronous players are essentially a specific ordering of 2-player interactions.

3   The Model

We play a variant of the one-shot ultimatum game [42], repeatedly, amongst pairs of individuals situated within a social network graph. The rules of the game, which are formally specified below, reflect the real-world scenario of multi-party sharing, namely determining access settings for content associated with multiple stakeholders [15,25,33]. These rules formally capture empirical evidence of concession behaviour in multi-party sharing [33], like being generally accommodating to the preferences of others to reach agreement [3,20,39].

Consider a social network graph $G = \{V, E\}$ where V is the set of users, represented as nodes in the graph. The set E of pairwise links between nodes represents relationships, or more generally, users with some connection who may both be party to the same content. Links may be weighted according to a weigh function W_{ij}, where weights between users i and j indicate strength of relation, or strength of social influence.

Each user i has an inherent, personal *comfort* C_i with sharing and an inherent *stubbornness* T_i that do not change over the lifespan of the game. Both are represented as value in $[0, 1]$. In the case of comfort, 0 indicates private and 1 public[1]; likewise for stubbornness, 0 is least stubborn and 1 most stubborn. Each

---

[1] Note that we abstract ourselves from the actual privacy settings or access control paradigm used by the online social network provider. For each social media infrastructure or privacy policy language used, a mapping could be defined that turns available settings into values in $[0, 1]$ and vice versa.

user is also perpetually endowed with two dynamic values – a "proposal" value $P_i(t)$, and a "response" value $R_i(t)$. These values represent the user's preferred settings when acting as the content owner ("poster") or when party to content posted by someone else, respectively, which is aligned with empirical evidence that shows that the perceptions and behaviours of users are significantly different when they are playing the role of poster or stakeholder [34]. Changes in these values over time are governed by the set of rules of the game, detailed as follows. We initialize the proposal value and the response values for each user as his comfort value, i.e., $P_i(0) = R_i(0) = C_i$. The intuition here is that, without the influence of peers (i.e., without playing the game) a user is inclined to both offer and accept the sharing level for a piece of content that most closely matches his comfort level.

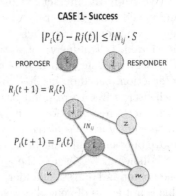

Fig. 1. Example of successful interaction and update rule

The game is played for some fixed number of iterations. At each iteration, a "proposer" is chosen at random. Intuitively, this is the owner/poster of a piece of content in which other users have a stake. A "responder" is selected at random from among his contacts, namely those users adjacent on the social network graph. The proposer offers his proposal value to the responder, i.e. the privacy level or disclosure setting for the co-owned content to be shared. The responder in turn accepts or declines this offer. Intuitively, the decision to accept or decline represents the responder's approval or disapproval of the proposed privacy setting. This decision is made based on the responder's *willingness to compromise*, which in turn relies primarily on two factors: (1) the *strength of influence* of the proposer on the responder, i.e., their relationship strength (possibly asymmetric) [10,11], and (2) the *sensitivity* of the content in question — if a user feels that an item is very sensitive for her, she will be less willing to approve sharing [30,38]. Conditions for acceptance and success of an interaction are given in the next definition and examples of successful and unsuccessful interactions are depicted in Figs. 1 and 2.

Definition 1 (Successful Interaction Conditions). *Let the strength of influence of user i on user j be represented by a value in $IN \in [0,1]$, with 0 indicating most weak and 1 most strong. Likewise, let the sensitivity of the content be denoted $S \in [0,1]$, with 0 most sensitive and 1 least sensitive. An interaction is successful, i.e., the responder j accepts the proposer's (i) proposal if*

$$|P_i(t) - R_j(t)| < IN(i,j) \times S$$

and a failure otherwise.

After each interaction, the involved players' proposal and response values are updated [42], as follows:

- *If the interaction is successful*, the proposer and responder do nothing. Specifically, P_i and R_j remain the same moving forward in time.
- *If the interaction is unsuccessful*, the proposer and responder move their proposal and response values, respectively, by some amount modulated by the *stubbornness* of each individual user *toward the midpoint* of the two as a way of conceding, so that future interactions are more likely to be successful.

$$P_i(t+1) = P_i(t) \times T_i + \frac{P_i(t) + R_j(t)}{2} \times (1 - T_i) \qquad (1)$$

and

$$R_j(t+1) = R_j(t) \times T_j + \frac{P_i(t) + R_j(t)}{2} \times (1 - T_j). \qquad (2)$$

The rules above capture notions of social influence and empirical evidence of multi-party access control decisions. In particular, informed by Fredkin's social influence theory [13], stating that strong ties are more likely to affect users' opinions and result in persuasion or social influence, in both Eqs. (1) and (2) users will move toward their peers values. This is consistent with empirical evidence about multi-party access control decisions that showed that both proposer and stakeholders are willing to collaborate and make concessions toward some compromise privacy setting [3,20,35,39]. The amount each party shifts (or concedes) may not be the same, as each party may be influenced by peers only to a certain point [36] driven by their stubbornness [2,40] and degree of selfishness [34].

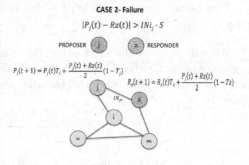

Fig. 2. Example of failed interaction

Of note, the proposer i does not change his response value and the responder j does not change his proposal value moving forward, i.e., $R_i(t+1) = R_i(t)$ and $P_j(t+1) = P_j(t)$. Likewise, all players in the game who were not involved in the interaction undergo no change in either proposal or response.

4 Theoretical Findings

In this section, we present our theoretical findings for the proposed Ultimatum Game. We demonstrate that, unless trivially impossible, the system converges towards a consensus state, wherein each individual's preferences are accounted for. In this state, both proposal and response values approach a constant c that is dependent on the stubbornness values associated with individual users, until all values are within a window of compromise which depends on the structure of the network.

4.1   Energy Conservation on Repeated Iterations

We first derive the following technical lemma on energy conservation, which will help determine conditions and value of convergence.

Lemma 1. *In an ultimatum game, let P_i and R_i be proposer's and offeror's value for user i, with P_i and R_i defined according to Eqs. (1) and (2), respectively. When $T_i \neq 1 \forall i$, the quantity $\sum_i \frac{P_i(t) + R_i(t)}{1 - T_i}$ is conserved.*

Proof. Consider a single Ultimatum game, with proposer i and responder j.

If the proposal is accepted, neither P_i nor R_j change, and no other proposal or response values are affected, so every value remains the same, and thus the weighted sum is unaffected.

If the proposal is rejected, then the new proposal value becomes $P'_i = T_i P_i + (1 - T_i)\frac{P_i + R_j}{2}$ and the new response value becomes $R'_j = T_j R_j + (1 - T_j)\frac{P_i + R_j}{2}$. Since no other values are changed, then:

$$\sum_i \frac{P'_i + R'_i}{1 - T_i} - \sum_i \frac{P_i + R_i}{1 - T_i} = \frac{P'_i - P_i}{1 - T_i} + \frac{R'_j - R_j}{1 - T_j} = \frac{P_i + R_j}{2} - P_i + \frac{P_i + R_j}{2} - R_j = 0$$

This means that regardless of which proposals are given or whether or not they is accepted, the quantity $\sum_i \frac{P_i(t) + R_i(t)}{1 - T_i}$ remains constant. ◇

We will show in the next subsection that the $P_i(t)$ and $R_i(t)$ converge to a given constant c. Using the relation obtained in Lemma 1, we posit that the constant c must be the unique constant for which this sum is conserved. Therefore

$$c = \frac{\sum_i \frac{P_i(0) + R_i(0)}{1 - T_i}}{\sum_i \frac{2}{1 - T_i}} \tag{3}$$

Next, we define the following vector $\mathbf{d}(t)$. We compute $|P_i(t) - c|$ and $|R_i(t) - c|$ for each i, and sort each difference in non-increasing order. We show that $\mathbf{d}(t)$ constructed in this way decreases in lexicographical order over time, and therefore $P_i(t)$ and $R_i(t)$ both approach c. The following Lemma holds.

Lemma 2. *At each time step t, the inequality $\mathbf{d}(t + 1) \leq_{lex} \mathbf{d}(t)$ is verified. In particular, at time $t + 1$, $\mathbf{d}(t + 1) <_{lex} \mathbf{d}(t)$ when the conditions of acceptance per Definition 1, are not met, and P_i is rejected by j.*

Proof. Consider a single ultimatum game taking place at time t, with proposer i and responder j. If the proposal P_i is accepted (i.e. acceptance condition per Definition 1 hold true), no changes are made to P_i and R_j. Since this proposal does not affect any other proposal or responder values, $\mathbf{d}(t + 1) = \mathbf{d}(t)$.

If the proposal is rejected, then let $a = \max\{|P_i(t) - c|, |R_j(t) - c|\}$ and $b = \min\{|P_i(t) - c|, |R_j(t) - c|\}$. Note that rejection means that $a > b$.

Since we sort these differences (including a and b) in non-increasing order, let k be the index of the last occurrence of a in $\mathbf{d}(t)$. We note that:

$$|P_i(t+1) - c| = \left| T_i P_i + (1 - T_i)\frac{P_i + R_j}{2} - c \right|$$

$$\leq T_i|P_i - c| + (1 - T_i)\left| \frac{P_i + R_j}{2} - c \right|$$

$$\leq \frac{1 + T_i}{2}|P_i - c| + \frac{1 - T_i}{2}|R_j - c|$$

Assuming $T_i < 1$, $|P_i(t+1) - c| < a$. Similarly, $|R_j(t+1) - c| < a$. Since no other values in $\mathbf{d}(t)$ change, this means that $\mathbf{d}_k(t+1) = \max\{|P_i(t+1) - c|, |R_j(t+1) - c|, \mathbf{d}_{k+1}(t)\}$. All of these possibilities are strictly smaller than $\mathbf{d}_k(t) = a$. Since none of indices preceding k are affected, the inequality $\mathbf{d}(t+1) <_{lex} \mathbf{d}(t)$ holds. $\diamond$

The lemma essentially shows that so long as there is a positive probability that a proposal will fail to be accepted, $\mathbf{d}(t)$ will converge towards $\mathbf{0}$, meaning that P_i and R_i will all converge to c.

Next, we formally identify conditions under which failure has to be possible. In this case, unlike Lemmas 1 and 2, the results are influenced by the structure of G, the sensitivity of content S and influence between players IN_{ij}.

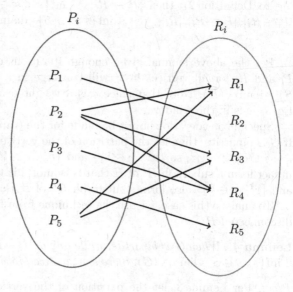

Fig. 3. Example of auxiliary graph

4.2 Convergence Results

We first create an auxiliary graph wherein we split apart the P_i and the R_i values for every user i. In this auxiliary graph, each P_i and R_i is associated with its own vertex. Because every game iteration involves one P_i and one R_j value (and never a P_i with a R_i or even another P_j value), this graph will be bipartite. An example of this type of graph is reported in Fig. 3.

Definition 2 (Auxiliary Graph). *Let (G, V) be a connected graph, wherein each $i \in G$ is associated with values (P_i, R_i). H is the auxiliary graph obtained by taking 2 copies of the vertices of G. Label the vertices by i^1 and i^2 respectively. Then, let $i^1 \sim_H j^2, i^2 \sim_H j^1 \iff i \sim_G j$. We will associate i^1 with P_i and i^2 with R_i.*

In the general case (i.e. G is connected and not bipartite), there is an odd cycle $i_1 i_2 \dots i_k i_1$ in G. This means that i_1^1 is connected to i_1^2 in H by the path $i_1^1 i_2^2 \dots i_k^1 i_1^2$. Because G is connected, this means that H is also connected.

We will use the notation $\text{diam}_{IN}(G)$ to denote the usual diameter of G with edge weights given by IN. The same is true for H, where the weight of an edge (i^1, j^2) is the same as the weight in G of (i, j).

Lemma 3. *When G is not bipartite, while any of $|P_i - R_j|, |P_i - P_j|$, or $|R_i - R_j| > \inf\{s : s \in S\} \cdot \text{diam}_{IN}(H)$, there is a positive probability that $\mathbf{d}(t+1) <_{lex} \mathbf{d}(t)$.*

Proof. Assume that $\mathbf{d}(t+1) = \mathbf{d}(t)$ with probability 1. By Lemma 2, this means that every possible ultimatum game (each edge that can be chosen with positive probability) results in acceptance. This means that for any i and j adjacent, $|P_i - R_j| < s \cdot IN(i, j)$ for any $s \in S$, so $|P_i - R_j| \le \inf\{s : s \in S\} \cdot w_{(i^1, j_2)}$.

Since there is a path between vertices i^1 and j^2 in H (H constructed according to Definition 2), then $|P_i - R_j| \le \inf\{s : s \in S\} \cdot d_{IN}(i^1, j^2)$, and thus $|P_i - R_j|, |P_i - P_j|, |R_i - R_j| \le \inf\{s : s \in S\} \cdot \text{diam}_{IN}(H)$ for any i and j. ◇

Per the above lemma, given enough iterations of the game, every value of P_i and R_i for all vertices in V will converge in a window of size $\inf\{s : s \in S\} \cdot \text{diam}_{IN}(H)$. Note that since c is a weighted average of these values (see Eq. 3), c is in the window.

Special considerations must be made for the (rare) case of G being bipartite. If G is bipartite, then let the partition of the vertices be V_1 and V_2. Note that if $i \sim j$ then $i^1 \sim j^2$, so $\{v^1 : v \in V_1\}$ and $\{v^2 : v \in V_2\}$ are connected in H, and in fact form a subgraph H' of H that is isomorphic to G. Similarly $\{v^2 : v \in V_1\}$ and $\{v^1 : v \in V_2\}$ also form a subgraph H'' of H that is isomorphic to G.

To analyze this case, we use the technique from Lemma 3 on each part of the disconnected H:

Lemma 4. *When G is bipartite, while any of $|P_i - R_j|, |P_i - P_j|$, or $|R_i - R_j| > 2 \cdot \inf\{s : s \in S\} \cdot \text{diam}_{IN}(G)$, there is a positive probability that $\mathbf{d}(t+1) <_{lex} \mathbf{d}(t)$.*

Proof. Per Lemma 3, let the partition of the vertices be into sets V_1 and V_2. Considering only the $\{P_i : i \in V_1\}$ and $\{R_i : i \in V_2\}$, we can use Lemmas 1 and 2 to define c' and $\mathbf{d}'$ to only consider those values. We can also define c'' and $\mathbf{d}''$ to be defined using only $\{R_i : i \in V_1\}$ and $\{P_i : i \in V_2\}$.

Because we consider only one of each P_i and R_i, algebraic manipulation shows that $c = \frac{c' + c''}{2}$. However, since $R_i(0) = P_i(0)$ for every i, we note that $c' = c''$, and thus both are equal to c. This means that vector $\mathbf{d}$ is simply a reordering of the entries of $\mathbf{d}'$ and $\mathbf{d}''$.

Using the same techniques used to prove Lemma 3, it is easy to show that if $\mathbf{d}'(t+1) = \mathbf{d}'(t)$ with probability 1, then all $P_i(t)$ and $R_i(t)$ in H' fall within a window of diameter $\inf\{s : s \in S\} \cdot \text{diam}_{IN}(H')$ that contains c.

In an identical manner, all $P_i(t)$ and $R_i(t)$ in H'' fall within another window of diameter $\inf\{s : s \in S\} \cdot \text{diam}_{IN}(H'')$ that also contains c. However, since both H' and H'' are isomorphic to G including edge weights, the union of these windows is a window of diameter less than $2 \cdot \inf\{s : s \in S\} \cdot \mathrm{diam}_{IN}(G)$. $\diamond$

By relying on the two lemmas above, we now derive the following theorem for general values of T_i:

Theorem 1. *For any graph G, one of the following will occur:*

1. *If $T_i \neq 1 \forall i \in V$, all P_i and R_i will eventually converge to a window of size $2 \cdot \inf\{s : s \in S\} \cdot \mathrm{diam}_{IN}(G)$ or $\inf\{s : s \in S\} \cdot \mathrm{diam}_{IN}(H)$ as appropriate around c.*
2. *If $\exists f$ such that $P_i(0) = R_i(0) = f \forall i : T_i = 1$, all P_i and R_i will eventually converge to a window of size $2 \cdot \inf\{s : s \in S\} \cdot \mathrm{diam}_{IN}(G)$ or $\inf\{s : s \in S\} \cdot \mathrm{diam}_{IN}(H)$ as appropriate around f.*
3. *Otherwise, consensus is impossible.*

Proof. The proof considers each case. For case 1: If $T_i \neq 1 \forall i \in V$, then by Lemmas 3 and 4, there is a positive probability of decrease if this window is larger than $2 \cdot \inf\{s : s \in S\} \cdot \mathrm{diam}_{IN}(G)$ or $\inf\{s : s \in S\} \cdot \mathrm{diam}_{IN}(H)$. This means that eventually the size of the window will decrease.

With respect to case 2: If $\exists f$ such that $P_i(0) = R_i(0) = f \forall i : T_i = 1$ Let $f' = \frac{\sum_{i:T_i \neq 1} \frac{P_i + R_i}{1 - T_i}}{\sum_{i:T_i \neq 1} \frac{2}{1 - T_i}}$ be the weighted average value over all the less stubborn players.

In the same manner as in Lemma 1, f' is conserved for any outcome of any ultimatum game between two vertices from the set $\{i : T_i \neq 1\}$. Note that the same is true for any outcome of of any ultimatum game between two vertices from the set $\{i : T_i = 1\}$, as well as a game with a vertex from each set where the ultimatum is accepted.

If we have a game with a vertex from each set where the ultimatum is not accepted, then without loss of generality, let $T_i \neq 1$. This means that $P_i(t+1)$ (or $R_i(t+1)$) is a weighted average of $P_i(t)$ (or $R_i(t)$) and f, so $f'(t+1)$ is a weighted average of f and $f'(t)$. This means that f' approaches f.

Using the same techniques as Lemmas 2, 3, and 4 with f instead of c, all P_i and R_i will eventually converge to a window of size $2 \cdot \inf\{s : s \in S\} \cdot \mathrm{diam}_{IN}(G)$ or $\inf\{s : s \in S\} \cdot \mathrm{diam}_{IN}(H)$ around f.

Finally, for case 3: if $\exists P_i(0) = R_i(0) \neq P_j(0) = R_j(0)$ for $T_i = T_j = 1$, then for any $t, P_i(t) = R_i(t) \neq P_j(t) = R_j(t)$, so trivially no consensus is possible. $\diamond$

In summary, if content sensitivity can be arbitrarily small, unless there is trivially no way to establish consensus, then all players will converge to a consensus based on their stubbornness values. The rate of convergence will actually depend on the topology of the network, and on how homogeneous users' comfort values and stubbornness levels are. We provide some insights on these dimensions in the next section.

5   Empirical Results

Our convergence results guide understanding of behavioral trajectories in a social network. However, some interesting and more practical issues are unaccounted for in our analysis, especially with respect to the effects of scale. That is, large social networks may have multiple stubborn users, users who interact very often, or those who interact very infrequently. Informed by our theoretical findings, we can further our understanding of these effects through controlled experiments, varying specific parameters of the game that we anticipate may play a significant role in real-world networks.

Through simulation, we explore the effects of specific personal and structural characteristics (e.g., stubbornness, degree) at the node level as they relate to short- and long-term evolution of privacy preferences for jointly managed content throughout the system. We study:

1. The role of the stubborn users, and their evolution in the network (e.g. how does an extremely opinionated user affect others? Does his/her behavior change over time?);
2. The role of high-degree users, and their relative rates of successful or failed interactions (i.e. are popular users more likely to experience successful interactions?); and
3. The short- vs. long-term nature of observed effects (i.e. is convergence to a fair value possible in the short term? if so, under which conditions)?

5.1   Local Influencers: Stubbornness and Connectivity

Two types of users are likely to affect the dynamics of our system. These are highly stubborn users (i.e. $T_i \simeq 1$), and highly popular users $deg(i) > avg(deg(n))$. Users with high stubbornness (who are slower to concede their preferences) are influential in their neighborhood. Recall that, per Theorem 1 in the extreme case of a fully-connected graph with exactly one perfectly stubborn user $(T_i = 1)$, given the conditions of our model described, all users' proposal and response values will converge to his comfort level.

In the case that a network has multiple stubborn users, each becomes a *local influencer*, with the speed and diameter of influence dependent on local connectivity patterns and proximity to "competing" stubborn users. In this way, stubborn users typically serve as centers of "communities", closely aligned with community structure detected by classic community structure detection algorithms.

We study this case through a simulation through the benchmark Karate Club network $(N = 34)$ [41]. The Karate Club network is used as a first network topology as it is well understood and its small size allows for explicit tracking and visualization at the individual node level. In addition, the close-knit peer group represents a micro-scale view of a larger social system.

Over this network structure, we start from baseline assumptions that: (1) Users' inherent privacy preferences; (2) Users' stubbornness scores T_i; (3) Influence scores over pairwise links $IN(i, j)$; and (4) Content sensitivity scores are all uniformly distributed in $[0, 1]$. We let zero represent least inclined to share, least stubborn, least influenced and least public (most sensitive), respectively.

Consider the following representative example (Fig. 4). User comfort levels, i.e., initial proposal and response values for all users are taken from a uniform random distribution in $[0, 1]$, with the exception that user 1 is seeded with a comfort of 0.1 (strict sharing) and user 34 with a comfort of 0.9 (public sharing). In addition, users 1 and 34 are seeded as perfectly stubborn, i.e., $T_1 = T_{34} = 1$. The left hand image visualizes initial comforts, equivalently initial proposal values, for all individuals in the network. Nodes are colored on a temperature scale, with blue representing 0 and red representing 1. The game is run to convergence (10K iterations of play), and the resultant final proposal values are reflected on the right; note that proposal and response values are equivalent in the limit. Consistent with our theoretical findings, we see convergence around stubborn users. In this case wherein multiple perfectly stubborn users are present within a single connected component of the graph, convergence is localized around each and specific diffusion of influence depends on the local connectivity patterns.

As such, high centrality (degree, betweenness, closeness) users may play an interesting and notable role in system-wide behaviors. Specifically, we suggest that users embedded in their local communities may have more rapid influence; in time-limited real-world scenarios of evolving social graphs, more rapid influence likely means wider influence as well. In addition, high-degree users "play" more often (are involved in more shared content and subsequent negotiations); in the framework we have described, highly connected users may be selected as responder any time a friend is selected as proposer. In sum, location, degree, number and extent of stubborn users are interrelated determinants of system-wide preferences, i.e., proposal and response values, at convergence.

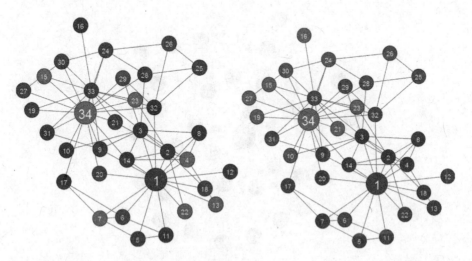

Fig. 4. Initial (left) and converged (right) proposal values on the karate club network.

5.2 Evolution of the Ultimatum Game at the Slow Time Scale

We have theoretically (Sect. 4) and experimentally described the behavior of the ultimatum game at the *long time horizon*. These analyses allowed for a formal understanding of limit behavior, but were not necessarily realistic in real-world, time-constrained scenarios. Here, we consider the implications of our findings at the shorter-term horizon, or for more sparse interactions.

Estimates indicate [5] that Facebook users share on average 0.35 photos per day, or 1 photo every 3 days (350 million photos per day, divided by 1 billion active users per day). In our small network of 34 users, we estimate 12 instances of sharing/interactions per day. Provided a static network structure, over the course of one month the game is played for 360 iterations. Figure 5 illustrates the influence of one stubborn user (user 3) after 500 iterations of play. We seeded user 3 with a sharing comfort of 0.9 and all other users with comfort 0.1. User 3 was seeded as perfectly stubborn, $T_3 = 1$, while other users' stubbornness values were taking from a uniform distribution on $[0, 1]$. We have proven that in this contrived but important extreme case, all users will eventually converge to proposal and response values at 0.9. However at the shorter horizon, notice the local influence of user 3, where variants in neighbors' final values are attributable to connectivity patters, number of interactions and inherent stubbornness.

Stubborn users, then, seem to be playing a winning strategy. In the long term, they pull other users toward their own preferences and exhibit greater influence regionally over time. However, we note one consequence for stubborn users, namely a greater expectation of failed interactions. As their peers move more quickly toward compromise and bring their own preferences in line with their neighbors, stubborn users are slower to narrow this gap. Accordingly, as

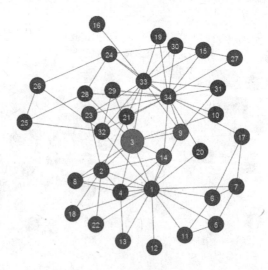

Fig. 5. Proposal values after 500 iterations of the ultimatum game. Observe node 3 and its local influence in its neighborhood.

pairs of less stubborn users reach preferences within the window of consensus and begin to increase their rate of successful interactions, all else (connectivity, preferences) being equal, stubborn users continue to fail further into the game. In addition to less stubborn users, it can also be said that users whose comfort level is nearer to the mean (in our case, the mean is fixed at 0.5) experience more successful interactions with their peers. That is, the expected value of the difference between their own proposal/response value and that of their neighbor is lesser than the expected value of that difference for a user with a preference nearer to either end of interval.

5.3 The Importance of Homophily

Our last observation brings us to an important consideration. The examples we have provided thus far have involved fixing personal preferences and stubbornness near extreme values in order to demonstrate effects. However, consider expected scenarios where connected users have generally similar preferences and are in general moderately stubborn. The social science literature on homophily provides evidence that real world social systems are well-modeled using 'birds of a feather' assumptions [23,27].

Consider the same network of Fig. 5, wherein initial preferences and stubborness are distributed in a uniform (and random) fashion from the interval [0.3, 0.7]. Figure 6 represents the preferences of all 34 users over 5000 rounds of play. Notice, in this framework, all players' values tend to converge to a common small range of values, with less than 0.2 separating the preferences of any two ·users in the network.

This tendency toward agreement in more homogeneous communities holds implications for prototypical real-world social networks wherein densely linked groups of users tend to be more 'similar' by some measure. It is an open question whether documented instances of homophily in social systems extends to privacy

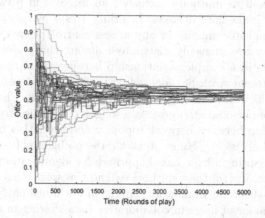

Fig. 6. Proposal values plotted for 34 players over 5000 iterations of the ultimatum game for multi-party content.

preferences as well, but our model would suggest that it may. Furthermore, anomalous cases for which this tendency is not observed may be indicative of areas for deeper investigation.

6   Related Work

Our work lies at the crossroad of game theory for modeling social interactions and multi-user access control.

There is a long history using the ultimatum game to model pairwise interactions amongst individuals seeking to rectify opposing forces of cooperation and selfishness [8,37]. In particular, in the Ultimatum Game, one player proposes a division of a sum of money between herself and a second player, who either accepts or rejects. Based on rational self-interest, responders should accept any nonzero offer and proposers should offer the smallest possible amount. Traditional, deterministic models of evolutionary game theory agree: in the one-shot anonymous Ultimatum Game, natural selection favors low offers and demands. However, experiments in real populations reveal a preference for fairness. When carried out between members of a shared social group (e.g., a village, a tribe, a nation, humanity) people offer "fair" splits close to 50-50, and offers of less than 30% are often rejected [14,28]. There are several theories as to why this difference between theoretically optimal and practical behaviors may exist, including reputation and memory effects [6], natural selection [26], empathy and perspective taking [24]. In [42], we study this phenomenon using a similar model to that presented in this paper, but in a general setting unrelated to privacy and access control. Accordingly, the formulation explored in [42] involves a more general rule set, leaning on notions of greed and charity, rather than consensus-formation.

With respect to privacy and related decision making processes, researchers from many communities have noted the trade-off between privacy and utility (e.g., [4,7,21,29,31,32]). The majority of this prior work tends to view the privacy/utility trade-off as mutually exclusive: an increase in privacy (resp. utility) results in an immediate decrease in utility (resp. privacy). We note that the interplay of multiple entities in any access control/privacy decision where privacy and utility are unevenly distributed among the players and context-dependency results in a complex relationship between these concepts [1,19]. A growing body of recent work has focused on multi-party access control mechanisms, some of which have used game-theoretical concepts. Chen et al. model users' disclosure of personal attributes as a weighted evolutionary game and discuss the relationship between network topology and revelation in environments with varying level of risk [9]. Hu et al. tackle the problem of multi-party access control in [17], proposing a logic-based approach for identification and resolution of privacy conflicts. In [18] these authors extend this work, this time proposing adopting a game-theoretic framework, specifically a multi-party control game to model the behavior of users in collaborative data sharing in OSNs. Another game-theoretic model is given in [35], in which automated agents negotiate on behalf of users access control settings in a multi-user scenario. Other very recent approaches to multi-party access control mechanism use a mediator [33] or a recommendation system [12] to suggest the optimal decision in one-shot multiparty access control scenarios. The primary difference between our work and previous ones on multi-party access control (whether game-theoretic or not) is our unique focus on the effects of one-time interactions to a given network, and the related consequences for users in the network over a number of interactions.

7 Conclusions and Future Work

In this paper, we presented a macro-model to describe how individuals' sharing decisions change over time, who are the most influential users and how they benefit from it, along with privacy gains and losses from a collective perspective. Through a carefully designed ultimatum game, informed by the body of work on multi-party access control, we were able to capture the most important dynamics underlying privacy decision making in online social networks. Our results show users' overall tendency to converge toward a self-adjusted environment, wherein successes and failures commensurate with users' stubbornness and underlying network dynamics.

This work is the first step toward a more systematic analysis of how people's privacy attitudes evolve over time, and change their personal information sharing patterns as a result. As such, we anticipate several extensions and possible avenues for research.

Further theoretical work may look into the system's convergence properties for nonvanishing content sensitivity and study time to convergence (within some bounds) in network topologies that reflect real-world social structure. Related to this, convergence in a practical sense will reflect agreement on a *discrete* privacy setting and accounting for this will impact these findings.

With respect to discretization of privacy settings, our model is thus far agnostic to the actual privacy settings or access control paradigm used by the online social network provider. We plan to define a mapping that converts available settings into values in $[0, 1]$ and vice versa. For instance, default Facebook settings go from private, to friends, to friends of friends, to public. Also, users may choose particular users or groups. In that case, the comfort value would be the distance between a user's desired privacy policy and the one she may finally accept, in a similar way to [35], in which the euclidean distance is used to compare the distance between two privacy policies to quantify the actual concession being made during an access control negotiation

Further empirical, simulated studies may look at larger network graphs and regimes of influence. That is, we have shown that stubborn users have disproportionate influence in their local neighborhood, but their global influence is dependent on their place in the network topology. We envision that these considerations may support a full taxonomy of users categorized in multiple dimensions including centrality, stubbornness and inherent privacy preferences. Ultimately, categorizing users in this way and developing a common language with which to discuss different user privacy behaviors will be very useful to further understand the interplay between local one-shot decisions and overall sharing dynamics at the social network in multi-party access control.

Finally, as more detailed data becomes available on instances of multi-party access control negotiations in the wild, especially longitudinal data about repeated negotiations over time, either through collected data from popular networking sites or through smaller and more targeted user studies, this data may be used to verify and parameterize the proposed model. We believe that the ultimatum game framework is a reasonable starting point, given its fundamental role in modeling social cooperation broadly and existing evidence on one-shot multi-party access control decisions. The update rules we have chosen are motivated by the psychology literature on in-group/out-group behaviors, peer pressure, and one-shot multi-party access control decisions. However, these rules and parameters thereof should be further researched in the specific context of repeated decisions on multi-party access control settings.

Acknowledgements. Portions of Dr. Squicciarini's and Mr. Semonsen's work were funded by Army Research Office under the auspices of Grant W911NF-13-1-0271. Portions of Dr. Squicciarini's work were additionally supported by a National Science Foundation grant 1453080. Dr. Such was supported by the EPSRC under grant EP/M027805/2.

References

1. Acquisti, A.: Privacy in electronic commerce and the economics of immediate gratification. In: Proceedings of the 5th ACM Electronic Commerce Conference, pp. 21–29. ACM Press (2004)
2. Arendt, D.L., Blaha, L.M.: Opinions, influence, and zealotry: a computational study on stubbornness. Comput. Math. Organ. Theor. **21**(2), 184–209 (2015)
3. Besmer, A., Richter Lipford, H.: Moving beyond untagging: photo privacy in a tagged world. In: Proceedings of the SIGCHI Conference on Human Factors in Computing Systems, pp. 1563–1572. ACM (2010)
4. Bhumiratana, B., Bishop, M.: Privacy aware data sharing: balancing the usability and privacy of datasets. In: Proceedings of the 2nd International Conference on PErvasive Technologies Related to Assistive Environments, p. 73. ACM (2009)
5. Brandwatch.com. Facebook statistics from Brandwatch (2016). https://www.brandwatch.com/blog/47-facebook-statistics-2016/
6. Bravo, G., Castellani, M., Squazzoni, F., Boero, R.: Reputation and judgment effects in repeated trust games (2008)
7. Brush, A.J., Krumm, J., Scott, J.: Exploring end user preferences for location obfuscation, location-based services, and the value of location. In: Proceedings of the 12th ACM International Conference on Ubiquitous Computing, pp. 95–104. ACM (2010)
8. Camerer, C.: Behavioral Game Theory: Experiments in Strategic Interaction. Princeton University Press, Princeton (2003)
9. Chen, J., Brust, M.R., Kiremire, A.R., Phoha, V.V.: Modeling privacy settings of an online social network from a game-theoretical perspective. In: 2013 9th International Conference Conference on Collaborative Computing: Networking, Applications and Worksharing (Collaboratecom), pp. 213–220. IEEE, October 2013

10. Cosley, D., Huttenlocher, D.P., Kleinberg, J.M., Lan, X., Suri, S.: Sequential influence models in social networks. In: ICWSM, vol. 10, p. 26 (2010)
11. Crandall, D., Cosley, D., Huttenlocher, D., Kleinberg, J., Suri, S.: Feedback effects between similarity and social influence in online communities. In: Proceedings of the 14th ACM SIGKDD International Conference on Knowledge Discovery and Data Mining, pp. 160–168. ACM (2008)
12. Fogues, R.L., Murukannaiah, P.K., Such, J.M., Singh, M.P.: Sharing policies in multiuser privacy scenarios: incorporating context, preferences, and arguments in decision making. ACM Trans. Comput. Hum. Interact. (TOCHI) **24**(1), 5 (2017)
13. Friedkin, N.E.: A Structural Theory of Social Influence, vol. 13. Cambridge University Press, Cambridge (2006)
14. Henrich, J.P.: Foundations of Human Sociality: Economic Experiments and Ethnographic Evidence from Fifteen Small-Scale Societies. Oxford University Press on Demand, Oxford (2004)
15. Hu, H., Ahn, G.-J.: Multiparty authorization framework for data sharing in online social networks. In: Li, Y. (ed.) DBSec 2011. LNCS, vol. 6818, pp. 29–43. Springer, Heidelberg (2011). doi:10.1007/978-3-642-22348-8_5
16. Hu, H., Ahn, G-J., Jorgensen, J.: Detecting and resolving privacy conflicts for collaborative data sharing in online social networks. In: Proceedings of the 27th Annual Computer Security Applications Conference, pp. 103–112. ACM (2011)
17. Hongxin, H., Ahn, G.-J., Jorgensen, J.: Multiparty access control for online social networks: model and mechanisms. IEEE Trans. Knowl. Data Eng. **25**(7), 1614–1627 (2013)
18. Hu, H., Ahn, G-J., Zhao, Z., Yang, D.: Game theoretic analysis of multiparty access control in online social networks. In: Proceedings of the 19th ACM Symposium on Access Control Models and Technologies (SACMAT), London, Ontario, pp. 93–102. ACM (2014)
19. John, L.K., Acquisti, A., Loewenstein, G.: Strangers on a plane: context-dependent willingness to divulge sensitive information. J. Consum. Res. **37**(5), 858–873 (2011)
20. Lampinen, A., Lehtinen, V., Lehmuskallio, A., Tamminen, S.: We're in it together: interpersonal management of disclosure in social network services. In: Proceedings of the SIGCHI Conference on Human Factors in Computing Systems, pp. 3217–3226. ACM (2011)
21. Liu, K., Terzi, E.: A framework for computing the privacy scores of users in online social networks. ACM Trans. Knowl. Discov. Data **5**(1), 6:1–6:30 (2010)
22. Marwick, A.E., Boyd, D.: I tweet honestly, i tweet passionately: Twitter users, context collapse, and the imagined audience. New Media Soc. **13**(1), 114–133 (2011)
23. McPherson, M., Smith-Lovin, L., Cook, J.M.: Birds of a feather: homophily in social networks. Ann. Rev. Sociol. **27**, 415–444 (2001)
24. Page, K.M., Nowak, M.A.: Empathy leads to fairness. Bull. Math. Biol. **64**(6), 1101–1116 (2002)
25. Rajtmajer, S., Squicciarini, A., Griffin, C., Karumanchi, S., Tyagi, S.: Constrained social-energy minimization for multi-party sharing in online social networks. In: Proceedings of the 2016 International Conference on Autonomous Agents & Multiagent Systems, pp. 680–688. International Foundation for Autonomous Agents and Multiagent Systems (2016)
26. Rand, D.G., Tarnita, C.E., Ohtsuki, H., Nowak, M.A.: Evolution of fairness in the one-shot anonymous ultimatum game. Proc. Natl. Acad. Sci. **110**(7), 2581–2586 (2013)
27. Ryan, B., Gross, N.: The diffusion of hybrid seed corn in two Iowa communities. Rural Sociol. **8**(1), 15–24 (1943)

28. Sanfey, A.G., Rilling, J.K., Aronson, J.A., Nystrom, L.E., Cohen, J.D.: The neural basis of economic decision-making in the ultimatum game. Science **300**(5626), 1755–1758 (2003)
29. Schlegel, R., Kapadia, A., Lee, A.J.: Eyeing your exposure: quantifying and controlling information sharing for improved privacy. In: Proceedings of the Seventh Symposium on Usable Privacy and Security, p. 14. ACM (2011)
30. Sleeper, M., Balebako, R., Das, S., McConahy, A.L., Wiese, J., Cranor, L.F.: The post that wasn't: exploring self-censorship on Facebook. In: Proceedings of the 2013 Conference on Computer Supported Cooperative Work, pp. 793–802. ACM (2013)
31. Solove, D.J.: Nothing to Hide: The False Tradeoff Between Privacy and Security. Yale University Press, New Haven (2011)
32. Stutzman, F., Hartzog, W.: Boundary regulation in social media. In: Proceedings of the ACM 2012 Conference on Computer Supported Cooperative Work, pp. 769–778. ACM (2012)
33. Such, J.M., Criado, N.: Resolving multi-party privacy conflicts in social media. IEEE Trans. Knowl. Data Eng. **28**(7), 1851–1863 (2016)
34. Such, J.M., Porter, J., Preibusch, S., Joinson, A.: Photo privacy conflicts in social media: a large-scale empirical study. In: Proceedings of the 2017 CHI Conference on Human Factors in Computing Systems, pp. 3821–3832. ACM (2017)
35. Such, J.M., Rovatsos, M.: Privacy policy negotiation in social media. ACM Trans. Auton. Adapt. Syst. **11**(4), 1–29 (2016)
36. van Baaren, R., Janssen, L., Chartrand, T.L., Dijksterhuis, A.: Where is the love? the social aspects of mimicry. Philos. Trans. R. Soc. B **364**(1528), 2381–2389 (2009)
37. Wallace, B., Cesarini, D., Lichtenstein, P., Johannesson, M.: Heritability of ultimatum game responder behavior. Proc. Natl. Acad. Sci. **104**(40), 15631–15634 (2007)
38. Wang, Y., Norcie, G., Komanduri, S., Acquisti, A., Leon, P.G., Cranor, L.F.: "I regretted the minute i pressed share": a qualitative study of regrets on Facebook. In: Proceedings of the Seventh Symposium on Usable Privacy and Security, p. 10. ACM (2011)
39. Wisniewski, P., Lipford, H., Wilson, D.: Fighting for my space: coping mechanisms for SNS boundary regulation. In: Proceedings of the SIGCHI Conference on Human Factors in Computing Systems, pp. 609–618. ACM (2012)
40. Yildiz, E., Acemoglu, D., Ozdaglar, A.E., Saberi, A., Scaglione, A.: Discrete opinion dynamics with stubborn agents (2011)
41. Zachary, W.W.: An information flow model for conflict and fission in small groups. J. Anthropol. Res. **33**, 452–473 (1977)
42. Zhu, Q., Rajtmajer, S., Belmonte, A.: The emergence of fairness in an agent-based ultimatum game. In: Preparation (2016)

The U.S. Vulnerabilities Equities Process: An Economic Perspective

Tristan Caulfield[1](✉), Christos Ioannidis[2], and David Pym[1,3]

[1] University College London, London, England
{t.caulfield,d.pym}@ucl.ac.uk
[2] Aston Business School, Birmingham, England
[3] The Alan Turing Institute, London, England
c.ioannidis@aston.ac.uk

Abstract. The U.S. Vulnerabilities Equities Process (VEP) is used by the government to decide whether to retain or disclose zero day vulnerabilities that the government possesses. There are costs and benefits to both actions: disclosing the vulnerability allows the vulnerability to be patched and systems to be made more secure, while retaining the vulnerability allows the government to conduct intelligence, offensive national security, and law enforcement activities. While redacted documents give some information about the organization of the VEP, very little is publicly known about the decision-making process itself, with most of the detail about the criteria used coming from a blog post by Michael Daniel, the former White House Cybersecurity Coordinator. Although the decision to disclose or retain a vulnerability is often considered a binary choice—to either disclose or retain—it should actually be seen as a decision about timing: to determine *when* to disclose. In this paper, we present a model that shows how the criteria could be combined to determine the optimal time for the government to disclose a vulnerability, with the aim of providing insight into how a more formal, repeatable decision-making process might be achieved. We look at how the recent case of the WannaCry malware, which made use of a leaked NSA zero day exploit, EternalBlue, can be interpreted using the model.

1 Introduction

Governments, for national security, military, law enforcement, or intelligence purposes, often require an ability to access electronic devices or information stored on devices that are protected against intrusion. One way this access can be achieved is through the exploitation of vulnerabilities in the device's software or hardware. To this end, governments acquire, through a number of different methods, knowledge of these vulnerabilities—which are usually unknown to the software vendor and users—and how they may be successfully exploited.

However, the role a government plays is dual: in addition to the national security and law enforcement purposes above, which may require the exploitation of vulnerabilities, the government is also responsible for defending its national

© Springer International Publishing AG 2017
S. Rass et al. (Eds.): GameSec 2017, LNCS 10575, pp. 131–150, 2017.
DOI: 10.1007/978-3-319-68711-7_8

assets in cyberspace. It has a responsibility to protect its own government and military networks, the nation's critical infrastructure, as well as the information assets of its businesses and citizens. When a government acquires knowledge of a vulnerability, this dual role presents a conflict. The government must decide between two competing national security interests: whether to retain the vulnerability, keeping it secret so it can be used to gain access to systems for intelligence purposes, or if it should instead be disclosed to the vendor, allowing it to be fixed so that the security of systems and software can be improved.

In the United States, this decision is now guided by the Vulnerabilities Equities Process (VEP), which the government uses to assess whether to retain or release each vulnerability it acquires. Publicly, relatively little is know the criteria used in this assessment. A Freedom of Information Act request from the Electronic Frontier Foundation (EFF) saw the release of a redacted version of a document [4] that describes how the VEP works within the government, but without any indication of how the decision to retain or disclose is made. A blog post [5] in April, 2014 by Michael Daniel, the White House Cybersecurity Coordinator, provided some insight, revealing a number of factors that are used in the decision-making process, but also that 'there are no hard and fast rules'.

The factors listed in the blog post are very high-level concepts, describing *what* decision-makers consider, but not *how* they do so. For example, some of these factors describe values, such as 'the extent of the vulnerable system's use in the Internet infrastructure' or 'the risks posed and the harm that could be done if the vulnerability is left unpatched', and yet no there is no indication of how they can be quantified or compared against each other. Given the lack of hard and fast rules, it is not unreasonable to assume that decisions are made on an ad hoc, case-by-case basis.

There has been some discussion and commentary about the VEP—and about how vague known information about it is. A June 2016 discussion paper by Schwartz and Knake [24] examines what is publicly known about the VEP and makes a number of recommendations to improve the process. Among these is the recommendation to 'make public the high-level criteria that will be used to determine whether to disclose ... or to retain [a] vulnerability' and that it is possible to 'formalize guidelines for disclosure decisions while preserving flexibility in the decision-making process'. Similarly, a September, 2016 EFF blog post [3] recommends that the government be more transparent about the VEP decision-making process, including the criteria used, and that the policy should be 'more than just a vague blog post'.

This paper aims to further understanding of how the different factors that are used in a VEP decision can be included in a more formal decision-making process. The intent is not to be normative: we do not aim to say, for example, how much potential harm is an acceptable trade-off for the benefits gained from exploiting a particular vulnerability. Instead, we look at different possible ways in which each of the factors may be evaluated or quantified, how the factors relate to each other, and how this information could be combined to make a decision. Specifically, we present the government's decision about whether or not to disclose a vulnerability as a timing problem, where the solution is the optimal amount of time to delay disclosing the vulnerability given the costs and benefits of doing so. We then look at how how the long delay before the disclosure of the EternalBlue vulnerability can be interpreted using this model.

Section 2, next, presents background information about vulnerabilities, exploits, and their increasing use by governments. Next, Sect. 3 introduces the VEP: its purpose and origins, the factors used in decision-making, and the discussion and debate surrounding the process. Section 4 examines each of the factors in detail, looking at how they might affect the disclosure decisions, and Sect. 5 looks at how the factors could be combined into a model to determine the optimal time of disclosure. Section 6 then looks at the WannaCry malware and the timing of the disclosure of the leaked vulnerability it used.

2   Background

During the development of a piece software, flaws—or *bugs*—may arise in its design or implementation which cause the software to behave differently than intended. A bug that can cause behaviour affecting the security of a system is called a *vulnerability*. An *exploit* is a technique or action (for example, a piece of software or a series of commands) that can be used to take advantage of the vulnerability. An *attack* is the use of an exploit to attempt this.

Creating software without bugs is a very difficult challenge and is not economically feasible for most software. As such, software often has to be updated after its initial release in order to fix bugs discovered later on. A vulnerability, once discovered, may be *disclosed*—either to the vendor directly, through an organization such as CERT, which coordinates disclosures with the vendor, or publicly. Once the developer is aware of the vulnerability, they may work to create a fix that removes the vulnerability. An updated version of the software containing the fix, called a *patch*, is then released; end-users of the software must then apply this patch to their systems to remove the vulnerability.

The sequence of events including the discover of a vulnerability, the creation of an exploit for it, the vulnerability's disclosure, and eventual patching is

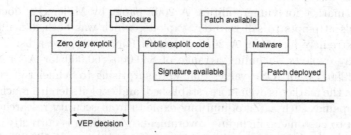

Fig. 1. The vulnerability timeline, showing the events that can occur from the discovery of a vulnerability to its eventual patching. This is a guideline: not all of these events will occur for every vulnerability, and the order in which they occur may differ.

known as the vulnerability timeline. Figure 1, adapted from [2], illustrates this sequence of events. Related to this is the notion of the window of exposure, discussed in [20], which is the time from the creation of an exploit until systems are patched during which systems are at risk from a vulnerability. This window can be reduced by improving the speed with which patches are produced and deployed. The VEP deals with the government's decision to disclose or retain *zero day* vulnerabilities; these are vulnerabilities that are unknown both publicly and to the software developer, so named because the developer and end users have had zero days to fix or mitigate the vulnerability. Disclosing the vulnerability allows a patch to be produced sooner, reducing the window of exposure.

The timeline includes an event for a signature becoming available, which indicates the the availability of methods to mitigate the vulnerability before the official patch has been released, including anti-virus or intrusion-detection signatures. There is also a distinction between the public release of exploit code and the development of malware. The former refers to code that utilizes the exploit—perhaps as a proof-of-concept or demonstration that the exploit works—but does not cause significant damage; the latter refers to more sophisticated and damaging uses of the exploit. However, publicly publishing proof-of-concept code can make it easier for more damaging exploits to be developed. All of the events in the timeline do not always occur for each vulnerability, and the ordering of events and the time between them is fluid. For example, there might not be a zero day exploit, or the patch could be released before any exploit is developed.

2.1   Increasing Use of Vulnerabilities

The exploitation of vulnerabilities by both governments and other parties is growing—and this is not surprising. The use of digital technologies in all aspects of life and business continues to grow at an astounding rate and more and more information is stored on electronic devices. Access to these devices and the information stored on them has value. For governments this could be the value of intelligence, the ability of law enforcement to conduct surveillance, or the ability to disrupt systems. For criminal actors, access to these systems can enable a host of different crimes, from theft or ransom of information to sabotage.

Evidence of the increasing importance of vulnerabilities to all parties is the rise of the market for vulnerabilities. A 2007 paper by Miller [12] documents the author's attempts to sell zero day exploits, which was both difficult to do and not extremely lucrative. A 2012 article in Forbes [8] gave a list of prices for zero day exploits, including a range of $30,000–$60,000 for Android, and $100,000–$250,000 for iOS—values that were surprising to Schneier at the time [22]. Today, the market is even more established, and exploits fetch a much higher price. Companies such as Zerodium buy exploits from security researchers and resell them to customers, including governments. Zerodium is currently offering researchers up to $200,000 for Android exploits and up to $1,500,000 for iOS exploits [25]. Other products that are less secure (so it is easier to find exploitable vulnerabilities) or less popular fetch lower prices. This increase in market price (and the expansion of the market itself) over the last decade is an indicator of the increasing demand for and importance of zero day exploits.

Often, exploits can be purchased with an exclusivity agreement, meaning that it will not be sold to anyone else. However, exclusivity agreements are no guarantee that the vulnerability will remain undiscovered by others. Other researchers, governments, or criminals may independently discover the same vulnerability. This is what causes the tension between the dual roles of the government: just because it it believes it is the only entity with access to a vulnerability does not mean it will not be used against assets it is charged to protect. Thus, every decision to retain a vulnerability instead of disclosing it so it can be fixed and patched increases the risk to systems the government aims to protect.

3   The Vulnerabilities Equities Process (VEP)

The Vulnerabilities Equities Process was created to address the tension between the offensive and defensive missions of the government. Schwartz and Knake [24] provide a thorough explanation of the background and origins of the VEP, and Healey [9] also gives a good overview; we will provide a brief summary here.

President George W. Bush signed a directive [14] in 2008 creating a government-wide Comprehensive National Cybersecurity Initiative (CNCI). This initiative required a number of government departments to develop a plan for coordinating the 'application of offensive capabilities to defend US information systems', which led to the production of the VEP document [4] in February, 2010.

A redacted version of the VEP document was obtained via a Freedom of Information Act request by the EFF. The document begins by stating its purpose:

> This document establishes policy and responsibilities for disseminating information about vulnerabilities discovered by the United States Government (USG) or its contractors, or disclosed to the USG by the private sector or foreign allies in Government Off-The-Shelf (GOTS), Commercial Off-The-Shelf (COTS), or other commercial information technology or industrial control products or systems (to include both hardware or software). This policy defines a process to ensure that dissemination decisions regarding the existence of a vulnerability are made quickly, in full consultation with all concerned USG organizations, and in the best interest of USG missions of cybersecurity, information assurance, intelligence, counterintelligence, law enforcement, military operations, and critical infrastructure protection.

The document also specifies conditions for whether or not a vulnerability is entered into the VEP: 'to enter the process a vulnerability must be both newly discovered and not publicly known' but that 'vulnerabilities discovered before the effective date of this process need not be put through the process'. The VEP document creates an Equities Review Board (ERB) that makes the decision about disclosing or retaining a vulnerability, establishes an Executive Secretariat, specifies how government agencies that come into possession of a vulnerability should notify the Executive Secretary, and how agency-designated Subject Matter Experts (SMEs) hold discussions to evaluate the course of action for each vulnerability.

In short, the VEP document specifies how the process of submitting vulnerabilities works, how the various stakeholders have inputs, and how the process is managed. It does not mention what inputs or factors are used when making a decision, nor how any such factors would be considered.

3.1   The Daniel Blog Post

Information about the VEP was first released under the Obama Administration in 2014, in response to allegations by Bloomberg News [18] that the NSA was aware of and had exploited the Heartbleed vulnerability in OpenSSL, which the NSA denied. The White House commented, saying that the NSA would have disclosed the vulnerability, had they known about it, and in most cases would disclose any vulnerability discovered to allow it to be fixed. Referring to the VEP: 'unless there is a clear national security or law enforcement need, this process is biased toward responsibly disclosing such vulnerabilities' [15,19].

Further information about the VEP came in the form of a blog post [5] by Michael Daniel, the White House Cybersecurity Coordinator, responding to the debate caused by the Heartbleed vulnerability. In it, Daniel discusses the trade-offs between disclosing and retaining a vulnerability— 'disclosing a vulnerability can mean that we forego an opportunity to collect crucial intelligence' but 'building up a huge stockpile of undisclosed vulnerabilities while leaving the Internet vulnerable and the American people unprotected would not be in our national security interest'.

Following this, Daniel provides the only public insight into the factors that are considered when deciding to retain or disclose a vulnerability:

> We have also established a disciplined, rigorous and high-level decision-making process for vulnerability disclosure. This interagency process helps ensure that all of the pros and cons are properly considered and weighed. While there are no hard and fast rules, here are a few things I want to know when an agency proposes temporarily withholding knowledge of a vulnerability:
> - How much is the vulnerable system used in the core internet infrastructure, in other critical infrastructure systems, in the U.S. economy, and/or in national security systems?
> - Does the vulnerability, if left unpatched, impose significant risk?
> - How much harm could an adversary nation or criminal group do with knowledge of this vulnerability?
> - How likely is it that we would know if someone else was exploiting it?
> - How badly do we need the intelligence we think we can get from exploiting the vulnerability?

- Are there other ways we can get it?
- Could we utilize the vulnerability for a short period of time before we disclose it?
- How likely is it that someone else will discover the vulnerability?
- Can the vulnerability be patched or otherwise mitigated?

These factors are weighed 'through a deliberate process that is biased toward responsibly disclosing the vulnerability'—but what this decision-making process is remains unknown.

3.2   Debate and Recommendations

The VEP document and the Daniel blog post have been analysed and criticized a number of times. Schwartz and Knake [24] explore the history of the VEP and what is known about it from various sources and make recommendations to improve the process.

Several of the recommendations concern the decision-making process and are of interest here. First, 'the principles guiding these decisions, as well as a high-level map of the process that will be used to make such decisions, can and should be public'. Next, 'make public the high-level criteria that will be used to determine whether to disclose to a vendor a zero day vulnerability in their product, or to retain the vulnerability for government use'. Finally, if a vulnerability is not disclosed, the process should 'ensure that any decision to retain a zero day vulnerability for government use is subject to periodic review' and that vulnerabilities should be 'disclosed to the responsible party once (1) the government has achieved its desired national security objectives or (2) the balance of equities dictate that the vulnerability should be disclosed'.

The EFF also makes recommendations about the VEP. In August, 2016, an entity naming itself 'The Shadow Brokers' released a collection of files containing code for exploiting vulnerabilities in various firewall products from vendors such as Cisco and Fortinet. These exploits were linked to the NSA and, crucially, were exploiting previously unknown zero day vulnerabilities. The exploit code was stolen in 2013 and the NSA was aware it had been exposed, but the vulnerabilities were never disclosed.

In response to this, the EFF wrote in [3]:

> We think the government should be far more transparent about its vulnerabilities policy. A start would be releasing a current version of the VEP without redacting the decision-making process, the criteria considered, and the list of agencies that participate, as well as an accounting of how many vulnerabilities the government retains and for how long. After that, we urgently need to have a debate about the proper weighting of disclosure versus retention of vulnerabilities.

Similarly, Mozilla discusses the VEP in response to the Shadow Brokers leak [6] and makes recommendations, including:

- All security vulnerabilities should go through the VEP and there should be public timelines for reviewing decisions to delay disclosure;
- All relevant federal agencies involved in the VEP must work together to evaluate a standard set of criteria to ensure all relevant risks and interests are considered;
- Independent oversight and transparency into the processes and procedures of the VEP must be created. All security vulnerabilities should go through the VEP and there should be public timelines for reviewing decisions to delay disclosure.

Common to these three sets of recommendations is the desire for greater insight into the decision-making process and the factors or criteria that are used. Additionally, the recommendations from Schwartz and Knake and Mozilla are both concerned with the timing for reviews of vulnerabilities that have been retained. Proposed legislation, the Protecting our Ability To Counter Hacking (PATCH) Act [7], would turn the VEP into law and allows for periodic review of vulnerabilities—meaning that a vulnerability could be used for a time and then disclosed. We agree with these recommendations and, in the next two sections, we examine the factors from the Daniel blog post—to better understand how they might influence the decisions made—and then present a model for a decision-making process that utilizes the different factors to determine the optimal time for disclosure.

4   Factors

The first step in improving understanding of the decision-making process is to focus on the factors involved and try to understand in greater detail what they mean and how they can be measured. The next step is then to examine how they affect the decision. The choice to retain a vulnerability gives a benefit to the government: it allows the collection of additional information for national security, intelligence, or law enforcement purposes; it also brings a cost: the increased risk of harm to its own networks, businesses, and individuals. The government aims to find the correct balance between these two, and each of the factors affects the outcome of this decision.

As discussed above, the VEP has two possible outcomes. First, a vulnerability may be disclosed; if this is the case, then the process ends with the disclosure. The other outcome is the decision to retain the vulnerability for use. If this is the case, then according to the process, the decision should be reviewed again at some point in the future and either disclosed or retained further. The VEP can be seen, then, as a timing problem: given the costs and benefits associated with disclosing or retaining a vulnerability, when is the best time time to disclose?

Each of the factors in the decision-making process can then be considered to have either an accelerating or a retarding effect on the time of disclosure. For example, if a factor reduces the risks or costs of non-disclosure, it will tend to delay disclosure; if it increases the risks, then it will move disclosure forward.

In this section, we will discuss each of the factors from the Daniel blog post, looking at what they mean and how they can be measured, and examining their impact on the costs and benefits to the government.

4.1   Extent of Use

How much is the vulnerable system used in the core internet infrastructure, in other critical infrastructure systems, in the U.S. economy, and/or in national security systems?

The meaning of this factor is straightforward, as is its measurement. Data about the number of units sold or deployed for a particular device or piece of software is not difficult to acquire or estimate. This factor is related to the risks and harm, below—where and how widely a device with a vulnerability is used will affect the potential risks and harms.

The extent of use may change over time. For example, end users might switch to newer devices or upgrade to newer versions of software that are not affected by the vulnerability.

Effect. This factor affects the decision to disclose in both directions, though not necessarily equally. First, a vulnerability in a widely-used device or piece of software can potentially cause harm to a larger group of individuals, businesses, or systems; this will have an accelerating effect on the time of disclosure. However, the opposite is also true: a vulnerability in a more widely-deployed system can potentially allow the government to access a greater number of systems, which would delay disclosure.

4.2   Risks and Harm

Does the vulnerability, if left unpatched, impose significant risk? How much harm could an adversary nation or criminal group do with knowledge of this vulnerability?

There are many potential ways in which exploitation of the vulnerability by others could cause harm. At a national level, there are potential harms from the compromise of government networks or the disruption of critical infrastructure. For businesses, harms can include direct monetary loss (from fraud, theft, sabotage, or ransomware) or loss of competitive ability (from industrial espionage), and also reputational harm caused by a breach. Harms to individuals include, for example, direct losses from crime, identity theft, and loss of privacy.

For each vulnerability, the risks of each of these harms will be different—it is unlikely, for example, that a vulnerability in an industrial control system will present much risk of identity theft to individuals, but the same vulnerability could present a large risk to infrastructure or businesses. The government must estimate how likely different harms are for each vulnerability, as well as the magnitude of those harms; this is related to the extent of use: where and how much devices with the vulnerability are used will affect the likelihood and impact.

Effect. That the government considers risks and harms instead of simply losses implies that they distinguish between the risk of discovery and use of the vulnerability by others and the 'lumpiness' of the harm. If the government has an aversion to substantial harm from single events, then its potential presence makes the decision to retain the vulnerability costlier, and will accelerate disclosure, even if the likelihood is low. If the risk of discovery and use is very high, even if the potential harm is modest in terms of impact on individuals or businesses, that will also accelerate the decision to disclose.

4.3 Detect Exploitation by Others

How likely is it that we would know if someone else was exploiting it?

This is hard to estimate without knowledge of the government's capabilities. A quote from [11] in the aftermath of the Shadow Brokers leak gives an indication that the NSA does have such an ability:

> After the discovery, the NSA tuned its sensors to detect use of any of the tools by other parties, especially foreign adversaries with strong cyber espionage operations, such as China and Russia.
> That could have helped identify rival powers' hacking targets, potentially leading them to be defended better. It might also have allowed U.S officials to see deeper into rival hacking operations while enabling the NSA itself to continue using the tools for its own operations.
> Because the sensors did not detect foreign spies or criminals using the tools on U.S. or allied targets, the NSA did not feel obligated to immediately warn the U.S. manufacturers, an official and one other person familiar with the matter said.

Effect. If the government has a high confidence in their ability to detect the exploitation of the vulnerability by others then this will have a delaying effect on disclosure time. From the quote above, this appears to be the case. If confidence in the ablity to detect is comparatively lower, then disclosure will happen sooner. Once use of the exploit has been detected, disclosure should follow immediately.

4.4 Is the Vulnerability Needed?

How badly do we need the intelligence we think we can get from exploiting the vulnerability? Are there other ways we can get it?

This factor is essentially the government's own estimation of the value of access to a device and the information it contains. If there are other vulnerabilities than can be exploited—or other methods entirely—with less cost or risk, then those other methods might be preferable.

Effect. The existence of other, less costly methods of obtaining the desired information will reduce the value of retaining this vulnerability and accelerate the timing of the disclosure. The availability of substitute methods depends on the nature of the information needed: concentrated info might be easier to acquire with other means, whilst broad-based information, spread over a number of sources, might not be possible to acquire without the exploitation of the vulnerability.

4.5 Discovery by Others

How likely is it that someone else will discover the vulnerability?

In a 2013 discussion about the government's approach to vulnerabilities [17], Hayden discussed the concept of 'Nobody But Us' (NOBUS) vulnerabilities, which the government believes others are unable to exploit:

> If there's a vulnerability here that weakens encryption but you still need four acres of Cray computers in the basement in order to work it you kind of think 'NOBUS' and that's a vulnerability we are not ethically or legally compelled to try to patch — it's one that ethically and legally we could try to exploit in order to keep Americans safe from others.

However, simultaneous discovery of a vulnerability may be relatively common. Schneier mentions several examples of simulatenous discovery [21]—including Heartbleed, which was discovered by both Google and Codenomicon. Studies of vulnerabilities in Microsoft software by Ozmnet [16] also suggest that simultaneous independent discovery is likely. More recently, a RAND report by Ablon and Bogart [1] followed a number of zero day exploits over time, and concluded that for a given collection of vulnerabilities, after one year 5.7% of them will have been discovered and disclosed by others. Another recent paper by Herr, Schneier, and Morris [10] studies a larger number of vulnerabilities and estimates that between 15% and 20% will be rediscovered within a year.

Different types of vulnerabilities probably experience different rates of independent discovery. If the government's ability to detect the use of known vulnerabilities by others is sufficient, they may be able to estimate how frequently simultaneous discovery occurs for different types of vulnerability.

Effect. If the vulnerability is likely to be discovered by others then it will accelerate disclosure. However, government confidence in a unique ability to discover or exploit some vulnerabilities will delay disclosure.

4.6   Can the Vulnerability Be Used?

Could we utilize the vulnerability for a short period of time before we disclose it?

This can be interpreted in different ways. First, it may simply not be possible to develop an exploit for a particular vulnerability—not every bug found in software can be successfully exploited. Or, alternatively, this may refer to the time it takes to develop and utilize an exploit for this vulnerability. If exploit development takes a long time, it is more likely that either the information needed will no longer be obtainable or no longer be of value, or that the vulnerability will be discovered and disclosed by another party. Another interpretation could be whether or not there is any benefit that can be gained by exploiting the vulnerability—perhaps the systems that could be accessed using the vulnerability have no intelligence or strategic value.

Effect. If the vulnerability cannot be utilized, then this will accelerate disclosure. If there is no benefit to be gained from retaining the vulnerability, then disclosing is the best option.

4.7   Can the Vulnerability Be Patched?

Can the vulnerability be patched or otherwise mitigated?

There are a few reasons why it may not be possible to patch a vulnerability: some types of devices or software (for example, industrial control systems, SCADA systems, PLCs, or embedded devices) are rarely—or never—updated, and older devices or software may no longer be supported by the vendor. However, many of these vulnerabilities can be mitigated, if known, through additional security measures. There are cases when a vulnerability cannot be patched or mitigated. For example, old Android phones stop receiving security updates, and little can be done to mitigate this—other than switching to a newer device. In this case. disclosure of a vulnerability will not help users of the older devices (unless it encourages them to upgrade), but can help increase the security of newer devices if they share the same code.

Effect. If a vulnerability can truly never be patched or mitigated in any way, then it can lead to a considerable delay in disclosure—because doing otherwise will reveal the vulnerability to potential exploitation when the system can not be defended. However, this is unlikely. The speed at which a patch can be created and deployed may also have an effect on the disclosure timing. If patch creation and deployment is fast, then systems can be made secure more quickly if someone else discovers the vulnerability, which will delay disclosure.

5   Modelling the Decision-Making Process

In considering whether to reveal the discovery of the vulnerability at any point in time, the government agency will consider the benefits and costs of the current situation—keeping the vulnerability undisclosed—and comparing them the the possible consequences after they have revealed the vulnerability to the public.

On one hand, retaining the vulnerability will allow the agency to access the information required for their purposes, and the longer the vulnerability persists—and the agency is able to exploit it undetected—the greater the potential accumulated benefit. On the other hand, if the vulnerability is not disclosed and remains unknown to the vendor and users, there is a chance that others will be able to exploit it, causing damage to the information assets the government is charged to protect. This constitutes the expected loss to the government. The model we present here should be seen as a formalization of a thinking process; there is no hard data to populate the model, but it shows how the factors would be considered when making a decision.

In a general form, assuming continuous time, the benefits and costs the government will receive from not disclosing the vulnerability until a particular time, T, can be expressed as

$$B_T = \int_0^T \texttt{Benefit}(t)dt \quad \text{and} \quad C_T = \int_0^T \texttt{Cost}(t)dt,$$

which represent the total benefits and costs received from now until time T.

The government's aim is to find the best time to disclose the vulnerability,

$$\max_T V_T = B_T - C_T,$$

where V_t is the value to the government of disclosing at time t. This is shown in Fig. 2, which shows the expected costs and benefits for disclosure at different times. The costs and benefits increase at different rates. The optimal time for disclosure maximizes the difference between costs and benefits. If the costs rise faster than the benefits, then the best action would be to disclose immediately.

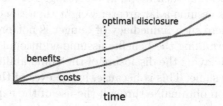

Fig. 2. Total costs and benefits over time. The optimal timing for disclosure maximizes the difference between benefits and costs.

We relate the different factors discussed in the previous section to these benefits and costs. Any benefit of the vulnerability depends on the ability to use it (Can the vulnerability be used? — F_{use}). The benefits that the government expects to receive at time t depend on the extent of use of software (F_{extent}). It is not necessarily known in advance if there will be any use for the exploit at a particular time t; this depends on whether or not there is information that is needed at that time, or if there is a system to which the government requires access at that time. A greater extent of use increases the probability of a need for the exploit. The perceived value gained from exploiting the vulnerability will increase the benefits (Is the vulnerability needed? — F_{value}).

The expected cost from not disclosing the vulnerability will be rising with the extent of use (F_{extent}) and the possible harm that can result from its exploitation by others (F_{harm}). The ability to detect its exploitation by others reduces the expected cost to the government (F_{detect}). Finally, the ability of others with high probability to exploit the vulnerability increases the cost of non-disclosure ($F_{discovery}$). The speed with which the patch can be developed and deployed (F_{patch}) reduces the expected cost of non-disclosure.

Immediate disclosure of the vulnerability upon discovery reduces the potential benefits to zero while minimizing the expected costs due to information assets damaged. However, this policy does not take into account the impact of the factors determining expected costs and benefits. Once such considerations are taken into account, the decision of when to disclose the vulnerability is equivalent to the solution of the problem to calculate the optimal timing for disclosure. In this context, the government is fully aware of both costs and benefits and their determinants and in effect decides when to exercise the 'option to disclose'. Intuitively, the decision will be such that at the time of the disclosure the marginal benefits from the retention vulnerability will be equal to the expected costs. Further delay in disclosing will result in the expected costs rising above the benefits. Although it is possible that such exact calculations cannot be made, the adoption of this equality as the organizing principle for the decision-making seems rational and more importantly, as it contains measurable quantities, it can be evaluated ex-post.

The analysis above is based on the assumption the the government is motivated 'equally/in a stable manner' by both benefits and costs. There are situations that call into question such a stable weighting. For example, in states of high alert, the benefits assume a far greater weight than the costs, compared to a normal situation where such immediacy of danger is not present. In this situation, the factors determining the benefits assume additional importance in the decision, resulting in delaying the disclosure of the vulnerability even though the expected costs are the same. This is because, in the eyes of the government, the value of the information obtainable through the use of the exploit is far higher.

5.1   Timing Rules

We look at two different timing rules using, for simplicity, a discrete-time model. The first considers no delay, so disclosure happens at time $t = 0$. The second considers some delay, with disclosure at time $t = T$. For both timing rules, we can consider the benefits as immediate benefits (received at $t = 0$) plus discounted expected future benefits, with the same done for costs: $B = B_0 + B^e$ and $C = C_0 + C^e$.

For the first case, immediate disclosure, the immediate benefits, B_0, are zero because the vulnerability is disclosed at time $t = 0$, so the government has no chance to gain from its exploitation. Additionally, there will be no future benefits, so $B^e = 0$. For the second case, where disclosure is delayed, the government sees an immediate benefit B_0. The value of B_0 is determined by two factors, the value of the information and the ability to use the exploit to gain it, and can be written $B_0 = f(F_{value}, F_{use})$. The expected future benefits, B^e, are

$$B^e = \sum_{t=1}^{T} d_b^t \, \mathbb{E}\left[B_t\right],$$

where d_b is the discount factor applied to future benefits and $\mathbb{E}\left[B_t\right]$ is the expected value of benefits at time t. These expected benefits depend on all of the factors and evolve according to the time-evolution of the underlying factors. For example, were the extent of use to expand in the future, the expected benefits would increase because the likelihood of being able to access needed information using the exploit increases. If, in the future, the information that can be collected by exploiting the vulnerability is not needed, the value of future benefits will decline. The total benefits for retaining the vulnerability until a time T is

$$B = B_0 + \sum_{t=1}^{T} d_b^t \, \mathbb{E}\left[B_t\right].$$

Next, we look at the costs of non-disclosure. Similarly to the benefits, these can be decomposed into two parts: the initial cost and the costs incurred during the time period before disclosure. For both immediate and delayed disclosure, the initial costs C_0 are zero, and for immediate disclosure, so are the expected future costs, C^e. In the case of delayed disclosure, the expected future cost C^e acquires a positive value and can be written as

$$C^e = \sum_{t=1}^{T} d_c^t \, \mathbb{E}\left[C_t\right]$$

where d_c is the discount factor applied to future costs, and $\mathbb{E}\left[C_t\right]$ is the expected value of costs at time t. The value of these expected costs will be influenced by the evolution over time of the factors mentioned above. These factors will affect both the probability of incurring the costs, which might be increasing with time, and the value of the losses which also might be functions of time.

Finally, the total costs until a time T can be written as

$$C = C_0 + \sum_{t=1}^{T} d_c^t \, \mathbb{E}\left[C_t\right].$$

5.2   Optimal Timing

The problem of the timing of disclosure can be reduced to the solution of

$$V_{T^*} = \max_{T}\left(B - C\right),$$

where V_{T^*} is the net benefit to the government when the vulnerability is disclosed at the optimal time, T^*. Substituting, we get

$$V_{T^*} = \max_T \left(B_0 + \sum_{t=1}^{T} d_b^t \, \mathbb{E}\,[B_t] - \left(C_0 + \sum_{t=1}^{T} d_c^t \, \mathbb{E}\,[C_t] \right) \right).$$

Each element B_0, B_t, C_0, C_t, of the equation is a function of the different factors and, in the case of B_t and C_t, also of time. We consider B_0 to be influenced primarily by the value of information needed, F_{value}, and the ability to use the vulnerability, F_{use}. At time $t = 0$, it is known which information is required and available through use of the exploit, and as such, its value can be determined. However, if it is not possible to use the exploit (F_{use}), then the value of B_0 is likely to be very low. Future expected benefits, B_t, are influenced by the same factors, but are also influenced by the extent of use F_{extent}. At some time in the future, it may be that there is information needed that is available through the use of the exploit. If the extent of use is larger, then it is more likely that such benefits will be available; if the extent is lower, it is less likely. For the initial costs, C_0, the value is always 0; none of the factors influence this. This is because costs accrue over time, and at $t = 0$, no time has passed. Expected future costs, C_t, are influenced by a host of factors. The extent of use, F_{extent}, will influence positively costs as a greater number of information assets are exposed to the possible exploit. These costs are increasing over time. As the risk and harm, F_{harm}, increases, so will the value of expected future costs. If the risk of discovery of the exploit by others increases $F_{discovery}$, this also increases the expected future costs, while the ability of the government to detect (F_{detect}) the use of the exploits by others will reduce such costs.

Table 1 shows how the different factors affect the benefits and costs, compared to immediate disclosure, and their influence on the timing of disclosure. This gives a general picture of how the factors affect timing. With a richer model of how the costs and benefits arise for each factor, it would be possible to have a deeper analysis of the timing problem.

Table 1. Influence of factors on the costs and benefits, compared to immediate disclosure, and how they affect the timing of disclosure. While F_{extent} influences both benefits and costs, it will likely have a greater influence on costs, moving disclosure forward.

Factor	F_{extent}	F_{harm}	F_{detect}	F_{value}	$F_{discovery}$	F_{use}	F_{patch}
Benefits	+			+		+	
Costs	+	+	−		+		−
Timing	−?	−	+	+	−	+	+

5.3 Making Decisions

The model above shows how each of the factors can affect the timing of disclosure. This is useful for understanding, in a general sense, how the decision of when to disclose depends on the interactions of the different factors, but would not be useful for actually making such a decision. To make decisions with this type of model, it must first be parametrized: estimates for how the expected values of each factor change over time are needed.

Only the government knows how it estimates and weights the different factors, and making accurate estimations is probably extremely difficult. However, given that the decision-making process is supposed to be 'biased toward responsibly disclosing vulnerabilities', any estimations *should* err on the side of caution by overestimating costs and risks, and underestimating the values of benefits. The same should be done for discount factors: by reducing d_b, the discount factor for benefits, compared to d_c, the discount factor for costs, future costs will outweigh potential future benefits, and move the timing decision forward.

Even if it is impossible to determine the exact time for optimal disclosure, having an estimate can still be useful. If retained vulnerabilities are periodically reviewed, the estimated optimal time of disclosure could be used to set an upper bound on the time before the next review. With conservative estimates for the factors, this would help ensure that retained vulnerabilities can be reconsidered (with updated information) in good time.

6 EternalBlue and WannaCry

Recent events have shown that the decisions the government makes about whether to disclose or retain vulnerabilities can have significant repercussions. The WannaCry malware, which severely affected businesses and hospitals around the world is an excellent example. The malware used a vulnerability from a NSA-developed exploit known as EternalBlue, which was leaked to the public by the ShadowBrokers on April 14, 2017.

The vulnerability used in the EternalBlue exploit would only have been considered under the VEP if it was discovered after the introduction of the VEP in 2010. According to the Washington Post [13], EternalBlue was used for 'more than 5 years', implying that it would have been considered under the VEP—for the following discussion, we will assume that this is the case.

In discussions about the VEP (for example, in [23]), there is a tendency to think of the VEP decision as a binary choice: either disclose or retain. We have argued that this should be viewed instead as a timing decision: not *if* a vulnerability should be disclosed, but *when*. When the EternalBlue exploit was leaked to the public in April, Microsoft had already created and released patches for the 0-day vulnerabilities in March—presumably after being informed by the NSA it they became aware of the ShadowBrokers leak. The initial decision here was to retain and use the vulnerability in EternalBlue, but to disclose it when it became clear it had leaked and could cause losses; it was a matter of timing.

While we do not know if the government makes decisions using the approach we described above, it can still be a useful tool for analysing the government's disclosure decisions. For the WannaCry/EternalBlue example, there different possible interpretations. The first case is that the decision was made using a correct model. This implies that the vulnerability was disclosed at the appropriate time, and that the benefits gained from the long-term retention of the vulnerability were valuable enough that they where not outweighed by the damages and costs that arose from the leaked vulnerability and resulting malware. The second case is that the timing of the disclosure was wrong because the model was missing a factor: the possibility of a vulnerability being leaked. From the Daniel blog post, we know that the risk of independent discovery is considered when making a decision, but it is unknown if this also includes the risk of leaks. If not, then the time of disclosure would have been after the optimal point. The final case is where the timing of the disclosure was wrong because the model's parameters were incorrect. First are the extent of use and patching factors: even though the patch was released by Microsoft before the WannaCry malware, many computer systems were still vulnerable, either because the patch had not yet been applied or because they were running older versions of Windows that were out of support and so did not receive the patch. If the rate at which patches can be developed and applied is overestimated, or the number of systems running software that is no longer supported is underestimated in the model, then the potential costs will be underestimated resulting in a non-optimal, later time of disclosure. Incorrectly underestimating the probability of a leak (possibly included in the discovery factor) would also cause such a delay in disclosure.

Without knowing how much value the government gained from use of the exploit, a detailed understanding of the factors used when making a decision and how they are calculated and weighted, it is impossible to know which, if any, of these cases is true. However, WannaCry caused a lot of damage and could have caused a lot more, had it not been stopped. It is unlikely that this was anticipated and accepted, and therefore unlikely that the first case is true.

The remaining two cases suggest some possible improvements to the decision-making process. First, if the risk of vulnerabilities leaking is not included, it needs to be added. Second, a better understanding of how systems are patched over time may be needed when deciding when to disclose. Many older machines running out of date software are still used in critical processes; the costs of attacks on these machines must be considered. It may also be beneficial to disclose before these machines become out-of-support or to reduce potential costs by sponsoring the creation of patches for out-of-support software still widely in use when the vulnerability is finally disclosed.

7   Conclusions

Government disclosure of vulnerabilities is important, but so is the ability of the government to conduct intelligence, offensive national security, and law enforcement tasks. It would be a mistake to immediately disclose every vulnerability discovered, but it would also be a mistake to disclose none. Recommendations for and proposed legislation about the VEP include periodic reviews of any retained vulnerabilities, allowing them to be used for a time before disclosure.

We have presented a model that shows how the different factors used in the decision can be combined to determine the optimal time to disclose. Understanding how the different factors affect the timing allows the decisions about vulnerabilities made by the government to be better interpreted. We have looked at the case of the WannaCry malware, which used a leaked NSA zero day vulnerability. The vulnerability was disclosed to Microsoft before the malware was created, but before that remained undisclosed for 5 or more years.

It is likely that the disclosure came after the optimal time, as many systems remained unpatched and were vulnerable to WannaCry. The government could have underestimated or ignored the risk of the vulnerability leaking, or overestimated the speed with which systems could be patched. In any case, future decisions should include or improve the estimation of these factors in order to better determine the optimal time of disclosure.

References

1. Ablon, L., Bogart, T.: Zero Days, Thousands of Nights: The Life and Times of Zero-Day Vulnerabilities and Their Exploits. RAND Corporation publication, Santa Monica (2017)
2. Beres, Y., Griffin, J., Shiu, S.: Security analytics: Analysis of security policies for vulnerability management. Technical report HPL-2008-121, HP Labs (2008)
3. Budington, B., Crocker, A.: NSA's failure to report shadow broker vulnerabilities underscores need for oversight, September 2016. https://www.eff.org/deeplinks/2016/09/nsas-failure-report-shadow-broker-vulnerabilities-underscores-need-oversight
4. Commercial and government information technology and industrial control product or system vulnerabilities equities policy and process. https://www.eff.org/files/2015/09/04/document_71_-_vep_ocr.pdf
5. Daniel, M.: Heartbleed: understanding when we disclose cyber vulnerabilities, April 2014. https://obamawhitehouse.archives.gov/blog/2014/04/28/heartbleed-understanding-when-we-disclose-cyber-vulnerabilities
6. Dixon-Thayer, D.: Improving government disclosure of security vulnerabilities, September 2016. https://blog.mozilla.org/netpolicy/2016/09/19/improving-government-disclosure-of-security-vulnerabilities/
7. Fidler, M., Herr, T.: PATCH: debating codication of the VEP, May 2017. https://lawfareblog.com/patch-debating-codification-vep
8. Greenberg, A.: Shopping for zero-days: a price list for hackers' secret software exploits, March 2012. https://www.forbes.com/sites/andygreenberg/2012/03/23/shopping-for-zero-days-an-price-list-for-hackers-secret-software-exploits/
9. Healey, J.: The U.S. Government and Zero-Day Vulnerabilities: From Pre-Heartbleed to Shadow Brokers. J. Int. Aff. (2016). https://jia.sipa.columbia.edu/online-articles/healey_vulnerability_equities_process
10. Herr, T., Schneier, B., Morris, C., Stock, T.: Estimating vulnerability rediscovery, March 2017. https://ssrn.com/abstract=2928758

11. Menn, J., Walcott, J.: Exclusive: Probe of leaked U.S. NSA hacking tools examines operative's 'mistake', September 2016. http://www.reuters.com/article/us-cyber-nsa-tools-idUSKCN11S2MF
12. Miller, C.: The legitimate vulnerability market: Inside the secretive world of 0-day exploit sales. In: Sixth Workshop on the Economics of Information Security (2007)
13. Nakashima, E., Timberg, C.: NSA officials worried about the day its potent hacking tool would get loose. Then it did, May 2017. https://www.washingtonpost.com/business/technology/nsa-officials-worried-about-the-day-its-potent-hacking-tool-would-get-loose-then-it-did/2017/05/16/50670b16-3978-11e7-a058-ddbb23c75d82_story.html
14. National Security Policy Directive 54. https://fas.org/irp/offdocs/nspd/nspd-54.pdf
15. ODNI Public Affairs Office. Statement on bloomberg news story that NSA knew about the "Heartbleed bug" aw and regularly used it to gather critical intelligence, April 2014. https://icontherecord.tumblr.com/post/82416436703/statement-on-bloomberg-news-story-that-nsa-knew
16. Ozment, A.: The likelihood of vulnerability rediscovery and the social utility of vulnerability hunting. In: Workshop on Economics and Information Security (2005)
17. Peterson, A.: Why everyone is left less secure when the NSA doesn't help fix security flaws, October 2013. https://www.washingtonpost.com/news/the-switch/wp/2013/10/04/why-everyone-is-left-less-secure-when-the-nsa-doesnt-help-fix-security-flaws/
18. Riley, M.: NSA said to have used heartbleed bug, exposing consumers, April 2014. https://www.bloomberg.com/news/articles/2014-04-11/nsa-said-to-have-used-heartbleed-bug-exposing-consumers
19. Sanger, D.E.: Obama lets N.S.A. exploit some internet flaws, officials say, April 2014. https://www.nytimes.com/2014/04/13/us/politics/obama-lets-nsa-exploit-some-internet-flaws-officials-say.html?_r=1
20. Schneier, B.: Managed security monitoring: Closing the window of exposure (2000). http://www.keystoneisit.com/window.pdf
21. Schneier, B.: Simultaneous discovery of vulnerabilities, February 2016. https://www.schneier.com/blog/archives/2016/02/simultaneous_di.html
22. Schneier, B.: The Vulnerabilities market and the future of security, June 2012. https://www.schneier.com/blog/archives/2012/06/the_vulnerabili.html
23. Schneier, B.: WannaCry and Vulnerabilities. June 2017. https://www.schneier.com/blog/archives/2017/06/wannacry_and_vu.html
24. Schwartz, A., Knake, R.: Government's Role in Vulnerability Dis- closure, June 2016. http://www.belfercenter.org/publication/governments-role-vulnerability-disclosure-creating-permanent-and-accountable
25. Zerodium: How to sell your 0day exploit to ZERODIUM, March 2017. https://zerodium.com/program.html

A Stackelberg Game Model for Botnet Data Exfiltration

Thanh Nguyen[(✉)], Michael P. Wellman, and Satinder Singh

University of Michigan, Ann Arbor, USA
{thanhhng,wellman,baveja}@umich.edu

Abstract. Cyber-criminals can distribute malware to control computers on a networked system and leverage these compromised computers to perform their malicious activities inside the network. Botnet-detection mechanisms, based on a detailed analysis of network traffic characteristics, provide a basis for defense against botnet attacks. We formulate the botnet defense problem as a zero-sum Stackelberg security game, allocating detection resources to deter botnet attacks taking into account the strategic response of cyber-criminals. We model two different botnet data-exfiltration scenarios, representing exfiltration on single or multiple paths. Based on the game model, we propose algorithms to compute an optimal detection resource allocation strategy with respect to these formulations. Our algorithms employ the double-oracle method to deal with the exponential action spaces for attacker and defender. Furthermore, we provide greedy heuristics to approximately compute an equilibrium of these botnet defense games. Finally, we conduct experiments based on both synthetic and real-world network topologies to demonstrate advantages of our game-theoretic solution compared to previously proposed defense policies.

1 Introduction

Cyber-criminals intent on denial-of-service, spam dissemination, data theft, or other information security breaches often pursue their attacks with *botnets*: collections of compromised computers (bots) subject to their control [14,23,30,31,33]. In 2014 testimony, the US Federal Bureau of Investigation cited over $9 billion of US losses and $110 billion losses globally due to botnet activities [7]. The estimated 500 million computers infected globally each year by botnet activities amounts to 18 victims per second.

The threat of botnets has drawn significant attention from network security researchers [1,5,6,10–13,32]. Much existing work focuses on detection mechanisms to identify compromised computers based on network traffic characteristics. For example, BotSniffer [13] searches for spatial-temporal patterns in network traffic characteristic of coordinated botnet behavior. Given some underlying detection capability, the defender faces the problem of how to effectively deploy its detection resources against potential botnet attacks. For example, Venkatesan et al. consider the problem of allocating a limited number of localized detection

S. Rass et al. (Eds.): GameSec 2017, LNCS 10575, pp. 151–170, 2017.
DOI: 10.1007/978-3-319-68711-7_9

resources on a network in order to maximally disrupt *data exfiltration* attacks, where the botnet aims to transfer stolen information out of the network [38]. Their first solution allocated resources statically, which could effectively disrupt one-time attacks but is vulnerable to adaptive attackers. They extended this method to randomize detector placement dynamically to improve robustness against adaptation [37]. In a related work, Mc Carthy et al. address the additional challenge of imperfect botnet detection [20].

Our work extends these prior efforts by formulating the botnet defense problem as a Stackelberg security game, thus accounting for the strategic response of attackers to deployed defenses. In our botnet defense game, the defender attempts to protect data within a computer network by allocating detection resources (*detectors*). The attacker compromises computers in the network to steal data, and attempts to exfiltrate the stolen data by transferring it outside the defender's network. We consider two formulations of data exfiltration: (i) *uni-exfiltration*, where the source bot routes the stolen data along a single path designated by the attacker; and (ii) *broad-exfiltration*, where each bot propagates the received stolen data to all other bots in the network.

We propose algorithms to compute defense strategies for these data exfiltration formulations: **ORANI** (**O**ptimal **R**esource **A**llocation for u**N**i-exfiltration **I**nterception) and **ORABI** (**O**ptimal **R**esource **A**llocation for **B**road-exfiltration **I**nterception). Both **ORANI** and **ORABI** employ the double-oracle method [21] to control exploration of the exponential strategy spaces available to attacker and defender. Our main algorithmic contributions lie in defining mixed-integer linear programs (MILPs) for the defender and attacker's best-response oracles. In addition, we introduce greedy heuristics to approximately implement these oracles. Finally, we conduct experiments based on both synthetic and real-world network topologies to evaluate solution quality as well as runtime performance of our game-theoretic algorithms, demonstrating significant improvements over previous defense strategies.

2   Related Work

Prior studies of botnet security tend to focus on designing botnet detection mechanisms [1,5,6,10–13,32] or advanced botnet designs against these detection mechanisms [29,39]. Some studies provide empirical and statistical analysis on related cyber-security implications such as the role of Internet service providers in botnet mitigation [35] or contagion in cyber attacks [2].

Recent work has introduced game-theoretic models and corresponding defense solutions for various botnet detection and prevention problems [4,17, 27,28]. In these models, cyber criminals intrude by compromising computers in a network. Users or owners of computers in the network defend by patching or replacing their computers based on alerts of potential security threats.

Stackelberg security games have been successfully applied to many real-world physical security problems [3,9,19,26,34]. Jain et al. address a problem in urban network security partially analogous to uni-exfiltration, as the attacker follows a single path to attack its best target in an urban road network [15]. Vaněk et al. tackle a problem of malicious packet prevention, where the attacker determines which entry point to access a network to attack a specific target assuming the corresponding traversing path is fixed [36]. In our botnet defense problem, cyber-criminals decide not only which computers to compromise but also create an overlay network over these bots to exfiltrate data from multiple targets in the network. The additional complexity of considering the exfiltration plan leads to a distinct and difficult security problem.

3   Game Model: Uni-exfiltration

Our game model for uni-exfiltration is built on the botnet model introduced by Venkatesan et al. [38]. Let $G = (V, E)$ represent a computer network where the set of nodes V comprises network elements such as routers and end hosts, and edges in E connect these nodes. We denote by V^c a set of *mission-critical nodes* in the network which contain sensitive data. Data exchange is governed by a routing algorithm fixed by the network system. For each pair of nodes (u, v), we denote by $P(u, v)$ the *routing path* between u and v. In our experiments, we assume that routing is via the shortest path.

We model the botnet defense problem as a *Stackelberg security game* (SSG) [16]. In such a game, the defender commits to a mixed (randomized) strategy to allocate limited security resources to protect important targets. The attacker then optimally chooses targets with respect to the distribution of defender allocations. In our context, the defender is the security controller of a computer network, with limited detection resources. The defender attempts to deploy its detectors in the most effective way to impede the attack chosen in response.

The attacker in the botnet exfiltration game is a cyber-criminal who attempts to steal sensitive network data. Compromising a mission-critical node $c \in V^c$ enables the attacker to steal data owned by c. Compromising other nodes in the network helps the attacker to relay the stolen data to a server S^a outside the network, which he controls. The attacker specifies a sequence of compromised nodes (bots) to relay stolen data. Routing between consecutive bots in the sequence follows fixed paths out of the attacker's control. We call this chain of ordered bots and nodes on routing paths between consecutive bots an *exfiltration path*, denoted by $\pi(c, S^a)$.

Definition 1 (Exfiltration Prevention). *Given a network $G = (V, E)$ and a set of mission-critical nodes V^c, data exfiltration from $c \in V^c$ is prevented by the defender iff there is a detector on the exfiltration path $\pi(c, S^a)$.*

Though the attacker's remote server S^a is located outside the network, we assume the defender is aware of which nodes in the network can relay data to S^a.

In our Stackelberg game model, the defender moves first by allocating detection resources, and the attacker responds with a plan for compromise and exfiltration to evade detection. The defender placement of detectors is randomized, so any attack plan succeeds with some probability.

Definition 2 (Strategy Space). *The strategy spaces of the players are as follows:*

Defender: *The defender has $K^d < |\mathbf{V}|$ detection resources available for deployment on network nodes. We denote by $\mathbf{D} = \{\mathbf{D}_i \mid \mathbf{D}_i \subseteq \mathbf{V}, |\mathbf{D}_i| \leq K^d\}$ the set of all pure defense strategies of the defender. Let $\mathbf{x} = \{x_i\}$ be a mixed strategy of the defender where $x_i \in [0,1]$ is the probability that the defender plays $\mathbf{D}_i$, and $\sum_i x_i = 1$.*

Attacker: *The attacker can compromise up to $K^a < |\mathbf{V}|$ nodes. We denote by $\mathbf{A} = \{\mathbf{A}_j = (\mathbf{B}_j, \mathbf{\Pi}_j) \mid \mathbf{B}_j \subseteq \mathbf{V}, |\mathbf{B}_j| \leq K^a, \mathbf{\Pi}_j = \{\pi_j(c, S^a) \mid c \in \mathbf{B}_j \cap \mathbf{V}^c\}\}$ the set of all pure strategies of the attacker. Each pure strategy $\mathbf{A}_j$ consists of: (i) $\mathbf{B}_j$: a set of compromised nodes; and (ii) $\mathbf{\Pi}_j$: a set of exfiltration paths over $\mathbf{B}_j$.*

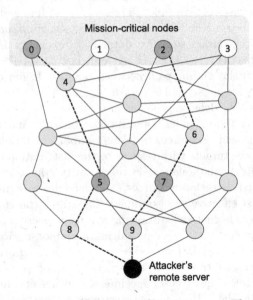

Fig. 1. An example scenario of the botnet exfiltration game. There are four mission-critical nodes, $\mathbf{V}^c = \{0, 1, 2, 3\}$. If $K^a = 4$, then a possible pure strategy of the attacker $\mathbf{A}_j$ can be: (i) a set of compromised nodes $\mathbf{B}_j = \{0, 2, 5, 7\}$; and (ii) a set of exfiltration paths $\mathbf{\Pi}_j = \{\pi_j(0), \pi_j(2)\}$ to exfiltrate data from stealing bots 0 and 2 to the attacker's server S^a. These exfiltration paths $\pi_j(0) = \mathbf{P}(0, 5) \cup \mathbf{P}(5, S^a)$ and $\pi_j(2) = \mathbf{P}(2, 7) \cup \mathbf{P}(7, S^a)$ relay stolen data via relaying bots 5 and 7 respectively, where $\mathbf{P}(0, 5) = (0 \rightarrow 4 \rightarrow 5)$, $\mathbf{P}(5, S^a) = (5 \rightarrow 8 \rightarrow S^a)$, $\mathbf{P}(2, 7) = (2 \rightarrow 6 \rightarrow 7)$ and $\mathbf{P}(7, S^a) = (7 \rightarrow 9 \rightarrow S^a)$ are routing paths fixed by the network system. Suppose $K^d = 1$. If the defender allocates its one detector to node 9, the attacker fails at exfiltrating data from node 2 since $9 \in \pi_j(2)$ but succeeds from node 0 since $9 \notin \pi_j(0)$.

A simple scenario of the botnet defense game is shown in Fig. 1. The model specification is completed by defining the payoff structure, which we take to be zero-sum.

Definition 3 (Game Payoff). *Each mission-critical node $c \in \mathbf{V}^c$ is associated with a value, $r(c) > 0$, representing the importance of data stored at that node. Successfully exfiltrating data from c yields the attacker a payoff $r(c)$, and the defender receives a payoff $-r(c)$. For prevented exfiltrations, both players receive zero.*

Note that the maximum achievable payoff for a defender is zero, obtained by preventing all exfiltration attempts. In general terms, let $U^a(\mathbf{D}_i, \mathbf{A}_j)$ denote the payoff to the attacker if the defender plays $\mathbf{D}_i$ and the attacker plays $\mathbf{A}_j$. Since the game is zero-sum, the defender payoff $U^d(\mathbf{D}_i, \mathbf{A}_j) \equiv -U^a(\mathbf{D}_i, \mathbf{A}_j)$. The payoff can be decomposed across mission-critical nodes,

$$U^a(\mathbf{D}_i, \mathbf{A}_j) \equiv \sum_{c \in \mathbf{V}^c} r(c)h(c), \tag{1}$$

where $h(c)$ indicates whether the attacker successfully exfiltrates the data of the mission-critical node $c \in \mathbf{V}^c$. This is determined as follows:

$$h(c) = \begin{cases} 1 & \text{if } c \in \mathbf{B}_j \text{ and } \mathbf{D}_i \cap \pi_j(c, S^a) = \emptyset \\ 0 & \text{otherwise.} \end{cases} \tag{2}$$

The expected utility for the attacker when the defender plays mixed-strategy $\mathbf{x}$ is

$$U^a(\mathbf{x}, \mathbf{A}_j) = \sum_i x_i U^a(\mathbf{D}_i, \mathbf{A}_j),$$

which is negated to obtain the expected defender payoff $U^d(\mathbf{x}, \mathbf{A}_j)$. A defender mixed strategy that maximizes $U^d(\mathbf{x}, \mathbf{A}_j)$ given the attacker plays a best response and breaks ties in favor of the defender constitutes a *Strong Stackelberg Equilibrium* (SSE) of the game.

4   ORANI: An Algorithm for Uni-exfiltration Games

In zero-sum games, the first mover's SSE strategy is also a maximin strategy [18]. Therefore, finding an optimal mixed defense strategy can be formulated as follows:

$$\max_{\mathbf{x}} U^d_* \tag{3}$$

$$\text{s.t. } U^d_* \leq U^d(\mathbf{x}, \mathbf{A}_j), \ \forall j \tag{4}$$

$$\sum_i x_i = 1, \ x_i \geq 0, \ \forall i, \tag{5}$$

where U^d_* is the defender's utility for playing mixed strategy $\mathbf{x}$ when the attacker best-responds. Constraint (4) ensures the attacker chooses an optimal action against $\mathbf{x}$, that is, $U^d_* = \min_j U^d(\mathbf{x}, \mathbf{A}_j) = \max_j U^a(\mathbf{x}, \mathbf{A}_j)$. Solving (3)–(5) is computationally expensive due to the exponential number of pure strategies of the defender and the attacker. To overcome this computational challenge,

---

Algorithm 1. ORANI Algorithm Overview

---
1 Initialize the sets of pure strategies: $\mathbf{A} = \{\mathbf{A}_j\}$ and $\mathbf{D} = \{\mathbf{D}_i\}$ for some j and i;
2 **repeat**
3      $(\mathbf{x}^*, \mathbf{a}^*) = \text{MaximinCore}(\mathbf{D}, \mathbf{A})$;
4      $\mathbf{D}_o = \text{DefenderOracle}(\mathbf{a}^*)$;
5      $\mathbf{A}_o = \text{AttackerOracle}(\mathbf{x}^*)$;
6      $\mathbf{A} = \mathbf{A} \cup \{\mathbf{A}_o\}, \mathbf{D} = \mathbf{D} \cup \{\mathbf{D}_o\}$
7 **until** *converge*;

ORANI applies the *double-oracle* method [15, 21]. Algorithm 1 presents a sketch of **ORANI**.

ORANI starts by solving a maximin sub-game of (3)–(5) by considering only small seed subsets $\mathbf{D}$ and $\mathbf{A}$ of pure strategies for the defender and attacker (Line 3). Solving this sub-game yields a solution $(\mathbf{x}^*, \mathbf{a}^*)$ with respect to the strategy subsets. **ORANI** iteratively adds new best pure strategies $\mathbf{D}_o$ and $\mathbf{A}_o$ to the current strategy sets $\mathbf{D}$ and $\mathbf{A}$ (Lines 4–6). These strategies $\mathbf{D}_o$ and $\mathbf{A}_o$ are chosen by the oracles to maximize the defender and attacker utility, respectively, against the current (in iteration) counterpart solution strategies $\mathbf{a}^*$ and $\mathbf{x}^*$. This iterative process continues until the solution *converges*: when no new pure strategy can be added to improve the defender and the attacker's utilities. At convergence, the latest solution $(\mathbf{x}^*, \mathbf{a}^*)$ an equilibrium of the game [21]. Following this general methodology, the specific contribution of **ORANI** is in defining MILPs representing the attacker and the defender oracle problems in botnet exfiltration games.

4.1   ORANI Attacker Oracle

The attacker oracle returns a pure strategy for the attacker maximizing utility against a given defender mixed strategy $\mathbf{x}^*$. Below, we present a MILP exactly representing the attacker oracle and show that the problem is NP-hard. We then provide a greedy heuristic to approximately solve the attacker oracle problem.

MILP Representation. We parameterize each pure strategy of the attacker as follows:

1. *bot* variables $\mathbf{z} = \{z_w \mid w \in \mathbf{V}\}$, indicate whether the attacker compromises node w ($z_w = 1$) or not ($z_w = 0$), and
2. *bot-chain* variables $\mathbf{q} = \{q_c(u, v) \mid c \in \mathbf{V}^c, u \in \mathbf{V}, v \in \mathbf{V} \cup \{S^a\} \setminus \{c, u\}\}$, represent exfiltration paths.

For each stealing bot c, $\{q_c(u, v)\}$ represents the bot chain to exfiltrate data from c to S^a. Note that the bot-chain variables employ compromised nodes only. This means that $q_c(u, v) = 0$ for all (c, u, v) such that $z_c = 0$ or $z_u = 0$ or $z_v = 0$.

Conversely, when $z_c = z_u = z_v = 1$, $q_c(u,v) = 1$ iff (u,v) are consecutive bots in the bot chain for c. This entails that the exfiltration path $\pi(c, S^a)$ includes the routing path $\mathbf{P}(u,v)$.

Given the attacker's pure strategy $(\mathbf{z}, \mathbf{q})$, we introduce *data-exfiltration* variables $\mathbf{h} = \{h_i(c)\}$ to describe the outcome of the attack. For stealing bot $c \in \mathbf{V^c}$ with $z_c = 1$, $h_i(c)$ indicates whether the attacker successfully exfiltrates from c when the defender plays $\mathbf{D}_i \in \mathbf{D}$. Specifically, $h_i(c) = 0$ if $\mathbf{D}_i$ includes a detector on the exfiltration path from node c to S^a. Otherwise, $h_i(c) = 1$. The attacker utility can be computed based on $\mathbf{h} = \{h_i(c)\}$,

$$U^a(\mathbf{x}^*, (\mathbf{z}, \mathbf{q})) = \sum_{\mathbf{D}_i \in \mathbf{D}} x_i \sum_{c \in \mathbf{V^c}} r(c) h_i(c).$$

The optimization problem for the attacker can now be formulated as a MILP (6)–(15). Variables $\mathbf{z}$ and $\mathbf{h}$ are constrained to be binary. Constraints (7)–(9) enforce that there is only a single exfiltration path from each mission-critical node $c \in \mathbf{V^c}$ to S^a if node c is compromised ($z_c = 1$). In particular, when $z_c = 1$, constraint (7) indicates that there is a single out-exfiltration path from node c and constraint (8) imposes that there is only a single in-exfiltration path to the attacker's server S^a. Otherwise, when c is not compromised ($z_c = 0$), there is no exfiltration path from c. Constraint (9) ensures, for each $c \in \mathbf{V^c}$, that the total number of in-exfiltration paths to any node v equals the total number of out-exfiltration paths from that node. Constraints (10) and (11) guarantee that exfiltration paths are determined using compromised nodes only (i.e., if either $z_u = 0$ or $z_v = 0$, then $q_c(u,v) = 0$). Constraint (12) ensures that the number of compromised nodes does not exceed the attacker's resource limit, K^a. Finally, constraint (13) enforces $h_i(c) = 0$ when $\mathbf{P}(u,v) \cap \mathbf{D}_i \neq \emptyset$ for some pair of consecutive bots (u,v) on the exfiltration path from c (i.e., such that $q_c(u,v) = 1$). Constraint (14) ensures $h_i(c) = 0$ when c is not compromised.

$$\max_{\mathbf{z}, \mathbf{q}, \mathbf{h}} U^a(\mathbf{x}^*, (\mathbf{z}, \mathbf{q})) \tag{6}$$

$$\text{s.t.} \sum_{u \in \mathbf{V} \cup \{S^a\} \setminus \{c\}} q_c(c, u) = z_c, \forall c \in \mathbf{V^c} \tag{7}$$

$$\sum_{u \in \mathbf{V}} q_c(u, S^a) = z_c, \forall c \in \mathbf{V^c} \tag{8}$$

$$\sum_{u \in \mathbf{V} \setminus \{v\}} q_c(u, v) = \sum_{w \in \mathbf{V} \cup \{S^a\} \setminus \{v, c\}} q_c(v, w), \forall c \in \mathbf{V^c}, v \in \mathbf{V} \setminus \{c\} \tag{9}$$

$$q_c(u, v) \leq z_u, \forall c \in \mathbf{V^c}, u \in \mathbf{V}, v \in \mathbf{V} \cup \{S^a\} \setminus \{c, u\} \tag{10}$$

$$q_c(u, v) \leq z_v, \forall c \in \mathbf{V^c}, u \in \mathbf{V}, v \in \mathbf{V} \setminus \{c, u\} \tag{11}$$

$$\sum_{w \in \mathbf{V}} z_w \leq K^a, z_w \in \{0, 1\}, \forall w \in \mathbf{V} \tag{12}$$

$$h_i(c) \leq 1 - q_c(u, v), \forall c \in \mathbf{V^c}, u \in \mathbf{V}, v \in \mathbf{V} \cup \{S^a\} \setminus \{u, c\}, \text{ and} \tag{13}$$

$\forall \mathbf{D}_i \in \mathbf{D}$ such that $\mathbf{P}(u,v) \cap \mathbf{D}_i \neq \emptyset$

$$h_i(c) \leq z_c, \forall c \in \mathbf{V^c}, \mathbf{D}_i \in \mathbf{D} \tag{14}$$

$$q_c(u,v) \in [0,1], h_i(c) \in \{0,1\}, \forall c, u, v, i \tag{15}$$

Theorem 1. *A solution to MILP (6)–(15) is an optimal pure strategy for the attacker against defender mixed strategy* $\mathbf{x}^*$.

Proof. Given a solution of (6)–(15), consider each mission-critical node $c \in \mathbf{V^c}$ such that $h_i(c) = 1$ for some i. This means that the attacker successfully exfiltrates data from c given defender pure strategy $\mathbf{D}_i$. There must exist a positive exfiltration path, $\pi^+(c)$, from c to S^a. That is $q_c(u,v) > 0$ for all consecutive bots (u,v) on $\pi^+(c)$. This conclusion results from the attacker strategy constraints in (7)–(9). Then an optimal pure strategy for the attacker consists of: (i) the set of compromised nodes u with $z_u = 1$; and (ii) the set of positive exfiltration paths $\{\pi^+(c)\}$ for any c which satisfies $h_i(c) = 1$ with some i.

Solving this MILP may take exponential time. In fact, the problem is NP-hard.

Proposition 1. *The attacker oracle problem for data uni-exfiltration is NP-hard.*

The proof is presented in Online Appendix B.[1] We introduce a greedy heuristic to approximately solve the problem.

Attacker Greedy Heuristic. Our greedy heuristic iteratively adds nodes to compromise until the resource limit K^a is reached. At each iteration, given the current set of compromised nodes $\mathbf{B^c}$ (which is initially empty), the greedy heuristic selects among uncompromised nodes $u \in \mathbf{V} \setminus \mathbf{B^c}$ the best next node for the attacker to compromise. A key step of the algorithm is to determine optimal exfiltration paths given the compromised set $\mathbf{B^c} \cup \{u\}$ and the defender strategy $\mathbf{x}^*$.

Overall, the problem of finding an optimal set of exfiltration paths for the attacker given a set of compromised nodes $\mathbf{B^c} \cup \{u\}$ and the defender's strategy $\mathbf{x}^*$ can be represented as a MILP which is a simplification of (6)–(15). In this MILP simplification, the bot variables $\mathbf{z} = \{z_w\}$ are no longer needed. Furthermore, the bot-chain and data-exfiltration variables can be limited to the current set of compromised nodes $\mathbf{B^c} \cup \{u\}$, rather than the whole node set $\mathbf{V}$. As a result, the total number of variables and constraints involved is reduced significantly.

4.2 ORANI Defender Oracle

The defender oracle attempts to find a new pure defense strategy which maximizes the defender utility against the current mixed attack strategy $\mathbf{a}^* = \{a_j^*\}$

---

[1] Link: http://hdl.handle.net/2027.42/137970.

returned by MaximinCore. Here, a_j^* is the probability that the attacker follows $\mathbf{A}_j$ such that $\sum_j a_j^* = 1, a_j^* \in [0, 1]$. We first present a MILP to exactly solve this defender oracle problem and then show that the problem is NP-hard.

MILP Representation. We parameterize each pure strategy of the defender using *detection* variables $\mathbf{z} = \{z_w\}$ where $w \in \mathbf{V}$. In particular, $z_w = 1$ if the defender deploys a detector on node w. Otherwise, $z_w = 0$. In addition, given that the attacker plays $\mathbf{A}_j$ and the defender plays $\mathbf{z}$, we introduce *data-exfiltration* variables $\mathbf{h} = \{h_j(c)\}$ where $c \in \mathbf{V^c} \cap \mathbf{B}_j$, implying whether the attacker successfully exfiltrates the data of c (i.e., $h_j(c) = 1$) or not ($h_j(c) = 0$). Given that the attacker plays $\mathbf{a}^*$ and the defender plays $\mathbf{z}$, the defender's utility can be now computed based on $\mathbf{h}$ as follows:

$$U^d(\mathbf{z}, \mathbf{a}^*) = - \sum_{\mathbf{A}_j \in \mathbf{A}} a_j^* \sum_{c \in \mathbf{V^c} \cap \mathbf{B}_j} r(c) h_j(c) \tag{16}$$

The problem of finding an optimal pure defense strategy which maximizes the defender's utility against the attacker's strategy $\mathbf{a}^*$ can be now formulated as the following MILP (17)–(20).

$$\max_{\mathbf{z}, \mathbf{I}} U^d(\mathbf{z}, \mathbf{a}^*) \tag{17}$$

$$\text{s.t. } h_j(c) \geq 1 - \sum_{w \in \pi_j(c, S^a)} z_w, \forall c \in \mathbf{V^c} \cap \mathbf{B}_j, \forall j \tag{18}$$

$$\sum_{w \in \mathbf{V}} z_w \leq K^d \tag{19}$$

$$z_w \in \{0, 1\}, h_j(c) \in [0, 1], \ \forall w \in \mathbf{V}, c \in \mathbf{V^c} \cap \mathbf{B}_j, \forall j \tag{20}$$

In (17)–(20), only $\mathbf{z} = \{z_w\}$ are required to be binary. Constraint (18) ensures that $h_j(c) = 1$ when the attacker successfully exfiltrates from an stealing bot $c \in \mathbf{V^c} \cap \mathbf{B}_j$ (i.e., the defender does not deploy a detector on the exfiltration path of that bot: $z_w = 0$ for all $w \in \pi_j(c, S^a)$). On the other hand, since the MILP attempts to maximize the defender's utility (Eq. 16) which is a monotonically decreasing function of $h_j(c)$, then any MILP solver will automatically force $h_j(c) = 0$ if possible given the bound constraint (20). Constraint (19) guarantees that the number of detection resources deployed does not exceed the limit K^d.

Finally, Proposition 2 shows the complexity of the defender oracle problem. Its proof is in Online Appendix C.

Proposition 2. *The defender oracle problem corresponding to data uni-exfiltration is NP-hard.*

Defender Greedy Heuristic. We introduce a greedy heuristic to approximately solve the defender oracle problem in polynomial time. Given the attacker's mixed strategy $\mathbf{a}^*$ and an initially empty set of monitored nodes $\mathbf{D^c}$, the greedy heuristic iteratively adds the next best node to monitor to the set $\mathbf{D^c}$ until $|\mathbf{D^c}| = K^d$. At each iteration, given the current set of monitored nodes $\mathbf{D^c}$, the greedy heuristic searches over all unmonitored nodes $u \in \mathbf{V} \setminus \mathbf{D^c}$ to find the best next node to monitor such that the defender's utility is maximized. Computing the defender's utility given a set of monitored nodes and the attacker's strategy $\mathbf{a}^*$ is possible in polynomial time (Eqs. 1 and 2), thus our defender greedy heuristic runs in polynomial time.

5   Data Broad-Exfiltration

In the botnet defense game model with respect to uni-exfiltration (Sect. 3), for each stealing bot, the attacker is assumed to only select a single exfiltration path from that bot to exfiltrate data. In this section, we study the botnet defense game model with respect to the alternative data broad-exfiltration. In particular, for every stealing bot, the attacker is able to broadcast the data stolen by this bot to all other compromised nodes via corresponding routing paths. Once receiving the stolen data, each compromised node continues to broadcast the data to all other compromised nodes, and so on. The game model for broad-exfiltration is motivated by the botnet models studied by Rossow et al. [25]. Overall, there is a higher chance that the attacker can successfully exfiltrate network data with broad-exfiltration compared to uni-exfiltration. In the following, we briefly describe the botnet defense game model with data broad-exfiltration and the corresponding algorithm, **ORABI**, to compute an optimal mixed defense strategy.

5.1   Game Model

In the botnet defense game model with data broad-exfiltration, the strategy space of the defender remains the same as shown in Sect. 3. On the other hand, since the attacker now can broadcast the data, we can abstractly represent each pure strategy of the attacker as a set of compromised nodes $\mathbf{A}_j \equiv \mathbf{B}_j$ only. Given a pair of pure strategies $(\mathbf{D}_i, \mathbf{B}_j)$, we need to determine payoffs the players receive. Note that in the case of broad-exfiltration, given $(\mathbf{D}_i, \mathbf{B}_j)$, the attacker succeeds in exfiltrating the stolen data from a stealing bot if there is an exfiltration path among all the possible exfiltration paths over the compromised set $\mathbf{B}_j$ from this bot to S^a which is not blocked by $\mathbf{D}_i$. Therefore, the players receive a payoff computed as in (1) where the binary indicator $h(c)$ for each mission-critical node $c \in \mathbf{V^c}$ is now determined as:

$$h(c) = \begin{cases} 1 & \text{if } \exists c \in \mathbf{B}_j \ \& \ \exists \pi_j(c, S^a) \\ & \text{s.t. } \mathbf{D}_i \cap \pi_j(c, S^a) = \emptyset \\ 0 & \text{otherwise} \end{cases}$$

In fact, when players plays $(\mathbf{D}_i, \mathbf{B}_j)$, we can determine if there is an exfiltration path from a stealing bot $c \in \mathbf{B}_j \cap \mathbf{V^c}$ which is not blocked by $\mathbf{D}_i$ by using depth or breath-first search over the compromised set $\mathbf{B}_j$, which runs in polynomial time. We next aim at computing an SSE of the botnet defense games with data broad-exfiltration. Based upon the double oracle methodology, we introduce a new algorithm, **ORABI**, which consists of new MILPs to exactly solve the resulting attacker and the defender's oracle problems. We also provide greedy heuristics to approximately solve these oracle problems in polynomial time. In the following, we briefly explain our MILPs in **ORABI**.

5.2   ORABI Attacker Oracle

MILP Representation. In solving the attacker oracle problem with respect to data broad-exfiltration, we can extend the MILP (6)–(15) for data uni-exfiltration as follows. First, each pure strategy of the attacker is now parameterized using only bot variables $\mathbf{z} = \{z_w\}$ for $w \in \mathbf{V}$. Second, although bot-chain variables $\{q_c(u, v)\}$ are not parts of the attacker's pure strategies anymore, we extend these variables $\mathbf{q} = \{q_{i,c}(u, v)\}$ for each pure strategy of the defender $\mathbf{D}_i$. For each mission-critical node $c \in \mathbf{V^c}$ and for each $\mathbf{D}_i \in \mathbf{D}$, $\{q_{i,c}(u, v)\}$ will determine if there is an exfiltration path which successfully exfiltrates stolen data from c given the attacker's pure strategy $\mathbf{z}$. Third, the path-exfiltration constraints in (7)–(11) and the data-exfiltration constraint in (13) are extended accordingly. Finally, the data-exfiltration and all other constraints and the objective are kept unchanged. Given the extended bot-chain variables $\mathbf{q} = \{q_{i,c}(u, v)\}$ and corresponding extended constraints, the resulting extension of (6)–(15) will search over all possible exfiltration paths with respect to the attacker's strategy $\mathbf{z}$ to find exfiltration paths which are not blocked by each $\mathbf{D}_i \in \mathbf{D}$. Thus, the extended MILP of (6)–(15) returns an optimal set of compromised nodes u with $z_u = 1$ for the attacker.

Finally, the attacker oracle problem with broad-exfiltration is NP-hard (Proposition 3 with proof is in the Online Appendix D). The resulting MILP involves a larger number of variables and constraints compared to the uni-exfiltration case due to the extension of bot-chain variables $\mathbf{q} = \{q_{i,c}(u, v)\}$ with respect to the defender's pure strategies $\{\mathbf{D}_i\}$. In the following, we apply the greedy approach for solving the attacker oracle problem in polynomial time.

Proposition 3. *The attacker oracle problem corresponding to data broad-exfiltration is NP-hard.*

Attacker Greedy Heuristic. The attacker greedy heuristic with respect to data broad-exfiltration is similar to the uni-exfiltration case. Nevertheless, given a mixed defense strategy $\mathbf{x}^*$ and a set of compromised nodes $\mathbf{B^c} \cup \{u\}$, we no longer need to find an optimal set of exfiltration paths as in the uni-exfiltration case. As shown in Sect. 5.1, we can compute the players' utility given $\mathbf{x}^*$ and $\mathbf{B^c} \cup \{u\}$ in polynomial time using depth or breadth-first search.

In addition to this heuristic, we propose a modification of the greedy approach which iteratively adds multiple new pure strategies as follows. Instead of starting the greedy search with an initial empty compromised set $\mathbf{B^c} = \emptyset$, we create $|\mathbf{V^c}|$ different compromised sets $\mathbf{B^c}$, each consists of a mission-critical node $c \in \mathbf{V^c}$

as a compromised seed node. Then for each initial compromised set $\mathbf{B}^c$ with one seed node, we run the greedy search. As a result, we obtain $|\mathbf{V}^c|$ different compromised sets or pure strategies for the attacker. In other words, we add $|\mathbf{V}^c|$ new pure strategies for the attacker at each iteration. We call this modified greedy approach as **greedy-multi** heuristic. Intuitively, by adding multiple new pure strategies, we expect **ORABI** with the greedy-multi heuristic for solving the attacker oracle problem would potentially converge to a solution close to the optimal one. Indeed, our experimental results confirm our conjecture.

5.3   ORABI Defender Oracle

MILP Representation. Although we can also extend the MILP (17)–(20) for uni-exfiltration to represent the defender oracle problem with broad-exfiltration, solving this extended MILP is impractical. Specifically, in the constraint (18) of the MILP (17)–(20), we need to iterate over all exfiltration paths to find if the defender's pure strategy $\mathbf{z}$ can block these exfiltration paths or not. Since each pure strategy of the attacker with uni-exfiltration only consists of a small set of exfiltration paths, it is straightforward to iterate over these exfiltration paths. On the other hand, in the broad-exfiltration case, given a pure strategy of the attacker which is now a set of compromised nodes, there is an exponential number of exfiltration paths over these nodes to relay the stolen data. Iterating over all these exfiltration paths is thus impractical.

Given this computational challenge, **ORABI** introduces a new MILP to solve the defender oracle problem. First, we continue to use *detection* variables $\mathbf{z} = \{z_w\}$ to represent a pure strategy of the defender in which $z_w = 1$ if the defender deploys a detector on node w. Otherwise, $z_w = 0$. Second, for each pure strategy of the attacker $\mathbf{B}_j$ and the defender's pure strategy $\mathbf{z}$, we introduce *relaying* variables $\mathbf{l} = \{l_j(u, v)\}$ where $u, v \in \mathbf{B}_j$ are two compromised nodes, indicating whether the attacker can successfully relay data via the routing path $\mathbf{P}(u, v)$. Specifically, the attacker successfully relays data from u to v (i.e., $l_j(u, v) = 1$) if the defender does not deploy a detector on the routing path $\mathbf{P}(u, v)$. Otherwise, $l_j(u, v) = 0$. Third, we introduce variables $\mathbf{s} = \{s_j^c(w)\}$ where $c \in \mathbf{V}^c \cap \mathbf{B}_j$ and $w \in \mathbf{B}_j$. By an abuse of variable name, we also call these new variables as *data-exfiltration* variables. In particular, for each stolen bot $c \in \mathbf{V}^c \cap \mathbf{B}_j$ and $w \in \mathbf{B}_j$, $s_j^c(w)$ indicates if the attacker successfully exfiltrates data of c to the compromised node w ($s_j^c(w) = 1$) or not ($s_j^c(w) = 0$). In other words, $s_j^c(w) = 1$ only when there is an exfiltration path from the stealing bot $c \in \mathbf{V}^c \cap \mathbf{B}_j$ to the compromised node w which is not blocked by the defender. Given $\mathbf{s}$, the defender's utility is computed as follows:

$$U^d(\mathbf{z}, \mathbf{a}^*) = -\sum_{\mathbf{B}_j} a_j^* \sum_{c \in \mathbf{V}^c \cap \mathbf{B}_j} s_j^c(S^a) r(c) \tag{21}$$

where $s_j^c(S^a) = 1$ indicates that the attacker successfully exfiltrates data of $c \in \mathbf{V}^c \cap \mathbf{B}_j$ to S^a. Otherwise, $s_j^c(S^a) = 0$. We now can formulate the defender oracle problem as the following MILP:

$$\max_{\mathbf{z,l,s}} U^d(\mathbf{z,a^*}) \tag{22}$$

$$\text{s.t. } l_j(u,v) \geq 1 - \sum_{w \in \mathbf{P}(u,v)} z_w, \forall u,v \in \mathbf{B}_j, u \neq v, \forall j \tag{23}$$

$$s_j^c(w) \geq s_j^c(w') + l_j(w',w) - 1, \tag{24}$$

$$\forall c \in \mathbf{V^c} \cap \mathbf{B}_j, w \in \mathbf{B}_j \cup \{S^a\} \setminus \{c\}, w' \in \mathbf{B}_j, w' \neq w, \forall j$$

$$s_j^c(c) \geq 1 - z_c, \forall c \in \mathbf{B}_j \cap \mathbf{V^c}, \forall j \tag{25}$$

$$\sum_{w \in \mathbf{V}} z_w \leq K^d, z_w \in \{0,1\}, \forall w \in \mathbf{V} \tag{26}$$

$$l_j(u,v), s_j^c(w) \in [0,1], \forall c \in \mathbf{V^c}, u,v,w \in \mathbf{B}_j, u \neq v, \forall j \tag{27}$$

which maximizes the defender's utility in Eq. 21. Constraint (23) ensures that the attacker can successfully relay data from compromised node u to compromised node v ($l_j(u,v) = 1$) if there is no detector of the defender on the routing path, i.e., $z_w = 0, \forall w \in \mathbf{P}(u,v)$. Constraint (24) guarantees that if the defender does not block the routing path $\mathbf{P}(w',w)$ (i.e., $l_j(w',w) = 1$), node w receives data broadcasted by node w' (i.e., $s_j^c(w) \geq s_j^c(w')$). Furthermore, constraint (25) implies that if the defender does not deploy a detector on a stealing bot $c \in \mathbf{B}_j \cap \mathbf{V^c}$, then the attacker can steal the data of c. In other words, $s_j^c(c) = 1$ if $z_c = 0$ for all $c \in \mathbf{B}_j \cap \mathbf{V^c}$. Finally, constraint (26) imposes the requirement of detection resource limit for the defender.

In our MILP (22)–(27), only the detection variables $\mathbf{z} = \{z_w\}$ are required to be binary. The relaying variables and the data-exfiltration variables will be forced to be equal to one by constraints (23)–(25) if the attacker can successfully exfiltrate the data. Otherwise, since the defender utility in (21) is monotonically decreasing with respect to the data-exfiltration variables, (22)–(27) will force these variables to be zero whenever possible. Thus, all the variables are either zero or one in the optimal solution of (22)–(27). Finally, the defender oracle problem with respect to broad-exfiltration is NP-hard (Proposition 4 with proof is in the Online Appendix E).

Proposition 4. *The defender oracle problem corresponding to data broad-exfiltration is NP-hard.*

Defender Greedy Heuristic. We also apply the greedy approach to solve the defender oracle problem in polynomial time. The idea is similar to the attacker greedy heuristic.

6   Experiments

We evaluate both solution quality and runtime performance of our algorithms compared with previously proposed defense policies. We conduct experiments based on two different datasets: (i) synthetic network topology—we use

JGraphT [22], a free Java graph library, to randomly generate scale-free graphs since many real-world network topologies exhibit the power-law property [8]; and (ii) real-world network topology—we derive different network topologies from the Rocket-fuel dataset [24]. Each data point in our results is averaged over 50 different samples of network topologies.

6.1 Synthetic Network Topology

Data Uni-exfiltration We compare six different algorithms: (i) **ORANI** – both exact oracles; (ii) **ORANI-AttG** – exact defender oracle and greedy attacker oracle; (iii) **ORANI-G** – both greedy oracles; (iv & v) CWP & ECWP – heuristics proposed in [37] to generate a mixed defense strategy based on the centrality values of nodes in the network; and (vi) Uniform – generating a uniformly-mixed defense strategy. We consider CWP, ECWP, and Uniform as the three baseline algorithms.

In the first four experiments (Figs. 2(a), (b), (c) and (d)), we examine solution quality of the algorithms with varying number of nodes, of defender resources, of attacker resources, and of mission-critical nodes respectively. In Figs. 2(a), (b), (c) and (d), the x-axis is the number of nodes, of defender resources, of attacker resources, and of mission-critical nodes in each graph respectively. In the later three figures, the number of nodes is 30. The y-axis is the averaged expected utility of the defender obtained by the evaluated algorithms. The data value associated with each mission-critical node is generated uniformly at random within $[0, 1]$. Intuitively, the higher averaged expected utility an algorithm gets, the better the solution quality of the algorithm is. Figures 2(a), (b), (c) and (d) show that all of our algorithms, **ORANI, ORANI-AttG, ORANI-G** defeat the baseline algorithms in obtaining a much higher utility for the defender. Moreover, when the number of defender resources increases, the defender's expected utility on average increases quickly and reaches the defender's highest utility of zero with just five defender resources. On the other hand, when the number of attacker resources increases, there is only a small decrease in the defender's expected utility on average. Finally, both **ORANI-AttG** and **ORANI-G** obtain a lower average utility of the defender compared to **ORANI** as expected. Yet, we show that the greedy heuristics help in significantly reducing the time of solving the double oracle problem.

In our fifth experiment (Fig. 2(e)), we examine the convergence of the double oracle used in **ORANI**. The x-axis is the number of iterations of adding new strategies for both players until convergence. In addition, the y-axis is the average of the defender's expected utility at each iteration with respect to the defender oracle, the attacker oracle, and the Maximin core. The number of nodes in the graph is set to 40. Figure 2(e) shows that **ORANI** converges quickly, i.e., after approximately 25 iterations. This result implies that there is only a small set of pure strategies of players involved in the game equilibrium despite an exponential number of strategies in total. In addition, **ORANI** can find this set of pure strategies after a small number of iterations.

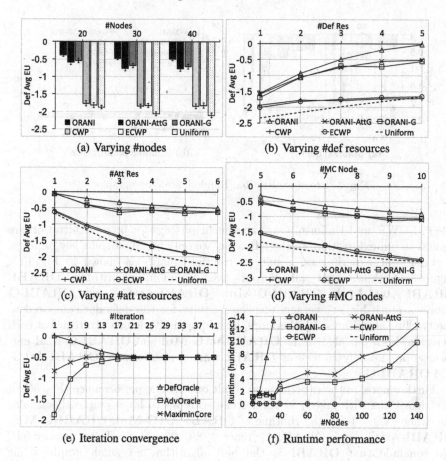

Fig. 2. Uni-Exfiltration: Random scale-free graphs

In our sixth experiment (Fig. 2(f)), we investigate runtime performance. In Fig. 2(f), the x-axis is the number of nodes in the graphs and the y-axis is the runtime on average in hundreds of seconds. As expected, the runtime of **ORANI** grows exponentially when $|\mathbf{V}|$ increases. In addition, by using the greedy heuristics, **ORANI-AttG** and **ORANI-G** run significantly faster than **ORANI**. For example, **ORANI** reaches 1333 seconds on average when $|\mathbf{V}| = 35$ while **ORANI-AttG** and **ORANI-G** reach 1266 and 990 seconds respectively when $|\mathbf{V}| = 140$.

Data Broad-Exfiltration. In the case of data broad-exfiltration, we compare eight algorithms: (i) **ORABI** – both exact oracles; (ii) **ORABI-AttG** – exact defender oracle and greedy attacker oracle; (iii) **ORABI-G** – both greedy oracles; (iv) **ORABI-AttG-Mul** – exact defender oracle and greedy-multi attacker oracle; (v) **ORABI-G-Mul** – both greedy-multi oracles; and (vi), (vii) and (viii)

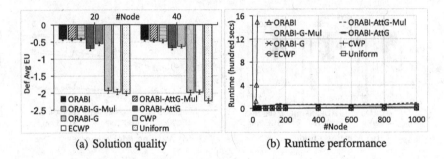

(a) Solution quality   (b) Runtime performance

Fig. 3. Broad-exfiltration: Random scale-free graphs

CWP, ECWP, and Uniform. Our experiment settings for broad-exfiltration are similar to uni-exfiltration. In the following, we only highlight some key findings.

First, our experimental result on solution quality is shown in Fig. 3(a). Figure 3(a) shows that all of our five evaluated algorithms, **ORABI**, **ORABI-AttG-Mul**, **ORABI-G-Mul**, **ORABI-AttG**, and **ORABI-G** obtain a much higher averaged expected utility for the defender compared to the baseline algorithms. Furthermore, by adding multiple new strategies at each iteration, both our algorithms **ORABI-AttG-Mul** and **ORABI-G-Mul** perform approximately as well as **ORABI** while outperforming **ORABI-AttG**, and **ORABI-G**.

Furthermore, in the experimental result on runtime performance (Fig. 3(b)), our algorithms with greedy heuristics can scale up to large graphs. For example, when $|V| = 1000$, the runtime of **ORABI-AttG-Mul**, **ORABI-G-Mul**, **ORABI-AttG**, and **ORABI-G** reaches 89, 20, 71, and 2 s respectively. We conclude that **ORABI** is the best algorithm for small graphs while **ORABI-AttG-Mul** and **ORABI-G-Mul** are proper choices for large-scale graphs.

Finally, we investigate the benefit to the attacker from broad-exfiltration compared to uni-exfiltration. We run **ORANI** and **ORABI** on the same set of 50 scale-free graph samples generated by uniformly at random with 20, 30, 40 nodes in each graph respectively. Among all the samples, there are only 58%, 72%, and 52% of the 20-node, 30-node, and 40-node graphs respectively for which the attacker obtains a strictly higher utility by using broad-exfiltration. This result shows that the attacker does not always benefit from broad-exfiltration. Indeed, despite broad-exfiltration, the data exchange between any pairs of compromised nodes must follow fixed routing paths specified by the network system, thus constraining the data exfiltration possibilities.

6.2   Real-World Network Topology

Our third set of experiments is conducted on real-world network topologies from the Rocket-fuel dataset [24]. Overall, the dataset provides router-level topologies of 10 different ISP networks: Telstra, Sprintlink, Ebone, Verio, Tiscali, Level3,

Exodus, VSNL, Abovenet, and AT&T. In this set of experiments, we mainly focus on evaluating the solution quality of our algorithms compared with the three baseline algorithms. For each of our experiments, we randomly sample fifty 40-node sub-graphs from every network topology using random walk. In addition, we assume that all external routers located outside the ISP can potentially route data to the attacker's server. Each data point in our experimental results is averaged over 50 different graph samples. The defender's averaged expected utility obtained by the evaluated algorithms is shown in Figs. 4 and 5 with respect to data uni-exfiltration and broad-exfiltration respectively.

Figures 4 and 5 show that all of our algorithms obtain higher defender expected utility than the three baseline algorithms. Further, the greedy

Dataset	ORANI	ORANI-AttG	ORANI-G	CWP	ECWP	Uniform
Telstra	-0.42	-0.44	-0.45	-1.9	-1.94	-2.38
Sprintlink	-0.43	-0.45	-0.45	-1.84	-1.89	-2.36
Ebone	-0.72	-0.73	-0.73	-1.71	-1.75	-2.32
Verio	-0.47	-0.47	-0.47	-1.84	-1.84	-2.25
Tiscali	-0.59	-0.62	-0.61	-1.97	-1.97	-2.2
Level3	-0.63	-0.64	-0.65	-1.85	-1.89	-2.25
Exodus	-0.68	-0.68	-0.68	-1.44	-1.47	-2.34
VSNL	-0.67	-0.68	-0.68	-1.69	-1.78	-2.3
Abovenet	-0.62	-0.64	-0.62	-1.77	-1.77	-2.3
AT&T	-0.31	-0.32	-0.33	-1.91	-1.96	-2.3

Fig. 4. Uni-exfiltration: Defender's average utility

Dataset	ORABI	ORABI-AttG-Mul	ORABI-G-Mul	ORABI-AttG	ORABI-G	CWP	ECWP	Uniform
Telstra	-0.41	-0.41	-0.41	-0.41	-0.42	-1.72	-1.78	-2.27
Sprintlink	-0.41	-0.41	-0.41	-0.43	-0.42	-1.72	-1.78	-2.21
Ebone	-0.71	-0.71	-0.71	-0.72	-0.73	-1.58	-1.66	-2.32
Verio	-0.47	-0.47	-0.47	-0.5	-0.5	-1.81	-1.85	-2.26
Tiscali	-0.51	-0.51	-0.51	-0.56	-0.56	-1.88	-1.95	-2.2
Level3	-0.67	-0.67	-0.67	-0.69	-0.68	-1.99	-2.03	-2.37
Exodus	-0.74	-0.74	-0.74	-0.75	-0.75	-1.58	-1.63	-2.37
VSNL	-0.73	-0.73	-0.73	-0.73	-0.73	-1.67	-1.76	-2.38
Abovenet	-0.67	-0.67	-0.68	-0.69	-0.68	-1.81	-1.88	-2.41
AT&T	-0.34	-0.34	-0.34	-0.35	-0.38	-1.88	-1.94	-2.28

Fig. 5. Broad-exfiltration: Defender's average utility

algorithms—**ORANI-AttG, ORANI-G,** and **ORABI-AttG, ORABI-G**—are shown to consistently perform well on all the ISP network topologies compared to the optimal ones—**ORANI** and **ORABI** respectively. In particular, the average expected defender utility obtained by **ORANI-G** is only ≈ 3% lower than **ORANI** on average over the 10 network topologies.

7   Summary

Many computer networks have suffered from botnet data exfiltration attacks, leading to a significant research emphasis on botnet defense. Our Stackelberg game model for the botnet defense problem accounts for the strategic response of cyber-criminals to deployed defenses. We propose two double-oracle based algorithms, **ORANI** and **ORABI**, to compute optimal defense strategies with respect to data uni-exfiltration and broad-exfiltration formulations, respectively. We also provide greedy heuristics to approximate the defender and the attacker best-response oracles. We conduct experiments based on both random scale-free graphs and 10 real-world ISP network topologies, demonstrating advantages of our game-theoretic solution compared to previous strategies.

Acknowledgment. This work was supported in part by MURI grant W911NF-13-1-0421 from the US Army Research Office.

References

1. Bacher, P., Holz, T., Kotter, M., Wicherski, G.: Know your enemy: tracking botnets. Technical report (2005)
2. Baldwin, A., Gheyas, I., Ioannidis, C., Pym, D., Williams, J.: Contagion in cyber security attacks. J. Oper. Res. Soc. **68**, 780–791 (2017)
3. Basilico, N., Gatti, N., Amigoni, F.: Leader-follower strategies for robotic patrolling in environments with arbitrary topologies. In: 8th International Conference on Autonomous Agents and Multiagent Systems, pp. 57–64 (2009)
4. Bensoussan, A., Kantarcioglu, M., Hoe, S.C.: A game-theoretical approach for finding optimal strategies in a botnet defense model. In: 1st Conference on Decision and Game Theory for Security, pp. 135–148 (2010)
5. Choi, H., Lee, H., Lee, H:, Kim, H.: Botnet detection by monitoring group activities in DNS traffic. In: 7th IEEE International Conference on Computer and Information Technology, pp. 715–720. IEEE (2007)
6. Cooke, E., Jahanian, F., McPherson, D.: The zombie roundup: understanding, detecting, and disrupting botnets. In: Workshop on Steps to Reducing Unwanted Traffic on the Internet (SRUTI), pp. 39–44 (2005)
7. Demarest, J.: Taking down botnets. Statement before the Senate Judiciary Committee, Subcommittee on Crime and Terrorism (2014)
8. Faloutsos, M., Faloutsos, P., Faloutsos, C.: On power-law relationships of the Internet topology. ACM SIGCOMM Comput. Commun. Rev. **29**(4), 251–262 (1999)
9. Fang, F., Nguyen, T.H., Pickles, R., Lam, W.Y., Clements, G.R., An, B., Singh, A., Tambe, M., Lemieux, A.: Deploying PAWS: field optimization of the protection assistant for wildlife security. In: 28th Conference on Innovative Applications of Artificial Intelligence, pp. 3966–3973 (2016)

10. Feily, M., Shahrestani, A., Ramadass, S.: A survey of botnet and botnet detection. In: 3rd International Conference on Emerging Security Information, Systems, and Technologies, pp. 268–273 (2009)
11. Gu, G., Perdisci, R., Zhang, J., Lee, W., et al.: BotMiner: clustering analysis of network traffic for protocol-and structure-independent botnet detection. In: 17th USENIX Security Symposium, pp. 139–154 (2008)
12. Gu, G., Porras, P.A., Yegneswaran, V., Fong, M.W., Lee, W.: BotHunter: detecting malware infection through IDS-driven dialog correlation. In: 16th USENIX Security Symposium, pp. 167–182 (2007)
13. Gu, G., Zhang, J., Lee, W.: BotSniffer: detecting botnet command and control channels in network traffic. In: 15th Annual Network and Distributed System Security Symposium (2008)
14. Holz, T., Engelberth, M., Freiling, F.: Learning more about the underground economy: a case-study of keyloggers and dropzones. In: Backes, M., Ning, P. (eds.) ESORICS 2009. LNCS, vol. 5789, pp. 1–18. Springer, Heidelberg (2009). doi:10. 1007/978-3-642-04444-1_1
15. Jain, M., Korzhyk, D., Vaněk, O., Conitzer, V., Pěchouček, M., Tambe, M.: A double oracle algorithm for zero-sum security games on graphs. In: 10th International Conference on Autonomous Agents and MultiAgent Systems, pp. 327–334 (2011)
16. Kiekintveld, C., Jain, M., Tsai, J., Pita, J., Ordó/ nez, F., Tambe, M.: Computing optimal randomized resource allocations for massive security games. In: 8th International Conference on Autonomous Agents and Multi-Agent Systems, pp. 689–696 (2009)
17. Kolokoltsov, V., Bensoussan, A.: Mean-field-game model for botnet defense in cyber-security. Appl. Math. Optim. **74**, 669–692 (2016)
18. Korzhyk, D., Yin, Z., Kiekintveld, C., Conitzer, V., Tambe, M.: Stackelberg vs. Nash in security games: an extended investigation of interchangeability, equivalence, and uniqueness. J. Artif. Intell. Res. **41**, 297–327 (2011)
19. Letchford, J., Vorobeychik, Y.: Computing randomized security strategies in networked domains. Appl. Advers. Reason. Risk Model. **11**, 06 (2011)
20. Mc Carthy, S.M., Sinha, A., Tambe, M., Manadhata, P.: Data exfiltration detection and prevention: virtually distributed POMDPs for practically safer networks. In: 7th Conference on Decision and Game Theory for Security, pp. 69–61 (2016)
21. McMahan, H.B., Gordon, G.J., Blum, A.: Planning in the presence of cost functions controlled by an adversary. In: 20th International Conference on Machine Learning, pp. 536–543 (2003)
22. Naveh, B., Contributors: JGraphT - a free java graph library (2009)
23. Peng, T., Leckie, C., Ramamohanarao, K.: Survey of network-based defense mechanisms countering the DoS and DDoS problems. ACM Comput. Surv. **39**(1), 3 (2007)
24. Rocketfuel: Rocketfuel: an ISP topology mapping engine (2002)
25. Rossow, C., Andriesse, D., Werner, T., Stone-Gross, B., Plohmann, D., Dietrich, C.J., Bos, H.: SoK: P2PWNED – modeling and evaluating the resilience of peer-to-peer botnets. In: IEEE Symposium on Security and Privacy, pp. 97–111 (2013)
26. Shieh, E., An, B., Yang, R., Tambe, M., Baldwin, C., DiRenzo, J., Maule, B., Meyer, G.: PROTECT: a deployed game theoretic system to protect the ports of the United States. In: 11th International Conference on Autonomous Agents and Multiagent Systems, pp. 13–20 (2012)
27. Soper, B., Musacchio, J.: A botnet detection game. In: 52nd Annual Allerton Conference on Communication Control and Computing, pp. 294–303. IEEE (2014)

28. Soper, B.C.: Non-zero-sum, adversarial detection games in network security. Ph.D. thesis, University of California, Santa Cruz (2015)
29. Stinson, E., Mitchell, J.C.: Towards systematic evaluation of the evadability of bot/botnet detection methods. In: 2nd USENIX Workshop on Offensive Technologies (2008)
30. Stone-Gross, B., Abman, R., Kemmerer, R.A., Kruegel, C., Steigerwald, D.G., Vigna, G.: The underground economy of fake antivirus software. In: 10th Workshop on the Economics of Information Security (2011)
31. Stone-Gross, B., Cova, M., Cavallaro, L., Gilbert, B., Szydlowski, M., Kemmerer, R., Kruegel, C., Vigna, G.: Your botnet is my botnet: analysis of a botnet takeover. In: 16th ACM Conference on Computer and Communications Security, pp. 635–647 (2009)
32. Strayer, W.T., Lapsely, D., Walsh, R., Livadas, C.: Botnet detection based on network behavior. In: Lee, W., Wang, C., Dagon, D. (eds.) Botnet Detection: Countering the Largest Security Threat. Advances in Information Security, vol. 36, pp. 1–24. Springer, Boston (2008)
33. Sweeney, P.J.: Designing effective and stealthy botnets for cyber espionage and interdiction: finding the cyber high ground. Ph.D. thesis, September 2014
34. Tambe, M. (ed.): Security and Game Theory: Algorithms, Deployed Systems, Lessons Learned. Cambridge University Press, Cambridge (2011)
35. Van Eeten, M., Bauer, J.M., Asghari, H., Tabatabaie, S., Rand, D.: The role of Internet service providers in botnet mitigation an empirical analysis based on spam data. In: 9th Workshop on the Economics of Information Security (2010)
36. Vaněk, O., Yin, Z., Jain, M., Bošanský, B., Tambe, M., Pěchouček, M.: Game-theoretic resource allocation for malicious packet detection in computer networks. In: 11th International Conference on Autonomous Agents and Multiagent Systems, pp. 905–912 (2012)
37. Venkatesan, S., Albanese, M., Cybenko, G., Jajodia, S.: A moving target defense approach to disrupting stealthy botnets. In: ACM Workshop on Moving Target, Defense, pp. 37–46 (2016)
38. Venkatesan, S., Albanese, M., Jajodia, S.: Disrupting stealthy botnets through strategic placement of detectors. In: IEEE Conference on Communications and Network Security (CNS), pp. 95–103 (2015)
39. Wang, P., Sparks, S., Zou, C.C.: An advanced hybrid peer-to-peer botnet. IEEE Trans. Dependable Secure Comput. **7**(2), 113 (2010)

Optimal Strategies for Detecting Data Exfiltration by Internal and External Attackers

Karel Durkota[1(✉)], Viliam Lisý[1], Christopher Kiekintveld[2], Karel Horák[1], Branislav Bošanský[1], and Tomáš Pevný[1,3]

[1] Deptartment of Computer Science, FEE, Czech Technical University in Prague, Prague, Czech Republic
{karel.durkota,viliam.lisy,karel.horak,branislav.bosansky,
tomas.pevny}@agents.fel.cvut.cz
[2] Computer Science Department, University of Texas at El Paso, El Paso, USA
cdkiekintveld@utep.edu
[3] Cisco Systems, Inc., Prague, Czech Republic

Abstract. We study the problem of detecting data exfiltration in computer networks. We focus on the performance of optimal defense strategies with respect to an attacker's knowledge about typical network behavior and his ability to influence the standard traffic. Internal attackers know the typical upload behavior of the compromised host and may be able to discontinue standard uploads in favor of the exfiltration. External attackers do not immediately know the behavior of the compromised host, but they can learn it from observations.

We model the problem as a sequential game of imperfect information, where the network administrator selects the thresholds for the detector, while the attacker chooses how much data to exfiltrate in each time step. We present novel algorithms for approximating the optimal defense strategies in the form of Stackelberg equilibria. We analyze the scalability of the algorithms and efficiency of the produced strategies in a case study based on real-world uploads of almost six thousand users to Google Drive. We show that with the computed defense strategies, the attacker exfiltrates 2–3 times less data than with simple heuristics; randomized defense strategies are up to 30% more effective than deterministic ones, and substantially more effective defense strategies are possible if the defense is customized for groups of hosts with similar behavior.

Keywords: Data exfiltration detection · Game theory · Network security

1 Introduction

A common type of cyber attack is a *data breach* which involves the unauthorized transfer of information out of a system or network in a process called *information exfiltration*. Information exfiltration is a major source of economic harm from cyber attacks, including the loss of credit card numbers, personal information,

© Springer International Publishing AG 2017
S. Rass et al. (Eds.): GameSec 2017, LNCS 10575, pp. 171–192, 2017.
DOI: 10.1007/978-3-319-68711-7_10

trade secrets, unreleased media content, and other sensitive data. Many recent attacks on high-profile companies (e.g., Sony Pictures and Target) have involved large amounts of data theft over long periods of time without detection [7]. Improving strategies for detecting information exfiltration is therefore of great importance for improving cybersecurity.

We focus on methods for detecting information exfiltration activities based on detecting anomalous patterns of behavior in user upload traffic. An important advantage of this class of detection methods is that it does not require knowledge of user data or the ability to modify this data (e.g., to use honey tokens). To better understand the strategic aspects of anomaly detection and information exfiltration we introduce a two-player game model that captures the defender and attacker decisions in a sequential game. While we focus mainly on the information exfiltration example, note that raising alerts based on detecting anomalous behavior is a common strategy for detecting cyber attacks, so our model and results are relevant beyond just information exfiltration.

One of the novel aspects of our game model is that we consider both insider and outsider threats. A recent McAfee report [1] states that that 40% of serious data breaches were caused by insiders trusted by the organization, while the remaining 60% are due to outside attackers infiltrating the enterprise. Since both types of attacks are prevalent we consider both cases. There are significant differences between insiders and outsiders for information exfiltration. One difference is that an insider *knows* his typical behavior already and can use this knowledge to evade detection, while an outsider must *learn* this behavior from observation. A second difference is that insiders may be able to *replace* their normal activity with malicious activity, while an outsider's actions will be observed *in addition* to the normal activity. We model both of these key differences and examine how they affect both attacker and defender behavior in information exfiltration.

We introduce a sequential game model in which the objective of the attacker is to exfiltrate as much data as possible before detection, and the objective of the defender is to minimize the data loss before detection. The defender monitors the amount of data uploaded to an external location (e.g., Dropbox or Google Drive) and raises an alert if the traffic exceeds a (possibly randomized) threshold in a give time period. Some network companies use only uploaded data volume as feature to detect the data exfiltration. In our paper we follow this approach, however, our algorithm allows using more features as well. The defender is constrained to policies that limit the expected number of false positives that will be generated. The attacker chooses the amount of data to exfiltrate in each time period. We model both insider and outsider attackers, and both additive and replacing attacks. In the additive attack the total traffic observed is the sum of the normal user activity and the attack traffic, while in the replacing attack only the attack traffic is observed by the defender. Outsider attackers also receive an observation of the user traffic in each time period that can be used to learn the behavior pattern (and therefore infer something about the likely threshold for detection).

Our first main contribution is the exfiltration game model that considers the differences between insider and outsider attackers. Our second contribution is a set of algorithms for approximating the optimal strategies for both defender and attacker players in these games. For outsider attackers, we use *Partially Observable Markov Decision Process* (POMDP) to model the learning process for the attacker. We also consider randomized policies for the defender, since it has been shown that static decision boundaries can be quickly learned [4] and randomizing can mitigate successful attacks [11]. Our third main contribution is an experimental analysis of a case study based on real-world data from a large enterprise with 5864 users connecting to a Google drive service for 12 weeks. We compute optimal strategies against different classes of attackers, and examine the characteristics of the strategies, the effects of randomization and attacker learning, and the robustness of strategies against different types of attackers. We show that with the computed defense strategies, the attacker exfiltrates 2–3 times less data than with simple heuristics; randomized defense strategies are up to 30% more effective than deterministic ones, and substantially more effective defense strategies are possible if the defense is customized for groups of hosts with similar behavior.

2 Related Work

Several previous works focus on detection and prevention of data exfiltration. A common approach is anomaly detection, e.g., a system can automatically learn the structure of standard database transactions on the level of SQL queries and raise alerts if a new transaction does not match this structure [5,10]. An alternative option is to create signatures of the sensitive data based on their content and detect if this content is sent out [14]. The signatures should be resilient against the addition of noise or encryption, such as wavelet coefficients for multimedia files, which are resilient against added noise. Data exfiltration can also be partially mitigated by introducing automatically generating honey-tokes, a bait documents that rise alarm when are opened or otherwise manipulated [2]. These works do not consider volume characteristics of the traffic as means of detecting data exfiltration and do not study the learning process of the external attacker, which are the focus of this paper. A commonly studied option of exfiltration is to use a covert channel and hide the communication in packet timing differences of DNS requests [19]. If the covert channel increases the volume of traffic to some service, the methods presented in this paper can help with its detection. More general data exfiltration motivations and best practices to protect the data are described in [13].

Data exfiltration and similar security problems have been previously studied in the framework of game theory. Liu et al. [12] propose a high level abstract model of insider threat in the form of partially observable stochastic game (POSG). They propose computing players' strategies using generic algorithms developed for this class of games, which have very limited scalability. The instance of the game they analyze in the case study focuses on data corruption and not exfiltration. Our work can also be seen as a special instance of POSG, but we provide more scalable algorithms to solve it and analyze the produced strategies in the context of data exfiltration.

Similar to our work, [9] investigates selecting thresholds for intrusion detection systems protecting distinct subsets of a network. The goal is to find optimal trade-offs between false positives and the likelihood of detection of an attack, which is simultaneously executed on a several subsets of the network. The attacker cannot decide what action to execute, only which systems to attack; nor he has an ability to learn the possible thresholds before the attack is conducted. McCarthy et al. [15] use POMDP to compute defender's optimal sequence of (imperfect) sensors to accumulate enough evidence whether data exfiltration over Domain Name System queries is happening in the network or not. However, unlike our paper, they assume non-adaptive attackers. Lisý et al. [11] investigated the effect of randomization of detection thresholds to strength of attacks and their overall cost to the defender. Our modeling of insider attacks is similar to this work. In contrast to our work, the attacker has perfect knowledge about the detector and the attacked system before the attack.

3   Game Theoretic Model

We model the problem of data exfiltration as a dynamic (sequential) game between the defender (network administrator) and the attacker trying to exfiltrate data over the network. We first discuss the basic setting of the game and focus on the fundamental differences between the insider and outsider and whether their activity is added to or replaces the normal traffic of the host. Then, we define the exact interaction between the attacker and the defender.

The **defender** monitors the volume of data uploaded by each network host[1] to a specific service over time, in time windows of constant length, e.g., 6 h. His action is to select a detection threshold θ from the set of available thresholds Θ. If the volume of the host's upload surpasses θ in a time window, an alarm is raised and the activity of the user is inspected by the administrator.

The **attacker** controls one of the users and tries to upload as much data as possible to the selected service before being detected. His actions are to choose the amount of data $a \in A \subseteq \mathbb{N}_0$ he exfiltrates in the next time window. If this amount (possibly) combined with the host's standard activity is below θ, the attacker immediately receives reward a and the defender suffers a penalty proportional to a. In this latter case, the attacker can act again in the following time window.

Since each **host** in a company may have different pattern of standard activity, the defender might want to set the threshold for each of them individually. However, this approach can be laborious in big companies and individual users rarely produce enough data for creating high quality models of their behavior. Therefore, it is common to create groups of hosts with similar behaviors and

---

[1] Hosts are non-strategic actors in the game considered to be part of the environment.

reason about these groups instead. In our models we refer to each group as a host type λ from a set of all types Λ. We assume both players know the probability $P(\lambda)$ that a randomly selected host in the network is of type λ. Each host type is characterized by its common activity pattern in the form of the probability $P(o|\lambda)$ that a host of type λ transfers the amount of data $o \in O \subseteq \mathbb{N}_0$ in a time interval. We call these amounts observations, since they are the information observed by the external attackers.

The standard host's activity can sometimes surpass the selected threshold even without any attacker's activity and the host is still inspected. These *false positives* take a lot of time for the administrator to investigate and are typically a key concern in designing IDS. To capture this constraint we require the defender's strategies to have an expected number of false positives bounded by a constant FP.

3.1   Outsider Vs. Insider

The outsider is an external attacker who compromises a host in the computer network to exfiltrate data. Although the outsider may know what types (groups) there are in the company (secretaries, IT admins, etc.), they often *do not know* which host type they compromised. However, they can observe the activity of the compromised host in each time window and update their belief about its type. Starting an aggressive exfiltration is likely not the best strategy, since once attacker surpasses a threshold, he is detected and the attack is stopped. However, conducing too much observation may cause that the host is disconnected or turned off before any exfiltration was conducted; or that the user's normal traffic surpasses the threshold, in which case the host is inspected and the attacker may be detected; or the data may become useless. We model this risk by discounting future rewards t time steps ahead with γ, where $\gamma \in (0,1)$. The outsiders must cautiously weigh how much to exfiltrate at the current time step versus how long to learn the host type to increase future rewards. Typically, he would first emphasize learning with little data exfiltration, and proceed to more aggressive exfiltration when he is more certain about the host type.

The insiders are the regular users of the network and they *know* their host type and the deployed defenses. If the defender sets a fixed threshold for each host type, an insider can exfiltrate exactly at that threshold (we assume that the amount of data has to surpass the threshold to trigger the alarm). Such a defense strategy is not optimal, and the defender should minimize the insiders certainty about the threshold by randomization of her choices.

3.2   Additivity Vs. Replacement

Consider a situation where the host uploads o MB and has set threshold θ. Then the attacker can exfiltrate at most $\theta - o$, if he does not want to surpass the threshold. Additivity is important mainly for the external attacker operating on the host without its user's knowledge. However, we allow additivity even for the insider in our model so that we can analyze the effect of incomplete knowledge of the external attacker with all other conditions equal. Assuming that the attacker completely replaces the existing host traffic with the exfiltration is more natural for the insider. However, even the external attacker can, in principle, throttle or to completely block the standard user's traffic to increase his own bandwidth for exfiltration. We analyze combinations of scenarios when the attacker is insider/outsider and when the user's normal traffic is and is not present.

3.3   Formal Definition of Game Model

We have introduced the following components: Λ is the set of host types and $P(\lambda)$ the probability of their occurrence; O (resp. A) is the set of possible amounts of data that the hosts (resp. attackers) can upload; $P(o|\lambda)$ describes host's standard activity; Θ is the set of thresholds the defender can choose; FP is the defender's maximal false positive rate; γ is the discount factor.

In our model, we assume that the network administrator models the user's normal traffic using discrete representation, e.g., histograms. In that case, the attacker and defender's action are also discrete, as they have no incentive to choose actions between the discrete values.

Defender's Strategy. We allow *mixed* (or randomized) strategies in form of $\sigma(\theta|\lambda)$, where the defender chooses a probability distribution of thresholds θ given host-type λ. As a special case, the defender may choose a *pure* strategy ψ : $\Lambda \to \Theta$, a threshold for each host-type λ. Let Ψ and Σ be the set of all pure and mixed strategies, respectively. A valid defender strategy σ must satisfy the false positive constraint $\sum_{\lambda \in \Lambda} \sum_{\theta \in \Theta} \sigma(\theta|\lambda) P(\lambda) FP(\theta|\lambda) \leq FP$, where $FP(\theta|\lambda) = \sum_{o \in O: o > \theta} P(o|\lambda)$ is type λ's amount of false positives if threshold is θ.

Attacker's Strategy. In the course of the attack, the attacker follows a policy which prescribes what action he should take when he played actions $a_1, \ldots, a_k$ and saw observations $o_1, \ldots, o_k$ so far [3]. We assume, that the defender chooses his threshold strategy first, and the attacker acts afterwards, knowing the defender's strategy (we will discuss it in section Solution Concept. In such a case, the attacker acts only against the nature, without adversarial actor, and Partial Observable Markov Decision Processes (POMDPs) can be used to reason about (approximately) optimal attacker's policies for the attacker. In the POMDP the attacker is not required to remember the entire history of his actions and observations. Instead, he can capture all relevant information he has acquired in the course of the interaction in the form of a belief state $b \in \Delta(\Lambda \times \Theta)$, which is a probability distribution over possible host types and threshold settings. We can then define attacker's policy based on his belief as $\pi : \Delta(\Lambda \times \Theta) \to A$. In the course of the interaction, the attacker keeps track of his belief b using a Bayesian update rule when he takes the last action and observation into account. Based on his current belief, he chooses action $\pi(b)$ to play. We denote the set of all belief-based policies as Π.

Note that the insider knows the host type from which he exfiltrates the data (it is his own host machine), therefore, there is no need to update belief based on the observation. Therefore, for the insider the attack policy is to choose an action from A. Mixed strategies are probability distribution among these choices.

Utilities. We define the attacker's expected utility as $u_a(\sigma, \pi)$, which is the discounted total expected amount of exfiltrated data using policy π against the defense strategy σ.

We define the defender's utility as $u_d(\sigma, \pi) = -Cu_a(\sigma, \pi)$, where $C > 0$. That means, that players have opposing objectives and their payoffs are proportional. Typically $C > 1$, which means that the defender suffers more than the attacker gains.

Solution Concept. Game theory provides a variety of solution concepts and algorithms for analyzing games with different characteristics. In zero-sum games and their slight generalizations, such as our payoff structure, many of these solution concepts lead to the same strategies. We use Kerckhoffs's principle, which assumes that the attacker knows the defender's algorithm or can conduct surveillance of the defender's behavior, therefore, knows his strategy. In game theory, Stackelberg equilibrium corresponds to such assumptions, where *leader* (the defender) acts first, by choosing strategy σ. Then, *follower* (the attacker), plays any *best response* strategy, which maximizes the attacker's utility against leader's strategy σ.

Definition 1 (best response). *Attacker plays* best response *if it maximizes the attacker's expected utility, taking the defender's strategy as given. Formally, $\pi \in BR_a(\sigma)$ iff $\forall \pi' \in \Pi : u_a(\sigma, \pi) \geq u_a(\sigma, \pi')$.*

In zero-sum games, all attacker's best responses have the same expected utility to the defender and the attacker, therefore, there is no need to distinguish between specific best responses. Because we use approximative algorithm to compute the attacker's policy, we focus on finding approximate ϵ-SE. The defender's strategy in ϵ-SE guarantees, that the defender's utility cannot be improved by a factor of more than $1 + \epsilon$ in the exact SE.

Definition 2 (ϵ-Stackelberg equilibrium (ϵ-SE)). *Let $\epsilon \in (0, 1]$. Solution profile (σ^*, π^*) where $\pi^* \in BR_a(\sigma^*)$ belongs to ϵ-SE, if $\forall \sigma \in \Sigma, \forall \pi \in BR_a(\sigma)$: $\frac{u_d(\sigma, \pi) - u_d(\sigma^*, \pi^*)}{|u_d(\sigma^*, \pi^*)|} \leq \epsilon$.*

Note, that we use *multiplicative* definition of approximate solution concept [6], rather then more typical *additive* approximation. In our opinion, the multiplicative approximation is slightly more reasonable for our domain. However, the algorithm can be easily modified to return additive ϵ-SE.

4   Algorithms

In this section, we present two algorithms. First algorithm computes exact SE against the insider. Second algorithm finds ϵ-SE against the outsider.

4.1   Optimal Defense Strategy Against Insiders

Since we assume that the insider knows from which user type he exfiltrates data (they have complete information), we can model the interaction between the attacker and a host type as a normal-form game, where the attacker chooses a probability distribution over actions A for each host type and the defender chooses probability distribution over thresholds Θ for each host type. We formalize the game between all host types and the attacker as one problem by extending the zero-sum normal-form linear program (LP) [17] with multiple host types and a false-positive constraint.

$$\min_{\sigma(\theta|\lambda)} U_a \tag{1a}$$

$$\text{s.t. } : (\forall \lambda \in \Lambda, \forall a \in A) : \sum_{\theta \in \Theta} u_a(\theta, a, \lambda)\sigma(\theta|\lambda) \leq U_{a,\lambda} \tag{1b}$$

$$\sum_{\lambda \in \Lambda} P(\lambda)U_{a,\lambda} \leq U_a \tag{1c}$$

$$(\forall \lambda \in \Lambda) : \sum_{\lambda \in \Lambda} \sigma(\theta|\lambda) = 1 \tag{1d}$$

$$(\forall \lambda \in \Lambda \forall \theta \in \Theta) : \sigma(\theta|\lambda) \geq 0 \tag{1e}$$

$$\sum_{\lambda \in \Lambda} \sum_{\theta \in \Theta} P(\lambda)\sigma(\theta|\lambda)FP(\theta|\lambda) \leq FP \tag{1f}$$

The variables in the LP are: $\sigma(\theta|\lambda), U_a$ and $U_{a,\lambda}$. The objective (1a) minimizes the attacker's expected utility U_a, which consists of expected utilities $U_{a,\lambda}$ of each type, weighed by its probability (1c). Constraints (1b) ensure that the attacker plays a best response in each host type against the given defense strategy; (1d) and (1e) ensures that the defender's strategy is proper probability distribution; and (1f) makes sure the strategy meets the false-positive rate.

In LP, we need to compute the attacker's payoff $u_a(\theta, a, \lambda)$ when the defender plays action θ and the attacker attacks host type λ with action a. For the attacker with replacement, we compute $u_a(\theta, a, \lambda)$ as follows:

$$u_a(\theta, a, \lambda) = \begin{cases} \frac{a}{1-\gamma} & \text{if } \theta \geq a \\ 0 & \text{otherwise,} \end{cases} \tag{2}$$

and for the attacker with additivity as follows:

$$u_a(\theta, a, \lambda) = \frac{aP(o+a \leq \theta|\lambda)}{1 - \gamma P(o+a \leq \theta|\lambda)} \tag{3}$$

where $P(o+a \leq \theta|\lambda) = \sum_{o \in O: a+o \leq \theta} P(o|\lambda)$ is the probability that the user's action o combined with the attacker's action a is below threshold θ for type λ. To compute the defender's pure strategy, we replace (2d) by $(\forall \lambda \in \Lambda \forall \theta \in \Theta) :$ $\sigma(\theta|\lambda) \in \{0,1\}$.

4.2   Optimal Defense Strategy Against External Attacker

The outsider observes the activity of the host in an attempt to learn and infer it's type. Due to the learning process, the strategies of the attacker are more complex, compared to the insider case, as the strength of the attack can vary over time. We can reason about attacker's behavior under this uncertainty using Partially Observable Markov Decision Processes (POMDPs) and his optimal, best-response strategy can be computed by algorithms for solving POMDPs. Originally, POMDPs were designed to reason about actions of a single decision maker. However, since the defender only decides the initial belief of the POMDP and the defender then has no influence on the dynamics of the system, we can extend the POMDP framework to solve our game-theoretic problem.

The POMDP framework assumes that in every time step, the player chooses an action and receives an observation from the environment as a result. Based on this observation he updates his belief over the possible current states of the environment. Additionally, in each time step the player obtains a reward which depend on the state of the environment and the action chosen. A solution of the POMDP is a policy which prescribe an action to use given every possible belief state. Here, we extend a well-established algorithm for solving POMDPs, Heuristic Search Value Iteration (HSVI) [18] to find an ϵ-Stackelberg Equilibrium, with key ideas inspired by [8]. The main idea of the algorithm is to iteratively compute the attacker's and defender's best response strategies, which will eventually converge to a Stackelberg equilibrium.

The structure of this section is as follows: first, we define POMDP models formally; then we explain the main ideas of the HSVI algorithm; and lastly, we present our contribution, the Adversarial HSVI algorithm, aimed to find ϵ-SE in our game.

POMDP Model. Let us now define a POMDP model formally, for a given defense strategy σ, as a tuple $\langle S, A, T, R, O, \gamma, \sigma \rangle$, where:

- S is set of states, where each state $s \in S$ is defined as $s = (\lambda_s, \theta_s)$, where λ is host-type and θ is the chosen detection threshold. We also define a terminal state s_T, which denotes that the attacker got detected and the attack was deflected.
- A is the set of attacker's actions;
- O is the set of observations about the traffic on host attacker tries to exfiltrate;
- $T(s, a, s')$ is the probability that action a in state s leads to new state s'. In our case, when additivity is considered $\forall s \in S \setminus \{s_T\} : T(s, a, s) = P(a + o \leq \theta_s)$, and $T(s, a, s_T) = 1 - P(a + o \leq \theta_s)$. If there is no additivity, then $\forall s \in S \setminus s_T : T(s, a, s') = 1_{a \leq \theta_s}$ and $T(s, a, s_T) = 1_{a > \theta_s}$, otherwise, where $1_A = 1$ if A is true and $1_A = 0$ otherwise is the indicator function.
- $R(s, a, s')$ is the immediate reward the attacker obtains for performing action a in state s. In our case $R(s, a, s') = a$ had the attacker not been detected yet, $R(s, a, s') = 0$ otherwise;
- $P(o|a, s)$ is the probability of observing $o \in O$ when action a is taken in state s. In our case $P(o|a, s = (\lambda, \theta)) = P(o|\lambda)$.
- $\gamma \in (0, 1)$ is the discount factor.

With $\mathscr{B}$ we denote the attacker's **belief space**, i.e. the set of all probability distributions over the states S. We derive the initial belief $b_0 \in \mathscr{B}$ according to the prior distribution over the host types $P(\lambda)$ and the strategy of the defender, i.e. $b_0(s) = P(\lambda)\sigma(\theta|\lambda)$ for state $s = (\lambda, \theta)$.

POMDP models are usually solved by approximating the optimal **value function** $v^* : \mathscr{B} \to \mathbb{R}$. This value function represents the utility $v^*(b)$ the attacker can obtain when the current distribution over the states is $b \in \mathscr{B}$ and he follows his optimal policy. We can then derive the optimal action to play in each belief state, i.e. the action $\pi(b)$, by solving the following equation

$$\pi(b) = \underset{a \in A}{\operatorname{argmax}} \left[\sum_{s \in S} \sum_{s' \in S} \Pr[s, s'|b, a] R(s, a, s') + \gamma \sum_{o \in O} \Pr[o|b, a] \cdot v^*(\tau(b, a, o)) \right] \quad (4)$$

where we account for the immediate rewards (expectation over $R(\cdot)$) as well as the expectation over future rewards (represented by the value function v^*). $\tau(b, a, o)$ stands for a Bayesian update of the belief b based on receiving the observation o when action a was used by the attacker.

HSVI Algorithm. We now provide an explanation of basic ideas of the HSVI algorithm, which we complement with illustrations in Fig. 1. For detailed explanation of the HSVI algorithm, we refer the reader to [18]. The algorithm maintains the upper and lower bounds on the optimal value function v^* for each point in the belief space $\mathscr{B}$, as depicted in Fig. 1a. The horizontal axis represents the belief space $\mathscr{B}$ and the vertical axis represents the expected utility the attacker can achieve (or lower and upper bounds on this utility, respectively). In each iteration, HSVI performs a single simulation of depth D, in the course of which the attacker plays D actions and obtains D observations. This simulation is conducted according to a forward-exploration heuristic, which aims to select beliefs which can be reached using the play starting from the initial belief b_0, and for which the approximation using the lower and upper bounds is excessively inaccurate. For these beliefs, we compute the optimal action of the attacker (see Eq. 4) and based on that we refine the bounds on v^*. In Fig. 1b we illustrate the way the lower and upper bounds get refined.

Let us use notation $LB(b)$ and $UB(b)$ to refer to values of the lower and upper bounds, respectively, in belief $b \in \mathscr{B}$. The original HSVI algorithm terminates, when $UB(b_0) - LB(b_0) < \epsilon_{hsvi}$, where ϵ_{hsvi} is the desired approximation error.

Finding ϵ–SE. Recall that the initial belief of the POMDP problem, $b_0(s) = P(\lambda)\sigma(\theta|\lambda)$, can be directly mapped to the defender's strategy σ (and vice versa). Therefore, we search such initial belief b_0 for the defender, that it meets maximal false positive constraint and minimizes the attacker's expected utility (POMDP upper bound value at b_0). In high lever, our approach iteratively alternates between selecting a promising initial belief b_0 (strategy for the defender) and solving POMDP at that belief b_0. In Fig. 1c we illustrate a subset of valid initial beliefs that meets the false-positive constraint.

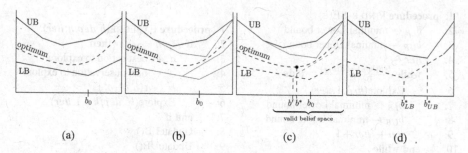

Fig. 1. Original and Adversarial HSVI algorithm: (a) initial upper bound (UB) and lower bound (LB) on the (unknown) optimal value function v^*. HSVI aims to minimizes the gap between UB and LB in the initial belief b_0. (b) After one HSVI iteration, tighter approximation using LB and UB is computed. (c) In Adversarial HSVI the defender chooses a new belief b' where LB has minimal value in every iteration. (d) A possible scenario when algorithm is converged and a conservative strategy for the defender, based on b^*_{UB}, is returned.

In detail, to find initial belief in ϵ-SE and the strategy σ of the defender, the Adversarial HSVI algorithm extends the original HSVI algorithm in two ways (the modified algorithm is presented in Fig. 2). First, instead of having fixed initial belief b_0, our algorithm chooses a new belief b' in every iteration. This belief, $b' = \arg\min_b LB(b)$, is chosen to minimize attacker's lower bound value. Second, we limit the depth D of the HSVI simulation by $\sqrt{iter}$, where $iter$ is the current iteration number. We do this to emphasize the exploration of the belief space first, and then focus on the computation of more accurate bounds later on (Fig. 2). The rest of the algorithm follows the ideas of the original HSVI algorithm. We refer the reader to Sects. 3.3 and 3.4 of [18] for details about the implementation of UpdateLB() and UpdateUB() procedures, and the forward exploration heuristic (lines 3–4 of the EXPLORE procedure).

Let $b_{LB} = \arg\min_b LB(b)$ and $b_{UB} = \arg\min_b UB(b)$ be the beliefs with minimal value of lower and upper bounds. We ensure that the algorithm finds ϵ-SE, by terminating when defender's and attacker's strategies have maximum relative error ϵ and we then return a secure strategy implied by belief b_{UB}. The attacker can guarantee that he will obtain at least $LB(b_{LB})$, while the defender can guarantee that he will not lose more than $UB(b_{UB})$. Based on these numbers, we compute an upper bound on the relative improvement of defender's strategy (i.e. if he plays b_{LB} instead of b_{UB}) as $\frac{UB(b_{UB})-LB(b_{LB})}{UB(b_{UB})}$.

Proposition 1. *Adversarial HSVI (Fig. 2) returns ϵ-SE.*

Proof. Without loss of generality, we assume the game is exactly zero-sum (i.e., $C = 1$). When the algorithm terminates and returns $\sigma(\theta|\lambda)$ induced by b_{UB}, we know that the best response of the attacker to the defender's strategy induced by b_{UB} cannot gain more than $UB(b_{UB})$, hence the defender's cost $-u_d(\sigma(\theta|\lambda), BR_a(\sigma(\theta|\lambda)) \leq UB(b_{UB})$. If the defender played any alternative strategy σ', we know that the attacker would always be able to exfiltrate at

```
 1: procedure FIND ε-SE(ε)
 2:     b_LB ← minimal Lower bound
 3:     b_UB ← minimal Upper bound
 4:     iter ← 1
 5:     while (UB(b_UB)−LB(b_LB))/(UB(b_UB)) > ε do
 6:         Explore(b_LB, 1, iter)
 7:         b_LB ← minimal Lower bound
 8:         b_UB ← minimal Upper bound
 9:         iter ← iter + 1
10:     end while
11:     return σ(θ|λ) induced by b_UB
12: end procedure
```

```
 1: procedure EXPLORE(b, depth, iter)
 2:     if depth < √iter then
 3:         a* ← best action to explore
 4:         o* ← best observation to explore
 5:         b' ← τ(b, a*, o*)
 6:         Explore(b', depth + 1, iter)
 7:     end if
 8:     UpdateLB()
 9:     UpdateUB()
10: end procedure
```

Fig. 2. Adversarial HSVI algorithm to find ε-SE.

least $LB(b_{LB})$ by definition of b_{LB}, hence $-u_d(\sigma', BR_a(\sigma')) \geq LB(b_{LB})$. If the termination condition is satisfied, $\frac{UB(b_{UB})-LB(b_{LB})}{UB(b_{UB})} \leq \epsilon$. Therefore, it is sufficient to show that the relative error of the computed strategy

$$\frac{u_d(\sigma', BR_a(\sigma')) - u_d(\sigma(\theta|\lambda), BR_a(\sigma(\theta|\lambda)))}{|u_d(\sigma(\theta|\lambda), BR_a(\sigma(\theta|\lambda)))|} \leq \frac{UB(b_{UB}) - LB(b_{LB})}{UB(b_{UB})}.$$

Since the defender's utility is always negative, we know $|u_d(\sigma(\theta|\lambda), BR_a(\sigma(\theta|\lambda)))| = -u_d(\sigma(\theta|\lambda), BR_a(\sigma(\theta|\lambda)))$. Hence, the above is equivalent to

$$1 - \frac{u_d(\sigma', BR_a(\sigma'))}{u_d(\sigma(\theta|\lambda), BR_a(\sigma(\theta|\lambda)))} \leq 1 - \frac{LB(b_{LB})}{UB(b_{UB})} \text{ and } \frac{-u_d(\sigma', BR_a(\sigma'))}{-u_d(\sigma(\theta|\lambda), BR_a(\sigma(\theta|\lambda)))} \geq \frac{LB(b_{LB})}{UB(b_{UB})}.$$

This is true, because from left to right in the last inequality, the nominator can only decrease and the denominator can only increase.

5   Real-World Data

From a large network security company we obtained anonymized data capturing the volumes of upload of 5864 active Google drive users uploaded during 12 weeks. For each user we computed the amount of data that the user uploads in 6 h windows. Next, we created histograms showing how often the user uploaded certain number of bytes per 6 h, which can be understood as user's upload probability distribution.

We used the Partitioning Around Medoids algorithm to find clusters of similar behavior of the users where similarity was measured by Earth Mover's Distance [16] metric. In Fig. 3 we present 7 histograms corresponding to user's average behavior in each cluster and their relative membership. The clusters (in order) contain 25.6%, 5.5%, 17.2%, 8.9%, 11.7%, 11.5% and 19.6% of the total users.

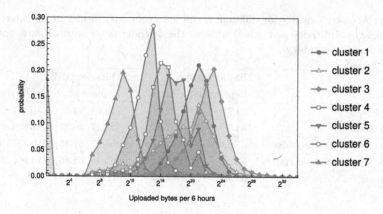

Fig. 3. Mean upload size histograms of the identified clusters of users.

6   Experiments

We now demonstrate our framework for a case study based on the real-world data. In all settings we choose: false positive rate $FP = 0.01$, the relative error of the strategy $\epsilon = 0.2$, and discount factor $\gamma = 0.9$. We chose the set of attacker actions $|A|$, the defender's thresholds $|\Theta|$, and the observations $|O|$ to be the set of $\{2^0, 2^2, 2^4, \ldots, 2^{34}\}$ bytes.

The structure of this section is as follows: In Sect. 6.1 we evaluate how much an optimal attacker can exfiltrate under various condition, in Sect. 6.2 we present a visualization of what optimal defense strategies look like, in Sect. 6.3 we evaluate defender strategies against different attacker models. In Sect. 6.4 we show how the presence of additivity influences the defense strategy, and finally, in Sect. 6.5 we present scalability results for computing the defense strategies.

6.1   Defender and Attacker Utilities

We now examine how much various attacker types can exfiltrate in our case study. In Table 1 we present a summary of the attacker's expected utilities (attacker maximizes and defender minimizes the value) for different types of attackers. Columns indicate whether the attacker is an insider or outsider and whether the attack is with replacement or with additivity. The rows indicate whether the defender plays a mixed or pure defense strategy, or a baseline defense. We present utilities against the outsider as minimal lower bound and minimal upper bound values from HSVI algorithm.

Note that the insider with replacement can exfiltrate up to 6 times more compared to insider with additivity. In in the case with additivity, the typical traffic of a user is added to the traffic of the attacker; hence, the attacker must choose a less aggressive strategy (i.e., upload less data) so that the total data upload does not exceed the threshold. Although the attacks with additivity are disadvantageous to the attacker, in some cases the additivity is unavoidable, e.g.,

Table 1. Attacker's utility for different scenarios: (columns) insider or outsider, with replacement or additivity and (rows) whether the defender plays optimal pure, optimal mixed or baseline strategy.

	Insider [MB]		Outsider [MB]	
	Replacement	Additivity	Replacement	Additivity
Mixed defense	23.71	3.56	(18.68, 24.81)	(3.11, 3.43)
Pure defense	33.58	5.03	(24.17, 29.87)	(3.78, 4.49)
Baseline single-quantile (mixed)	65.32	12.31	(54.85, 57.58)	(10.39, 10.84)
Baseline single-threshold (mixed)	68.86	14.54	(65.45, 68.86)	(13.96, 14.56)
One cluster (mixed)	63.29	12.95	N/A	N/A

when different detectors detect whether the user runs standard processes (which generate a standard traffic). Next, we see that the user type uncertainty caused around a 7%–12% decrease in the utility (computed from the upper bounds). To verify that this outcome does not rely on the fact that some of the host-types have higher prior probability than the others, we additionally ran experiments with uniform prior probability of the users, and the outsider had about 16% lower utility compared to the insider. Finally we note that if the defender must choose a pure strategy (e.g., due to practical deployment reasons) the amount of data exfiltrated by the attacker can be 24%–42% higher compared to randomized strategies.

We compare our strategies against two baseline approaches: (i) *single-quantile* and (ii) *single-threshold*. In (i), the defender sets for each host type a threshold at quantile $(1 - FP)$ of their upload probability distribution. Since we have discrete thresholds, the defender's strategy randomizes between two consecutive thresholds to reach the exact $(1 - FP)$ false positive rate. The experimental results show that the attacker can exploit this straightforward strategy and can exfiltrate about 3-times more data than against the optimal solution. The main reason is that this strategy chooses high thresholds for the users with large data upload (e.g., cluster 3) in order to satisfy the false positive constraint. In (ii) the defender chooses a single threshold for all host types such that the false positive rate requirement is satisfied. Although the strategy is quite different, it also performs poorly. The utility is even worse than the single-quantile strategy. This strategy is exactly contrary to the previous one: it sets the threshold for passive users (e.g. cluster 7) is too high, and attackers easily exfiltrate from them.

Additionally, we show that it is worth developing different defense strategies for different user types. We computed the optimal defense strategy where all users belong to one cluster (instead clustering them into 7 clusters), and utilities were 2x–3x worse than the optimal defense strategy where users were clustered. Since there is only one cluster, the attacker does not need to learn the cluster of the attacked host. Therefore, there is no difference between insiders and outsiders in this case.

6.2 Defender's and Attacker's Strategies

Insiders. In this section we present the computed optimal strategies of the defender. Figure 4a shows the defender's (cumulative) probability of selecting thresholds (x-axis) for each host type against the insider with replacement. The cumulative distributions show the probability that an attacker exfiltrating data at a certain rate is detected. In Fig. 4b we present the attacker's expected utility for different attacks on different host types when the defender is using the strategy depicted in Fig. 4a. The defender's strategy is computed in such a way that it makes the attacker indifferent between intervals of actions (e.g., for host type 1 the attacker is indifferent between actions 2^{22} through 2^{30}), which is typical for stable strategies. The attacker's best response is to choose any of the attack actions that have the highest expected utility. We also note that the host types with the highest activity (e.g., host type 3 and 1) result in the highest

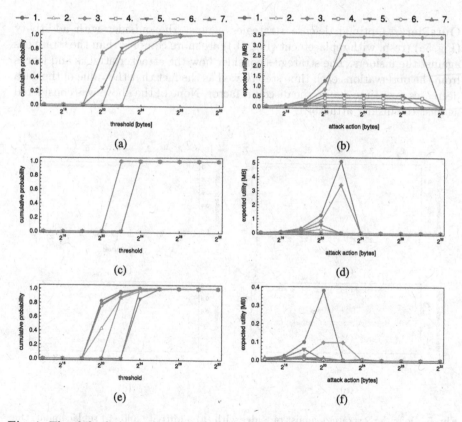

(a)   (b)

(c)   (d)

(e)   (f)

Fig. 4. The defender's strategies and the attacker's expected utilities for individual attack actions for: (a,b) mixed defense strategy against insider with replacement; (c,d) pure defense strategy against insider with replacement; and (e,f) mixed defense strategy against insider with additivity.

expected reward for the attacker. Therefore, we suggest that these hosts should be monitored thoroughly to avoid potentially large data loss.

Figure 4c shows the optimal strategy of the defender when restricted to pure strategies. For host types 3, 6 and 7, corresponding to the largest and the two smallest mean upload sizes, the defender chooses threshold 2^{22}. The threshold of 2^{24} is chosen for all the other types. The expected utility of the attacker depicted in Fig. 4b has peaks up to 5 MB, since with pure defense strategies it is impossible to make the attacker indifferent between multiple actions. By randomizing non-trivially between multiple thresholds the defender can significantly increase his expected utility.

The defense strategy against the insider with additivity (Fig. 4e) is quite different to the previous ones. With the additivity, the optimal defense strategy lowers the thresholds of the most active users (host types 1 and 3) to restrict their large loss, and increases the thresholds of the less active users to compensate the false-positives.

Outsiders. Optimal defense strategies against the outsider with additivity (Fig. 5a) (resp. with replacement (Fig. 5b)) are more complex than the strategies against the insiders. The strategies consider how the attacker attacks and learns from the observations each time step as well as the fact that the value of the data decreases over time due to the discount factor. None of the above was considered against the insider attacker.

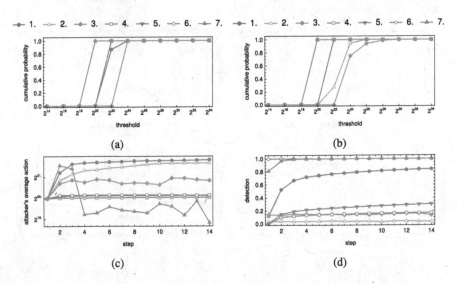

Fig. 5. Defender's strategy agaist outsider with (a) additivity and (b) replacement; the (near) optimal attacker's respone to (a) characterized by (c) the average action played at certain time step and (d) the probability of reaching the time step when attacking the given host type.

To explain how the defense strategy takes into account these aspects we first examine the attack strategy of the outsider with additivity. In Fig. 5c we show the attacker's average action in every time step *given* that the observations are drawn from a certain host type and that the attacker was not detected until the previous time step. First, the attacker plays safe action 2^{20} and after the observation, the attacker strengthens his attack as he learns about the host type their possible thresholds. Since the defender knows that the first attack action is 2^{20}, he prefers lowering the threshold for host types 6 and 7 (at 2^{20}), which causes the defender to almost certainly detect the attacker during his first attack action on 31.1% of hosts (see Fig. 5d). Not only does it increase the detection probability, it also prevents the attacker from obtaining information at the beginning, when it is most valuable due to the discounting. This generates a lot of false positives, so for highly active users the strategy uses higher thresholds (host types 2 and 3). The attacker will exfiltrate from these hosts aggressively in the later phase of the game but the loss will be less important by that time due to discounting.

This example shows how sophisticated the outsider's strategies can be as they must consider a complex behavior of the attacker and all possible sequences of observations and attacks. To minimize the loss, the defender aims to detect the attacker as soon as possible. In Fig. 5d we show the probability that the attacker is detected until given time step. Using the optimal defense strategies, host types 1, 6 and 7 (56.7% of the users) detect the attackers until his third time step with higher probability than 0.5.

6.3   Different Attacker Models

Computing a defense strategy against the insiders can be done using linear programming, which is computationally more efficient than the Adversarial HSVI algorithm used for outsiders. It rises a question of whether strategies against the insiders applied against the outsiders are significantly worse than the strategies optimized against the outsiders. In Table 2 we show the expected attacker's utilities of various defense strategies against different attacker models. The attacker

Table 2. Discounted expected amount of data that the attacker can exfiltrate if defense strategy (row) is optimized against the attacker in the "Strategy against" column, played against the different type of attackers (columns). The intervals for the outsiders represent lower and upper bounds of the optimal value.

	Strategy against	Insider, additive	Insider, replacement	Outsider, additive	Outsider, replacement
Mixed	insider, add	3.56 MB	26.58 MB	(3.56, 4.84) MB	(23.51, 29.5) MB
	insider, rep	5.91 MB	23.71 MB	(5.59, 8.33) MB	(23.71, 32.72) MB
	outsider, add	3.69 MB	27.59 MB	(2.67, 3.32) MB	(20.24, 27.59) MB
	outsider, rep	3.71 MB	27.59 MB	(2.42, 3.4) MB	(18.59, 26.35) MB
Pure	insider, add	5.03 MB	33.59 MB	(3.58, 4.43) MB	(24.54, 29.95) MB
	insider, rep	5.03 MB	33.59 MB	(3.58, 4.43) MB	(24.54, 29.96) MB
	outsider, add	5.03 MB	33.59 MB	(3.72, 4.49) MB	(24.55, 29.95) MB
	outsider, rep	5.03 MB	33.59 MB	(3.58, 4.42) MB	(24.17, 29.87) MB

of type (columns) plays a best response against the defense strategy (rows), where each defense strategy was optimized against attackers listed in column "Strategy against". For example, if we apply insider with additivity (resp. with replacement) to the outsider with additivity (resp. replacement), than the loss is between 24% and 45%. However, if we compute a defense strategy against the insider with replacement and apply it against the outsider with additivity, than the defender can lose up to 150% (comparing upper bounds) more than if the appropriate strategy is used, which is significantly worse. Therefore, it is beneficial for the network administrator to apply appropriate mixed defense strategies against different attacker models. The pure strategies do not have such big utility difference between various attacker models, due to the high similarity of all defense strategies. However, all of them have quite high loss compared to mixed strategies.

6.4 Effect of the Additivity

We now analyze how the uncertainty of the attacker's behavior affects the defender's strategy for the choice of thresholds. We created users with behavior of a normal distribution with varying standard deviation parameter. In Fig. 6 we show how the defense strategy against the insider with additivity changes given that the standard deviation of the user's behavior increases. Note that in the case where the user's behavior is constant (low standard deviation), it is optimal to choose a pure strategy with threshold at the user's mean behavior. If the attacker chooses any non-zero action, the sum of the observation and action will exceed the threshold and attacker is detected. If the defender's behavior is spread, the pure strategy is ineffective. It would have to be set at quantile $1 - FP$ due to the false-positive rate, and the attacker has a single best response action with highest expected reward. By mixing the thresholds, the defender can decrease the attacker's expected utility for a specific action and make the expected utility equal for an interval of actions (similarly, as was done for host type 3 in Fig. 4f).

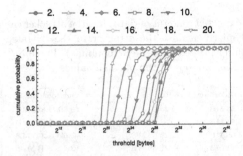

Fig. 6. Optimal defense strategies for users with standard activity of Normal distribution with $\mu = 2^{20}$ and standard deviations $\sigma = 2^i$, where i is given parameter.

This suggests that the users with close to constant behavior can be considered as the safest, as any exfiltration can be easily detected. For the users with uncertain behavior, the optimal defense strategies is to randomize among several thresholds, which forces the attacker to attacker weaker.

6.5   Algorithm Scalability

In Fig. 7a we present runtimes (note the logarithmic scale) for different numbers of host-types. All experiments were run on an Intel Xeon E5-2650 2.6 GHz with time limit 2 h. Strategies against the insider were computed in under 1 s, as they require single linear program computation. Adversarial HSVI, which iteratively improves the solution runs for between ten seconds and two hours, depending on the parameters of the problem. Even if problem was smaller (see outsider with two host types), the runtime could take longer than for seven host types. The reason is that the algorithm can temporarily get stuck in sequences of solutions with no or very little improvements. This is similar with the original HSVI. In Fig. 8 we show the relative error ϵ in each iteration for one and four host types. Our algorithm suffers with plateaus even more than HSVI, as the defender chooses initial belief with the lowest lower bound point every iteration. Despite the fact that Adversarial HSVI with $|\Lambda| = 4$ has 4 times more states, it is able to escape the plateau earlier than with $|\Lambda| = 1$. In Fig. 7b we show that the algorithm scales exponentially (note logarithmic y axis) with increasing number of actions, thresholds and observations.

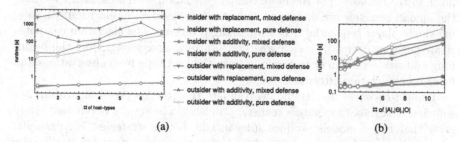

Fig. 7. (a) Runtime for increasing number of host types. (b) Runtime for increasing number of actions, thresholds, and observations.

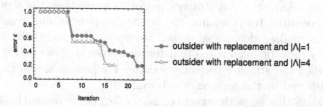

Fig. 8. ϵ error progress during computation strategies for outsider with replacement with 1 and 4 number of types.

All our case study strategies were computed with $|A| = |\Theta| = |O| = 15$, error $\epsilon = 20\%$ and number of host types $|\Lambda| = 7$ within 2 h. This demonstrates that the algorithm can be used to solve practically large problems. Moreover, HSVI and therefore Adversarial HSVI algorithms are easy to parallelize, which further improves the applicability of the presented approach.

7 Conclusion

Since computer networks, deployed defenses, and attacks are becoming more complex, developing effective decision support tools is critical for improving security. It is particularly difficult to consider the impact of all possible attacker's counteractions when the network administrator applies new defenses. Game theory provides a means to model these interactions and algorithms to compute the optimal strategies of the involved parties. We use the framework of game theory to model the problem of data exfiltration as a sequential game between the attacker and the network administrator. Sequential modeling allows us to model the decreasing value of data and the increasing chance of detection over time, as well as the development of attacker's knowledge about the network and user behavior and evolving attack strategy.

We propose two algorithms for computing (near) optimal defender strategies and bounds on their performance. For the case that the attacker does not need to learn the behavior of the attacked host, we specify a linear programming formulation for computing the optimal strategies. This situation typically corresponds to attacks by insiders, such as employees, who know the standard behavior of the hosts in the network. For the more complex situation with an attacker learning the upload behavior, we developed an algorithm based on recent results in solving single-player sequential decision-making problems. The algorithm computes strategies that optimally weigh whether to attack aggressively from the beginning and risk detection or to carefully learn the host type from observations and focus on exfiltration afterward.

Using real-world user traffic, we validate that the proposed algorithms are sufficiently scalable to analyze realistic problems. The results of our case study show that richer models produce substantially better strategies. For example, when facing the external attacker that does not replace the original traffic, the simple heuristic defense strategies let the attacker exfiltrate three times more data than the strategy optimized against a perfectly informed attacker using the linear program. Similarly, this strategy performs worse than the strategy optimized by Adversarial HSVI against the learning opponent. Our results further show that randomized defense strategies are up to 30% more effective in preventing data exfiltration compared to deterministic strategies. This is especially important when the attacker keeps the existing traffic intact, and the amounts of data transferred by the hosts vary substantially.

The attackers that know the exact behavior of the compromised host can exfiltrate by 7%–12% more data than the external attackers who have to learn it. Furthermore, a substantially more effective mitigation of data exfiltration is possible if the users are clustered into groups with similar behavior and a different detection strategy is used for each group. In our case study, the optimal strategy without the clustering allows the attacker to exfiltrate approximately three times more data than the optimal strategy using the clusters. Finally, we show that regardless of any other considered assumptions, if the attacker can replace the standard traffic of the compromised host, he can exfiltrate up to 6 times more data than the attacker who merely mixes his exfiltration traffic into the host's typical behavior. Therefore, monitoring the presence of the standard traffic (e.g., by expecting fake pre-scheduled transfers) may be a very effective countermeasure for decreasing the possible harm of data exfiltration.

Acknowledgments. This research was supported by the Czech Science Foundation (grant no. 15-23235S).

References

1. Grand theft data data exfiltration study: Actors, tactics, and detection. Technical report, McAfee, Inc. (2015). https://www.mcafee.com/us/resources/reports/rp-data-exfiltration.pdf
2. Bowen, B.M., Hershkop, S., Keromytis, A.D., Stolfo, S.J.: Baiting inside attackers using decoy documents. In: Chen, Y., Dimitriou, T.D., Zhou, J. (eds.) SecureComm 2009. LNICST, vol. 19, pp. 51–70. Springer, Heidelberg (2009). doi:10.1007/978-3-642-05284-2_4
3. Braziunas, D.: Pomdp solution methods. University of Toronto, Technical Report (2003)
4. Comesana, P., Pérez-Freire, L., Pérez-González, F.: Blind newton sensitivity attack. IEE Proc.-Inf. Secur. **153**(3), 115–125 (2006)
5. Fadolalkarim, D., Sallam, A., Bertino, E.: Pandde: provenance-based anomaly detection of data exfiltration. In: Proceedings of the Sixth ACM Conference on Data and Application Security and Privacy, pp. 267–276. ACM (2016)
6. Feder, T., Nazerzadeh, H., Saberi, A.: Approximating nash equilibria using small-support strategies. In: Proceedings of the 8th ACM Conference on Electronic Commerce, pp. 352–354. ACM (2007)
7. Wikipedia foundation: List of data breaches. https://en.wikipedia.org/wiki/List_of_data_breaches
8. Horák, K., Bošanský, B.: A point-based approximate algorithm for one-sided partially observable pursuit-evasion games. In: Zhu, Q., Alpcan, T., Panaousis, E., Tambe, M., Casey, W. (eds.) GameSec 2016. LNCS, vol. 9996, pp. 435–454. Springer, Cham (2016). doi:10.1007/978-3-319-47413-7_25
9. Laszka, A., Abbas, W., Sastry, S.S., Vorobeychik, Y., Koutsoukos, X.: Optimal thresholds for intrusion detection systems. In: Proceedings of the Symposium and Bootcamp on the Science of Security, pp. 72–81. ACM (2016)
10. Lee, S.Y., Low, W.L., Wong, P.Y.: Learning fingerprints for a database intrusion detection system. In: Gollmann, D., Karjoth, G., Waidner, M. (eds.) ESORICS 2002. LNCS, vol. 2502, pp. 264–279. Springer, Heidelberg (2002). doi:10.1007/3-540-45853-0_16
11. Lisý, V., Kessl, R., Pevný, T.: Randomized operating point selection in adversarial classification. In: Calders, T., Esposito, F., Hüllermeier, E., Meo, R. (eds.) ECML PKDD 2014. LNCS, vol. 8725, pp. 240–255. Springer, Heidelberg (2014). doi:10.1007/978-3-662-44851-9_16

12. Liu, D., Wang, X., Camp, J.: Game-theoretic modeling and analysis of insider threats. Int. J. Crit. Infrastruct. Prot. **1**, 75–80 (2008)
13. Liu, S., Kuhn, R.: Data loss prevention. IT Prof. **12**(2), 10–13 (2010)
14. Liu, Y., Corbett, C., Chiang, K., Archibald, R., Mukherjeé, B., Ghosal, D.: Sidd: a framework for detecting sensitive data exfiltration by an insider attack. In: 42nd Hawaii International Conference on System Sciences, HICSS 2009, pp. 1–10. IEEE (2009)
15. Mc Carthy, S.M., Sinha, A., Tambe, M., Manadhata, P.: Data exfiltration detection and prevention: virtually distributed POMDPs for practically safer networks. In: Zhu, Q., Alpcan, T., Panaousis, E., Tambe, M., Casey, W. (eds.) GameSec 2016. LNCS, vol. 9996, pp. 39–61. Springer, Cham (2016). doi:10.1007/978-3-319-47413-7_3
16. Rubner, Y., Tomasi, C., Guibas, L.J.: The earth mover's distance as a metric for image retrieval. Int. J. Comput. Vis. **40**(2), 99–121 (2000)
17. Shoham, Y., Leyton-Brown, K.: Multiagent Systems: Algorithmic, Game-theoretic, and Logical Foundations. Cambridge University Press (2008)
18. Smith, T., Simmons, R.: Heuristic search value iteration for pomdps. In: Proceedings of the 20th Conference on Uncertainty in Artificial Intelligence, pp. 520–527. AUAI Press (2004)
19. Zander, S., Armitage, G., Branch, P.: A survey of covert channels and countermeasures in computer network protocols. IEEE Commun. Surv. Tutorials **9**(3), 44–57 (2007)

Building Real Stackelberg Security Games for Border Patrols

Victor Bucarey[1]([✉]), Carlos Casorrán[1,2]([✉]), Óscar Figueroa[3], Karla Rosas[1], Hugo Navarrete[1], and Fernando Ordóñez[1]([✉])

[1] Departamento de Ingeniería Industrial, Universidad de Chile, Santiago, Chile
vbucarey@ing.uchile.cl, fordon@dii.uchile.cl
[2] Département d'Informatique, Université Libre de Bruxelles, Brussels, Belgium
ccasorra@ulb.ac.be
[3] Carabineros de Chile, Santiago, Chile

Abstract. We present a decision support system to help plan preventive border patrols. The system represents the interaction between defenders and intruders as a Stackelberg security game (SSG) where the defender pools local resources to conduct joint preventive border patrols. We introduce a new SSG that constructs defender strategies that pair adjacent precincts to pool resources that are used to patrol a location within one of the two precincts. We introduce an efficient formulation of this problem and an efficient sampling method to construct an implementable defender strategy.

The system automatically constructs the Stackelberg game from geographically located past crime data, topology and cross border information. We use clustering of past crime data and logit probability distribution to assign risk to patrol areas. Our results on a simplified real-world inspired border patrol instance show the computational efficiency of the model proposed, its robustness with respect to parameters used in automatically constructing the instance, and the quality of the sampled solution obtained.

Keywords: Stackelberg games · Security application · Border patrol

1 Introduction

Securing national borders is a natural concern of a country to defend it from the illegal movement of contraband, drugs and people. The European Union (EU) created the European Border and Coast Guard in October 2016 in response to the recent increase in migrant flows into the EU [1]. In the United States the Department of Homeland Security states as a primary objective that of "protecting [the] borders from the illegal movement of weapons, drugs, contraband, and people, while promoting lawful entry and exit" claiming it is "essential to homeland security, economic prosperity, and national sovereignty" [2].

The task of patrolling the border requires monitoring vast stretches of land 24/7. Given the size of the problem and resource constraints, the global border

S. Rass et al. (Eds.): GameSec 2017, LNCS 10575, pp. 193–212, 2017.
DOI: 10.1007/978-3-319-68711-7_11

monitoring is constructed by pooling and coordinating resources from different locations. The European Border and Coast Guard lists as one of its prime objectives "organizing joint operations and rapid border interventions to strengthen the capacity of the member states to control the external borders, and to tackle challenges at the external border resulting from illegal immigration or cross-border crime".

In this paper we consider the problem of patrolling a border in the presence of strategic adversaries that aim to cross the border, taking into account the defender patrolling strategies. We consider a Stackelberg game where the defender acts as the leader executing a preventive border patrol, which is observed prior to the optimal response by the strategic adversary, which acts as the follower. Due to the size of the border patrol problem the defender coordinates local resources to achieve a global defender strategy. Stackelberg Games, introduced by [3], are an example of bilevel optimization programs [4] where top level decisions are made by a player – the leader – that takes into account an optimal response of a second player – the follower – to a nested optimization problem.

Recent research has used Stackelberg games to model and provide implementable defender strategies in real life security applications. The Stackelberg games used in this context are referred to as Stackelberg security games (SSG). In these games, a defender aims to defend targets from a strategic adversary by deploying limited resources to protect them. The defender deploys resources to maximize the expected utility, anticipating that the adversary attacks a target that maximizes his own utility. Examples of Stackelberg security games applications include assigning Federal Air Marshals to transatlantic flights [5], determining U.S. Coast Guard patrols of port infrastructure [6], police patrols to prevent fare evasion in public transport systems [7], as well as protecting endangered wildlife [8].

One of the challenges that has to be addressed in solving SSGs is problem size. When the defender action is to allocate limited resources to various targets, the set of possible defender actions is exponential in the number of resources and targets. In [9] a relaxation of the SSG is formulated which determines the frequency with which each target is protected. This polynomial formulation (in the number of targets and security resources) is shown to be exact when there are no constraints on what constitutes a feasible defender action, but it is only an approximation in the general case.

In this work we tackle a border patrol problem, suggested by the Chilean National Police Force (known as Carabineros de Chile), where precincts pair up to jointly patrol border outposts in the presence of strategic attackers that observe these patrols before attacking. Combining resources from adjacent precincts provides overall coverage without excessively tasking each precinct. We use a mixed integer formulation for an SSG on a network, where the decision variables are two coverage distributions, one over the edges of the network and one over the targets that need to be protected. Further, we provide an approximate but computationally efficient sampling method that, given these optimal coverage distributions, recovers an implementable strategy for the defender, i.e., a valid pairing of precincts, and a set of targets to protect. We conduct computational experiments to measure the performance of our formulation and further experiments are carried out to measure the quality of the implementable defender strategies–recovered through the approximate sampling method– with respect to the optimal coverage probabilities returned by our formulation. In addition, we describe a case study that tackles the real-life border patrol problem presented by Carabineros de Chile. We develop a parameter generation methodology to construct Stackelberg games that model the border patrol setting and we carry out a sensitivity analysis to study the effect that perturbations in our parameter generation methodology can have on the solutions provided by our software.

The rest of the paper is as follows. In Sect. 2, we present the problem formulation considered. In Sect. 3 we describe the sampling method proposed to retrieve an implementable patrolling strategy from the optimal solution obtained. In Sect. 4 we describe the border patrol case study. In Sect. 5, we provide computational experiments that evaluate the efficiency of our sampling strategy and measure the performance of our problem formulation. Finally, we present our conclusions and discuss future work in Sect. 6.

2   Problem Formulation and Notation

In this section we first introduce the general framework of Stackelberg games, the notation and a review of benchmark models. Then, we present a Stackelberg game that seeks to select coordinated defender strategies given heterogeneous resources when facing strategic attackers.

2.1   Stackelberg Security Games

We consider a general Bayesian Stackelberg game, where a leader is facing a set K of followers, as introduced in [10]. In this model the leader knows the probability π_k of facing follower $k \in K$. We denote by I the finite set of pure strategies for the leader and by J be the finite set of pure strategies for each of the followers. A mixed strategy for the leader consists in a vector $\mathbf{x} = (x_i)_{i \in I}$, such that x_i is the probability with which the leader plays pure strategy i. Analogously, a mixed strategy for follower $k \in K$ is a vector $\mathbf{q}^k = (q_j^k)_{j \in J}$ such that q_j^k is the probability with which follower k plays pure strategy j. The payoffs for the agents are represented in the payoff matrices $(R^k, C^k)_{k \in K}$, where $R^k \in \mathbb{R}^{|I| \times |J|}$ gives the leader's reward matrix when facing follower $k \in K$ and $C^k \in \mathbb{R}^{|I| \times |J|}$ is the reward matrix for follower $k \in K$. The R_{ij}^k (C_{ij}^k) entry gives the reward for the leader (follower) when the leader takes action i and the k-th follower takes action j. With these payoff matrices, given a mixed strategy $\mathbf{x}$ for the leader and strategy $\mathbf{q}^k$ for follower k, the expected utility for follower k is given by $\sum_{i \in I} \sum_{j \in J} C_{ij}^k x_i q_j^k$ while the expected utility for the leader is given by $\sum_{k \in K} \pi_k \sum_{i \in I} \sum_{j \in J} R_{ij}^k x_i q_j^k$.

The objective of the game is for the leader to commit to a payoff-maximizing strategy, anticipating that every follower will best respond by selecting a payoff-maximizing strategy of their own. The solution concept used in these games is the Strong Stackelberg Equilibrium (SSE), introduced in [11]. In an SSE, the leader selects the strategy that maximizes his payoff given that every follower selects a best response breaking ties in favor of the leader when a follower is indifferent between several strategies. Therefore, without loss of generality the SSE concept can consider pure strategies as best response for each follower.

The problem of finding an SSE can be formulated as the following Mixed Integer Linear Program (MILP), referred to as (D2) [10]:

$$(D2) \quad \text{Max} \quad \sum_{k \in K} \pi^k f^k \tag{1}$$

$$\text{s.t.} \quad \mathbf{x}^\top \mathbf{1} = 1, \ \mathbf{x} \geq 0, \tag{2}$$

$$\mathbf{q}^{k^\top} \mathbf{1} = 1, \ \mathbf{q}^k \in \{0,1\}^{|J|} \qquad \forall k \in K \tag{3}$$

$$f^k \leq \sum_{i \in I} R_{ij}^k x_i + M(1 - q_j^k) \qquad \forall k \in K, \forall j \in J \tag{4}$$

$$0 \leq s^k - \sum_{i \in I} C_{ij}^k x_i \leq M(1 - q_j^k) \qquad \forall k \in K, \forall j \in J, \tag{5}$$

Constraints (2) and (3) indicate that the leader selects a mixed strategy and each follower responds with a pure strategy. The constant M in Constraints (4) and (5) is a large positive constant relative to the highest payoff value that renders the constraints redundant if $q_j^k = 0$. In Constraint (4), f^k is a bound on the leader's reward when facing the follower of type $k \in K$. This bound is tight for the strategy $j \in J$ selected by that follower. In Constraint (5), s^k is a bound on follower k's expected payoff. This bound is tight for the best response strategy for that follower. Together, Constraints (4) and (5) ensure that the leader's strategy and each follower's strategies are mutual best responses. The objective function maximizes the leader's expected reward.

The rewards in a Stackelberg game in a security setting only depend on whether the attack on a target is successful (if it is unprotected) or not (if the target is protected). Thus, we denote by $D^k(j|c)$ the utility of the defender when an attacker of type $k \in K$ attacks a covered target $j \in J$ and by $D^k(j|u)$ the utility of the defender when an attacker of type $k \in K$ attacks an unprotected target $j \in J$. Similarly, the utility of an attacker of type $k \in K$ when successfully attacking an unprotected target $j \in J$ is denoted by $A^k(j|u)$ and that attacker's utility when attacking a covered target $j \in J$ is denoted by $A^k(j|c)$. We express as $j \in i$ the condition that defender strategy i patrols target j. The relationship between the payoffs in a security game and in a general game are as follows:

$$R_{ij}^k = \begin{cases} D^k(j|c) \text{ if } j \in i \\ D^k(j|u) \text{ if } j \notin i \end{cases} \tag{6}$$

$$C_{ij}^k = \begin{cases} A^k(j|c) \text{ if } j \in i \\ A^k(j|u) \text{ if } j \notin i. \end{cases} \tag{7}$$

2.2 SSG for Border Patrol

We now adapt the SG formulation above to a problem where the defender first pairs up the resources from different precincts to form m combined patrols and then decides where to deploy these combined patrols. Let V be the set of police precincts and let $E \subset V \times V$ be the set of edges representing the set of possible precinct pairings, forming an adjacency graph $G = (V, E)$. We denote by $\delta(v) \subset E$ the set of edges incident to precinct $v \in V$, similarly for any $U \subset V$, $\delta(U) \subset E$ denotes the edges between U and $V \setminus U$, and $E(U) \subset E$ denotes the edges between precincts in U. We can then represent the possible combinations of m precincts pairs as the set of matchings of size m, which is given by:

$$\mathcal{M}_m := \left\{ \mathbf{y} \in \{0,1\}^{|E|} : \sum_{e \in E} y_e = m, \ \sum_{e \in \delta(v)} y_e \leq 1 \ \ \forall v \in V \right\}.$$

For every precinct $v \in V$, let J_v be the set of targets to patrol that are inside that precinct. Note that $\{J_v\}_{v \in V}$ is a partition of the set of targets J, i.e., $\cup_{v \in V} J_v = J$ and $J_u \cap J_v = \emptyset$ for all $u \neq v$. The set of defender strategies selects the m precinct pairings and for each pairing, further selects a target within the precincts in the pairing where the resource team for that given pairing is deployed. The combined patrol from the pairing of precincts u and v can only be deployed to a target in $J_u \cup J_v$. For each edge $e = (u,v) \in E$ we define $J_e = J_u \cup J_v$. It follows that the set I of defender strategies can be expressed as

$$I = \left\{ (\mathbf{y}, \mathbf{w}) \in \{0,1\}^{|E|} \times \{0,1\}^{|J|} : \begin{array}{l} \mathbf{y} \in \mathcal{M}_m, \\ \sum_{j \in \cup_{v \in U} J_v} w_j \leq \sum_{e \in E(U) \cup \delta(U)} y_e \ \ \forall U \subseteq V, \\ \sum_{j \in J} w_j = m \end{array} \right\}.$$

$$(8)$$

For $(y,w) \in I$, the variable y_e indicates whether edge e is selected for a precinct pairing and w_j indicates whether target j is patrolled. The first condition indicates that the coverage provided to any subset of targets is bounded by the coverage on all incident edges to this subset of targets. The second condition enforces that total target coverage is equal to the required number of resources. The set of pure strategies for an attacker of type $k \in K$ consist in, for each $k \in K$, selecting a single $j \in J$ with probability 1. In our border context $q_j^k = 1$ indicates that a criminal of type $k \in K$ attempts to penetrate the border through target $j \in J$.

The Border Patrol problem can be formulated as (D2) given in (1)–(5) by explicitly considering the exponentially-many defender pure strategies in the set (8) with defender and attacker rewards given by (6) and (7), respectively. This formulation however is computationally challenging as the resulting MILP considers variables $x_{(y,w)}$ for each $(y,w) \in I$, of which there is an exponential number in terms of targets and edges. We derive a compact formulation following the formulation presented in [9]. Our formulation is based on the observation that if rewards are given by (6) and (7) then the utility of each player only depends on c_j, the coverage at a target $j \in J$. Where this coverage can be expressed as

$$c_j = \sum_{(y,w) \in I \,:\, w_j = 1} x_{(y,w)} \qquad\qquad j \in J. \qquad (9)$$

We further consider the variables

$$z_e = \sum_{(y,w) \in I \,:\, y_e = 1} x_{(y,w)} \qquad\qquad e \in E, \qquad (10)$$

$$g_{ej} = \sum_{(y,w) \in I \,:\, y_e = 1, w_j = 1} x_{(y,w)} \qquad\qquad e \in E, j \in J_e \qquad (11)$$

where z_e is the coverage on edge $e \in E$ and can be obtained by summing over the pure strategies that assign coverage to that edge. Similarly, g_{ej} represents the combined coverage on edge $e \in E$ and target $j \in J$ and can be obtained by summing over the pure strategies where edge e and target j receive coverage.

Given a graph $G = (V, E)$ with V the set of precincts and E the feasible pairings between precincts, we propose the following SSG formulation for a border patrol (BP) problem:

(BP)

$$\text{Max} \quad \sum_{k \in K} \pi^k f^k \qquad\qquad (12)$$

$$\text{s.t.} \quad \mathbf{q}^{k\top} \mathbf{1} = 1, \; \mathbf{q}^k \in \{0,1\}^{|J|} \qquad\qquad \forall k \in K, \; (13)$$

$$\sum_{j \in J} c_j = \sum_{e \in E} z_e = m, \qquad\qquad (14)$$

$$\sum_{e \in \delta(v)} z_e \le 1 \qquad\qquad \forall v \in V, \; (15)$$

$$\sum_{e \in E(U)} z_e \le \frac{|U| - 1}{2} \qquad\qquad \forall U \subseteq V, |U| \ge 3, |U| \text{ odd} \; (16)$$

$$\sum_{e \in E : j \in J_e} g_{e,j} = c_j \qquad\qquad \forall j \in J \; (17)$$

$$\sum_{j \in J_e} g_{e,j} = z_e \qquad\qquad \forall e \in E \; (18)$$

$$f^k \le D^k(j|c)c_j + D^k(j|u)(1 - c_j)$$
$$\qquad + (1 - q_j^k) \cdot M \qquad\qquad \forall j \in J, \forall k \in K, \; (19)$$

$$0 \le s^k - A^k(j|c)c_j$$
$$\qquad - A^k(j|u)(1 - c_j) \le (1 - q_j^k) \cdot M \qquad \forall j \in J, \forall k \in K \; (20)$$

$$\mathbf{c} \in [0,1]^{|J|}, \; \mathbf{z} \in [0,1]^{|E|}, \qquad\qquad (21)$$

$$\mathbf{s}, \mathbf{f} \in \mathbb{R}^K, \; \mathbf{g} \in [0,1]^{|E||J|}. \qquad\qquad (22)$$

Constraints (13) ensure that the each attacker $k \in K$ attacks a single target $j \in J$ with probability 1. Constraint (14) indicates that the defender uses all his resources in a feasible solution and that in order to form his resources he pairs up precincts without exceeding the number of resources he wants to deploy. Constraint (15) indicates that a precinct's contribution to a pairing cannot exceed 1. Constraints (16) correspond to the *Odd Set Inequalities*, as introduced in [12], and together with (14) and (15) enforce that the coverage probabilities on the edges belong to the convex hull of the matching polytope of size m. Constraints (17) and (18) enforce the conservation between marginal coverages in nodes and edges. Finally, Constraints (19) and (20) are the same as in the formulation introduced in [9] and ensure that c and q are mutual best responses. The objective function in (BP), maximizes the defender's expected utility.

2.3   Discussion

To ensure the correctness of the formulation (BP) we need to be able to recover the variables $x_{(y,w)}$ for $(y,w) \in I$–that represent the probability distribution over the defender pure strategies–from an optimal solution c, z, q, s, f, g to (BP). In particular, we need to find variables $x \in [0,1]^{|I|}$ that satisfy constraint (10). Note that the odd set inequalities (16) are necessary. We give an example of z variables for which there does not exist a probability distribution over I that would satisfy (10). Consider the example in Fig. 1, which shows a non-negative z that satisfies constraints (14) and (15) for $m = 2$ but violates the odd set inequalities (for the set $U = \{3,4,5\}$). We observe that this solution cannot be expressed as a convex combination of pure matchings of size 2, making it impossible to retrieve an implementable defender strategy x.

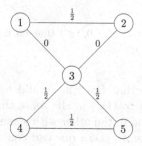

Fig. 1. Variables $z \in [0,1]^{|E|}$ that satisfy (14) and (15) but violate (16) for $m = 2$

Similarly, Constraints (17) and (18) also play a vital role in that they establish a link between the coverage variables on the edges and on the targets. This becomes much more apparent if one applies Farkas' Lemma [13] on the linear system defined by (17), (18) and $g \geq 0$ to understand which conditions on c and

z guarantee feasibility of the system. Applying Farkas provides the following necessary conditions on c and z which offer a more direct interpretation:

$$\sum_{e \in E : e \in \overline{E}(U)} z_e \geq \sum_{j \in \cup_{u \in U} J_u} c_j \quad \forall U \subseteq V, \tag{23}$$

$$\sum_{j \in J : j \in \cup_{u \in V : e \in \delta(u) \cap E'} J_u} c_j \geq \sum_{e \in E'} z_e \quad \forall E' \subseteq E. \tag{24}$$

Constraint (23) states that given a subset of nodes, the coverage provided on all targets inside these nodes cannot exceed the weight of the edges incident to these nodes. Constraint (24) indicates that given a fixed set of edges E', the weights on those edges is a lower bound on the coverage of the targets in nodes to which those edges are incident. Figure 2 shows an example of variables $\mathbf{z} \in [0,1]^{|E|}$ and $\mathbf{c} \in [0,1]^{|J|}$ that satisfy all constraints in (BP) except (17) and (18). The numbers on the top of the nodes represent total coverage on targets in that node, $\sum_{j \in J_u} c_j$, and the numbers on the edges represent the coverage probabilities on the edges, z_e. The solution in Fig. 2 violates (23) for $U = \{1, 2\}$. It is also not possible to find in this example an implementable defender strategy $\mathbf{x} \in [0,1]^{|I|}$ related to these variables $\mathbf{z}$ and $\mathbf{c}$.

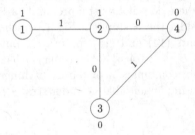

Fig. 2. Variables $\mathbf{z} \in [0,1]^{|E|}$ and $\mathbf{c} \in [0,1]^{|J|}$ that do not satisfy (17) and (18) with $m = 2$

It can in fact be proven that (BP) is a valid formulation for the SSG by showing that it is equivalent to (D2) in the sense that a feasible solution from one leads to a feasible solution in the other with same objective value and viceversa. Given a feasible solution to (D2), one can construct a feasible solution to (BP), with same objective value, through conditions (9)–(11). Conversely, given a feasible solution to (BP), a feasible solution to (D2), with same objective value, can be obtained relying on the fact that the cardinality constrained matching polytope is integral, [14]. The formal proof is omitted here.

3   Sampling Method

In this section we consider a two-stage approximate sampling method to recover an implementable mixed strategy $\mathbf{x} \in [0,1]^{|I|}$ which complies with the optimal probabilities on targets and edges, (c^*, z^*) as given by (BP). The quality of the proposed sampling method is later tested in Sect. 5.

Given z^*, the coverage vector over $|E|$ of size m, we discard all the edges in E such that $z_e = 0$. We then select m of the remaining edges according to a uniform random variable $U(0, m)$. This, in itself, could provide edges that do not form a matching. Therefore, let M be the set of m edges we have sampled. Now, solve the following optimization problem:

$$\text{Max} \sum_{e \in M} z_e^* y_e$$
$$\text{s.t.} \quad \mathbf{y} \in \mathcal{M_m}$$

Out of all matchings of size m, the objective function guarantees that we pick a maximum weight matching. The optimization problem either returns an optimal solution, in which case the edges in M admit a matching of size m, or, the problem is infeasible and such a matching cannot be constructed. If the problem is infeasible, we sample a new edge which we add to the set M and we re-optimize the optimization problem above. The sampled matching respects the optimal coverages if our algorithm returns the required matching after one iteration. Otherwise, the matching will deviate from the optimal coverages. We proceed in this iterative fashion until we construct a matching of size m. Note that this algorithm will produce a matching in at most $|E| - m$ iterations, as we know that such a matching exists in the original graph.

Having obtained M^*, the sampled matching of size m, the second stage of our sampling consists in sampling an allocation of resources to targets that satisfies the optimal target coverage probability returned by our formulation. To do so, we discard targets j that belong to precincts which are not paired. For each target j that belongs to a paired pair of precincts, say u and v, we normalize their coverage probability by the weight of the total coverage provided by the optimal coverage vector c^* in the two areas u and v that are paired and denote it by $\bar{c}_j^*$:

$$\bar{c}_j^* = \frac{c_j^*}{\sum_{j \in J_u \cup J_v} c_j^*} \qquad \forall (u, v) = e \in M^*.$$

This way, we ensure that one resource is available per paired pairs of precincts. The defender's coverage on targets is composed by sampling over the newly constructed $\bar{c}^*$.

4   Case Study: Carabineros de Chile

In this section, we describe a realistic border patrol problem proposed by Carabineros de Chile. In this problem, Carabineros considers three different types of crime, namely, drug trafficking, contraband and illegal entry. In order to minimize the free flow of these types of crime across their borders, Carabineros organizes both day shift patrols and night shift patrols along the border, following different patterns and satisfying different requirements.

We are concerned with the specific actions that Carabineros can take during night shift patrols. The region is divided into several police precincts. Due to the vast expanses and harsh landscape at the border to patrol and the lack of manpower, for the purpose of the defender actions under consideration, a number of these precincts are paired up when planning the patrol. Furthermore, Carabineros have identified a set of locations along the border of the region that can serve as vantage points from where to conduct surveillance with technical equipment such as night goggles and heat sensors (Figs. 3 and 4). A night shift action consists in deploying a joint detail with personnel from two paired precincts to conduct vigilance from 22h00 to 04h00 at the vantage point located within the territory of the paired precincts. Due to logistical constraints, for a given precinct pair, Carabineros deploys a joint detail from every precinct pair to a surveillance location once a week.

Fig. 3. A Carabinero conducts surveillance

Fig. 4. Harsh border landscape

Carabineros requires a schedule indicating the optimal deployment of details to vantage points for a given week. Figure 5 depicts a defender strategy in a game with $m = 3$ pairings, $|V| = 7$ precincts and $|J| = 10$ locations. Table 1 shows a tabular representation of the implemented strategy for that week.

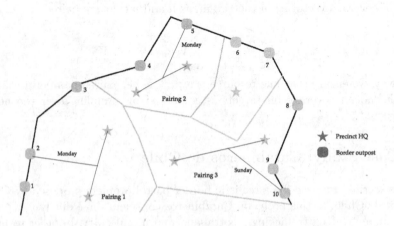

Fig. 5. Feasible schedule for a week

Table 1. Tabular representation for the feasible schedule in Fig. 5

Pairing/outpost	1	2	3	4	5	6	7	8	9	10
Pairing 1		M								
Pairing 2					M					
Pairing 3										Su

Therefore, we have an adjacency graph $G(V, E)$ where V is the set of police precincts and E are the edges that represent valid pairing of precincts. Further, the set of vantage points that need to be protected, corresponds to the set of targets J. Furthermore, the vantage points are partitioned among the different vertices of G, such that for a given $u \in V$, J_u contains all the vantage points inside precinct u. The set of attacker types is given by $K = \{\text{Drugs, Contraband, Illegal entry}\}$. In this setting, the pairings among precincts is fixed at the beginning of each month. Therefore, the game is separable into different standard SSG within every pair of paired precincts and one can use a standard SSG formulation such as the one presented in [9] to solve the different subproblems. Within each subproblem, the defender has a single resource to allocate to one of the different vantage points on a given day of the week. Given a coverage strategy over the targets, an adversary of type $k \in K$ plays the game with probability π^k and tries to cross the border through the vantage point $j \in J$ and on the day of the week that maximizes his payoff. It remains to construct the payoffs of the game for the problem described. To that end, Carabineros supplied us with arrest data in the region between 1999 and 2013 as well as other relevant data discussed next. In the following section, we discuss a payoff generation methodology.

4.1   Payoff Estimation

An accurate estimation of the payoffs for the players is one of the most crucial factors in building a Stackelberg model to solve a real-life problem. For each target in the game, we need to estimate 12 different values corresponding to a reward and penalty for Carabineros and the attacker for each type of crime $k \in K$.

We tackle this problem in several steps. First, we use QGIS [15], an open source geographic information system, to determine what we call action areas around each vantage point provided by Carabineros, based on the visibility range from each outpost. Such an action area represents the range of a detail stationed at a vantage point, i.e., the area within which the detail will be able to observe and intercept a criminal.

Further, consider, for each type of crime $k \in K$, a network $\mathcal{G}^k(\mathcal{V}^k, \mathcal{E}^k)$ that models that type of crime's flow from some nodes outside the border to some nodes inside the border, crossing the border precisely through the action areas previously defined. As nodes of origin for the different types of crime, we consider cross border cities. As destination nodes we consider the locations inside Chile where Carabineros has performed an arrest of that type of crime. In order to have a more manageable sized network, we consider a clustering of these destination nodes. We later show that our methodology is robust versus changes in the number of cluster nodes.

Specifically, for a crime of type $k \in K$, let us define $S^k \subset \mathcal{V}^k$ as the nodes of origin situated outside the borders, $F^k \subset \mathcal{V}^k$ as the nodes of destination and J as the set of action areas along the border. Each destination node, $f \in F^k$, resulting from a clustering procedure is then assigned a demand $b(f)$ which corresponds to the number of destination nodes which are contracted into f. We use the k-means model to cluster crime data. For each $k \in K$, the edge set $\mathcal{E}^k$ is constructed as follows. All nodes of origin are linked to all action areas. These areas are then linked to all of the destination nodes for crime $k \in K$. Figure 6 is a representation of such a network. The nodes to the left represent the points of origin of crime and the three nodes to the right are clusters of destination nodes for those crime flows. Note that crime enters the country through the four action areas marked as squares along the border.

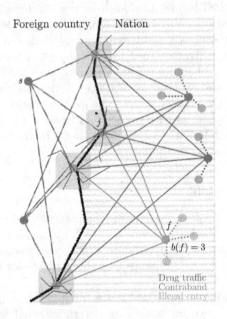

Fig. 6. Three crime flow networks, one per type of crime

We propose the following attractiveness parameter for a given action area $j \in J$ for a criminal of type $k \in K$ attempting to move from node $s \in S^k$ to node $f \in F^k$ through action area j:

$$U_{sf}^j = \frac{\text{Kilometers of roads inside action area j}}{d_{sj} + d_{jf}},$$

where d_{uv} is the distance in kilometers between nodes $u \in \mathcal{V}^k$ and $v \in \mathcal{V}^k$. This attractiveness parameter is proportional to the total length of roads inside a given action area and it is inversely proportional to how much an attacker moving from s^k to f^k has to travel in order to cross the border through area j.

We model the flow of crime $k \in K$ through a single route from $s \in S^k$ to $f \in F^k$ passing through $j \in J$ as follows:

$$x(s, j, f, k) = \frac{e^{\lambda U_{sf}^j}}{\sum_{s' \in S^k} \sum_{j' \in J} e^{\lambda U_{s'f}^{j'}}} \cdot b(f).$$

The flow of crime $k \in K$ through a route (s, j, f) is expressed as a proportion with respect to the flow of crime $k \in K$ through all routes leading into the same destination point $f \in F^k$. The parameter $\lambda \in \mathbb{R}_+$ provides a proxy of how the defender expects crime to behave. A value of $\lambda = 0$ means that crime $k \in K$ between any node of origin and destination distributes itself evenly among the different action areas. A high value of λ, however, is consistent with a flow of that type of crime through those action areas $j \in J$ with a higher attractiveness parameter U_{sf}^j. It follows that the total flow of crime of type $k \in K$ through $j \in J$ can be computed by summing over all origin nodes $s \in S^k$ and all destination nodes $f \in F^k$:

$$x(j, k) = \sum_{s \in S^k} \sum_{f \in F^k} \frac{e^{\lambda U_{sf}^j}}{\sum_{s' \in S^k} \sum_{j' \in J} e^{\lambda U_{s'f}^{j'}}} \cdot b(f) \quad \forall j \in J, \forall k \in K.$$

Based on this parameter, we propose the following values for the players' payoff values:

$$A^k(j|u) = x(j, k) \cdot AG(k) \qquad \forall j \in J, \forall k \in K,$$
$$A^k(j|c) = -x(j, k) \cdot OC(k) \qquad \forall j \in J, \forall k \in K,$$
$$D^k(j|c) = 0 \qquad \forall j \in J, \forall k \in K,$$
$$D^k(j|u) = -x(j, k) \cdot AG(k) \qquad \forall j \in J, \forall k \in K,$$

where $AG(k)$ denotes the average gain of successfully committing crime $k \in K$, and $OC(k)$ the opportunity cost of being captured while attempting to perpetrate a crime $k \in K$. Note that the reward Carabineros perceives when capturing a criminal is 0, irrespective of the crime. Carabineros is only penalized when a crime is successfully perpetrated on their watch. These values were calculated following open source references [16–18] and where then vetted by Carabineros to ensure that our estimates were realistic.

4.2 Building Software for Carabineros

We provide Carabineros with a graphical user interface developed in PHP to determine optimal weekly schedules for the night shift actions for a set of border precincts in the XV region of Chile. The software provided for Carabineros is divided into two parts: a first part devoted to the parameter generation of the game according to the indications of the previous section, and a second part, which solves for the optimal deployment of resources. We discuss the two parts separately.

Parameter Estimation Software. The objective of the parameter estimation software is to construct the payoff matrices for the SSG. This software allows for the matrices to be updated when new criminal arrests are recorded in Carabineros' database. The input for this software is a *csv* data file with arrest data which is uploaded to the software. The main screen of the software shows a map of the region to the left and the following options to the right:

1. *Crimes:* Shows all criminal arrests in the area, color-coded according to the type of crime.
2. *Nodes of origin:* Shows the nodes of origin used in the networks constructed to determine the crime flow through the action areas.
3. *Cluster:* Clusters the criminal arrest points and constructs the crime flow networks joining nodes of origin, action areas and the clustered arrest points, which are the destination nodes for the different types of crime. It displays the payoff matrices for the different action areas.
4. *Input file and update:* Allows to upload a *csv* data file with arrest data. One then re-clusters to obtain new destination points and to construct the new crime flow networks that lead to new payoff matrices.

Deployment Generation Software. The deployment generation software is the part of the software that optimizes the SSGs and returns an implementable patrols strategy for Carabineros. The user is faced with a screen that on the left shows a map of the region where the different action areas are color-coded along the border, and on the right shows different available user options. Clicking on an action area reveals the payoff values for that area. The values can be modified on-screen although this is discouraged. The user can additionally select the number of resources in a given paired pair of precincts. Increasing the number of resources can be used to model that a joint detail can perform a night-shift patrol as many times during a week as the number of resources he has. Further, the user can select the number of weekly schedules that are to be sampled from the optimal target coverage distribution, allowing him to change the weekly schedule to a monthly schedule. Once all parameters are set, clicking on solve returns the desired patrol schedule such as the one shown in Table 1.

Once a patrol strategy has been returned, the user can perform several actions. If the patrol is not to the planner's liking, he can re-sample based on the optimal coverage distribution returned by the optimization. This produces a different patrol strategy that still complies with the same coverage distribution over targets. The user can further impose different types of constraints on each paired pair of precincts to model different requirements such as forcing a deployment on a given day of the week or to a particular target. Similarly, the user can forbid a deployment on a given day of the week, or forbid deployment to a given target. Further, the user can ensure that at least one of a subset of targets is protected or that deployment to a given target happens on at least one out of a subset of days. Solving the game under these constraints and sampling will produce a deployment strategy that complies with the user's requirements.

4.3   Robustness of Our Approach

We study the robustness of the solutions produced by our software to variations in the payoff matrices. Specifically, we study the robustness of our method against variations of two key parameters in the payoff generation methodology: λ, which models the defender's belief on how crime flows across the border and $b(f)$, which indicates the number of nodes clustered into a given destination node f. Equivalently, one can consider variations in a vector $h = (h_1, h_2, h_3)$ which determines the number of cluster nodes for the three types of crime considered. We study the effects of variations in the parameter λ and in the vector of cluster nodes h separately.

As a base case, we generate payoffs for the players by setting $\overline{\lambda} = 50$ and $\overline{h} = (6, 6, 6)$. This appears reasonable given the size of the problem and distribution and number of arrests per type of crime in the studied region. Let $\lambda \in \Lambda = \{0.5\overline{\lambda}, 0.75\overline{\lambda}, 1.25\overline{\lambda}, 1.5\overline{\lambda}\}$ and $h \in H = \{(h_1, h_2, h_3) \in \mathbb{N}^3 : h_t = \overline{h}_t \pm s, t \in \{1, 2, 3\}, s \in \{0, 1, 2, 3\}\}$. We denote by $c(\lambda, h)$, the optimal coverage probabilities on the targets when the payoffs have been defined according to λ and h. Given two vectors $p, q \in \mathbb{R}_+^{|J|}$, we consider the usual distance function between them:

$$d(p, q) = \sqrt{\sum_{j \in J} (p_j - q_j)^2}.$$

We identify $\lambda^* \in \arg\max\{d(c(\overline{\lambda}, \overline{h}), c(\lambda, \overline{h}))\}$ and $h^* \in \arg\max\{d(c(\overline{\lambda}, \overline{h}), c(\overline{\lambda}, h))\}$ and plot $c(\overline{\lambda}, \overline{h})$, $c(\lambda^*, \overline{h})$ and $c(\overline{\lambda}, h^*)$.

Figure 7 shows the optimal coverage probabilities $c(\overline{\lambda}, \overline{h})$, $c(\lambda^*, \overline{h})$ and $c(\overline{\lambda}, h^*)$ for a game with five paired police precincts and twenty targets. One can see that the optimal probabilities are very robust towards variations in the number of clusters. As one could expect, they are less robust to variations in the parameter λ. Recall that a low value of λ constructs the payoff matrices under the assumption that crime distributes itself uniformly among the different action areas $j \in J$. It is therefore understandable that the optimal coverage probabilities reflect this by trying to cover the targets uniformly. On the other hand the optimal coverage probabilities tend to be more robust for higher values of λ.

5   Computational Experiments

In this section we run computational experiments to explore the quality of the two-stage sampling method described in Sect. 3 that recovers an implementable

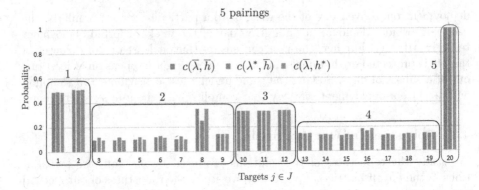

Fig. 7. Robustness of the solution method to variations in the parameters λ and h

defender strategy given an optimal solution to (BP). Further, we analyze the performance of the proposed formulation (BP) against solving the game, that results from explicitly enumerating all the defender pure strategies, with formulation (D2).

5.1   Performance of the Alternative Sampling Method

To evaluate the performance of the proposed alternative sampling method, consider an optimal solution to (BP). In particular, (c^*, z^*) are the optimal coverage distributions over targets and edges. Repeated executions of the sampling method will lead to estimates on said distributions $(\hat{c}, \hat{z})$. In this section, we describe how to construct these estimates and study how close the estimated coverage distributions are to the optimal distributions.

Consider z^* and construct $\hat{z}$ as follows. Sample $i = 1, \ldots, N$ matchings of size m according to the first stage of the sampling method. In our experiments, $N = 1000$. For each edge $e \in E$, its estimated coverage is given by $\hat{z}_e = \frac{1}{N} \sum_{i=1}^{N} z_e^i$, where $z_e^i \in \{0, 1\}$ depending on whether or not edge $e \in E$ was sampled in sampling $i \in \{1, \ldots, N\}$.

We use the Kullback-Leibler divergence [19] to measure the closeness of the two distributions z^* and $\hat{z}$ over instances with n nodes where $n \in \{5, 25, 50, 100\}$. For each instance size, we generate 30 estimations $\hat{z}$ and plot the results as box diagrams as shown in Fig. 8.

Observe that the Kullback-Leibler distance between z^* and $\hat{z}$ is very small, below 0.2 over all instances, which is a good indicator that $\hat{z}$ is a good estimator for z^*. in particular we observe that the larger the set of nodes in an instance the better an estimator $\hat{z}$ appears to be. Further, for instances with 100 nodes, most of the $\hat{z}$ have a Kullback-Leibler distance to z^* which is below 0.02. We performed the same analysis to measure the closeness of the optimal coverage distribution over targets c, to an estimated distribution $\hat{c}$, obtained from N samplings in the second stage of our sampling method, but omit it here due to space limitations. Our analysis reveals that $\hat{c}$ is a good estimator for c^*.

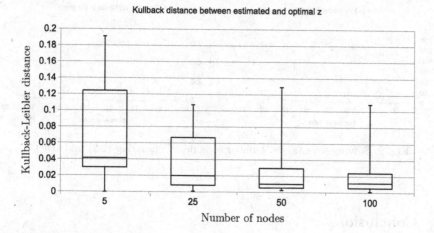

Fig. 8. Kullback-Leibler distance between z^* and $\hat{z}$ over instances of different size

5.2   Performance of (BP)

We study the performance of the proposed formulation (BP) on randomly generated instances against using the formulation (D2) to solve the Stackelberg game
that results from explicitly enumerating all the defender pure strategies. The
instances we consider are generated as follows. We consider random graphs with
n nodes, where $n \in \{5, 6, \ldots, 22\}$ and edges such that the graphs are connected
and that, in average, each node has degree three. Further, we consider four tar-
gets inside each node. The set of targets, J, is thus of size $|J| = 4n$. We consider
$|K| = 3$. We then uniformly generate payoff values for the defender and each
attacker type by considering for each player, rewards $D^k(j|c)$ and $A^k(j|u)$ for
all $k \in K$ and $j \in J$ in the range $[0, 100]$ and penalties $D^k(j|u)$ and $A^k(j|c)$ for
all $k \in K$ and $j \in J$ in the range $[-100, 0]$.

In Fig. 9, we show the running time of the different solution methods over
the generated instances. On the left hand side, we consider instances where the
number of pairings is 2. On the right hand side, we consider instances where
the number of pairings is 3. For these last instances, we only consider graphs
with up to 20 nodes. In both plots, for each instance size, we record the average
solving time of 30 randomly generated instances. Our compact formulation (BP)
outperforms (D2). The set of leader strategies grows exponentially and (D2) can
only explicitly enumerate these strategies for very small graphs of less than 12
nodes. For graphs that both methods can handle, (BP) solves instances much
faster than (D2). Our compact formulation (BP) scales much better than (D2)
being able to comfortably handle instances on graphs with up to 20–21 nodes.

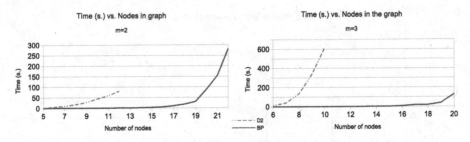

Fig. 9. Solving time (s) vs. number of nodes. Comparing (D2) and (BP)

6 Conclusions

In this paper, we have studied a special type of SSG played on a network. In this game, the defender has to commit to a mixed strategy which consists of two distribution strategies, one over the edges of the network, representing pairings between nodes, and one over the targets of the game which are inside the nodes. The defender can pair m nodes and protect m targets. Further, coverage on a target can only occur if the node in which the target is contained, is incident to a covered edge.

We have provided a compact formulation for the SSG network problem presented and also provided a sampling method to recover an implementable defender strategy given the optimal coverage distributions. In addition, we have described a real-life border patrol problem and have presented a parameter generation methodology that takes into account past crime data and geographical and social factors to construct payoffs for the Stackelberg game. Robustness tests have shown that the solutions our software provides are fairly robust to the networks we generate as well as to minor changes in the flow of crime along the border. Computational tests have shown that the two-stage sampling method we describe, provides implementable strategies that do not deviate much from the optimal coverage distributions. Further computational tests have shown (BP) to have smaller solution times and better scaling capabilities than the extensive formulation (D2) on randomly generated security instances.

There are many promising lines of future work. First, from a Mathematical Programming perspective, we intend to develop decomposition approaches for (BP)–which has an exponential number of the so-called odd set inequalities–to allow it to efficiently solve instances on larger graphs. Second, from a modeling perspective several enhancements could be addressed. In the model presented, a single security resource is available to patrol a target in a pairing of precincts, as it happens in the border patrol problem studied. A natural extension is to consider that different pairings of precincts have different numbers of security resources available to patrol. Further, our payoff estimation methodology could be enhanced in different ways. Temporal weighing of crime data would increase the relative importance of the more recent crimes. Our estimation methodology currently builds the attractiveness of the action areas for a certain type of criminal based on road density around the action area and distances to be traveled by the criminals from source to destination. Other environmental factors such as maximum altitude or availability of shelter along a route or distance of settlements from a route could be taken into account to compute a more realistic attractiveness of an outpost. The research question that remains is verifying whether these modeling enhancements can lead to a better payoff estimation and, thus, to a better representation of the game. Finally, we plan to evaluate the proposed patrol planner following deployment. This evaluation should include both a comparison of crime rate data before and after the deployment of this system and the expert validation that Carabineros de Chile will undertake.

Acknowledgements. Casorrán wishes to acknowledge the FNRS for funding his PhD research through a FRIA grant. This work is also partially supported by the Interuniversity Attraction Poles Programme P7/36 "COMEX" initiated by the Belgian Science Policy Office. Ordóñez acknowledges the support by CONICYT through Fondecyt grant 1140807 and the Complex Engineering Systems Institute, ISCI (CONICYT: FB0816).

References

1. Council of the EU: European border and coast guard: final approval (2016). http://www.consilium.europa.eu/en/press/press-releases/2016/09/14-european-border-coast-guard/. Accessed Feb 2017
2. Department of Homeland Security: Border security (2017). https://www.dhs.gov/border-security. Accessed Feb 2017
3. von Stackelberg, H.: Principios de teoria economica. Oxford University Press, New York (1952)
4. Bracken, J., McGill, J.T.: Mathematical programs with optimization problems in the constraints. Oper. Res. **21**(1), 37–44 (1973)
5. Jain, M., Tsai, J., Pita, J., Kiekintveld, C., Rathi, S., Tambe, M., Ordóñez, F.: Software assistants for randomized patrol planning for the lax airport police and the federal air marshal service. Interfaces **40**(4), 267–290 (2010)
6. Shieh, E., An, B., Yang, R., Tambe, M., Baldwin, C., DiRenzo, J., Maule, B., Meyer, G.: Protect: a deployed game theoretic system to protect the ports of the United States. In: Richland, S. (ed.) Proceedings of the 11th International Conference on Autonomous Agents and Multiagent Systems, AAMAS 2012, Richland, SC, vol. 1, pp. 13–20. International Foundation for Autonomous Agents and Multiagent Systems (2012)
7. Yin, Z., Jiang, A.X., Tambe, M., Kiekintveld, C., Leyton-Brown, K., Sandholm, T., Sullivan, J.P.: Trusts: scheduling randomized patrols for fare inspection in transit systems using game theory. AI Mag. **33**(4), 59–72 (2012)
8. Yang, R., Ford, B., Tambe, M., Lemieux, A.: Adaptive resource allocation for wildlife protection against illegal poachers. In: International Conference on Autonomous Agents and Multiagent Systems (AAMAS) (2014)
9. Kiekintveld, C., Jain, M., Tsai, J., Pita, J., Ordóñez, F., Tambe, M.: Computing optimal randomized resource allocations for massive security games. In: Proceedings of The 8th International Conference on Autonomous Agents and Multiagent Systems, AAMAS 2009, Richland, SC, vol. 1, pp. 689–696. International Foundation for Autonomous Agents and Multiagent Systems (2009)

10. Paruchuri, P., Pearce, J.P., Marecki, J., Tambe, M., Ordóñez, F., Kraus, S.: Playing games for security: an efficient exact algorithm for solving Bayesian Stackelberg games. In: Proceedings of the 7th International Joint Conference on Autonomous Agents and Multiagent Systems, AAMAS 2008, Richland, SC, vol. 2, pp. 895–902. International Foundation for Autonomous Agents and Multiagent Systems (2008)

11. Leitman, G.: On generalized Stackelberg strategies. J. Optim. Theor. Appl. **26**(4), 637–643 (1978)

12. Edmonds, J.: Maximum matching and a polyhedron with 0, l-vertices. J. Res. Nat. Bur. Stand. B **69**(1965), 125–130 (1965)

13. Farkas, J.: Theorie der einfachen ungleichungen. Journal für die reine und ange-wandte Mathematik **124**, 1–27 (1902)

14. Schrijver, A.: Combinatorial Optimization - Polyhedra and Efficiency. Springer, Heidelberg (2003)

15. QGIS: QGIS Geographic Information System. Open Source Geospatial Foundation (2009)

16. Comisión Económica para América Latina y el Caribe: Costo económico de los delitos, niveles de vigilancia y políticas de seguridad ciudadana en las comunas del gran santiago (2000). http://www.cepal.org/es/publicaciones/7258-costo-economico-de-los-delitos-niveles-de-vigilancia-y-politicas-de-seguridad

17. Aduanas de Chile: Qué tributos deben pagar las importaciones? (2016). https://www.aduana.cl/importaciones-de-productos/aduana/2007-02-28/161116.html

18. Ministerio del Trabajo y Previsión Social: Reajusta monto del ingreso mínimo men-sual, de la asignación familiar y maternal del subsidio familiar, para los periodos que indica (2016). http://www.leychile.cl/Navegar?idLey=20763

19. Kullback, S., Leibler, R.A.: On information and sufficiency. Ann. Math. Stat. **22**(1), 79–86 (1951)

Strategic Defense Against Deceptive Civilian GPS Spoofing of Unmanned Aerial Vehicles

Tao Zhang[(✉)] and Quanyan Zhu

Department of Electrical and Computer Engineering, Tandon School of Engineering,
New York University, Brooklyn, NY 11201, USA
{tz636,qz494}@nyu.edu

Abstract. The Global Positioning System (GPS) is commonly used in civilian Unmanned Aerial Vehicles (UAVs) to provide geolocation and time information for navigation. However, GPS is vulnerable to many intentional threats such as the GPS signal spoofing, where an attacker can deceive a GPS receiver by broadcasting incorrect GPS signals. Defense against such attacks is critical to ensure the reliability and security of UAVs. In this work, we propose a signaling game framework in which the GPS receiver can strategically infer the true location when the attacker attempts to mislead it with a fraudulent and purposefully crafted signal. We characterize the necessary and sufficient conditions of perfect Bayesian equilibrium (PBE) of the game and observe that the equilibrium has a PLASH structure, i.e., pooling in low types and separating in high types. This structure enables the development of a game-theoretic security mechanism to defend against the civil GPS signal spoofing for civilian UAVs. Our results show that in the separating part of the PLASH PBE, the civilian UAV can infer its true position under the spoofing attack while in the pooling portion of the PLASH PBE, the corresponding equilibrium strategy allows the civilian UAV to rationally decide the position that minimizes the deviation from its true position. Numerical experiments are used to corroborate our results and observations.

Keywords: Game theory · Signaling game · GPS spoofing · Cybersecurity

1 Introduction

The unmanned aerial vehicle (UAV) is the next generation of aerial platform in various domains. Apart from the military applications, the civilian UAVs are anticipated to play an essential role in commercial applications including business to business (B2B) and business to consumer (B2C) purposes, especially for the delivery systems with logistics services and supply chain support. Prime Air, for example, is a delivery system, currently in development by Amazon, using fully autonomous GPS-guided UAVs to provide rapid parcel delivery (Fig. 1 shows an

© Springer International Publishing AG 2017
S. Rass et al. (Eds.): GameSec 2017, LNCS 10575, pp. 213–233, 2017.
DOI: 10.1007/978-3-319-68711-7_12

Fig. 1. Illustration of a GPS-guided UAV conducts delivery mission between two locations. The attacker in the lower-right corner indicates that the mission is under threat.

example), showing a great potential to improve the efficiency and safety of the overall supply chain system [1].

Emerging applications that primarily depend on autonomous UAV requires a dependable and trustworthy navigation system. Global Positioning System (GPS) is the most common and popular navigation sensor used in the navigation system of UAVs to achieve high-performance flights. In military applications, GPS signals are encrypted to prevent unauthorized use and imitation. However, the current civilian GPS signal is transparent and easily accessible worldwide, which makes the civilian GPS-guided infrastructures vulnerable to different types of GPS spoofing attacks.

It has been shown by researchers in recent literature [22] that civilian UAVs can be easily spoofed. For example, in 2002 researchers from Los Alamos National Laboratory have successfully performed an simplistic GPS spoofing attack [24]. In 2012, Humphreys et al. have shown the spoofing of a UAV by sending the false positional data to its GPS receiver and thus misled the UAV to crash into the sand [7].

Therefore, it is imperative to develop an appropriate defense mechanism to make the civilian GPS dependable for UAVs. Cryptography is one prospective approach. However, the encryption of civilian GPS signals requires high level of secrecy, expense, and scalability. It will create a significant computational and communication overhead when widely used, which can be impractical and limit the scope of its applications. Moreover, the cryptographic keys can be leaked to or stolen by a stealthy adversary who launches an advanced persistent threat (APT) attacks that exploit zero-day exploits and human vulnerabilities. Therefore, an alternative protection mechanism is needed to build a trust mechanism that allows UAV to mitigate the risk of UAV by anticipating the spoof attacks.

To this end, we propose a two-player game-theoretic framework to capture the strategic behaviors of the spoofer and the GPS receiver in which the spoofer aims to inject a counterfeit signal to the UAV to mislead its command and control while the receiver aims to decide whether to estimate the true signal upon receiving the signal. In the two-player game, the receiver does not know the true signal while the adversary knows the correct signal and is able to generate a counterfeit one. To capture the information asymmetry, we use a continuous-kernel signaling game model in which the receiver does not completely know its current location but can form a belief given the received GPS signal. The location of the UAV can be taken as the private information of the sender and hence it is taken as the type of the sender, which is a continuous variable unknown to the receiver. This treatment aligns with the literature in the games of incomplete information. The objective of the receiver is to estimate the correct location based on the received signal and the risk of trusting it. The spoofer, on the other hand, designs a deceptive scheme to manipulate the UAV to move toward an adversarial direction. The spoofer can act stealthily by carefully crafting a signal that takes into account the response of the receiver. The equilibrium analysis of the two-stage game with information asymmetry provides a fundamental understanding of the risk of a UAV under spoofing attacks and yields a strategic trust mechanism that can defend against a rational attacker.

Our results show that the perfect Bayesian equilibrium (PBE) of the game is pooling in low types and separating in high types (PLASH), known as a PLASH PBE. In the separating part of the PLASH PBE, the UAV can strategically infer its true position under the spoofing attack; while in the pooling part of the PLASH PBE, the civilian UAV could not infer its true position exactly, but the corresponding equilibrium strategy enables the civilian UAV to rationally decide the position that minimizes the deviation from its true position. When the deception cost is small enough relative to the level of deviation of aimed by the spoofer, the PLASH PBE becomes a fully pooling PBE (PPBE); while the deception cost is sufficiently large compared to the level of deviation, the PLASH PBE becomes a fully separating PBE (SPBE). These two PBEs coincide with the intuition that the spoofer prefers pooling (resp. separating) strategy when the deception cost is low (resp. high). The main contributions of this paper are summarized as follows:

(i) We model the deceptive spoofing using a continuous-kernel signal game framework and capture the information asymmetry between the sender and the receiver through the private type.
(ii) We develop a risk-based defense mechanism in which the GPS receiver can strategically trust the received messages by taking into account the spoofing threat that a civilian UAV is subject to.
(iii) We characterize the PLASH perfect Bayesian equilibrium (PBE) of the signaling game between the GPS spoofer and the UAV, which has implications in developing defense mechanisms.

1.1   Related Work

There have been a number of approaches based on cryptography proposed to defend against GPS spoofing attacks. For example, spreading code encryption (SCE) [6,18] is currently the only cryptographic technique in widespread use, exclusively in military applications [21]. Techniques based on SCE have provided a very high degree of resistance to the GPS spoofing attacks; however, the high level of secrecy, expense, and scalability of such approach makes it impractical for the civilian GPS [21]. Kuhn et al. [12] have used short sequences of spread spectrum security codes to modify the GPS signal to suit the civilian application; however, the modification in the standard signal protocols makes it impractical to be widely use [21]. Other cryptographic techniques include the navigation message authentication (NMA) [18,25,27], which allows both the uncertified and certified GPS receivers to read navigation messages with different levels of security; however, it has shown that NMA can be fully circumvented by powerful spoofers [6,16].

There has also been a significant amount of work on GPS spoofing defense techniques based on signaling processing [5]. For example, receiver autonomous integrity monitoring (RAIM) is the most widely used approach to detecting GNSS spoofing attacks [8,13]; RAIM is successful in any spoofing attacks that confined to one or two aberrant satellites, but fails when the attacks are confined to the entire constellation [21]. Another line of anti-spoofing work lies in the correlation with other GNSS sources. For example, the external sources of position and timing information such as inertial measurement unit (IMU) is one of the possible sources for the verification of the GPS position data [8,13]. These techniques can accumulate errors due to the inaccuracy of external sources compared to the GPS signal, thereby causing a quick drift from the accurate information. There are also anti-spoofing techniques using machine learning. For example, Wang et al. [23] have developed a machine learning classifier to detect time synchronization attack in cyber-physical systems.

Game theory has been widely applied in the intrusion detection systems [31], and the cyber security systems in various fields, including wireless networks [10,20], mobile networks [19], and control systems [17,29,30]. Signaling game has attracted attention in the field of cyber security [2,3,28]. Xu et al. [28], for example, have proposed an impact-aware defense mechanism using a cyber-physical signaling game. Casey et al. [2] provided a game-theoretical model to simultaneously study systems properties and human incentives.

In this work, we use the signaling game to capture the strategic interactions between the sender and the receiver. The GPS receiver does not have complete location information and the spoofer aims to send signals to mislead the UAV to another location. The game-theoretic defense provides an algorithmic solution that can be implemented on the embedded system in the UAV against GPS spoofings.

1.2 Organization

This paper is organized as follows. Section 2 presents the problem statement and develops a signaling game model. In Sect. 3, we analyze signaling game, define the PLASH PBE, and provide the necessary and sufficient conditions of the equilibrium. The numerical results are shown in Sect. 4. Finally, Sect. 5 concludes the paper.

2   Problem Statement

In this section, we formulate the game-theoretic model for UAV spoofing. First, we describe the dynamic state-space control model of the UAV and show that the UAV can be manipulated by controlling the source of the position information. Then, we describe the GPS signal spoofing attack model. Finally, we develop a signaling game model for the strategic defence mechanism.

2.1   State-Space Model of UAV

Consider an autonomous UAV that conducts a delivery mission from the origin to the destination as shown in Fig. 1. Suppose that the navigation of the UAV is fully supported by the GPS, and there is no other infrastructure such as radar that can provide navigation information. For each specific mission, the UAV flies along a prescribed flight path. Without loss of generality, we assume that the UAV flies at the same altitude; thus we focus on the 2-dimensional (2-D) navigation model with longitude and latitude.

Let $t = [t_x, t_y]$, $v = [v_x, v_y]$ and $\lambda = [\lambda_x, \lambda_y]$ be position, velocity and acceleration of the UAV, respectively, where J_x and J_y are the x and y components of $J \in \{t, v, \lambda\}$. Note that we use t to denote the position, which is referred as the *type* in the signaling game or the incomplete information of the game. The linear state-space model for the UAV plant is described as:

$$\dot{\chi}_z = \Lambda \chi_z + B \lambda_z,$$

where $\dot{\chi}_z = \begin{bmatrix} v_z \\ \lambda_z \end{bmatrix}$, $\chi_z = \begin{bmatrix} t_z \\ v_z \end{bmatrix}$, for $z \in \{x, y\}$, $\Lambda = \begin{bmatrix} 0 & 1 \\ 0 & 0 \end{bmatrix}$, $B = \begin{bmatrix} 0 \\ 1 \end{bmatrix}$. Thus, the state χ is driven by an acceleration λ, which is the control input. The control objective of the UAV is to track a prescribed flight path. Let $\tilde{t} = [\tilde{t}_x, \tilde{t}_y]$, $\tilde{v} = [\tilde{v}_x, \tilde{v}_y]$, and $\tilde{\lambda} = [\tilde{\lambda}_x, \tilde{\lambda}_y]$ be the prescribed reference position, velocity, and acceleration, respectively. Similarly, the double integrator dynamics of the prescribed reference model is $\dot{\tilde{\chi}}_z = \Lambda \tilde{\chi}_z + B \tilde{\lambda}_z$, where $\dot{\tilde{\chi}}_z = \begin{bmatrix} \tilde{v}_z \\ \tilde{\lambda}_z \end{bmatrix}$, $\tilde{\chi}_z = \begin{bmatrix} \tilde{t}_z \\ \tilde{v}_z \end{bmatrix}$, for $z \in \{x, y\}$. We model the controller of the UAV by a Proportional-Derivative (PD) compensator $\lambda_z = -K(\chi_z - \tilde{\chi}_z)$, where $K = [K_p, K_d]$ is the gain matrix with $K_p, K_d > 0$ such that the closed-loop control system is stable. Thus, the continuous-time linear state space model of the UAV can be written as:

$$\begin{bmatrix} \dot{\chi}_z \\ \dot{\tilde{\chi}}_z \end{bmatrix} = \begin{bmatrix} \Lambda - BK & BK \\ 0 & \Lambda \end{bmatrix} \begin{bmatrix} \chi_z \\ \tilde{\chi}_z \end{bmatrix} + \begin{bmatrix} 0 \\ B \end{bmatrix} \tilde{\lambda}_z. \tag{1}$$

We consider the case when GPS is the only source of navigation information. Suppose the UAV receives a GPS signal indicating a current position $t = (t_x, t_y)$ that shows a deviation of the UAV from the prescribed flight path. The controller adjusts the velocity v and the acceleration λ according to the state space model (1) as: $v_z = (\Lambda + BK)t_z + BK\tilde{t}_z$, and $\lambda = (\Lambda - BK)v_z + BK\tilde{v}_z$, for $z \in \{x, y\}$.

As shown in Sect. 2.2, a GPS spoofer aims to mislead the UAV to a wrong destination via creating a reset flight path by GPS signaling spoofing. The GPS spoofer starts a spoofing attack by sending a fake GPS signal indicating a wrong position $t' = (t'_x, t'_y)$ that shows a fake deviation. The reset flight path is determined based on the first spoofing signal. In this paper, we only consider that once the reset flight path is determined, it is fixed during the entire delivery mission. If the UAV is naive, its controller completely accept $t' = (t'_x, t'_y)$. The corresponding v'_z and λ'_z are then obtained; the GPS spoofer continues spoofing the GPS signal based on the first spoofed signal to lead the UAV to fly on the reset flight path towarding the wrong destination while making the controller believe it is the original prescribed flight path. We model the communication between the GPS spoofer and the UAV by a signaling game, and show that the strategic acceptance of $t' = (t'_x, t'_y)$ will significantly reduce or completely avoid the damage that might be caused by the spoofing attack.

2.2   GPS Signal Spoofing

In this paper, we consider a GPS signal spoofer located from a distance as shown in Fig. 2. At time τ during one mission, the spoofer starts to launch an spoofing attack. The spoofer is capable of capturing the authentic navigation message for the UAV from all visible GPS satellites and sends the counterfeit navigation message to the UAV as shown in Fig. 2. The navigation message from GPS satellites does not directly reveal the 2D position; instead, the message contains the time and the orbital information of the GPS satellites for computing the 2D position by the GPS receiver of the UAV via 2D trilateration. The spoofer aims to make the GPS receiver of the UAV report the current location as the simulated position $t' = [t'_x, t'_y]$ while the true position is $t = [t_x, t_y]$.

Starting from time τ, the spoofer continuously sends the UAV the counterfeit navigation messages such that the UAV would be deceived to fly along the reset flight path as shown in Fig. 3. The deviation between the true path and the reset path depends on the simulated position chosen by the adversary at time τ.

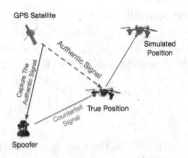

Fig. 2. Illustration of a GPS spoofing attack targeting a GPS-guided UAV.

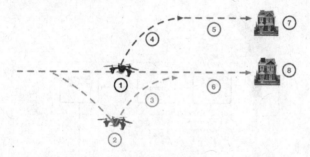

Fig. 3. Illustration of a complete GPS spoofing procedure. 1: True position of the UAV; 2: Counterfeit GPS signal makes the UAV think that its current position is deviated from the original path; 3: UAV control system adjusts the velocity and acceleration to return to the original path; 4: Actual move of the UAV; 5: Reset path; 6: Original path; 7: Wrong destination; 8: Correct destination.

2.3 Signaling Game

In this sub-section, we propose a game-theoretic cyber-security mechanism to capture the receiver's uncertainties on the received GPS signals, which can be either the true locations or the counterfeit ones. The analysis of the game yields a defense mechanism that allows the UAV to strategically minimize its risk and deal with the GPS signal spoofing without terminating the mission or resorting to other costly navigation infrastructures.

Signaling games are a class of the incomplete information games, in which one player has more information than the other. Specifically, the more informed player strategically decides to signal the private information called *type*, which is unknown to the opponent; the less informed player decides how to respond to the signal received [9,15]. In this paper, we model the communications between the GPS spoofer and the UAV by the signaling game and propose a game-theoretic approach to dealing with the GPS deception.

In our scenario, the role of GPS spoofer is the signal sender, denoted as S, and the role of GPS receiver of the UAV is the signal receiver, denoted as R. It is clear that the GPS spoofer is the more informed player and the UAV is the less informed counterpart. To capture the information asymmetry, we use the signaling game framework in which the navigation message (thus the position information) is only known to S. The position is viewed as type $t = [t_x, t_y] \in T$, where t_x and t_y are the latitude and longitude, respectively, in the form of decimal degrees, and $T = [t_x^m, t_x^M] \times [t_y^m, t_y^M]$ is the 2D location space with t_z^m and t_z^M are the minimum and maximum values of $z \in \{x, y\}$, respectively, which are determined based on the mission of the UAV. Note that the position or the type t takes a continuum of values in set T. Hence the game is a continuous-kernel signal game.

Let $m \in M$ be the navigation message sent by S. We denote $\Omega(m) = [\Omega_x(m), \Omega_y(m)] : M \to T$ as the 2D trilateration function to compute the 2D position. The output of the computation is $t' = [t'_x, t'_y] = \Omega(m)$ is the position

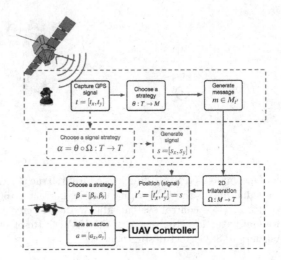

Fig. 4. Illustration of the signaling game model. The procedure represented by the solid blue line is equivalent to the procedure represented by the dashed blue line, i.e., the strategy θ generates a message m that tells R the position $t' = \Omega(m) = s$, where s is the signal generated by the signal strategy α.

claimed in message m. This process is illustrated in Fig. 4. The procedure of 2D trilateration is a pure mathematical computation and there is no strategic activity involved; thus, we can equivalently regard the action of generating message m as the action of generating a signal $s = [s_x, s_y] \in T$, i.e., choosing $s = t'$ means is equivalent to generating a message m that indicates $t' = [t'_x, t'_y] = \Omega(m)$.

The signaling game is played at τ, which is chosen by the spoofer, S. Since the choice of τ contains no strategic activity, we assume that τ is chosen according to a uniform distribution. Suppose a UAV, R, is flying at position $t = [t_x, t_y]$ at time τ. Here, we assume that t_x and t_y are drawn independently according to a uniform distribution over a credible interval to the receiver. After capturing the authentic navigation message for R from the GPS satellites, S generates a counterfeit message $m \in M$ leading to $t' = \Omega(m)$ or, equivalently, generates a signal $s = t'$. Then, S sends message m to R (equivalently sends signal s to R). Sender S tells the truth if $s = t$; otherwise, $s = t'$, for $t' \neq t$. Once s is observed by the receiver, R can strategically estimate the true location t by taking an action $a = [a_x, a_y] \in A$. It is natural to take $A = T$. The receiver then estimates the position of the UAV based on its belief and the received message. The navigation system of the UAV then adjusts the direction and speed according to the estimated position.

S has the cost function $C^S(a, t, s) = C^A(a, t) + k_1 C^D(t, s) : A \times T \times T \to \mathbb{R}$, where $C^A(a, t) : A \times T \to \mathbb{R}$ is the action-related cost, and $C^D(t, s) : T \times T \to \mathbb{R}$ is the deception cost, and $k_1 > 0$ is a constant scaling the intensity of the deception cost. The signal s (thus the message m) is only cost relevant to S in C^D. R has the cost function $C^R(a, t) : A \times T \to \mathbb{R}$. The goal of S is to choose a message to minimize the cost function by anticipating the action of R, while the goal of R is to take an action to minimize the cost function based on the belief about the true type after observing the signal s.

Suppose that the true type is $t = [t_x, t_y]$. S chooses the message m claiming $t' = \Omega(m)$ based on the pure strategy, which is a measurable function $\theta(t) = [\theta_x(t_x), \theta_y(t_y)] : T \to M$. Equivalently, we define a measurable function $\alpha(t) = [\alpha_z(t_z), \alpha_z(t_z)] := T \to T$ as the signal strategy, based on which S chooses the signal s. The aforementioned relationship between s and m yields $\alpha(t) = t'$. The interpretation is that the signal strategy $\alpha(t)$ indicates the position S wants R to believe. R chooses its action $a = [a_x, a_y]$ using a pure strategy $\beta(\Omega(m)) : M \to T$. Based on the action, the strategically chosen position is sent to the UAV control system. The signaling game model is illustrated in Fig. 4.

Due to the fact that no GPS satellite is in a geostationary orbit, all the GPS satellites are moving all the time with respective to the ground; thus, there exists a message subspace M_t such that for each pair of different messages m_i, $m_j \in M_t$, we have $\Omega(m_i) = \Omega(m_j) = t$. Thus, every message $m \in M_t$ gives $\Omega(m) = t$. Clearly, $M = \cup_t M_t$ and $|M_t| = \infty$. Therefore, S can send an infinite number of messages for any strategy $\theta(t)$. Equivalently, we can claim that for every specific signal strategy $\alpha(t) = t'$, there is an infinite number of messages $m \in M_{t'}$ that S can choose.

3   Signaling Game Analysis

In this section, we define the cost functions of the sender S and the receiver R and analyze the solution of the signaling game based on the perfect Bayesian equilibrium (PBE).

3.1   Cost Function and Strategy

Let $C^A(a, t) = \| a - t - L \|^2$ and $C^D(t, m) = \| s - t \|^2 + \rho \| s \|^2$. The cost function of S is defined as:

$$
\begin{aligned}
C^S(a, t, s) &= C^A(a, t) + k_1 C^D(t, s) \\
&= \| a - t - L \|^2 + k_1 \left(\| s - t \|^2 + \rho \| s \|^2 \right) \\
&= \left[(a_x - t_x - l_x)^2 + k_1 \left((s_x - x)^2 + \rho s_x^2 \right) \right] \\
&\quad + \left[(a_y - t_y - l_y)^2 + k_1 \left((s_y - y)^2 + \rho s_y^2 \right) \right],
\end{aligned}
\tag{2}
$$

where $L = (l_x, l_y)$ with $l_x, l_y > 0$ represents the malignity of S that models the conflict of interests between S and R. Therefore, the optimal action that minimizes the cost function of R leads to a strictly positive C^A, $\rho \| s \|^2$ with $\rho > 0$ models the other cost including message generation cost and transmission cost, and $k_1 > 0$ parameterizes the intensity of the cost C^D.

The cost function of R is defined as:

$$
C^R(a, t) = k_2 \| a - t \|^2 = k_2 (a_x - t_x)^2 + k_2 (a_y - t_y)^2,
\tag{3}
$$

where $k_2 > 0$ is a constant. Let $C^{S,z} = (a_z - t_z - l_z)^2 + k_1((s_z - t_z)^2 + \rho s_z^2)$, for $z \in \{x, y\}$, and let $C^{R,x} = k_2(a_x - t_x)^2$ and $C^{R,y} = k_2(a_y - t_y)^2$. Therefore, R chooses an action $a = (a_x, a_y)$ to solve the following problem

$$\min_{a \in A} C^R(a, t) := C^{R,x} + C^{R,y}. \tag{4}$$

S aims to choose a message m to solve the following problem

$$\min_{s \in T} C^S(a, t, s) := C^{S,x} + C^{S,y}. \tag{5}$$

Since t_x and t_y are generated independently. Thus, $\min_s C^{S,x}$ and $\min_s C^{S,y}$ are independent to each other and can be solved independently and so are $\min_{a_x} C^{R,x}$ and $\min_{a_y} C^{R,y}$. Therefore, $\min_{a \in A} C^R(a, t) = \min_{a_x} C^{R,x} + \min_{a_y} C^{R,y}$, and $\min_{s \in T} C^S(a, t, m) = \min_{s_x} C^{S,x} + \min_{s_y} C^{S,y}$. Then, (4) and (5) are equivalent to the following

$$\min_{a_z} C^{R,z}(a_z, t_z) = k_2(a_z - t_z)^2, \tag{6}$$

and $\min_{s_z} C^{S,z}(a_z, t_z, s_z) = C^{A,z}(a_z, t_z) + k_1 C^{D,z}(t_z, s_z)$, where $C^{A,z}(a_z, t_z) = (a_z - t_z - l_z)^2$ and $C^{D,z}(t_z, s_z) = (s_z - t_z)^2 + \rho s_z^2$, for $z \in \{x, y\}$ (hereafter). The function $C^{A,z}(\cdot, \cdot)$ and $C^{R,z}(\cdot, \cdot)$ are double differentiable at both arguments with $C_{12}^{A,z} < 0 < C_{11}^{A,z}$ and $C_{12}^{R,z} < 0 < C_{11}^{R,z}$; thus, $C^{A,z}$ and $C^{R,z}$ are convex in action a_z and super-modular in (a_z, t_z). Let $a_{R,z}^*(t_z) := \arg\min_{a_z} C^{R,z} = t_z$ and $a_{S,z}^*(t_z) := \arg\min_{a_z} C^{S,z} = t_z + l_z$, respectively, be the most preferred action (taken by R) for R and S with $\frac{da_{J,z}^*(t_z)}{dt_z} > 0$ for $J \in \{R, S\}$; and $a_{R,z}^*(t_z) < a_{S,z}^*(t_z)$ that coincides with the existence of conflict of interest. $C^{D,z}(\cdot, \cdot)$ is double differentiable for both arguments and $C_{12}^{D,z} < 0 < C_{11}^{D,z}$, which implies that given a type t_z, a larger s_z leads to a larger deception cost.

Based on the pure strategy $\alpha(t)$, S chooses a signal $s(t) = (s_x(t_x), s_y(t_y))$ and sends a corresponding message m. After observing the signal s_z, R updates its posterior belief about t_z, denoted as $g_z(t_z|s_z)$, using Bayes' rule. Using the pure strategy $\beta(s) = (\beta_x(s_x), \beta_y(s_y))$, R takes an action $a = (a_x, a_y)$. Let $p_z(t_z)$ be the prior belief of R about type t_z. Let $q^{S,z}(s_z)|t_z)$ and $q^{R,z}(a_z|s_z)$ be the probability distributions induced by $\alpha_z(t_z)$ and $\beta_z(s_z)$, respectively, which satisfy

$$\int_{s_z \in T} q^{S,z}(s_z|t_z) ds_z = 1, \quad \int_{a_z} q^{R,z}(a_z|s_z) da_z = 1.$$

Our solution concept to deal with the GPS signal deception in the signaling game model is the perfect Bayesian equilibrium, which is defined as follows.

Definition 1. *The strategy profile $(\alpha(t), \beta(s(t))$ with the belief $g_z(t_z|s(t))$ of the signaling game is a the perfect Bayesian equilibrium (PBE) if*

– *(Consistent belief)* for all s_z,

$$g_z(t_z|s_z) = \begin{cases} \frac{p_z(t_z)q^{S,z}(s_z|t_z)}{\int_{\hat{t}_z} p_z(\hat{t}_z)q^{S,z}(s_z|\hat{t}_z)d\hat{t}_z} & \text{if } \int_{\hat{t}_z} p_z(\hat{t}_z)q^{S,z}(s_z|\hat{t}_z)d\hat{t}_z > 0, \\ any\ distribution & otherwise. \end{cases}$$

– *(Sequential rationality)*

$$\alpha(t) \in \arg\min_{s \in T} C^S(\beta(s_z), t_z, s_z),$$

$$\beta_z(s_z) \in \arg\min_{a_z} \int_{t_z} g_z(t_z|s_z)C^{R,z}(a_z, t_z)dt_z.$$

Remark 1. There are two pure strategy equilibria. One is the separating PBE (SPBE), in which S chooses strategies for different types and the other one is the pooling PBE (PPBE), in which S uses the same strategy for different types.

3.2 Equilibrium Analysis

In this section, we characterize the equilibrium of the signaling game model. In our scenario of GPS signal deception, S aims to lead R to believe the type that is actually deviated from the true type. In this paper, we focus on the pure PBE strategy, and consider the case when $\frac{da_z(t_z)}{dt_z} \geq 0$.

First, we consider if there exists a SPBE. In any differentiable SPBE, the cost function $C^{S,z}$ and the signal strategy α_z have to satisfy the following necessary first-order condition for optimality based on the sequential rationality:

$$C_1^{S,z}(a_{R,z}^*(t_z), t_z, \alpha_z(t_z))\frac{da_{R,z}^*(t_z)}{dt_z} + C_3^{S,z}(a_{R,z}^*(t_z), t_z, \alpha_z(t_z))\frac{d\alpha_z(t_z)}{dt_z} = 0. \quad (7)$$

However, since $\frac{d\alpha_z(t_z)}{dt_z} \geq 0$ and $C_1^{S,z}(a_{R,z}^*(t_z), t_z, \alpha_z(t_z)) = 2(a_{R,z}^*(t_z) - t_z - l_z) = -2l_z$ is independent of $\alpha_z(t_z)$, there is no strategy such that $C_1^{S,z}\frac{da_{R,z}^*(t_z)}{dt_z} = 0$ when $C_3^{S,z} = 0$. Instead, we rearrange (7) and obtain the following differential equation:

$$\frac{d\alpha_z(t_z)}{dt_z} = -\frac{C_1^{S,z}(a_{R,z}^*(t_z), t_z, \alpha_z(t_z))\frac{da_{R,z}^*(t_z)}{dt_z}}{C_3^{S,z}(a_{R,z}^*(t_z), t_z, \alpha_z(t_z))} = \frac{l_z}{k_1((1+\rho)\alpha_z(t_z) - t_z)},$$

to circumvent the case when $C_3^{S,z} = 0$. Let $\alpha^*(t) = \arg\min_s C^D$ be the signal strategy of choosing a signal $s^*(t) = (s_x^*(t_x), s_y^*(t_y))$ that minimizes the deception function. Then, $s_z^*(t_z) = \frac{t_z}{1+\rho} < t_z$ with $\frac{ds_z^*(t_z)}{dt_z} > 0$. We summarize the property of the strategy $\alpha_z(t_z)$ in any separating regime of the type space in the following lemma.

Lemma 1. *We say that in the type space $(t_z^s, t_z^l) \subset [t_z^m, t_z^M]$, the signaling game has a monotone SPBE with strategy $\alpha_z(t_z)$ if for each $t_z \in (t_z^s, t_z^l)$, $\alpha_z(t_z) > s_z^*(t_z)$, and*

$$\frac{d\alpha_z(t_z)}{dt_z} = \frac{l_z}{k_1((1+\rho)\alpha_z(t_z) - t_z)}. \tag{8}$$

Proof. See the proof in Appendix A.1.

Based on Lemma 1, we can conclude the following theorem.

Theorem 1. *There exists a unique SPBE portion $[\hat{t}, t_z^M] \subseteq [t_z^m, t_z^M]$ with initial condition $\alpha_z^*(t_z^M) = t_z^M$, where $\alpha_z^*(t_z)$ is the solution to (8).*

Proof. See the proof in Appendix A.2.

Since $\frac{d\alpha_z(t_z)}{dt_z} \geq 0$, $\frac{d\alpha_z^*(t_z)}{dt_z} > 0$, which means that in any separating region, the SPBE strategy of S is strictly increasing; thus, according to (8), we must have $\alpha_z^*(t_z) > \frac{t_z}{1+\rho} = s_z^*(t_z)$. Since S tells the truth if the type is t_z^M at the time τ (when S launches a spoofing attack), i.e., $\alpha_z(t_z^M) = t_z^M$, if t_z^M is in the separating region, $\alpha_z^*(t_z^M) = t_z^M$, which satisfies $\alpha_z^*(t_z^M) = t_z^M > \frac{t_z^M}{1+\rho}$. We summarize the existence of a full SPBE in the following corollary.

Corollary 1. *Let $\alpha_z^*(t_z)$ be the unique separating signal strategy given the initial condition $\alpha_z^*(t_z^M) = t_z^M$. There exists a single SPBE in the entire type space $[t_z^m, t_z^M]$, if $\alpha_z^*(t_z^m) = t_z^m$, which depends on the values of l_z and k_1.*

Proof. See the proof in Appendix A.2.

Corollary 1 shows that for certain values of l_z and k_1 there exists a unique single SPBE in the entire type space $[t_z^m, t_z^M]$. However, when there is no single SPBE existing, we are interested in a class of pooling strategy. For the separating region, Theorem 1 shows that there exists a continuous and increasing separating signal strategy function $\alpha_z^*(t_z)$ that solves (8) with initial condition $\alpha_z^*(t_z^M) = t_z^M$ for all $t_z \in [\hat{t}_z, t_z^M]$, where $\hat{t}_z \in (t_z^m, t_z^M)$ has a well-defined unique SPBE signal strategy $\alpha_z^*(\hat{t}_z) = t_z^m$. In this case, the maximal feasible interval of separating types is $[\hat{t}_z, t_z^M]$, while for all $t_z \in [t_z^m, \hat{t}_z]$, $\alpha_z(t_z) = t_z^m$. Before analyzing the pooling strategy, we first define the following equilibrium by introducing a boundary type $\bar{t}_z \in [\hat{t}_z, t_z^M]$.

Definition 2. *Let $t^m = (t_x^m, t_y^m)$ and $t^* = (\alpha_x^*(t_x), \alpha_x^*(t_x))$. A strategy θ and the corresponding signal strategy α_z is a PLASH (Pooling in Low types And Separating in High types) strategy if there exists a boundary type $\bar{t}_z \in [\hat{t}_z, t_z^M]$ such that:*

1. *(Pooling strategy) $\theta(t) \in M_{t^m}$ and $\alpha_z(t_z) = t_z^m$ for all $t_z \in [t_z^m, \bar{t}_z)$,*
2. *(Separating strategy) $\theta(t) \in M_{t^*}$ and $\alpha_z(t_z) = \alpha_z^*(t_z)$ for all $t_z \in [\bar{t}_z, t_z^M]$.*

In the pooling type interval, any type $t_z \in [t_z^m, \bar{t}_z]$ induces the equal deception cost since the signal strategy $\alpha_z(t_z) = t_z^m$ is chosen for all $t_z \in [t_z^m, \bar{t}_z]$. Therefore, we can regard the communication in $[t_z^m, \bar{t}_z]$ as a cheap talk [26]. However, as shown in Sect. 2.3, all the message $m \in M_{t_z^m}$ give the same value of signal $s_z = \Omega_z(m)$; then, it is possible for S to choose the same signal strategy $\alpha_z(t_z)$ but different message-related strategy θ so that R can choose distinct actions for different types in the pooling interval $[t_z^m, \bar{t}_z]$. Let $[t_z', t_z''] \subseteq [t_z^m, \bar{t}_z]$. Suppose that based on the message m, R only knows that t_z lies in $[t_z', t_z'']$ for each type $t_z \in [t_z', t_z'']$. Let $\hat{a}_z(t_z', t_z'')$ be defined as follows:

$$\hat{a}_z(t_z', t_z'') = \arg\max_{a_z} \int_{t_z'}^{t_z''} C^{R,z}(a_z, t_z) dt_z = \frac{t_z' + t_z''}{2}.$$

Thus, R takes the same action $\hat{a}_z(t_z', t_z'')$ for each type $t_z \in [t_z', t_z'']$. Therefore, it is possible for R to choose $\hat{a}_z(t_z', t_z'')$ for different intervals $[t_z', t_z''] \subseteq [t_z^m, \bar{t}_z]$.

Indeed, Crawford and Sobel [4] has shown that there exists a pooling-partition for $[t_z^m, \bar{t}_z]$. Specifically, for a boundary type $\bar{t}_z$, $[t_z^m, \bar{t}_z]$ can be partitioned into multiple pooling sub-intervals, which can be represented by a strictly increasing sequence $[t_z^0, t_z^1, ..., t_z^N]$, where $t_z^0 = t_z^m$ and $t_z^N = \bar{t}_z$. Thus, for all $n \in \{1, 2, ..., N-1\}$, the cost for S satisfies

$$C^{S,z}(\hat{a}_z(t_z^{n-1}, t_z^n), t_z^n, s_z(t_z^n)) = C^{S,z}(\hat{a}_z(t_z^n, t_z^{n+1}), t_z^n, s_z(t_z^n)). \tag{9}$$

Note that the deception cost is the same for every type $t_z \in [t_z^m, \bar{t}_z]$, (9) implies $C^{A,z}(\hat{a}_z(t_z^{n-1}, t_z^n), t_z^n) = C^{A,z}(\hat{a}_z(t_z^n, t_z^{n+1}), t_z^n)$. The interpretation is that, for each $t_z \in (t_z^{n-1}, t_z^n)$, S sends the same message $m_n \in M_{t_z^m}$, and R takes the same action $\hat{a}_z(t_z^{n-1}, t_z^n)$. S can send either m_n or m_{n+1} for the connecting type t_z^n. Note that $\alpha_z(t_z)$ is the same for all types $t_z \in [t_z^m, \bar{t}_z]$, but $m_n \neq m_j$ for $n \neq j$ and $m_j \in M_{t_z^m}$; thus S uses the same signal for all types $t_z \in [t_z^m, \bar{t}_z]$ but different messages for types in different pooling sub-intervals and all the messages are chosen from the set $M_{t_z^m}$.

The necessary and sufficient conditions for the existence of PLASH equilibrium are summarized in the following theorem.

Theorem 2 *(Necessary condition).* *In any PLASH equilibrium, there exists a boundary type $\bar{t}_z \in [\hat{t}_z, t_z^M]$ such that the pooling interval $[t_z^m, \bar{t}_z]$ can be partitioned into multiple pooling sub-intervals, denoted by a strictly increasing sequence $[t_z^0, t_z^1, ..., t_z^N]$ with $\frac{0}{z} = t_z^m$ and $t_z^N = \bar{t}_z$, such that*

$$C^{A,z}(\hat{a}_z(t_z^{n-1}, t_z^n), t_z^n) = C^{A,z}(\hat{a}_z(t_z^n, t_z^{n+1}), t_z^n), \forall n \in \{1, ..., N-1\} \tag{10}$$

$$C^{S,z}(\hat{a}_z(t_z^{N-1}, \bar{t}_z), \bar{t}_z, t_z^m) = C^{S,z}(a_{R,z}^*(\bar{t}_z), \bar{t}_z, \alpha_z^*(\bar{t}_z)), if \bar{t}_z < t_z^M. \tag{11}$$

(Sufficient condition). *Given any boundary type and a pooling-partition shown in (10) and (11), and*

$$C^{S,z}(\hat{a}_z(t_z^{N-1}, \bar{t}_z), t_z^M, t_z^m) \leq C^{S,z}(a_{R,z}^*(t_z^M), t_z^M, t_z^M), if \bar{t}_z = t_z^M. \tag{12}$$

There exists a PLASH equilibrium.

In any PLASH equilibrium, both players must play on the equilibrium. Specifically, R chooses strategy $\beta_z(\Omega_z(m_n))$ and takes the action $\hat{a}_z(t_z^{n-1}, t_z^n)$ for any $m_n \in M_{t_z^m}$ with $\theta(t) = m_n$ and $t = [t_x, t_y]$ for all $t_z \in (t_z^{n-1}, t_z^n)$; while for any $t_z \in (\bar{t}_z, t_z^M]$, R chooses $\beta_z(\Omega_z(\theta(t))) = \alpha_{R,z}^*(t_z)$ with $t = [t_x, t_y]$. S chooses the signaling strategy $\alpha_z(t_z) = \alpha_z^*(t_z)$ for all $t_z \in (\bar{t}_z, t_z^M]$, and chooses $\alpha_z(t_z) = t_z^m$ for all $t_z \in [t_z^m, \bar{t}_z]$, and sends message $m_n \in M_{t_z^m}$ for any $t_z \in (t_z^{n-1}, t_z^n)$; for $t_z \in (t_z^{j-1}, t_z^j)$, S sends message $m_j \neq m_n$, but $\Omega_z(m_j) = \Omega_z(m_n) = t_z^m$.

Remark 2. In the separating PBE regime, S chooses the signal strategy $\alpha_z^*(t_z)$, which induces action $a_{R,z}^*$ of R; thus, the signal strategy $\alpha_z^*(t_z)$ reveals the true type; yet this signal strategy is costly since $\alpha_z^*(t_z) > s_z^*(t_z)$, which means that it does not minimize the deception cost $C^{D,z}$. However, if S chooses the least costly strategy $\alpha_z(t_z) = s_z^*(t_z)$, it would cause adverse inferences from R since R expects a certain degree of deception at separating PBE and rationally infers the true type.

4   Numerical Experiments

In this section, we simulate a simple scenario of GPS spoofing and construct a signaling game model in which the UAV plays the receiver (R) and the GPS spoofer plays the sender (S). In the numerical experiments, we set the minimum value and the maximum value of latitude or longitude as $t_z^m = 1$ and $t_z^M = 10$, respectively, and set the constant parameters $\rho = 1$ and $k_2 = 1$. The differential equation (8) becomes

$$\frac{d\alpha_z(t_z)}{dt_z} = \frac{l_z}{k_1(2\alpha_z(t_z) - t_z)}. \tag{13}$$

Let $c = \frac{l_z}{k_1}$, $w = 2\alpha_z(t_z) - t_z$, then $w' = 2\alpha_z'(t_z) - 1$; thus $\frac{d\alpha_z(t_z)}{dt_z} = \frac{w'+1}{2}$; substituting w to (13) yields $\frac{w}{2c-w}dw = dt_z$, which can be integrated and yield the solution form $t_z + \sigma = -w - 2c\ln(2c - w)$, where σ is a constant to be found. We assume that when the UAV reaches the maximum value of latitude (longitude), the spoofer does not spoof on the value of latitude (longitude). Therefore, we have the initial condition $\alpha_z^*(10) = 10$, and then can determine $\sigma = -20 - 2c\ln(2c - 10)$. Thus, the solution of (13) α_z^* satisfies

$$\frac{e^{\frac{-10k_1}{l_z}}k_1}{2l_z - 10k_1}\left(\frac{2l_z}{k_1} - 2\alpha_z^*(t_z) + t_z\right) = e^{\frac{-k_1}{l_z}\alpha_z^*(t_z)} \tag{14}$$

The solutions of (14) are shown in Fig. 5e–f. Since $\alpha_z^*(\hat{t}_z) = 1$, the value of $\hat{t}_z$ can be determined as

$$\hat{t}_z = \frac{2 - 10\frac{k_1}{l_z}}{\frac{k_1}{l_z}}e^{\frac{9k_1}{l_z}} + 2 - 2\frac{l_z}{k_1} \tag{15}$$

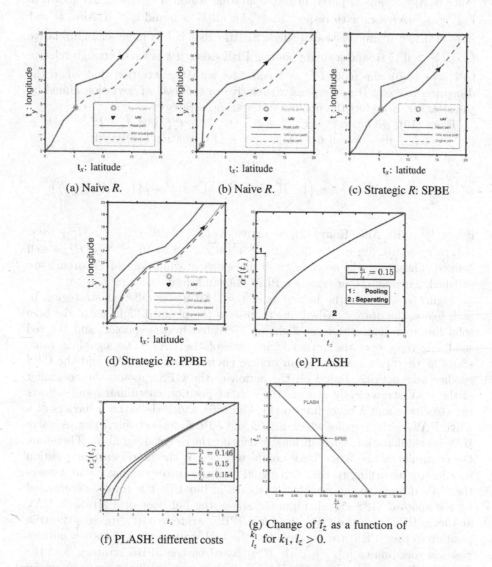

Fig. 5. 5a–d: Examples of UAV scenarios at PLASH equilibrium. The orange circle represents the place where both players take actions. (a) naive UAV (R) at the region where SPBE exists; (b) naive UAV at the region where PPBE exists; (c) strategic UAV at SPBE; (d) strategic UAV at PPBE. 5e–f: Examples of UAV scenarios at PLASH equilibrium (e): PLASH strategies of the GPS spoofer: PLASH (f): PLASH strategies of the GPS spoofer with different deception costs (relative to the malignity of the sender). 5g: Change of $\bar{t}_z$ as a function of $\frac{k_1}{l_z}$ for $k_1, l_z > 0$. PLASH equilibrium exists for all $1 < \bar{t}_z < 2$ (above the red line). (Color figure online)

Since $\alpha_z^*(\hat{t}_z) > \frac{\hat{t}_z}{2}$ is required in the separating region, $1 \leq \hat{t}_z < 2$. As shown in Fig. 5g, $\hat{t}_z$ decreases with respect to $\frac{k_1}{l_z}$, for all $k_1 > 0$ and $l_z > 0$. Also, $\hat{t}_z = 1$ if $\frac{k_1}{l_z} \approx 0.154$; it implies that a single SPBE exists if k is large enough relative to l_z ($\frac{k_1}{l_z} > 0.154$), and a single pooling PBE exists if k is small enough relative to l_z ($\frac{k_1}{l_z} \to 0$); the plot of $\hat{t}_z = 1$ coincides with the intuition that when the deception is cheap (resp. expensive) relative to the level of deviation aimed by the attacker, S prefers the pooling (separating) strategy.

From (10), we have: $t_z^n - t_z^{n-1} + 4l_z = t_z^{n+1} - t_z^n$; thus, $\bar{t}_z - t_z^{N-1} = t_z^n - t_z^{n-1} + 4(N-n)l_z$, for all $n \in \{1, 2, ..., N-1\}$. Equation (11) yields:

$$\left(\frac{t_z^{N-1} - \bar{t}_z}{2} - l_z\right)^2 - l_z^2 = k_1\left((\alpha_z^*(\bar{t}_z) - \bar{t}_z)^2 - (1 - \bar{t}_z)^2 + \rho\left(1 - (\alpha_z^*(\bar{t}_z))^2\right)\right),$$

if $\bar{t}_z < t_z^M = 10$. Also, from (12), we arrive at $t_z^{N-1} \geq 10 + 2l_z - 2\sqrt{l_z^2 + 18k_1}$, if $\bar{t}_z = t_z^M = 10$. Since $10 + 2l_z - 2\sqrt{l_z^2 + 18k_1} < t_z^M = 10$, $t_z^{N-1} < t_z^M$ is well defined. Thus, both the necessary and the sufficient conditions of Theorem 2 are satisfied. Therefore, there exists a PLASH equilibrium.

Figure 5a–d shows the behaviors of the UAV under different strategies. In each figure, the orange dashed line represents the planned flight path, the blue solid line represents the reset flight path created by the spoofer, and the red solid line represents the actual flight path of the UAV. The signaling game starts at the place marked by an orange circle, where the UAV and the GPS spoofer take actions. Based on the action of the GPS spoofer, the controller of the UAV strategically accepts the current position coordinates and adjusts the velocity v and λ according to (1). Figure 5a and b show the behaviors of a naive UAV in the regions where SPBE and PPBE, respectively, exist. A naive UAV is credulous, i.e., unconditionally trusting the received signal, s_z. Therefore, the controller of the naive UAV completely accept the literal current position coordinates according to the GPS signal, and the corresponding v and λ make the UAV deviate to the reset path (shown in blue) that is totally determined by the spoofed GPS signal. Figure 5c shows the behavior of a strategic UAV at the SPBE. Since the GPS spoofer's SPBE strategy $\alpha_z^*(t_z)$ reveals the true position in the SPBE, the controller of the UAV can obtain the correct current position coordinates $(a_{R,x}^*(t_x), a_{R,y}^*(t_y))$ based on the SPBE strategy, and the corresponding v and λ keep the UAV fly on the original flight path. Figure 5d shows the behavior of a strategic UAV at the PPBE. In the PPBE, the GPS spoofer plays the PPBE strategy $\alpha_z(t_z) = t_z^m$. However, in the PPBE region the spoofer can send different navigation messages $m_z \in M_{t_z^m}$ that induce the same value of signal $s_z = t_z^m$ (position coordinates) due to the existence of multiple pooling sub-intervals. The controller of the strategic UAV takes the current position coordinates as $(\hat{a}_x(t_x^{n-1}, t_x^n), \hat{a}_y(t_y^{n-1}, t_y^n))$ when the UAV is in the region (t_z^{n-1}, t_z^n), the corresponding v and λ make the UAV fly on a path shown in solid orange in Fig. 5d. As can be seen, the strategy of the UAV in the multiple pooling region cannot always obtain the exactly true position but performs better than being credulous.

5   Conclusion

Civilian UAVs primarily guided by GPS have been shown to be readily spoofable by researchers. Failing to detect and defend the civil GPS spoofing could cause a significant hazard in the national airspace and sabotage the businesses primarily based on UAVs. Thus, it is critical to design a security mechanism. We have proposed a signaling game-based defense mechanism against the civil GPS spoofing attacks for the civilian UAVs. Our focus is on the case when the position information is spoofed while the velocity and the time are assumed to be accurate. However, our method can be further extended to the spoofing of the velocity and time information.

We have defined a perfect Bayesian equilibrium (PBE) pooling in low types and separating in high types (PLASH). We have also shown that there can be a unique full separating PBE if the deception cost is sufficiently small compared to the malice of the GPS spoofer. A full pooling PBE can exist if the deception cost is sufficiently large. We have also shown that the pooling portion of the PLASH can be partitioned into multiple pooling subintervals such that the GPS spoofer chooses messages to for different pooling subintervals.

The simulation results have shown that in the separating portion of the PLASH, the GPS spoofer chooses a strategy that yields the optimal action of the UAV that reveals the true position and completely defends the spoofing. In the pooling portion, the UAV cannot exactly infer its true position, but the equilibrium action can reduce the deviation between the estimated position and the true position, thus mitigating the potential loss caused by the spoofing.

Acknowledgement. This research is partially supported by NSF grants CNS-1544782, CNS-1720230 and the DOE grant DE-NE0008571.

A   Appendix

A.1   Appendix A: Proof of Lemma 1

Proof. Since we require $\frac{d\alpha_z(t_z)}{dt_z} \geq 0$, the strategy $\alpha_z(t_z)$ in the separating portion must satisfy $\alpha_z(t_z) > s_z^*(t_z) = \frac{t_z}{1+\rho}$. Suppose that α_z is constant on some interval $\Phi \subseteq (t_z^s, t_z^l))$, then there exists some type $t_z \in \Phi$ such that S can send a signal $s_z(t_z + \delta)$ with $\delta > 0$ indicating a slightly higher type $t_z + \delta \in \Phi$ without inducing the additional deception cost, which contradicts the hypothesis of separating equilibrium in Lemma 1; therefore, α_z is strictly increasing on (t_z^s, t_z^l); thus, $\alpha_z \in (t_z^m, t_z^M)$ for any $t_z \in (t_z^s, t_z^l)$.

The incentive compatibility of SPBE requires that for any $t_z \in (t_z^s, t_z^l)$, $\alpha_z(t_z) \in \arg\min_{s_z} C^{S,z}(t_z, t_z, s_z)$. (8) is obtained by differentiating $C^{S,z}(t_z, t_z, s_z)$, which can be done only if $\alpha_z(t_z)$ is differentiable. In order to prove that $\alpha_z(t_z)$ on (t_z^s, t_z^l), we first prove that $\alpha_z(t_z) > \arg\min_s C^{D,z}$ and $\alpha_z(t_z)$ is continuous for all $t_z \in (t_z^s, t_z^l)$.

We prove $\alpha_z(t_z) > s_z^* = \frac{t_z}{1+\rho}$ for all $t_z \in (t_z^s, t_z^l)$ in two steps as follows.

Step 1: Suppose $\alpha_z(\bar{\tau}_z) = s_z^*(\bar{\tau}_z) = \frac{\bar{\tau}_z}{1+\rho}$ for some $\bar{\tau}_z \in (t_z^s, t_z^l)$. Then, $C_2^{D,z}(t_z, \alpha_z(\bar{\tau}_z)) = 0$. Let $\delta > 0$ be a position constant with small enough $|\delta|$. Let $U(\delta)$ be the expected change in the cost for type $\bar{\tau}_z - \delta \in (t_z^s, t_z^l)$ by changing from $\alpha_z(\bar{\tau}_z - \delta)$ to $\alpha_z(\bar{\tau}_z)$. Then,

$$
\begin{aligned}
U(\delta) =& C^{S,z}(\bar{\tau}_z, \bar{\tau}_z - \delta, s_z^*(\bar{\tau}_z)) - C^{S,z}(\bar{\tau}_z - \delta, \bar{\tau}_z - \delta, \alpha_z(\bar{\tau}_z - \delta)) \\
=& [C^{A,z}(\bar{\tau}_z, \bar{\tau}_z - \delta) - C^{A,z}(\bar{\tau}_z - \delta, \bar{\tau}_z - \delta)] \\
& + k_1 [C^{D,z}(\bar{\tau}_z - \delta, s_z^*(\bar{\tau}_z)) - C^{D,z}(\bar{\tau}_z - \delta, \alpha_z(\bar{\tau}_z - \delta))].
\end{aligned}
$$

Since $C^{A,z}(\bar{\tau}_z, \bar{\tau}_z - \delta) < C^{A,z}(\bar{\tau}_z - \delta, \bar{\tau}_z - \delta)$ and $C^{D,z}(\bar{\tau}_z - \delta, s_z^*(\bar{\tau}_z)) \leq C^{D,z}(\bar{\tau}_z - \delta, \alpha_z(\bar{\tau}_z - \delta))$, $U(\delta) < 0$, which implies that S strictly prefers to use the strategy $\alpha_z(\bar{\tau}_z)$ when the type is $\bar{\tau}_z - \delta$; this means that S uses the strategy $\alpha_z(\bar{\tau}_z)$ for both type $\bar{\tau}_z - \delta$ and type $\bar{\tau}_z$, which contradicts the hypothesis of SPBE for $\bar{\tau}_z$. Thus, $\alpha_z(\bar{\tau}_z) \neq s_z^*(\bar{\tau}_z)$.

Step 2: Suppose there exists a $\hat{\tau}_z \in (t_z^s, t_z^l)$ such that $\alpha_z(\hat{\tau}_z) < s_z^*(\hat{\tau}_z) < \hat{\tau}_z$. From (8), we have $\frac{d\alpha_z(\hat{\tau}_z)}{d\hat{\tau}_z} < 0$. Thus, the strict monotonicity of $\alpha_z(t_z)$ gives that $\alpha_z(\hat{\tau}_z - \delta) > \alpha_z(\hat{\tau}_z)$ for all $\delta > 0$. Then for small enough $\delta > 0$, we have $C^{D,z}(\hat{\tau}_z - \delta, \alpha_z(\hat{\tau}_z)) < C^{D,z}(\hat{\tau}_z - \delta, \alpha_z(\hat{\tau}_z - \delta))$. Also, we have $C^{A,z}(\hat{\tau}_z, \hat{\tau}_z - \delta) < C^{A,z}(\hat{\tau}_z - \delta, \hat{\tau}_z - \delta)$. As a result, $C^{S,z}(\hat{\tau}_z, \hat{\tau}_z - \delta, \alpha_z(\hat{\tau}_z)) < C^{S,z}(\hat{\tau}_z - \delta, \hat{\tau}_z - \delta, \alpha_z(\hat{\tau}_z - \delta))$. Therefore, S prefers to use the same strategy $\alpha_z(\hat{\tau}_z)$ for $\hat{\tau}_z - \delta$ as for $\hat{\tau}_z$, which contradicts the hypothesis of SPBE for $\hat{\tau}_z$. Thus, Step 1 and 2 yield that $\alpha_z(t_z) > s_z^*(t_z)$.

Now we prove the continuity of $\alpha_z(t_z)$ on $t_z \in (t_z^s, t_z^l)$. Suppose that there exists a discontinuity point at some $t_z \in (t_z^s, t_z^l)$. Let $\alpha_z(t_z) > \lim_{t_z \to t_z^-} = \hat{\alpha}_z$. Then,

$$
\lim_{\delta \to 0+} [C^{A,z}(t_z - \delta, \alpha_z(t_z - \delta)) - C^{A,z}(t_z - \delta, \alpha_z(t_z))] = 0.
$$

Since α_z is strictly increasing and $s_z^*(t_z) \leq \hat{\alpha}_z < \alpha_z(t_z)$, we also have

$$
\lim_{\delta \to 0} [C^{D,z}(t_z - \delta, \alpha_z(t_z - \delta) - C^{D,z}(t_z - \delta, \alpha_z(t_z))] = C^{D,z}(t_z, \hat{\alpha}_z) - C^{D,z}(t_z, \alpha_z(t_z)) < 0.
$$

Therefore, the cost of $\alpha_z(t_z - \delta)$ is less than $\alpha_z(t_z)$; thus, S prefers to use the same strategy $\alpha_z(t_z - \delta)$ for t_z as for $t_z - \delta$ for small enough $\delta > 0$, which contradicts the hypothesis of SPBE. Similar proof for the case $\alpha_z(t_z) < \lim_{t_z \to t_z^+} = \hat{\alpha}_z$ can show that S prefers to use the same strategy $\alpha_z(t_z + \delta)$ for t_z as for $t_z + \delta$ for small enough $\delta > 0$, contradicting the SPBE. Therefore, $\alpha_z(t_z)$ is continuous on (t_z^s, t_z^l).

Based on the same argument of the Proposition 2 in the Appendix of Mailath's work in [14] (also see the proof of [9]), α_z is differentiable. Therefore, Lemma 1 is proved.

A.2   Appendix B: Proof of Theorem 1

In this part, we prove that there exists a unique solution on $[\hat{t}, t_z^M]$ to (8) with initial condition $\alpha_z^*(t_z^M) = t_z^M = \hat{s}_z(t_z^M)$ and $\frac{d\alpha_z^*(t_z)}{dt_z} > 0$.

Proof. **Step 1: Local uniqueness and existence**
Let $B_z(t_z, s_z)$ be the inverse initial value problem and let $\eta_z(s_z)$ be the solution of $B_z(t_z, s_z)$. Then,

$$\eta_z' = B_z(\eta_z, s_z) = -\frac{C^{S,z}(\eta_z, \eta_z, s_z)_3}{C^{S,z}(\eta_z, \eta_z, s_z)_1}, \text{ with } \eta_z(s_z^*(t_z^M)) = t_z^M. \quad (16)$$

From the definition of $C^{S,z}$, B_z is Lipschitz continuous on $T \times T$. Then, from the existence and uniqueness theorems [11], we can find some $\delta > 0$ such that $\hat{s}_z(t_z) - \delta \geq s_z^*(t_z) = \frac{t_z}{1+\rho}$ and there exists a unique solution $\hat{\eta}_z$ to (16) on $[\hat{s}_z(t_z^M) - \delta, \hat{s}_z(t_z^M))$, and $\hat{\eta}_z$ is continuously differentiable on $[\hat{s}_z(t_z^M) - \delta, \hat{s}_z(t_z^M))$. From the definition of $\hat{s}_z(t_z^M)$, we have $B_z(t_z^M, \hat{s}_z(t_z^M)) > 0$, $B_z(t_z^M, s_z^*(t_z^M)) = 0$ and $\hat{s}_z^{-1}(t_z^M) = \frac{1}{\hat{s}_z(\hat{s}_z^{-1}(t_z^M))} > 0$; δ can be small enough such that $s_z < \hat{s}_z(\hat{\eta}_z(s_z))$ for all $s_z \in (\hat{s}_z(t_z^M) - \delta, \hat{s}_z(t_z^M)))$; and thus $\hat{\eta}_z'(s_z) > 0$. Let $\hat{\alpha}_z = \hat{\eta}_z^{-1}$ be a solution to 8 on $(\check{t}_z, t_z^M]$ for some $\check{t}_z < t_z^M$ with $\frac{d\hat{\alpha}_z}{dt_z} > 0$. Since the solution $\hat{\eta}_z$ to the inverse initial value problem is locally unique, the solution to the initial value problem (8) is locally unique.

Step 2: Suppose $\hat{\alpha}_z$ is the a solution to (8) with initial condition $\alpha_z^*(t_z^M) = t_z^M = \hat{s}_z(t_z^M)$ and $\frac{d\alpha_z^*(t_z)}{dt_z} > 0$, on $(t_z', t_z^M]$. Let $\bar{\alpha}_z = \lim_{t_z \to t_z'} \hat{\alpha}_z$. As been proved above, $\hat{\alpha}_z > s_z^*(t_z)$ for all $(t_z', t_z^M]$, and $\bar{\alpha}_z \geq s_z^*(t_z)$. Suppose $\bar{\alpha}_z = s_z^*(t_z)$. Then, $C_3^{S,z} = 0$, which yields $\lim_{t_z \to t_z'} = \infty$. Let $\zeta = \sup_{t_z \in [t_z', t_z^M]} (s_z^*(t_z))' = \frac{1}{1+\rho} < \infty$. Since $\hat{\alpha}_z'(t_z^M) > 0$ exists, there exists a $t_z'' > t_z'$ such that $\alpha_z(t_z'') > \zeta$ for all $t_z \in [t_z', t_z'']$. Let $\epsilon > 0$ such that $\hat{\alpha}_z(t_z'') > s_z^*(t_z'') + \epsilon$. Since $\bar{\alpha}_z = \lim_{t_z \to t_z'} \hat{\alpha}_z$, it follows

$$\bar{\alpha}_z = \hat{\alpha}_z(t_z'') + \lim_{t_z \to t_z'} \int_{t_z}^{t_z''} \alpha_z'(\tau)d\tau > s_z^*(t_z'') + \epsilon + \int_{t_z'}^{t_z''} \alpha_z'(\tau)d\tau$$

$$> s_z^*(t_z'') + \int_{t_z'}^{t_z''} (s_z^*(\tau))' d\tau + \epsilon = s_z^*(t_z') + \epsilon,$$

which contradicts that $\bar{\alpha}_z = s_z^*(t_z)$. Therefore, we have $\bar{\alpha}_z > s_z^*(t_z)$.

If the solution $\hat{\alpha}_z(t_z)$ is well defined on $(t_z', t_z^M]$ with $\lim_{t_z \to t_z'} \hat{\alpha}_z(t_z) > t_z^M$, then $-\frac{C_1^{S,z}}{C_3^{S,z}}$ is Lipschitz continuous and bounded in a neighborhood of $(\bar{\alpha}_z, t_z')$. According to the existence and uniqueness theorems, there exists a unique differentiable solution $\hat{\alpha}_z$ to (8) on $(t_z' - \epsilon', t_z^M]$ for some $\epsilon' > 0$ with $\lim_{t_z \to t_z' - \epsilon'} \hat{\alpha}_z(t_z) > s_z^*(t_z' - \epsilon')$ for $t_z' \in (t_z^m, t_z^M)$.

Clearly, $\hat{t}_z = \sup\{\hat{\tau}_z : \hat{\alpha}_z \text{ is well defined on } (\hat{\tau}_z, t_z^M]\}$, and setting $\hat{\alpha}_z(\hat{t}_z) = t_z^m$ finishes the proof of Theorem 1.

References

1. Amazon.com: Amazon prime air (2017). https://www.amazon.com/Amazon-Prime-Air/b?node=8037720011. Accessed 12 Apr 2017
2. Casey, W., Morales, J.A., Nguyen, T., Spring, J., Weaver, R., Wright, E., Metcalf, L., Mishra, B.: Cyber security via signaling games: toward a science of cyber security. In: Natarajan, R. (ed.) ICDCIT 2014. LNCS, vol. 8337, pp. 34–42. Springer, Cham (2014). doi:10.1007/978-3-319-04483-5_4
3. Casey, W.A., Zhu, Q., Morales, J.A., Mishra, B.: Compliance control: managed vulnerability surface in social-technological systems via signaling games. In: Proceedings of the 7th ACM CCS International Workshop on Managing Insider Security Threats, pp. 53–62. ACM (2015)
4. Crawford, V.P., Sobel, J.: Strategic information transmission. Econom. J. Econom. Soc. 1431–1451 (1982)
5. Curry, C., et al.: SENTINEL Project-Report on GNSS Vulnerabilities. Chronos Technology Ltd., Lydbrook (2014)
6. Hein, G., Kneissl, F., Avila-Rodriguez, J.A., Wallner, S.: Authenticating GNSS: proofs against spoofs, Inside GNSS 2(5), 58–63 (2007). part 2
7. Humphreys, T.: Cockrell school researchers demonstrate first successful spoofing of UAVs (2012). https://www.engr.utexas.edu/features/humphreysspoofing. Accessed 5 Apr 2017
8. Infrastructure, Transportation: Vulnerability assessment of the transportation infrastructure relying on the global positioning system (2001)
9. Kartik, N.: Strategic communication with lying costs. Rev. Econ. Stud. 76(4), 1359–1395 (2009)
10. Kashyap, A., Basar, T., Srikant, R.: Correlated jamming on mimo gaussian fading channels. IEEE Trans. Inf. Theor. 50(9), 2119–2123 (2004)
11. Khalil, H.K.: Nonlinear Systems. Prentice Hall, Upper Saddle River (2002)
12. Kuhn, M.G.: An asymmetric security mechanism for navigation signals. In: Fridrich, J. (ed.) IH 2004. LNCS, vol. 3200, pp. 239–252. Springer, Heidelberg (2004). doi:10.1007/978-3-540-30114-1_17
13. Ledvina, B.M., Bencze, W.J., Galusha, B., Miller, I.: An in-line anti-spoofing device for legacy civil GPS receivers. In: Proceedings of the 2010 International Technical Meeting of the Institute of Navigation, pp. 698–712 (2001)
14. Mailath, G.J.: Incentive compatibility in signaling games with a continuum of types. Econom. J. Econom. Soc. 1349–1365 (1987)
15. Noe, T.H.: Capital structure and signaling game equilibria. Rev. Financ. Stud. 1(4), 331–355 (1988)
16. Papadimitratos, P., Jovanovic, A.: GNSS-based positioning: attacks and countermeasures. In: Military Communications Conference, MILCOM 2008, pp. 1–7. IEEE (2008)
17. Pawlick, J., Zhu, Q.: Strategic trust in cloud-enabled cyber-physical systems with an application to glucose control. IEEE Trans. Inf. Forensics Secur. (2017)
18. Pozzobon, O.: Keeping the spoofs out: signal authentication services for future GNSS. Inside GNSS 6(3), 48–55 (2011)
19. Raya, M., Manshaei, M.H., Félegyházi, M., Hubaux, J.P.: Revocation games in ephemeral networks. In: Proceedings of the 15th ACM Conference on Computer and Communications Security, pp. 199–210. ACM (2008)
20. Sagduyu, Y.E., Berry, R., Ephremides, A.: MAC games for distributed wireless network security with incomplete information of selfish and malicious user types.

In: International Conference on Game Theory for Networks, GameNets 2009, pp. 130–139. IEEE (2009)

21. Schmidt, D., Radke, K., Camtepe, S., Foo, E., Ren, M.: A survey and analysis of the GNSS spoofing threat and countermeasures. ACM Comput. Surv. (CSUR) 48(4), 64 (2016)

22. Shepard, D.P., Bhatti, J.A., Humphreys, T.E., Fansler, A.A.: Evaluation of smart grid and civilian UAV vulnerability to GPS spoofing attacks. In: Proceedings of the ION GNSS Meeting, vol. 3 (2012)

23. Wang, J., Tu, W., Hui, L.C., Yiu, S., Wang, E.K.: Detecting time synchronization attacks in cyber-physical systems with machine learning techniques. In: 2017 IEEE 37th International Conference on Distributed Computing Systems (ICDCS), pp. 2246–2251. IEEE (2017)

24. Warner, J.S., Johnston, R.G.: A simple demonstration that the global positioning system (GPS) is vulnerable to spoofing. J. Secur. Adm. 25(2), 19–27 (2002)

25. Wesson, K., Rothlisberger, M., Humphreys, T.: Practical cryptographic civil GPS signal authentication. Navigation 59(3), 177–193 (2012)

26. Wikipedia: Cheap talk — wikipedia, the free encyclopedia (2017). https://en.wikipedia.org/w/index.php?title=Cheap_talk&oldid=771947821. Accessed 4 Apr 2017

27. Wullems, C., Pozzobon, O., Kubik, K.: Signal authentication and integrity schemes for next generation global navigation satellite systems. In: Proceedings of the European Navigation Conference GNSS (2005)

28. Xu, Z., Zhu, Q.: A cyber-physical game framework for secure and resilient multi-agent autonomous systems. In: 2015 IEEE 54th Annual Conference on Decision and Control (CDC), pp. 5156–5161. IEEE (2015)

29. Xu, Z., Zhu, Q.: A game-theoretic approach to secure control of communication-based train control systems under jamming attacks. In: Proceedings of the 1st International Workshop on Safe Control of Connected and Autonomous Vehicles, pp. 27–34. ACM (2017)

30. Zhu, Q., Basar, T.: Game-theoretic methods for robustness, security, and resilience of cyberphysical control systems: games-in-games principle for optimal cross-layer resilient control systems. IEEE Control Syst. 35(1), 46–65 (2015)

31. Zhu, Q., Fung, C., Boutaba, R., Basar, T.: A game-theoretical approach to incentive design in collaborative intrusion detection networks. In: International Conference on Game Theory for Networks, GameNets 2009, pp. 384–392. IEEE (2009)

A Game Theoretical Model for Optimal Distribution of Network Security Resources

Ziad Ismail[1][✉], Christophe Kiennert[2], Jean Leneutre[1], and Lin Chen[3]

[1] Télécom ParisTech, Université Paris-Saclay, 46 Rue Barrault, 75013 Paris, France
`ismail.ziad@telecom-paristech.fr`
[2] Télécom SudParis, 9 Rue Charles Fourier, 91011 Evry, France
[3] University of Paris-Sud 11, 15 Rue Georges Clemenceau, 91400 Orsay, France

Abstract. Enforcing security in a network always comes with a trade-off regarding budget constraints, entailing unavoidable choices for the deployment of security equipment over the network. Therefore, finding the optimal distribution of security resources to protect the network is necessary. In this paper, we focus on Intrusion Detection Systems (IDSs), which are among the main components used to secure networks. However, configuring and deploying IDSs efficiently to optimize attack detection and mitigation remain a challenging task. In particular, in networks providing critical services, optimal IDS deployment depends on the type of interdependencies that exists between vulnerable network equipment. In this paper, we present a game theoretical analysis for optimizing intrusion detection in such networks. First, we present a set of theoretical preliminary results for resource constrained network security games. Then, we formulate the problem of intrusion detection as a resource constrained network security game where interdependencies between equipment vulnerabilities are taken into account. Finally, we validate our model numerically via a real world case study.

Keywords: Intrusion detection · Optimization · Non-cooperative game theory

1 Introduction

As the amount of network communications keeps growing and the complexity of architectures keeps increasing, designing secure networks has become more challenging. One critical aspect of network security is optimizing the distribution of security resources given a constrained defense budget. In addition to firewalls,

This research was initially supported by the MSSTB project, in collaboration with Airbus Defence & Space CyberSecurity and Cogisys, through the Program *Investissement d'Avenir* funded by the French public financial organization *Caisse des Dépôts et des Consignations*, and later by the Cyber CNI Chair of Institut Mines-Télécom held by Télécom Bretagne and supported by Airbus Defence and Space, Amossys, BNP Paribas, EDF, Orange, La Poste, Nokia, Société Générale, and the Regional Council of Brittany, and acknowledged by the Center of Excellence in Cybersecurity.

© Springer International Publishing AG 2017
S. Rass et al. (Eds.): GameSec 2017, LNCS 10575, pp. 234–255, 2017.
DOI: 10.1007/978-3-319-68711-7_13

reverse proxies, or application level countermeasures, Intrusion Detection Systems (IDSs) allow network administrators to substantially refine security management by analyzing data flows dynamically. However, analyzing all the traffic in the network can be complex and costly. Therefore, an optimal IDS deployment strategy to maximize the overall probability of detecting attacks is needed.

In general, based on the data they store, some equipment in a network will be more attractive to attack than others. The interdependencies of equipment vulnerabilities need also to be taken into account. For example, accessing a user workstation is generally not very useful for an attacker unless if it allows him to get access to sensitive equipment more easily. Therefore, it is important to take into account such sequence of attacks in realistic approaches, as the actions of an attacker are not limited to independent atomic attacks.

In addition to classic security approaches, approaches based on game theory were recently used to study and analyze network security problems [1], and more specifically intrusion detection [2]. One of the first game theoretical approaches for intrusion detection was proposed by Alpcan and Basar in [3]. The authors describe and solve a static nonzero-sum imperfect information game where the attacker targets subsystems in the network and the defender tries to optimize the sensitivity of the IDS in each subsystem. This work was later extended in [4] with a zero-sum stochastic game formulation that aims to take into account the uncertainty of attack detection. The authors analyze the equilibria in the case of perfect and imperfect information, and compare the performances of various Q-learning schemes in the case of imperfect information.

Chen and Leneutre [2] consider the intrusion detection problem under budget constraints in a network comprised of independent nodes with different security assets. Nguyen et al. [5] address the same problem, but take into account node interdependencies, both in terms of vulnerabilities and security assets, modeled using linear influence networks [6]. Following the formalism introduced in [7], Nguyen et al. formulate the problem as a two-player zero-sum stochastic game where the states of the game are characterized by the state of each node, either compromised or healthy. Though we also take node interdependencies into account in this paper, formulating the problem as a static game allows us to manipulate more complex utility functions in order to remain as realistic as possible while keeping the solution tractable.

Another approach for the resource allocation problem consists in finding the optimal sampling rate of the IDS on each link in the network under budget constraints. Kodialam and Lakshman in [8] describe the problem as an attacker injecting malicious packets from a fixed entry node and trying to reach a target node without being detected. They formulate the problem as a zero-sum static game, where the attacker aims at choosing the path that minimizes the detection probability over all possible paths from the entry node to the target node. This work was later extended in [9,10] where the sampling rate problem under budget constraints and in the case of fragmented malicious packets are addressed respectively.

The paper proceeds as follows. In Sect. 2, we present a class of security games which we refer to as Resource Constrained Network Security (RCNS) games. The aim of this section is to present a generic framework that will serve as a basis for the analysis of different types of security games. In Sect. 3, we define our game theoretic model, which is as a subclass of RCNS games, for optimizing the allocation of defense resources in a network, focusing on intrusion detection in which the equipment interdependent vulnerabilities are taken into account. We pay a particular attention to the evaluation of the model parameters, as they are chosen in order to be naturally derived from information security risk assessment methods and correspond to what a chief information security officer would expect to find. We analyze the behavior of the attacker and the defender at the Nash Equilibrium (NE). In Sect. 4, we validate our model numerically via a case study. Finally, we conclude the paper in Sect. 5.

2   Resource Constrained Network Security Games

In this section, we introduce a new class of security games which we will refer to as Resource Constrained Network Security (RCNS) games. Before giving the definition of a RCNS game, we will introduce a number of simple intermediary games. In the remaining of this section, we will refer to a network as a set of interconnected nodes that could also be security-wise interdependent. The nodes can refer to the set of equipment in the network or the set of services running on equipment. Therefore, allocating a set of defense resources on a node refers to the set of defense resources used to monitor the node for any sign of security intrusion. This abstraction of the notion of a network node will allow us to cover a wide spectrum of use cases for applying our formal model.

2.1   Attack/Defense Game

Let $\mathcal{N}$ be a network consisting of $\mathcal{T}$ nodes.

Definition 1 (AD game). *A simple Attack/Defense (AD) game is a static game played on a node i in the network $\mathcal{N}$ between two players: an attacker and a defender. The attacker's actions are restricted to {Attack/Not attack} while the defender's actions are restricted to {Defend/Not defend}.*

An AD game is a simple game played between the attacker and the defender. It is *restricted* in the sense that the actions of each player are restricted to a single node in the network. The strategic form of a general AD game is given in Table 1.

Assumption 1. In an AD game, we can have $u_i \leq t_i$, $s_i' \leq u_i'$, $r_i - s_i \leq t_i - u_i$, and $r_i' - t_i' \geq s_i' - u_i'$.

Definition 2 (Realistic AD game). *A realistic AD game is an AD game satisfying Assumption 1.*

Table 1. Strategic form of the AD game for node i

	Defend	Not defend
Attack	r_i, r_i'	t_i, t_i'
Not attack	s_i, s_i'	u_i, u_i'

We suppose that a realistic AD game satisfies $u_i \leq t_i$ since the attacker will get a higher payoff when attacking a node that is not defended. Similarly, we have $s_i' \leq u_i'$ since the defender is better off defending a node when that node is under attack. Moreover, the difference in payoff for the attacker between the Attack/Not attack actions is higher when the defender chooses not to defend, which translates to $r_i - s_i \leq t_i - u_i$. Similarly, on the defender's side, we have $r_i' - t_i' \geq s_i' - u_i'$. We also note that in general, the attacker's payoffs $r_i, s_i, t_i,$ and u_i are nonnegative real numbers and the defender's payoffs $r_i', s_i', t_i',$ and u_i' are nonpositive real numbers.

Let $(p_i, 1 - p_i)$ and $(q_i, 1 - q_i)$ be the mixed strategy Nash equilibrium of the attacker and the defender for choosing the actions {Attack/Not attack} and {Defend/Not defend} respectively. Given the strategic form of the game shown in Table 1, the utility function $u_A^i(p_i, q_i)$ of the attacker can be written as $u_A^i(p_i, q_i) = \alpha_i p_i + \sigma_i q_i + \gamma_i p_i q_i + \delta_i$, where $\alpha_i = t_i - u_i$, $\sigma_i = s_i - u_i$, $\gamma_i = r_i - s_i - t_i + u_i$, and $\delta_i = u_i$. Similarly, the utility function $u_D^i(p_i, q_i)$ of the defender can be written as $u_D^i(p_i, q_i) = \alpha_i' p_i + \sigma_i' q_i + \gamma_i' p_i q_i + \delta_i'$, where $\alpha_i' = t_i' - u_i'$, $\sigma_i' = s_i' - u_i'$, $\gamma_i' = r_i' - s_i' - t_i' + u_i'$, and $\delta_i' = u_i'$. We have the following lemma, which follows directly from Assumption 1:

Lemma 1. *In a realistic AD game, we have $\alpha_i \geq 0$, $\gamma_i \leq 0$, $\sigma_i' \leq 0$, and $\gamma_i' \geq 0$.*

2.2   Network Security Game

Let $n = |T|$ be the number of nodes in the network $\mathcal{N}$. We define a network security game as follows:

Definition 3 (NS game). *A Network Security (NS) game is a game in which the attacker and the defender play n independent AD games on each node of the network $\mathcal{N}$.*

We also refer to a NS game where Assumption 1 holds in each of the n AD games as a *realistic NS game*. The NS game can be as well viewed as a game played between n attackers and n defenders where the attackers and the defenders do not cooperate with each other.

Since a NS game is just a set of AD games played in parallel between the attacker and the defender, the utility of the attacker can be expressed as $U_A(\mathbf{p}, \mathbf{q}) = \sum_{i \in T} u_A^i(p_i, q_i)$, where $u_A^i(p_i, q_i)$ is the utility the attacker gets from playing the AD game on node i, $\mathbf{p} = (p_1, ..., p_n) \in [0, 1]^n$, and

$\mathbf{q} = (q_1, ..., q_n) \in [0, 1]^n$. Similarly, the utility of the defender can be expressed as $U_D(\mathbf{p}, \mathbf{q}) = \sum_{i \in T} u_D^i(p_i, q_i)$.

2.3 Resource Constrained Network Security Game

In a NS game, the choices of actions in the AD game played on node i is independent of any other AD game played on node $j \neq i$. However, in realistic interactions between a defender and an attacker targeting the network, the choice of an action on a node depends on the choices of actions on other nodes as well. For example, given two target nodes, the attacker may assess the success likelihood of his attack and its potential payoff and decide to attack only one of these nodes. In practice, one of the main factors that play a role in the attacker's decision process is the set of attack resources at his disposal. Similarly, a constrained defense budget will influence the defender's allocation of security resources on network nodes. This observation leads us to define the class of resource constrained network security games.

Definition 4 (RCNS game). *A Resource Constrained Network Security (RCNS) game is a non-cooperative two player, static, complete information game between an attacker and a defender. The game features a set T of n targets. Let $\mathbf{p} = (p_1, ..., p_n) \in [0, 1]^n$ and $\mathbf{q} = (q_1, ..., q_n) \in [0, 1]^n$ be the strategies of the attacker and the defender, where p_i and q_i refer to the attack and defense resources allocated on node i respectively. The game features the resource constraints $\sum_{i \in T} p_i \leq P \leq 1$ and $\sum_{i \in T} q_i \leq Q \leq 1$.*

A RCNS game can be seen as a NS game where the allocation of attack and defense resources p_i and q_i on node i refer to the mixed strategy NE of an AD game played on node i. In fact, for the NS game, we have $U_A(\mathbf{p}, \mathbf{q}) = \sum_{i \in T} u_A^i(p_i, q_i) = \sum_{i \in T} \alpha_i p_i + \sigma_i q_i + \gamma_i p_i q_i + \delta_i$. Similarly, for the defender, we have $U_D(\mathbf{p}, \mathbf{q}) = \sum_{i \in T} \alpha_i' p_i + \sigma_i' q_i + \gamma_i' p_i q_i + \delta_i'$. By just looking at the shape of $U_A(\mathbf{p}, \mathbf{q})$ and $U_D(\mathbf{p}, \mathbf{q})$, it is as if we have a game in which the attacker and the defender are trying to find strategies $\mathbf{p} = (p_1, ..., p_n) \in [0, 1]^n$ and $\mathbf{q} = (q_1, ..., q_n) \in [0, 1]^n$ respectively. This is similar to what we have defined in the RCNS game in Definition 4. However, while p_i and q_i for each node i in the NS game are defined as probabilities, these variables refer to the attack and defense resources allocated on node i in the RCNS game respectively. Therefore, p_i and q_i differ only semantically in these two types of games. In addition, in a RCNS game, we have constraints related to the set of resources available to each player.

Definition 5 (Realistic RCNS game). *A realistic RCNS game is a RCNS game where $u_i \leq t_i$, $s_i' \leq u_i'$, $r_i - s_i \leq t_i - u_i$, and $r_i' - t_i' \geq s_i' - u_i'$, $\forall i \in T$, and there exists at least one $j \in T$ s.t. $\alpha_j + \gamma_j q_j > 0$, $q_j \in [0, 1]$.*

We can notice that the first set of conditions in Definition 5 are similar to the set of conditions in the definition of realistic AD games. In a realistic RCNS game, we assume that there exists at least one target node $j \in T$ s.t. $\alpha_j + \gamma_j q_j > 0$. Otherwise, by analyzing the utility of the attacker, we can notice that he will not have any incentive to attack any target. Therefore, the conditions defined in a realistic RCNS game ensure that the attacker will *play along* by giving him an incentive to allocate a set of his attack resources to target nodes in the network. We note that in a realistic RCNS game, we have $\alpha_i \geq 0$ and $\gamma_i \leq 0$, $\forall i$.

2.3.1 Nash Equilibrium Analysis

Many network security games, such as [2,11,12], can be formulated as RCNS games. The resource constraints $\sum\limits_{i \in T} p_i \leq P$ and $\sum\limits_{i \in T} q_i \leq Q$ represent constraints on players' budgets. In the rest of this section, we present a necessary condition for the existence of a NE in this type of games. In particular, we show that when $\gamma_i < 0$ and $\gamma_i' > 0$, at least the attacker has to use all his resources for a NE to exist.

Theorem 1. *A necessary condition for $(\mathbf{p}^*, \mathbf{q}^*)$ to be a Nash equilibrium in a realistic RCNS game where $\gamma_i < 0$ and $\gamma_i' > 0$ is $\sum\limits_{i \in T} p_i^* = P$.*

Proof. We consider a realistic RCNS game. We have $\gamma_i \leq 0$ and $\gamma_i' \geq 0$. First, we analyze the case where $\gamma_i = 0$. If $\gamma_i = 0$, then the hypothesis $t_i \geq u_i$ implies $r_i \geq s_i$. In this case, the attacker will always decide to attack node i since the payoff is higher independently from the behavior of the defender. This case being of no interest, we will suppose for the rest of this section that $\gamma_i < 0$. Similarly, we can show that when $\gamma_i' = 0$, the defender always gets a higher payoff by choosing not to defend. In the rest of this section, we suppose $\gamma_i' > 0$.

Let T_{S_d} be the set of targets on which the defender will allocate defense resources. For example, in a network, the defender monitors a subset of the network nodes to detect intrusions. Similarly, let T_{S_a} denote the target set that will be attacked by the attacker. In general, we note that $T_{S_d} \cap T_{S_a} \neq \varnothing$.

The conditions for the existence of a NE vary according to the hypothesis made on $\sum\limits_{i \in T} p_i$ and $\sum\limits_{i \in T} q_i$. In the general case where $\sum\limits_{i \in T} p_i \leq P$ and $\sum\limits_{i \in T} q_i \leq Q$, if a NE $(\mathbf{p}^*, \mathbf{q}^*)$ exists, $\mathbf{p}^*$ is a best response strategy to the defender strategy and $\mathbf{q}^*$ is a best response strategy to the attacker strategy. Since the utility of the attacker is linear with respect to the attacker's strategy $\mathbf{p}$, if a solution to the attacker's optimization problem exists, then an optimal solution at an extreme point of the feasible set defined by $\sum\limits_{i \in T} p_i \leq P$ exists (when $\sum\limits_{i \in T} p_i = P$). A similar analysis can be conducted for the case of the defender.

Case 1: $\sum\limits_{i \in T} p_i = P$ and $\sum\limits_{i \in T} q_i = Q$

From the definitions of T_{S_a} and T_{S_d}, the constraints on the attack and defense resources become $\sum\limits_{i \in T_{S_a}} p_i = P$ and $\sum\limits_{i \in T_{S_d}} q_i = Q$. From the Karush-Kuhn-Tucker (KKT) conditions, there exists $\lambda > 0$ s.t. $\dfrac{\partial U_A}{\partial p_i} = \lambda$ and $\lambda' > 0$ s.t.

$\frac{\partial U_D}{\partial q_i} = \lambda'$. We have $\frac{\partial U_A}{\partial p_i} = \alpha_i + \gamma_i q_i$. Therefore, $\alpha_i + \gamma_i q_i > 0 \Rightarrow q_i < -\frac{\alpha_i}{\gamma_i} \Rightarrow$

$Q < \sum_{i \in T_{s_d}} \frac{-\alpha_i}{\gamma_i}$. Since $\alpha_i \geq 0$ and $\gamma_i < 0$, we have $\sum_{i \in T_{s_d}} \frac{-\alpha_i}{\gamma_i} \geq 0$. Similarly,

considering $\frac{\partial U_D}{\partial q_i} = \sigma'_i + \gamma'_i p_i$, we have $P > \sum_{i \in T_{s_a}} \frac{-\sigma'_i}{\gamma'_i}$. Since $\sigma'_i \leq 0$ and $\gamma'_i > 0$,

we have $\sum_{i \in T_{s_a}} \frac{-\sigma'_i}{\gamma'_i} \geq 0$. We have already established that if a NE solution

exists, it must exist at least when $\sum_{i \in T} p_i = P$ and $\sum_{i \in T} q_i = Q$. Therefore,

from the results above, the necessary conditions for the existence of a NE are

$Q < \sum_{i \in T_{s_d}} \frac{-\alpha_i}{\gamma_i}$ and $P > \sum_{i \in T_{s_a}} \frac{-\sigma'_i}{\gamma'_i}$.

Case 2: $\sum_{i \in T} p_i = P$ and $\sum_{i \in T} q_i < Q$

Similarly to Case 1, we can verify that the conditions for the existence of a

NE are $Q < \sum_{i \in T_{s_d}} \frac{-\alpha_i}{\gamma_i}$ and $P = \sum_{i \in T, T \neq T_{s_a}} \frac{-\sigma'_i}{\gamma'_i}$.

Case 3: $\sum_{i \in T} p_i < P$ and $\sum_{i \in T} q_i \leq Q$

We have $\frac{\partial U_A}{\partial p_i} = 0$. Therefore, $q_i = -\frac{\alpha_i}{\gamma_i} \Rightarrow \sum_{i \in T} q_i = -\sum_{i \in T} \frac{\alpha_i}{\gamma_i}$. However,

from the first case, we have $Q < \sum_{i \in T_{s_d}} \frac{-\alpha_i}{\gamma_i} \leq \sum_{i \in T} \frac{-\alpha_i}{\gamma_i} = \sum_{i \in T} q_i$. Therefore,

$Q < \sum_{i \in T} q_i$ which contradicts the fact that $\sum_{i \in T} q_i \leq Q$. As a result, the scenario

in which $\sum_{i \in T} q_i \leq Q$ and $\sum_{i \in T} p_i < P$ does not admit a NE. $\square$

Table 2 exhibits the possible scenarios for the existence of a NE with respect
to the assumptions about the resources of the attacker and the defender. In
particular, given the conditions that P and Q must satisfy, a NE cannot be
found when $\sum_{i \in T} q_i < Q$ and $\sum_{i \in T} p_i < P$.

Table 2. Conditions for the existence of the NE in a realistic RCNS game

	Conditions
$\sum_{i \in T} q_i = Q$, $\sum_{i \in T} p_i = P$	$Q < \sum_{i \in T_{s_d}} \frac{-\alpha_i}{\gamma_i}$, $P > \sum_{i \in T_{s_a}} \frac{-\sigma'_i}{\gamma'_i}$
$\sum_{i \in T} q_i < Q$, $\sum_{i \in T} p_i = P$	$Q < \sum_{i \in T_{s_d}} \frac{-\alpha_i}{\gamma_i}$, $P = \sum_{i \in T, T \neq T_{s_a}} \frac{-\sigma'_i}{\gamma'_i}$
$\sum_{i \in T} q_i \leq Q$, $\sum_{i \in T} p_i < P$	Impossible

2.3.2 Stackelberg Equilibrium Analysis

In a Stackelberg game, a leader chooses his strategy first. Then, the follower, informed by the leader's choice, chooses his strategy. In this section, we analyze the scenario where the defender is the leader and the follower is the attacker. In this case, the defender tries to anticipate the attacker's strategy and chooses a strategy that minimizes the potential impact of attacks on the system.

Stackelberg games are generally solved by backward induction and the solution is known as Stackelberg Equilibrium (SE). We start by computing the best response strategy of the follower as a function of the leader's strategy. Then, according to the follower's best response, we compute the best strategy of the leader.

The attacker solves the following optimization problem:

$$\mathbf{p}(\mathbf{q}) = \underset{\mathbf{p} \in [0,1]^n}{\operatorname{argmax}} U_A(\mathbf{p}, \mathbf{q}) \; s.t. \sum_{i \in \mathcal{T}} p_i \leq P$$

On the other hand, the defender solves the following optimization problem:

$$\mathbf{q}(\mathbf{p}) = \underset{\mathbf{q} \in [0,1]^n}{\operatorname{argmax}} U_D(\mathbf{p}(\mathbf{q}), \mathbf{q}) \; s.t. \sum_{i \in \mathcal{T}} q_i \leq Q$$

Assumption 2. The attacker's resource allocation strategy on a node i depends only on the defender's strategy on that node.

As a result of Assumption 2, we have $p_i(\mathbf{q}) = p_i(q_i) \; \forall i \in \mathcal{T}$. In the rest of this section, we suppose that Assumption 2 holds. In what follows, we present necessary conditions for the existence of a Stackelberg equilibrium in a realistic RCNS game. In particular, we have the following theorem:

Theorem 2. *In a realistic RCNS game, the necessary conditions for the existence of a Stackelberg equilibrium are as follows, $\forall i \in \mathcal{T}$:*
If $\alpha_i' = \gamma_i' = 0$, $\forall j \in \mathcal{T}$ s.t. $\gamma_j' = \alpha_j' = 0$, we have $\sigma_i' = \sigma_j'$. Otherwise, if $\alpha_i' \neq 0$ or $\gamma_i' \neq 0$, $\exists \tau' \geq 0$ s.t. the strategy of the attacker p_i have the following form:

$$p_i = p_i^0 \left| \frac{\alpha_i'}{\alpha_i' + \gamma_i' q_i} \right| + \frac{\tau' - \sigma_i'}{\alpha_i' + \gamma_i' q_i}(q_i + D_i)$$

where $p_i^0 = p_i(0)$ and $D_i =$
$$\begin{cases} 0 & \text{if } \gamma_i' = 0, \alpha_i' \neq 0 \\ 0 & \text{if } \gamma_i' > 0, \alpha_i' \geq 0, q_i \neq \frac{-\alpha_i'}{\gamma_i'} \\ 0 & \text{if } \gamma_i' > 0, \alpha_i' \leq 0, q_i \in \left[0, \min\left(\frac{-\alpha_i'}{\gamma_i'}, Q \right) \right[\\ \frac{2\alpha_i'}{\gamma_i'} & \text{if } \gamma_i' > 0, \alpha_i' \leq 0, q_i \in \left] \min\left(\frac{-\alpha_i'}{\gamma_i'}, Q \right), Q \right] \end{cases}$$

Proof. Let $\mathbf{p}(\mathbf{q})$ be the strategy of the attacker. Next, we establish the conditions that $\mathbf{p}(\mathbf{q})$ must satisfy for a Stackelberg equilibrium for the RCNS game to exist in the presence of constraints on the attack and defense budgets.

From Assumption 2, we have $p_i(\mathbf{q}) = p_i(q_i) \; \forall i \in \mathcal{T}$. The utility of the defender is therefore given by:

$$U_D(\mathbf{p}(\mathbf{q}), \mathbf{q}) = \sum_{i \in \mathcal{T}} \alpha'_i p_i(q_i) + \sigma'_i q_i + \gamma'_i p_i(q_i) q_i + \delta'_i$$

We have the following constraint $\sum_{i \in \mathcal{T}} q_i \leq Q$. From the KKT conditions, there

exists $\tau' \geq 0$ s.t. $\dfrac{\partial U_D}{\partial q_i} = \tau'$. Therefore, we have $\dfrac{\partial p_i}{\partial q_i}(\alpha'_i + \gamma'_i q_i) + \gamma'_i p_i + \sigma'_i = \tau'$.
Let $p_i(0) = p_i^0$.

Case 1: $\gamma'_i = 0$ and $\alpha'_i = 0$

In this case, $\tau' = \sigma'_i$. However, if there are two nodes i and j in which $\gamma'_i = \alpha'_i = 0$, $\gamma'_j = \alpha'_j = 0$ and $\sigma'_i \neq \sigma'_j$, then a Stackelberg equilibrium does not exist.

Case 2: $\gamma'_i = 0$ and $\alpha'_i \neq 0$

In this case, we have $\dfrac{\partial p_i}{\partial q_i} = \dfrac{\tau' - \sigma'_i}{\alpha'_i} \Rightarrow p_i = \dfrac{\tau' - \sigma'_i}{\alpha'_i} q_i + p_i^0$

Case 3: $\gamma'_i > 0$ and $q_i \neq \dfrac{-\alpha'_i}{\gamma'_i}$

In this case, we have $\dfrac{\partial p_i}{\partial q_i} + \dfrac{\gamma'_i}{\alpha'_i + \gamma'_i q_i} p_i = \dfrac{\tau' - \sigma'_i}{\alpha'_i + \gamma'_i q_i}$. This first order differential equation has a unique solution s.t. $p_i(0) = p_i^0$ and is given by:

$$p_i = p_i^0 e^{F(0) - F(q_i)} + \int_0^{q_i} \dfrac{\tau' - \sigma'_i}{\alpha'_i + \gamma'_i x} e^{F(x) - F(q_i)} dx$$

where $F(x) = \displaystyle\int \dfrac{\gamma'_i}{\alpha'_i + \gamma'_i t} dt = \log(|\alpha'_i + \gamma'_i x|)$.

Therefore, $p_i^0 e^{F(0) - F(q_i)} = p_i^0 \left| \dfrac{\alpha'_i}{\alpha'_i + \gamma'_i q_i} \right|$ and $\displaystyle\int_0^{q_i} \dfrac{\tau' - \sigma'_i}{\alpha'_i + \gamma'_i x} e^{F(x) - F(q_i)} dx =$

$\displaystyle\int_0^{q_i} \dfrac{\tau' - \sigma'_i}{\alpha'_i + \gamma'_i x} \left| \dfrac{\alpha'_i + \gamma'_i x}{\alpha'_i + \gamma'_i q_i} \right| dx$

Case 3.1: $\alpha'_i \geq 0$

In this case, we have $\displaystyle\int_0^{q_i} \dfrac{\tau' - \sigma'_i}{\alpha'_i + \gamma'_i x} \left(\dfrac{\alpha'_i + \gamma'_i x}{\alpha'_i + \gamma'_i q_i} \right) dx = \dfrac{\tau - \sigma'_i}{\alpha'_i + \gamma'_i q_i} q_i$

Case 3.2: $q_i \in \left[0, \min\left(\dfrac{-\alpha'_i}{\gamma'_i}, Q \right) \right[$ and $\alpha'_i \leq 0$

Similar to Case 3.1.

Case 3.3: $q_i \in \left] \min\left(\dfrac{-\alpha'_i}{\gamma'_i}, Q \right), Q \right]$ and $\alpha'_i \leq 0$

In this case, we have:

$$\int_0^{q_i} \frac{\tau' - \sigma_i'}{\alpha_i' + \gamma_i'x} \cdot \frac{|\alpha_i' + \gamma_i'x|}{\alpha_i' + \gamma_i'q_i} dx = \int_0^{\frac{-\alpha_i'}{\gamma_i'}} \frac{\tau' - \sigma_i'}{\alpha_i' + \gamma_i'x} \cdot \frac{(-\alpha_i' - \gamma_i'x)}{\alpha_i' + \gamma_i'q_i} dx$$

$$+ \int_{\frac{-\alpha_i'}{\gamma_i'}}^{q_i} \frac{\tau' - \sigma_i'}{\alpha_i' + \gamma_i'x} \cdot \frac{(\alpha_i' + \gamma_i'x)}{\alpha_i' + \gamma_i'q_i} dx = \frac{\tau - \sigma_i'}{\alpha_i' + \gamma_i'q_i}\left(q_i + 2\frac{\alpha_i'}{\gamma_i'}\right)$$

Combining the 3 cases completes the proof.   □

Theorem 3. $\forall i \in \mathcal{T}$ s.t. $\alpha_i' \neq 0$ and $\gamma_i' = 0$. If the conditions in Theorem 2 are satisfied, a necessary condition for the uniqueness of the players' strategies on node i at the Stackelberg equilibrium is that $\exists \tau \geq 0$ s.t.:

$$\begin{cases} \Gamma(\alpha_i')\big((\alpha_i - \tau)(\tau' - \sigma_i') - \alpha_i'(\gamma_i p_i^0 - \sigma_i)\big) \leq 0 \\ \Gamma(\alpha_i')\alpha_i'\gamma_i(\tau' - \sigma_i') > 0 \end{cases}$$

where $\Gamma : \mathbb{R} \to \{1, -1\}$ s.t. $\Gamma(x) = 1$ if $x > 0$ and -1 otherwise.

Proof. The utility function of the attacker is given by: $U_A(\mathbf{p}, \mathbf{q}) = \sum_{i \in \mathcal{T}} \alpha_i p_i + \sigma_i q_i + \gamma_i p_i q_i + \delta_i$. To find the Stackelberg equilibrium, the attacker solves the following maximization problem:

$$\mathbf{p}(\mathbf{q}) = \underset{\mathbf{p} \in [0,1]^{N_c}}{\text{argmax}} \, U_A(\mathbf{p}, \mathbf{q}) \, s.t. \sum_{i \in \mathcal{T}} p_i \leq P$$

Let $\Gamma : \mathbb{R} \to \{1, -1\}$ s.t. $\Gamma(x) = 1$ if $x > 0$ and -1 otherwise.

Case 1: $\gamma_i' = 0, \alpha_i' \neq 0$

From Theorem 2, we know that a necessary condition for the existence of a Stackelberg equilibrium is that $\exists \tau' \geq 0$ s.t. $p_i = p_i^0 + \frac{\tau' - \sigma_i'}{\alpha_i'} q_i$.

Case 1.1: $\tau' = \sigma_i'$

In this case, the attacker's strategy p_i on node i is independent from the defender strategy q_i. Therefore, the strategy of the defender on node i has no influence on the attacker's strategy on that node. In this case, we may have an unlimited number of Stackelberg equilibriums. We note that if $\forall i \in \mathcal{T}$, $\tau' = \sigma_i'$, the study of this type of games is not interesting.

Case 1.2: $\tau' \neq \sigma_i'$

In this case, we have $q_i = \frac{\alpha_i'(p_i - p_i^0)}{\tau' - \sigma_i'}$. From the KKT conditions, there exists $\tau \geq 0$ s.t. $\frac{\partial U_A}{\partial p_i} = \tau$. Therefore, we have $2p_i\alpha_i'\gamma_i(\tau' - \sigma_i') + (\tau' - \sigma_i')\big((\alpha_i - \tau)(\tau' - \sigma_i') - \alpha_i'\gamma_i p_i^0 + \alpha_i'\sigma_i\big) = 0$. We have $p_i \in [0, 1]$ $\forall i \in \mathcal{T}$. Therefore, a necessary condition for the existence of a unique strategy on node i at the Stackelberg equilibrium in this case is that $\Gamma(\alpha_i')\big((\alpha_i - \tau)(\tau' - \sigma_i') - \alpha_i'(\gamma_i p_i^0 - \sigma_i)\big) \leq 0$ and $\Gamma(\alpha_i')\alpha_i'\gamma_i(\tau' - \sigma_i') > 0$.   □

Theorem 4. $\forall i \in \mathcal{T}$ s.t. $\gamma_i' > 0$ and $\alpha_i' \neq 0$, there exists at most two possible couple of strategies $(\mathbf{p_i^*}, \mathbf{q_i^*})$ and $(\mathbf{p_i^\dagger}, \mathbf{q_i^\dagger})$ at the Stackelberg equilibrium on each node i.

Proof. There are 3 possible cases to analyze.

Case 1: $\gamma_i' > 0$, $\alpha_i' \geq 0$, $q_i \neq \dfrac{-\alpha_i'}{\gamma_i'}$, and $p_i \neq \dfrac{\tau' - \sigma_i'}{\gamma_i'}$

In this case, we have $p_i = \dfrac{\alpha_i' p_i^0}{\alpha_i' + \gamma_i' q_i} + \dfrac{(\tau' - \sigma_i') q_i}{\alpha_i' + \gamma_i' q_i}$. We have a constraint on the attack budget $\sum\limits_{i \in \mathcal{T}} p_i \leq P$. Therefore, from the KKT conditions, $\exists \tau \geq 0$ s.t. $\dfrac{\partial U_A}{\partial p_i} = \tau$. Therefore, we have:

$$\alpha_i + \alpha_i' \gamma_i \left(\frac{p_i^0 - p_i}{\gamma_i' p_i - (\tau' - \sigma_i')} \right) + (\sigma_i + \gamma_i p_i) \left(\frac{\alpha_i'(\tau' - \sigma_i') - \alpha_i' \gamma_i' p_i^0}{\left(\gamma_i' p_i - (\tau' - \sigma_i') \right)^2} \right) = \tau$$

which can be written as $A_i p_i^2 + B_i p_i + C_i = 0$ where $A_i = \gamma_i'^2 (\alpha_i - \tau) - \alpha_i' \gamma_i' \gamma_i$, $B_i = 2(\tau' - \sigma_i')(\alpha_i' \gamma_i - \gamma_i'(\alpha_i - \tau))$, and $C_i = (\tau' - \sigma_i')((\alpha_i - \tau)(\tau' - \sigma_i') - \alpha_i' \gamma_i p_i^0 + \alpha_i' \sigma_i) - \alpha_i' \gamma_i' \sigma_i p_i^0$. This quadratic equation has at most 2 solutions, which concludes the proof for this case.

Case 2: $\gamma_i' > 0, \alpha_i' \leq 0, q_i \in \left[0, \min\left(\dfrac{-\alpha_i'}{\gamma_i'}, Q \right) \right[$, and $p_i \neq \dfrac{\tau' - \sigma_i'}{\gamma_i'}$

Similar to Case 2.

Case 3: $\gamma_i' > 0, \alpha_i' \leq 0, q_i \in \left] \min\left(\dfrac{-\alpha_i'}{\gamma_i'}, Q \right), Q \right]$, and $p_i \neq \dfrac{\tau' - \sigma_i'}{\gamma_i'}$

Similary to Case 1, from the partial derivative of U_A w.r.t. p_i, we can find that the strategy of the attacker is the solution of the quadratic equation $A_i p_i^2 + B_i p_i + C_i' = 0$ where $C_i' = (\tau' - \sigma_i')((\alpha_i - \tau)(\tau' - \sigma_i') - \alpha_i' \gamma_i p_i^0 - \gamma_i(\tau' - \sigma_i') \dfrac{2\alpha_i'}{\gamma_i'} - \alpha_i' \sigma_i) - \alpha_i' \gamma_i' \sigma_i p_i^0$. $\qquad \square$

Lemma 2. A realistic RCNS game can have an infinite number of Stackelberg equilibriums if $\exists i \in \mathcal{T}$ s.t. $\gamma_i' = 0, \alpha_i' \neq 0$, and $\tau' = \sigma_i'$. Otherwise, a realistic RCNS game can have at most 2^n Stackelberg equilibriums.

Lemma 2 follows directly from Theorems 3 and 4.

2.3.3 Maximin Strategy

In this section, we will be interested in analyzing the *maximin* strategy of the attacker. For space limitations, we will omit the analysis of the *maximin* strategy of the defender, which can be analyzed similarly.

A player's *maximin* strategy is a strategy in which he tries to maximize the worst payoff he can get for any strategy played by the other player. The attacker's *maximin* strategy is therefore given by $\mathbf{p} = \underset{\mathbf{p'}}{\text{argmax}} \min\limits_{\mathbf{q}} U_A(\mathbf{p'}, \mathbf{q})$.

We will study the attacker's *maximin* strategy under different constraints on the attacker's and defender's budgets $\sum_{i \in T} p_i$ and $\sum_{i \in T} q_i$ respectively.

Theorem 5. *For each strategy of the attacker, there exists a sensible target set $\mathcal{R}_D$ that will be of interest to the defender.*

Proof. For a given attacker strategy $\mathbf{p}$, the defender tries to compute $\min_{\mathbf{q}} U_A(\mathbf{p}, \mathbf{q}) = \min_{\mathbf{q}} \left(\sum_{i \in T} \alpha_i p_i + \delta_i + q_i(\sigma_i + \gamma_i p_i) \right)$. In the case of unconstrained defense budget, there exists a sensible target set $\mathcal{R}_D$ where $\forall i \in \mathcal{R}_D$, we have $q_i = 1$ and $\sigma_i + \gamma_i p_i < 0$, and $\forall j \in T \backslash \mathcal{R}_D$, we have $q_j = 0$ and $\sigma_j + \gamma_j p_j \geq 0$. In case of constrained defense budget $\sum_{i \in T} q_i = Q$, the sensible target set $\mathcal{R}_D$ is defined s.t. $\forall \{i, k\} \in \mathcal{R}_D$, $\sigma_i + \gamma_i p_i = \sigma_k + \gamma_k p_k$ and $i = \operatorname*{argmin}_{j \in T}(\sigma_j + \gamma_j p_j)$. $\square$

Theorem 6. *In the case of unconstrained defense budget, for a given sensible target set $\mathcal{R}_D$, there exists either 1 or an infinite maximin strategies for the attacker.*

Proof. Let ζ be the set of targets i s.t. $\alpha_i + \gamma_i = 0$. Let $\mathbb{1}_{expr} = 1$ if *expr* is true and 0 otherwise. In the case of unconstrained attacker budget, if $\zeta = \varnothing$, there exists a unique attacker *maximin* strategy where the attack resource on node i is determined by analyzing $r_i - t_i$ and σ_i. This can be found easily by analyzing the attacker's payoff $\alpha_i p_i + \delta_i + (\sigma_i + \gamma_i p_i) q_i$ on each target i. Otherwise, if $\zeta \neq \varnothing$, there exists an infinite number of attacker *maximin* strategies yielding at least a payoff of $\sum_{j \in \zeta} \delta_j + \sigma_j \mathbb{1}_{\sigma_j < 0}$ for targets in ζ. $\square$

In the rest of this section, we will analyze the attacker's *maximin* strategy in the presence of constraints on the defender's budget.

Let S be a large positive number. By analyzing the attacker's utility function $U_A(\mathbf{p}, \mathbf{q})$, we have the following lemma:

Lemma 3. *If $\sum_{i \in T} q_i = Q$ and in the absence of constraints on the attacker's budget, finding a maximin strategy for the attacker is equivalent to solving the following Mixed Integer Quadratic Program (MIQP):*

$$\max_{\mathbf{p}, \mathbf{q}, \mathbf{y}, b} U_A(\mathbf{p}, \mathbf{q})$$
$$s.t. \quad (y_i - 1)S \leq b - \sigma_i - \gamma_i p_i \leq 0$$
$$q_i \leq y_i S$$
$$\sum_{i \in T} q_i = Q$$
$$y_i \in \{0, 1\}, \ p_i \in [0, 1], \ q_i \in [0, Q], \ b \in \mathbb{R}$$

Lemma 4. *In the presence of constraints on the defender budget $\sum_{i \in T} q_i = Q$, for any sensible target set $\mathcal{R}_D$, assuming that the defender will focus on defending only one target in $\mathcal{R}_D$ will not change the impact of the defender's strategy on the maximin strategy of the attacker.*

Proof. If $\sum_{i \in \mathcal{T}} q_i = Q$, the defender will allocate his resources on the set of target i with the lowest $\sigma_i + \gamma_i p_i$. In addition, we have $\sigma_j + \gamma_j p_j = \sigma_m + \gamma_m p_m$, $\forall \{j, m\} \in \mathcal{R}_D$. By analyzing the attacker's utility function, we can notice that instead of setting $q_j \neq 0 \; \forall j \in \mathcal{R}_D$, the attacker can pick $m \in \mathcal{R}_D$ and set $q_m = Q$ without that changing the attacker's payoff.   □

Lemma 5. *In the presence of constraints on the attacker and defender budgets (resp. $\sum_{i \in \mathcal{T}} p_i = P$ and $\sum_{i \in \mathcal{T}} q_i = Q$), finding a maximin strategy for the attacker is equivalent to solving the following Mixed Integer Linear Program (MILP):*

$$\max_{\mathbf{y}, \mathbf{x}, b} \sum_{i \in \mathcal{T}} \left(\alpha_i \sum_{j \in \mathcal{T}} x_{ji} + \delta_i + (\sigma_i y_i + \gamma_i x_{ii}) Q \right)$$
$$s.t. \; (y_i - 1)S \leq b - \sigma_i - \gamma_i \sum_{j \in \mathcal{T}} x_{ji} \leq 0$$
$$\sum_{i \in \mathcal{T}} y_i = 1$$
$$y_i P \leq \sum_{j \in \mathcal{T}} x_{ij} \leq P$$
$$\sum_{i \in \mathcal{T}} x_{ij} \leq P$$
$$\sum_{i \in \mathcal{T}} \sum_{j \in \mathcal{T}} x_{ij} = P$$
$$y_i \in \{0, 1\}, \; x_{ij} \in [0, P], \; b \in \mathbb{R}$$

Proof. From Lemma 4, we can assume that the defender will defend 1 target with a resource Q. Let $y_i \in \{0, 1\} \; \forall i \in \mathcal{T}$. The *maximin* strategy of the attacker can then be found by maximizing $\sum_{i \in \mathcal{T}} \alpha_i p_i + \delta_i + (\sigma_i + \gamma_i p_i) y_i Q$ w.r.t. $\mathbf{p}$, $\mathbf{y}$, and b s.t. $(y_i - 1)S \leq b - \sigma_i - \gamma_i p_i \leq 0$, $\sum_{i \in \mathcal{T}} y_i = 1$, $\sum_{i \in \mathcal{T}} p_i = P$, and $b \in \mathbb{R}$. We can linearize this Mixed Integer Quadratic Program through the change of variables $x_{ij} = y_i p_j \; \forall \{i, j\} \in \mathcal{T}$.   □

3   Intrusion Detection Game

3.1   Game Model and Parameters

In this section, we introduce an intrusion detection game, which is a specific case of a RCNS game. We consider a heterogeneous network comprised of n interdependent equipment referred to as nodes in the remaining of this paper. The network can be represented as a weighted directed graph $\mathcal{G} = (\mathcal{T}, \mathcal{E}, \Theta)$, where $\mathcal{T} = \{1, ..., n\}$ is the set of network nodes, and $\mathcal{E}$ is a particular subset of $\mathcal{T}^2$ and referred to as the edges of $\mathcal{G}$. In particular, an edge (i, j) exists between node i and node j if compromising node i makes it easier for the attacker to compromise node j. Finally, a weight $\theta_i^j \in \Theta$, $\theta_i^j \in \left]0, 1\right]$, is associated to each edge $(i, j) \in \mathcal{E}$, quantifying the vulnerability dependency from node i to node j.

We model the intrusion detection problem as a non-cooperative static game with two players, an attacker and a defender. We assume that both players are rational. The objective of the attacker is to compromise targets in the network without being detected, whereas the defender's objective is to distribute monitoring resources on network nodes in order to detect attacks. For each node $i \in \mathcal{T}$, the attacker and the defender actions are limited to *Attack/Not Attack* and *Monitor/Not Monitor* respectively. The attacker's strategy is represented by a vector $\mathbf{p} = (p_1, ..., p_n) \in [0,1]^n$, where p_i is the probability of targeting node i. Similarly, the defender's strategy is represented by a vector $\mathbf{q} = (q_1, ..., q_n) \in [0,1]^n$, where q_i is the probability of monitoring node i. The resource constraints on the attacker and the defender budgets are P and Q respectively. Therefore, we have $\sum_{i=1}^{n} p_i \leq P$ and $\sum_{i=1}^{n} q_i \leq Q$, where $P \leq 1$ and $Q \leq 1$.

We associate to each node $i \in \mathcal{T}$ the following parameters:

- The security asset $W_i \geq 0$ representing the importance of services provided by node i to the network. Security assets are assumed to be independent, since the existing correlations between security assets may have already been taken into account through a formal risk analysis evaluation process.
- The intrinsic vulnerability $V_i^0 \in [0,1]$ quantifying local vulnerabilities of services on node i.
- The detection probability $a_i \in [0,1]$ representing the probability of detecting an attack on node i considering the current configuration of the defense system.

We assume that the costs of attacking and monitoring a node $i \in \mathcal{T}$ are proportional to the security asset W_i. In addition, these costs are affected by the intrinsic vulnerability V_i^0 on node i. In particular, the cost of attacking node i is inversely proportional to V_i^0, while the cost of monitoring node i is proportional to V_i^0. Therefore, the costs to attack and monitor node i are given by $C_a(1 - V_i^0)W_i$ and $C_m V_i^0 W_i$ respectively, where C_a and $C_m \in [0,1]$. Let $C_a^i = C_a(1 - V_i^0)$ and $C_m^i = C_m V_i^0$. Finally, we introduce a dependency parameter $\beta \in [0,1]$. β is used to assess the impact of interdependencies between network nodes in the utilities of the attacker and the defender. For example, $\beta = 0$ is equivalent to the case where interdependencies between network nodes are not taken into account in the model.

3.2 Utility Functions

Let $\Gamma^-(i)$ and $\Gamma^+(i)$ refer to the set of predecessors and the set of successors of node i in the network graph $\mathcal{G}$ respectively. The effect of interdependencies on node i is defined as $\Delta_i = \beta \sum_{j \in \Gamma^-(i)} \theta_j^i W_j p_j (1 - a_j q_j)$. Δ_i is the sum of the effect of interdependencies on node i from all its predecessors j that have been attacked (hence the p_j factor) without being detected (hence the $(1 - a_j q_j)$ factor) while taking into account the vulnerability dependency $\theta_j^i \in]0,1]$ from node j to i.

Table 3 presents the payoff matrix for both players in strategic form for a node $i \in \mathcal{T}$. A successful (i.e. undetected) attack on node i, which happens with probability $1 - a_i$, gives the attacker and the defender the payoffs $W_i(1 - a_i)$ and

Table 3. Payoff matrix in strategic form for node i

	Monitor	Not monitor
Attack	$W_i(1-2a_i-C_a(1-V_i^0))+\Delta_i,$ $W_i(2a_i-1-C_mV_i^0)-\Delta_i$	$W_i(1-C_a(1-V_i^0))+\Delta_i,$ $-W_i-\Delta_i$
Not attack	Δ_i , $-C_mV_i^0W_i-\Delta_i$	Δ_i , $-\Delta_i$

$-W_i(1-a_i)$ respectively. However, if the attack is detected, which happens with probability a_i, the payoffs for the attacker and the defender are given by $-W_ia_i$ and W_ia_i respectively. We take into account the impact of interdependencies between vulnerable network nodes. For example, even though the attacker can choose not to attack node i directly, he can benefit from the impact of attacks on the set of nodes whose compromise can affect his state on node i (e.g. in terms of information or privileges the attacker could decide to make use of).

The utilities U_A and U_D of the attacker and the defender respectively are as follows:

$$U_A(\mathbf{p},\mathbf{q}) = \sum_{i=1}^{n} \Big(p_iq_i(W_i(1-2a_i-C_a^i)+\Delta_i) + (1-p_i)q_i\Delta_i + p_i(1-q_i)(W_i(1$$
$$-C_a^i)+\Delta_i) + (1-p_i)(1-q_i)\Delta_i \Big) = \sum_{i=1}^{n} p_iW_i(1-2a_iq_i-C_a^i)+\Delta_i$$

$$U_D(\mathbf{p},\mathbf{q}) = \sum_{i=1}^{n} q_iW_i(2a_ip_i-C_m^i)-p_iW_i-\Delta_i$$

3.3 Solving the Game

3.3.1 Node Distribution

The values of the security assets and the impact of the interdependencies between nodes can affect the strategies of the attacker and the defender. In this section, we identify the set $\mathcal{T}_S$ of sensible targets that are attractive to the attacker and needs therefore to be monitored by the defender. Let $\mathcal{T}_U$ refer to the set of unattractive nodes that will not be the target of attacks. Therefore, we have $\mathcal{T} = \mathcal{T}_S \cup \mathcal{T}_U$. Let $\lambda_i = (1 - C_a^i + \beta\sum_{j\in\Gamma^+(i)} \theta_i^j)$ and $\mu_i = a_i(2 + \beta\sum_{j\in\Gamma^+(i)} \theta_i^j)$, $\forall i \in \mathcal{T}$.

Definition 6. *The sensible target set $\mathcal{T}_S$ and the set $\mathcal{T}_U$ are defined as follows:*

$$\begin{cases} W_i\lambda_i > \xi \ \forall i \in \mathcal{T}_S \\ W_i\lambda_i < \xi \ \forall i \in \mathcal{T}_U \end{cases} \text{ where } \xi = \dfrac{\sum\limits_{k\in\mathcal{T}_S}\left(\frac{\lambda_k}{\mu_k}\right)-Q}{\sum\limits_{k\in\mathcal{T}_S}\left(\frac{1}{W_k\mu_k}\right)}.$$

The case where $W_i\lambda_i = \xi$ does not need to be taken into account. In fact, this case happens with very low probability. Therefore, should this case happen, and since these values rely on estimations, replacing for instance W_i with a slightly different estimation $W_i + \epsilon$ or $W_i - \epsilon$ would be enough to solve the problem.

For the rest of this paper, we suppose that network nodes are numbered according to the following rule: $i < j \Leftrightarrow W_i \lambda_i \geq W_j \lambda_j$.

Lemma 6. *Given a network comprised of n nodes, T_S is uniquely determined and consists of n_S nodes with the highest $W_i \lambda_i$ values. The set T_S can be determined using Algorithm 1.*

Proof. We need to prove that T_S consists of the d highest $W_i \lambda_i$ values, where $d = n_S$ and the cases where $d < n_S$ and $d > n_S$ cannot be achieved. First, it is easy to prove that if $i \in T_S$, then $\forall j < i, j \in T_S$. We prove that $d = n_S$ with a proof by contradiction. Let us suppose that $d < n_S$, we have: $W_{n_S} \lambda_{n_S} \sum_{k=1}^{n_S} \left(\frac{1}{W_k \mu_k} \right) -$

$\sum_{k=d+1}^{n_S} \frac{\lambda_k}{\mu_k} > \sum_{k=1}^{d} \frac{\lambda_k}{\mu_k} - Q$. Noticing that $W_{n_S} \lambda_{n_S} \leq W_i \lambda_i, \forall i \leq n_S$ and $d < n_S$, we

have: $W_{d+1} \lambda_{d+1} \sum_{k=1}^{d} \left(\frac{1}{W_k \mu_k} \right) \geq W_{n_S} \lambda_{n_S} \sum_{k=1}^{d} \left(\frac{1}{W_k \mu_k} \right) = W_{n_S} \lambda_{n_S} \sum_{k=1}^{n_S} \left(\frac{1}{W_k \mu_k} \right)$

$- W_{n_S} \lambda_{n_S} \sum_{k=d+1}^{n_S} \left(\frac{1}{W_k \lambda_k} \frac{\lambda_k}{\mu_k} \right) \geq W_{n_S} \lambda_{n_S} \sum_{k=1}^{n_S} \left(\frac{1}{W_k \mu_k} \right) - \sum_{k=d+1}^{n_S} \left(\frac{\lambda_k}{\mu_k} \right) > \sum_{k=1}^{d} \left(\frac{\lambda_k}{\mu_k} \right) -$

Q. However, from Definition 6, we have $W_{d+1} \lambda_{d+1} \leq \frac{\sum_{k=1}^{d} \left(\frac{\lambda_k}{\mu_k} \right) - Q}{\sum_{k=1}^{d} \left(\frac{1}{W_k \mu_k} \right)}$. This con-

tradiction shows that it is impossible to have $d < n_S$. Similarly, we can show that it is impossible to have $d > n_S$. Therefore, $d = n_S$ and T_S is uniquely determined. $\qquad \square$

Algorithm 1. FindSensibleTargetSet

Data: The set of nodes T
Result: The sensible target set T_S
begin

    $W_i' \longleftarrow SortInDescendingOrder(W_{\sigma(i)} \lambda_{\sigma(i)})$
    $n_S \longleftarrow n$

    while $n_S \geq 1$ & $W_{n_S}' \leq \dfrac{\sum_{k=1}^{n_S} \frac{\lambda_k}{\mu_k} - Q}{\sum_{k=1}^{n_S} \left(\frac{1}{W_k' \mu_k} \right)}$ **do**

      | $n_S \longleftarrow n_S - 1$
    end
    $T_S = \{\sigma(i) \in T : i \in [\![1, n_S]\!]\}$
end

Theorem 7. *A rational attacker has no incentive to attack any node $i \in \mathcal{T}_U$.*

Proof. For space limitations, we only provide a sketch of the proof. The proof consists of showing that regardless of the defender's strategy $\mathbf{q}$, for any $\mathbf{p} \in [0,1]^n$ s.t. $\exists i \in \mathcal{T}_U$, $p_i > 0$, we can construct another strategy $\mathbf{p}'$ s.t. $p_i' = 0$, $\forall i \in \mathcal{T}_U$ and $U_A(\mathbf{p}, \mathbf{q}) < U_A(\mathbf{p}', \mathbf{q})$. If $\mathcal{T}_U = \varnothing$, the theorem holds. We focus in our proof on the case where $\mathcal{T}_U \neq \varnothing$. We consider a vector $\mathbf{q}^0 = (q_1^0, q_2^0, ..., q_N^0)$ s.t.:

$$
q_i^0 = \begin{cases} \dfrac{Q - \sum\limits_{k \in \mathcal{T}_S} \left(\frac{\lambda_k}{\mu_k} \right)}{W_i \mu_i \sum\limits_{k \in \mathcal{T}_S} \left(\frac{1}{\mu_k W_k} \right)} + \dfrac{\lambda_i}{\mu_i} & \forall i \in \mathcal{T}_S \\[6pt] 0 & \forall i \in \mathcal{T} - \mathcal{T}_S \end{cases}
$$

It holds that $\sum\limits_{i \in \mathcal{T}_S} q_i^0 = Q$, and $q_i^0 \geq 0$, $\forall i$. Let $\mathbf{q} = (q_1, ..., q_n)$ denote a defender strategy s.t. $\sum\limits_{i \in \mathcal{T}_S} q_i \leq Q$. By the pigeonhole principle, it holds that $\exists m \in \mathcal{T}_S$ s.t. $q_m \leq q_m^0$.

We consider an attacker strategy $\mathbf{p} = (p_1, ..., p_n)$ satisfying $\sum\limits_{i \in \mathcal{T}_U} p_i > 0$, i.e. the attacker attacks at least one target outside the sensible target set $\mathcal{T}_S$ with nonzero probability. We construct another attacker strategy profile $\mathbf{p}'$ based on $\mathbf{p}$ s.t.:

$$
p_i' = \begin{cases} p_i & i \in \mathcal{T}_S \text{ and } i \neq m \\ p_m + \sum\limits_{j \in \mathcal{T}_U} p_j & i = m \\ 0 & i \in \mathcal{T}_U \end{cases}
$$

After some algebraic operations, it is possible to show that $U_A(\mathbf{p}, \mathbf{q}) < U_A(\mathbf{p}', \mathbf{q})$. Therefore, the attacker is always better off attacking nodes in the sensible target set $\mathcal{T}_S$. □

Theorem 7 shows that the attacker only needs to attack nodes that belong to $\mathcal{T}_S$ in order to maximize his utility. Therefore, the defender has no incentive to monitor nodes that do not belong to $\mathcal{T}_S$. As a consequence, valuable defense resources would be wasted by monitoring nodes in $\mathcal{T}_U$. Therefore, a rational defender only needs to monitor nodes in $\mathcal{T}_S$.

3.3.2   NE Analysis

A strategy profile $(\mathbf{p}^*, \mathbf{q}^*)$ is a Nash Equilibrium of the intrusion detection game if each player cannot improve his utility by deviating from his strategy unilaterally. Let $\sum\limits_{i \in \mathcal{T}} p_i^* = P$ and $\sum\limits_{i \in \mathcal{T}} q_i^* = Q$. In this case, the attacker/defender uses all his resources to attack/defend the network. The game can be seen as a resource allocation problem, in which each player's objective is to maximize his/her utility given the action of the other player. The strategies of the attacker and the defender at the NE are as follows:

$$\forall i \in \mathcal{T}_S, \ p_i^* = \frac{P - \sum\limits_{k \in \mathcal{T}_S} \left(\frac{C_m^k}{\mu_k} \right)}{W_i \mu_i \sum\limits_{k \in \mathcal{T}_S} \left(\frac{1}{W_k \mu_k} \right)} + \frac{C_m^i}{\mu_i} \quad \text{and} \quad q_i^* = \frac{Q - \sum\limits_{k \in \mathcal{T}_S} \left(\frac{\lambda_k}{\mu_k} \right)}{W_i \mu_i \sum\limits_{k \in \mathcal{T}_S} \left(\frac{1}{W_k \mu_k} \right)} + \frac{\lambda_i}{\mu_i}$$

$$\forall i \in \mathcal{T}_U, \ p_i^* = 0 \text{ and } q_i^* = 0$$

The necessary conditions for the obtained result to be a NE are:

$$\begin{cases} W_i(2a_i p_i^* - C_m^i) + \beta W_i a_i p_i^* \sum\limits_{j \in \Gamma^+(i)} \theta_i^j \geq 0 \\ W_i(1 - 2a_i q_i^* - C_a^i) + \beta W_i(1 - a_i q_i^*) \sum\limits_{j \in \Gamma^+(i)} \theta_i^j \geq 0 \end{cases} \Rightarrow \begin{cases} P \geq \sum\limits_{i \in \mathcal{T}_S} \left(\frac{C_m^i}{\mu_i} \right) \\ Q \leq \sum\limits_{i \in \mathcal{T}_S} \left(\frac{\lambda_i}{\mu_i} \right) \end{cases}$$

In this case, the attacker and the defender focus on attacking and monitoring a subset $\mathcal{T}_S$ of nodes in the network. These nodes yield the maximum payoff for the attacker and therefore need to be monitored.

If $\sum\limits_{i \in \mathcal{T}} p_i^* < P$ and $\sum\limits_{i \in \mathcal{T}} q_i^* < Q$, both the attacker and the defender do not use all the available resources to attack and defend the network respectively. According to Theorem 1, in a realistic instance of this game, no NE exists.

4   Numerical Analysis

We consider a network comprised of $n = 10$ nodes. The type of the nodes and the values of some of the model parameters are depicted in Tables 4 and 5. The nodes in both tables are already sorted and numbered according to decreasing $W_i \lambda_i$ values as described in Sect. 3.

Table 4. Node types and individual parameters

Number	Node type	W_i	V_i^0	a_i
1	Business App. A	0.75	0.6	0.7
2	Intranet Portal	0.75	0.6	0.6
3	Mailing Server	0.75	0.3	0.6
4	Webmail Server	0.4	0.3	0.1
5	Business App. B	0.5	0.6	0.7
6	Intranet Common Services	1	0.6	0.1
7	Storage Area Network	1	0	0.1
8	Office Server	0.4	0.3	0.7
9	Authority Station	0.1	1	0.8
10	User Station	0.1	1	0.8

Table 5. Node interdependencies θ_i^j

i \ j	1	2	3	4	5	6	7	8	9/10
1	0	1	0	0	0	0.5	1	0	0
2	0	0	0	0	0	0.9	0.9	0	0
3	0	0	0	0	0	0.8	1	0	0
4	0	1	1	0	0	0.9	1	0	0
5	0	1	0	0	0	0.5	1	0	0
6	0	0	0	0	0	0	0	0	0
7	0	0	0	0	0	0	0	0	0
8	0	0	0	0	0	0.5	0.9	0	0
9	0.8	0.9	0.3	0.1	0.8	0.9	0	0.3	0
10	0.5	0.5	0.2	0.1	0.5	0.9	0	0.2	0

• We study the NE strategies of both players in two different scenarios. In the first scenario, we consider a typical network in which the attack and defense costs are relatively high compared with the security assets of the nodes (i.e. $C_a = C_m = 0.1$). In addition, the use of the interdependencies between nodes in the attack process is not considered of high criticality (i.e. $\beta = 0.5$). In this scenario, the attacker may not be tempted to fully exploit the node interdependencies in his attack. The resource constraints for the attacker and the defender are set to $P = 0.8$ and $Q = 0.9$ respectively, which means that the budget of the defender is slightly superior to the budget of the attacker. In the second scenario, the values of nodes security assets outweigh attack and defense costs (i.e. $C_a = C_m = 0.001$), and exploiting the interdependencies between nodes can play a significant role in the attack process (i.e. $\beta = 1$). In addition, due to the security requirements of such critical networks, the detection rate a_i on each node i is assumed to be $a_i \geq 0.5$. Finally, we consider that the attack and defense resource constraints are set to $P = 1$ and $Q = 1$ respectively.

The NE strategies of the attacker and the defender are depicted in Table 6. In both scenarios, the attacker/defender uses all his available resources to attack/defend. We note that both players focus on a sensible target set comprised of nodes 1, 2, 3, and 4 in the first scenario, and nodes 1, 2, 3, 4, and 5 in the second scenario. It is interesting to note that nodes 9 and 10 are not sensitive nodes despite having many dependencies stemming from them, as they have low security assets values to be worth attacking or defending. On the contrary, nodes 6 and 7 are not part of the sensible target set despite their relatively high security assets and the absence of dependencies stemming from them. In the second scenario, the sensible target set increased by one node (node 5). This is most probably due to the fact that the attacker has additional available resources and that node 4 had its detection probability a_i raised from 0.1 to 0.5, hence discouraging the attacker from spending too many resources to attack this node.

Table 6. Nash equilibrium for scenarios 1 and 2

Scenario 1	Scenario 2
$p_1^* = 0.0712, q_1^* = 0.3135$	$p_1^* = 0.1377, q_1^* = 0.3762$
$p_2^* = 0.0931, q_2^* = 0.2088$	$p_2^* = 0.1903, q_2^* = 0.2127$
$p_3^* = 0.0758, q_3^* = 0.1915$	$p_3^* = 0.1901, q_3^* = 0.2126$
$p_4^* = 0.5599, q_4^* = 0.1862$	$p_4^* = 0.2754, q_4^* = 0.1897$
$p_5^* = 0, q_5^* = 0$	$p_5^* = 0.2065, q_5^* = 0.0088$
$p_6^* = 0, q_6^* = 0$	$p_6^* = 0, q_6^* = 0$
$p_7^* = 0, q_7^* = 0$	$p_7^* = 0, q_7^* = 0$
$p_8^* = 0, q_8^* = 0$	$p_8^* = 0, q_8^* = 0$
$p_9^* = 0, q_9^* = 0$	$p_9^* = 0, q_9^* = 0$
$p_{10}^* = 0, q_{10}^* = 0$	$p_{10}^* = 0, q_{10}^* = 0$
$U_A = 0.898, U_D = -0.953$	$U_A = 1.736, U_D = -1.737$

The Security Information and Event Management (SIEM) software used in this industrial case study defines a metric to quantify the overall security of the network. This metric, which cannot be described in detail due to confidentiality reasons, consists in assessing, for each node, the types of attacks that can be mitigated given the current IDS configuration while taking into account the interdependencies between nodes in the evaluation process. After applying the optimal allocation of defense resources obtained at the NE, which translates in practice in configuring more efficient IDSs on critical nodes, we were able to notice a significant improvement of the overall security of the network, hence confirming the validity of our approach.

Sensitivity to θ_i^j. We analyze the impact of θ_i^j estimation errors on the identity of nodes that belong to the sensible target set $\mathcal{T}_S$. In both scenarios, nodes 8 to 10, due to their low security assets, remain in the set $\mathcal{T}_U$ even with a 20% estimation error on the values of each θ_i^j. In our model, the importance of a node is quantified by the value $W_i\lambda_i$, where λ_i mainly depends on β and the interdependencies θ_i^j. Therefore, inaccurate assessment of the interdependencies can have a significant impact on the results when the values of β and W_i are high. In our case study, when nodes 1, 2 and 3 have slightly erroneous interdependencies evaluations, we do not note any change in the sets $\mathcal{T}_S$ and $\mathcal{T}_U$. However, at the NE, we observe a small increase and decrease in the attacker and defender utilities respectively. For example, if on node 2, which has a relatively high security asset ($W_2 = 0.75$), $\sum_{j\in\Gamma^+(2)} \theta_2^j$ was overestimated by 0.4 (i.e. a 16% estimation error), U_A increases by 10% and U_D decreases by 5%. On the other hand, overestimating $\sum_{j\in\Gamma^+(5)} \theta_5^j$ by 0.1 (i.e. a 4% error) in scenario 1 is enough to include node 5 in $\mathcal{T}_S$. However, the impact of the error on U_A and U_D remains very low ($<1\%$).

Similarly, underestimating $\sum_{j \in \Gamma^+(5)} \theta_5^j$ by 0.1 in scenario 2 leads to the exclusion of node 5 from $\mathcal{T}_S$. At the NE, the attacker leverages this situation and targets node 5. However, it is interesting to note that the impact on the players' utilities remains inferior to 1% in this case as well. This shows that in some cases, an approximate construction of the sensible target set $\mathcal{T}_S$ does not necessarily entail a sudden substantial utility gain (*resp.* loss) for the attacker (*resp.* defender).

These observations demonstrate that our model is robust enough to deal with slight inaccuracies in the evaluation of interdependencies parameters. However, given the number of parameters θ_i^j to evaluate in large networks, important estimation errors on these parameters could have a significant impact on the strategies of the attacker and the defender, hence justifying the need for a more formal and rigorous evaluation method of these parameters.

5   Conclusion

In this paper, we introduced a set of security games that we refer to as Resource Constrained Network Security (RCNS) games and studied the necessary conditions for the existence of NE, Stackelberg equilibrium, and *maximin* strategies for this type of games. We then presented a game theoretical model for optimizing the allocation of monitoring resources to detect attacks in a network while taking into account nodes' vulnerabilities interdependencies. Finally, we validated our model via a real world case study. Our numerical study showed that the result of the analysis is sensitive to the values of parameters quantifying the interdependencies between network nodes. Therefore, elaborating a rigorous evaluation method for these parameters will be the subject of future work. In addition, we plan to investigate the impact of imperfect information in the general framework of RCNS games on the existence and uniqueness of equilibrium solutions.

References

1. Manshaei, M.H., Zhu, Q., Alpcan, T., Basar, T., Hubaux, J.P.: Game theory meets network security and privacy. ACM Comput. Surv. **45**(3), 25–39 (2013)
2. Chen, L., Leneutre, J.: A game theoretical framework on intrusion detection in heterogeneous networks. IEEE Trans. Inf. Forensics Secur. **4**(2), 165–178 (2009)
3. Alpcan, T., Basar, T.: A game theoretic approach to decision and analysis in network intrusion detection. In: Proceedings of the 42nd IEEE Conference on Decision and Control (CDC), vol. 3 (2003)
4. Alpcan, T., Basar, T.: An intrusion detection game with limited observations. In: Proceedings of the 12th International Symposium on Dynamic Games and Applications (2006)
5. Nguyen, K., Alpcan, T., Basar, T.: Stochastic games for security in networks with interdependent nodes. In: Proceedings of the International Conference on Game Theory for Networks (GameNets) (2009)

6. Miura-Ko, R., Yolken, B., Bambos, N., Mitchell, J.: Security investment games of interdependent organizations. In: Proceedings of the 46th Annual Allerton Conference on Communication, Control, and Computing (2008)
7. Sallhammar, K., Helvik, B., Knapskog, S.: Incorporating attacker behavior in stochastic models of security. In: Proceedings of the 2005 International Conference on Security and Management (2005)
8. Kodialam, M., Lakshman, T.: Detecting network intrusions via sampling: a game theoretic approach. In: IEEE INFOCOM (2003)
9. Otrok, H., Mohammed, N., Wang, L., Debbabi, M., Bhattacharya, P.: A game-theoretic intrusion detection model for mobile ad hoc networks. Comput. Commun. 31(4), 708–721 (2008)
10. Otrok, H., Mehrandish, M., Assi, C., Debbabi, M., Bhattacharya, P.: Game theoretic models for detecting network intrusions. Comput. Commun. 31(10), 1934–1944 (2008)
11. Zheng, D., Yu, F., Boukerche, A.: Security and quality of service (qos) co-design using game theory in cooperative wireless ad hoc networks. In: Proceedings of the Second ACM International Symposium on Design and Analysis of Intelligent Vehicular Networks and Applications (2012)
12. Djebaili, B., Kiennert, C., Leneutre, J., Chen, L.: Data integrity and availability verification game in untrusted cloud storage. In: Proceedings of the 5th International Conference on Decision and Game Theory for Security (2014)

Game-Theoretic Goal Recognition Models with Applications to Security Domains

Samuel Ang[1]($\boxtimes$), Hau Chan[2], Albert Xin Jiang[1], and William Yeoh[3]

[1] Department of Computer Science, Trinity University, San Antonio, TX 78212, USA
{sang,xjiang}@trinity.edu
[2] Department of Computer Science and Engineering, University of Nebraska-Lincoln, Lincoln, NE 68588, USA
hchan3@unl.edu
[3] Department of Computer Science and Engineering, Washington University in St. Louis, St. Louis, MO 63130, USA
wyeoh@wustl.edu

Abstract. Motivated by the goal recognition (GR) and goal recognition design (GRD) problems in the artificial intelligence (AI) planning domain, we introduce and study two natural variants of the GR and GRD problems with strategic agents, respectively. More specifically, we consider game-theoretic (GT) scenarios where a malicious adversary aims to damage some target in an (physical or virtual) environment monitored by a defender. The adversary must take a sequence of actions in order to attack the intended target. In the GTGR and GTGRD settings, the defender attempts to identify the adversary's intended target while observing the adversary's available actions so that he/she can strengthens the target's defense against the attack. In addition, in the GTGRD setting, the defender can alter the environment (e.g., adding roadblocks) in order to better distinguish the goal/target of the adversary.

We propose to model GTGR and GTGRD settings as zero-sum stochastic games with incomplete information about the adversary's intended target. The games are played on graphs where vertices represents states and edges are adversary's actions. For the GTGR setting, we show that if the defender is restricted to playing only stationary strategies, the problem of computing optimal strategies (for both defender and adversary) can be formulated and represented compactly as a linear program. For the GTGRD setting, where the defender can choose K edges to block at the start of the game, we formulate the problem of computing optimal strategies as a mixed integer program, and present a heuristic algorithm based on LP duality and greedy methods. Experiments show that our heuristic algorithm achieves good performance (i.e., close to defender's optimal value) with better scalability compared to the mixed-integer programming approach.

In contrast with our research, existing work, especially on GRD problems, has focused almost exclusively on decision-theoretic paradigms, where the adversary chooses its actions without taking into account the fact that they may be observed by the defender. As such an assumption is unrealistic in GT scenarios, our proposed models and algorithms fill a significant gap in the literature.

© Springer International Publishing AG 2017
S. Rass et al. (Eds.): GameSec 2017, LNCS 10575, pp. 256–272, 2017.
DOI: 10.1007/978-3-319-68711-7_14

1   Introduction

Discovering the objective of an agent based on observations of its behavior is a problem that has interested both artificial intelligence (AI) and psychology researchers for many years [7,23]. In AI, this problem is known as *goal recognition* (GR) or, more generally, *plan recognition* [25]. Plan and goal recognition problems have been used to model a number of applications ranging from software personal assistants [16–18]; robots that interact with humans in social settings such as homes, offices, and hospitals [8,26]; intelligent tutoring systems that recognize sources of confusion or misunderstanding in students through their interactions with the system [6,12,14,15]; and security applications that recognize the plan or goal of terrorists [5].

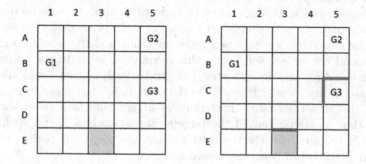

Fig. 1. Example Problem (left) and with Blocked Actions in Red (right).

One can broadly summarize the existing research in GR as one that primarily focuses on developing better and more efficient techniques to recognize the plan or the goal of the user given a sequence of observations of the user's actions. For example, imagine a scenario shown in Fig. 1 (left), where an agent is at cell $E3$, it can move in any of the four cardinal directions, and its goal is one of three possible goals G1 (in cell $B1$), G2 (in cell $A5$), and G3 (in cell $C5$). Additionally, assume that it will move along a shortest path to its goal. Then, if it moves left to cell $E2$, then we can deduce that its goal is G1. Similarly, if it moves right to cell $E4$, then its goal is either G2 or G3.

Existing research has focused on agent GR models that are non-strategic or partially strategic: The agent's objective is to reach its goal with minimum cost, and the agent does not explicitly reason about its interaction with the observer. However, when the observer's recognition of the agent's goal affects the agent in some way, then it is in the agent's best interest to be *fully strategic* – to *explicitly* reason about how the agent's choice affects the observer's recognition. As a result, the observer will need to take into account the agent's strategic reasoning when making decisions.

1.1   Game-Theoretic Goal Recognition Problems in Security Domains

Naturally, GT settings with strategic agents are common in many real-world (physical and cyber) security scenarios between an adversary and a defender. The adversary has a set of targets of interests and would be equally happy in attacking one of them. In physical security domains, the adversary must make a sequence of physical movements to reach a target; in cyber security domains, this could be a sequence of actions achieving necessary subgoals to carry out the attack. In any case, the defender is trying to recognize the adversary's goal/target. We coined this the game-theoretic goal recognition (GTGR) problem.

Let us describe the security games of interests using Fig. 1. Consider the security scenario in Fig. 1 (left), where an agent (i.e., terrorist) wants to reach its intended target and carry out an attack, while we, the observer (the defender) try to recognize the agent's goal as early as possible. Suppose once we recognize the agent's goal, we will strengthen the agent's target to defend against the attack. The more time we have between recognition and the actual attack, the less successful the attack will be. In this scenario, it is no longer optimal for the agent to simply choose a shortest path to its goal, as that could allow the observer to quickly identify its goal. On the other hand, the agent still wants to reach its goal in a reasonably short time, as a very long path could allow the observer time to strengthen all the targets. So, an optimal agent would need to explicitly reason about the tradeoffs between the cost of its path (e.g., path length) and the cost of being discovered early.

1.2   Game-Theoretic Goal Recognition Design Problems in Security Domains

So far we have been discussing the defender's task on recognizing goals. However, the task could become very difficult in general. For instance, going back to our security example in Fig. 1, if the agent moves up to $D3$, the observer cannot make any informed deductions. In fact, if the agent moves along any one of its shortest paths to goal G3, throughout its entire path, which is of length 4, we cannot deduce whether its goal is either G2 or G3! This illustrates one of the challenges with this approach, that is, there are often a large number of ambiguous observations that can be a result of a large number of goals. As such, it is difficult to uniquely determine the goal of the agent until a long sequence of observations is observed.

The work of [9, 10] proposed an orthogonal approach to *modify the underlying environment of the agent*, in such a way that *the agent is forced to reveal its goal as early as possible*. They call this problem the goal recognition design (GRD) problem. For example, if we block the actions $(E3, up), (C4, right), (C5, up)$ in our example problem, where we use tuples (s, a) to denote that action a is blocked from cell s, then the agent can make at most 2 actions (i.e., right to E4 then up to D4) before its goal is conclusively revealed. Figure 1 (right) shows the blocked actions. This problem finds itself relevant in many of the same applications of GR because, typically, the underlying environment can be easily modified.

As such, in addition to studying the GTGR problem, we consider the GTGRD problem where the observer can modify the underlying environment (i.e., adding K roadblocks) as to restrict the actions of the agent.

1.3 Related Work

GR and its more general forms, plan recognition and intent recognition, have been extensively studied [25] since their inception almost 40 years ago [23]. Researchers have made significant progress within the last decade through synergistic integrations of techniques ranging from natural language processing [3,27] to classical planning [20–22] and deep learning [15]. The closest body of work to ours is the one that uses game-theoretic formulations, including an adversarial plan recognition model that is defined as an imperfect information two-player zero-sum game in extensive form [13], a model where the game is over attack graphs [1], and an extension that allows for stochastic action outcomes [4]. The main difference between these works and ours is that ours focuses on goal recognition instead of plan recognition.

While GR has a long history and extensive literature, the field of GRD is relatively new. Keren et al. introduced the problem in their seminal paper [9], where they proposed a decision-theoretic STRIPS-based formulation of the problem. In the original GRD problem, the authors make several simplifying assumptions: (1) the observed agent is assumed to execute an optimal (i.e., cost-minimal) plan to its goal; (2) the actions of the agent are deterministic; and (3) the actions of the agent are fully observable. Since then, these assumptions have been independently relaxed, where agents can now execute boundedly-suboptimal plans [10], actions of the agents can be stochastic [28], and actions of the agents can be only partially observable [11]. Further, aside from all the decision-theoretic approaches above, researchers have also modeled and solved the original GRD problem using answer set programming [24]. The key difference between these works and ours is that ours introduced a game-theoretic formulation that can more accurately capture interactions between the observed agent and the observer in security applications.

1.4 Our Contributions

As a result of the strategic interaction in the GTGR and GTGRD scenarios, the concept of cost-minimal plan (the solution concept in GR problem) and worst-case distinctiveness (the solution concept in GRD problem) are no longer a suitable solution concept since it does not reflect the behavior of strategic agents. Instead, our objective here is to formulate game-theoretic models of the agent's and observer's interactions under GR and GRD settings. More specifically, we propose to model GTGR and GRGRD settings as zero-sum stochastic games with incomplete information where the adversary's target is unknown to the observer. For the GTGR setting, we show that if the defender is restricted to playing only stationary strategies, the problem of computing optimal strategies (for both defender and adversary) can be formulated and represented compactly as a linear program. For the GTGRD setting, where the defender can choose K edges to block at the start of the game, we formulate the problem of computing optimal strategies as a mixed integer program, and present a heuristic algorithm based on LP duality and greedy methods. We perform experiments to show that our heuristic algorithm achieves good performance (i.e., close to defender's optimal value) with better scalability compared to the mixed-integer programming approach.

2   Preliminary: Stochastic Games

In our two-player zero-sum single-controller stochastic game G, we have a finite set S of states, and an initial state $s_0 \in S$. The first player acts as an adversary attempting to reach some target within the environment, while second player acts as the observer of the environment. Given a state $s \in S$, there exist finite action sets J_s and I_s for the adversary and the observer respectively. Given a state $s \in S$ and $j \in J_s$, a single-controller transition function $\chi(s, j)$ deterministically maps state and action to a new state. Given a state $s \in S$, $j \in J_s$, $i \in I$, and intended target of the adversary θ, we define a reward function $r(s, i, j, \theta) \in \mathbb{R}$. Since this is a zero-sum game, without loss of generality, we define r as the reward for the observer and the additive inverse of the reward for the adversary. We consider a two-player zero-sum single-controller stochastic game where observer has incomplete information. In particular, the game consists of a collection of zero-sum single-controller stochastic games $\{G_\theta\}_{\theta \in B}$ and a probability distribution $P \in \Delta(B)$ over B. For our setting, we assume that each stochastic game G_θ could have different reward function r^θ, but all of the games $G'_\theta s$ have the same sets of states, actions, and transition rules. The game is played in stages over some finite time. First, a game G_θ is drawn according to P. The adversary is informed of θ while the observer does not know θ, but rather a set of states B of which θ is a part of. At each stage of game t with current state $s_t \in S$, the adversary selects $j_t \in J_s$ and the observer selects $i_t \in I$, and s_{t+1} is reached according to $\chi(s_t, j_t)$. However, we assume that the adversary does not know i_t, and both of the players do not know $r^\theta(s_t, i_t, j_t)$. Note that observer can infer the action of the adversary given the new state since our transition function is deterministic. Hence, the observer knows j_t, i_t, and s_{t+1}.

The strategies of the players can be based on their own history of the previous states and strategies. In addition, player 1 can condition his strategies based on θ. We consider a finite timestep to be at most T. Let $h_t^1 = (s_0, j_0, s_1, j_1, ..., j_{t-1}, s_t)$ and $h_t^2 = (s_0, j_0, i_0, s_1, ..., j_{t-1}, i_{t-1}, s_t)$ to denote a possible history of length t of player 1 and player 2 where $j_k \in J_{s_k}$ and $i_k \in I$ for $k = 1, ..., t$. Let $H_{s_t}^1$ and $H_{s_t}^2$ be the set of all possible histories of length t ended up at state s_t. Then, the sets of deterministic strategies for player 1 and player 2 are therefore $\prod_{t=0 \leq T, s_t \in S, h_{s_t}^1 \in H_{s_t}^1} J_{s_t}$ and $\prod_{t=0 \leq T, s_t \in S, h_{s_t}^2 \in H_{s_t}^2} I$, respectively. Indeed, for each possible history, the players need to select some actions. Naturally, the players mixed strategies are distributions over the deterministic strategies.

Definition 1. *Given $\theta \in B$, $0 \leq t \leq T$, $s_t \in S$, $h^1_{s_t} \in H^1_{s_t}$, player 1's behavioral strategy $\sigma_1(\theta, h^1_{s_t}, j_{s_t})$ returns the probability of playing $j_{s_t} \in J_{s_t}$ such that $\sum_{j_{s_t} \in J_{s_t}} \sigma_1(\theta, h^1_{s_t}, j_{s_t}) = 1$. (Player 2's behavioral strategy σ_2 is defined similarly and does not depend on θ).*

Definition 2. *A behavioral strategy σ is stationary if and only if it is independent of any timestep t and depends only on the current state (i.e., $\sigma_1(\theta, h^1_s, j_s) = \sigma_1(\theta, \bar{h}^1_s, j_s)$ such that h^1_s and $\bar{h}^1_s$ have the same last state and σ_2 can be defined similarly).*

Given a sequence $\{(s_t, i_t, j_t)\}^T_{t=1}$ of actions and states, the total reward for player 2 is $r_T = \sum^T_{t=1} r^\theta(s_t, i_t, j_t)$. Thus, the expected reward $\gamma_T(P, s_0, \sigma_1, \sigma_2) = \mathbf{E}_{P,s_0,\sigma_1,\sigma_2}[r_T]$ is the expectation of r_T over the set of stochastic games $\{G_\theta\}_{\theta \in B}$ given the the fixed initial state s_0 under P, σ_1, and σ_2, respectively.

Definition 3. *The behavioral strategy σ_2 is a best response to σ_1 if and only if for all σ'_2, $\gamma_T(P, s_0, \sigma_1, \sigma_2) \geq \gamma_T(P, s_0, \sigma_1, \sigma'_2)$. The behavioral strategy σ_1 is a best response to σ_2 if and only if for all σ'_1, $\gamma_T(P, s_0, \sigma_1, \sigma_2) \leq \gamma_T(P, s_0, \sigma'_1, \sigma_2)$.*

For two-player zero-sum games, the standard solution concept is the max-min solution: $\max_{\sigma_2} \min_{\sigma_1} \gamma_T(P, s_0, \sigma_1, \sigma_2)$. One can also define min-max solution $\min_{\sigma_1} \max_{\sigma_2} \gamma_T(P, s_0, \sigma_1, \sigma_2)$. For zero-sum games, the max-min value, min-max value, and Nash equilibrium values all coincide [2]. For simultaneous-move games this can usually be solved by formulating a linear program. In this work, we will be focusing on computing the max-min solution.

3 Game Model

We begin by describing our settings and introducing the GTGR and GTGRD models.

3.1 Game-Theoretic Goal Recognition Model

Consider a deterministic environment such as the one in the introduction. We can model the environment with a graph in which the nodes correspond to the states and the edges connect neighboring states. Given the environment and the graph, as in many standard GR problems, the agent wants to plan out a sequence of moves (i.e., determining a path) to reach its target location of the graph. The target location is unknown to the observer, and the observer's goals are to identify the target location based on the observed sequence of moves and to make preventive measure to protect the target location.

We model this scenario as a two-player zero-sum game, between the agent/ adversary and the observer. Given the graph $G = (L, E)$ of the environment, the adversary is interested in a set of potential targets $B \subseteq L$ and has a starting position $s_0 \in L \setminus B$. The adversary's aim is to attack a specific target $\theta \in B$, which is chosen at random according to some prior probability distribution P.

The observer does not know the target θ, and only the adversary knows its target θ. However, the observer knows the set of possible targets B and the adversary's starting position s_0. For any $s \in L$, we let $\nu(s)$ is the set of neighbors of s in the graph G.

The sequential game is played over several timesteps where both players move simultaneously. Each timestep, the observer selects a potential target in B to protect, and the agent moves to a neighboring node. We consider the zero-sum scenario: With each timestep, the adversary and the observer will lose and gain a value d, respectively. In addition, if the observer protects the correct target location θ, an additional value of q will be added to the observer and subtracted from the adversary. The game ends when the attacker reaches its target θ, a value of u^θ will be added to the adversary's overall score, and u^θ will be subtracted from the observer's overall score. Notice that during the play of the game, the adversary does not observe the observer's action(s), and the players do not know of their current scores.

Because of the potentially stochastic nature of the adversary's moves at each timestep, and the uncertainty of adversary's target in the system, our setting is most naturally modeled as a *stochastic game with incomplete information* as defined in Sect. 2. More specifically, the set of states is L with an initial state s_0. Given a state $s \in S$, $\nu(s)$ is the action set for the adversary and B is the action set for the observer. Given a state $s \in S$ and $j \in \nu(s)$, the single-controller transition function $\chi(s, j) = j$. Indeed, the transition between states are controlled by the adversary only and is deterministic: From state s, where $s \neq \theta$, given attacker action $j \in \nu(s)$, the next state is j. The state θ is terminal: Once reached, the game ends. Given a state $s \in S$, $j \in \nu(s)$, and $i \in B$, we define the reward function $r^\theta(s, i, j) \equiv r(s, i, j, \theta)$ from the observer's point of view as

$$
r(s, i, j, \theta) = \begin{cases} d & j \neq \theta \ \& \ i \neq \theta \\ d + q & j \neq \theta \ \& \ i = \theta \\ d - u^\theta & j = \theta \ \& \ i \neq \theta \\ d + q - u^\theta & j = \theta \ \& \ i = \theta. \end{cases} \tag{1}
$$

While, in theory, the game could go on forever if the adversary never reaches his target θ, because of the per-timestep cost of d, any sufficiently long path for the adversary would be dominated by the strategy of taking the shortest path to θ. Eliminating these dominated strategies allows us to set a finite bound for the duration of the game, which grows linearly in the shortest distance to the target that is furthest away. Even in games where the value of d is set to 0, the defender could potentially play a uniformly random strategy that imposes a cost of $\frac{q}{|B|}$ per timestep. Therefore, an adversary strategy taking forever would achieve a value of $-\infty$ against the uniformly random defender strategy. In any Nash equilibrium the attacker will always reach their target in finite time.

We call this the game-theoretic goal recognition (GTGR) model. All of the definitions in Sect. 2 follow immediately for our games.

3.2   Game-Theoretic Goal Recognition Design Model

As mentioned in the introduction, we also consider the game-theoretic goal recognition design (GTGRD) model. Formally, before the game starts, we allow the observer to block a subset of at most K actions from the game. In our model, that corresponds to blocking at most K edges from the graph. In one variant of the model, blocking an edge effectively removes that edge, i.e. the adversary can no longer take that action. In another variant, blocking an edge does not prevent the adversary from taking the action, but the adversary would incur a cost by taking that action. After placing the blocks, the game proceeds as described in Sect. 3.1.

4   Computation

4.1   Game-Theoretic Goal Recognition Model

With the game defined, we are interested in computing the solution of the game: What is the outcome of the game when both players behave rationally? Before defining rational behavior, we first need to discuss the set of strategies. In a sequential game, a pure strategy of a player is a deterministic mapping from the current state and the player's observations/histories leading to the state, to an available action. For the adversary, such observations/histories include its own sequence of prior actions and its target θ; the observer's observations/histories include the adversary's sequence of actions and the observer's sequence of actions. A mixed strategy is a randomized strategy, specified by a probability distribution over the set of pure strategies. The strategies are defined more formally in Sect. 2 and Definition 1.

As mentioned earlier, we are interested in computing the max-min solution, which is equivalent to the max-min value, min-max value, and Nash equilibrium value of the game. For simultaneous-move games this can usually be solved by formulating a linear program. However, for our sequential game, each pure strategy need to prescribe an action for each possible sequence of observations leading to that state and, as a result, the sets of pure strategies are exponential for both players.

To overcome this computational challenge, we focus on *stationary strategies*, which depend only on the current state (for the adversary, also on θ) and not on the history of observations (see Definition 2). While for stochastic games with complete information, it is known that there always exist an optimal solution that consists of stationary strategies [2], it is an open question whether the same property holds for our setting, which is an incomplete-information game. Nevertheless, there are some heuristic reasons that stationary strategies are at least good approximations of optimal solutions: The state (i.e., adversary's location) already captures a large amount of information about the strategic intention of the adversary.

An intuitively optimal non-stationary strategy in which the observer assigns resources to the target with maximal probability, determined through observing the actions of the adversary, presents additional challenges. An optimal strategy of this nature would require information regarding adversary's strategy from the beginning of the game, so as to determine the likelihood of a given action assuming a particular target for the adversary. Making such assumptions is a straightforward process when restricting the observer to stationary strategies. Later in this paper we will demonstrate how given a stationary strategy for the observer, there exists a best response strategy for the adversary that is also stationary.

Restricting to stationary strategies, randomized strategies now correspond to a mapping from state to a distribution over actions. We have thus reduced the dimension of the solution space from exponential to polynomial in the size of the graph. Furthermore, our game exhibits the single-controller property: The state transitions are controlled by the adversary only. For complete information stochastic games with a single controller, a *linear programming* (LP) formulation is known [19]. We adapt this LP formulation to our incomplete information setting.

We define $V(\theta, s)$ to be a variable that represents the expected payoff to the observer at state s and with adversary's intended target θ. We use $P(\theta)$ to denote the prior probability of $\theta \in B$ being the adversary's target such that $\sum_{\theta \in B} P(\theta) = 1$. The observer's objective is to find a (possibly randomized) strategy that maximizes his expected payoff given the prior distribution over the target set B, the moves of the adversary, and the adversary's starting location. The following linear program computes the utility of the observer in a max-min solution assuming both players are playing a stationary strategy.

$$\max_{V, \{f_i(s)\}_{i,s}} \sum_{\theta} P(\theta) V(\theta, s_o) \tag{2}$$

$$V(\theta, s) \leq \sum_{i \in B} r(s, i, j, \theta) f_i(s) + V(\theta, j) \quad \forall \theta \in B, \forall s \mid s \neq \theta, \forall j \in \nu(s) \tag{3}$$

$$V(\theta, s) = 0 \qquad\qquad\qquad\qquad\qquad\qquad \text{when } s = \theta \tag{4}$$

$$\sum_i f_i(s) = 1 \qquad\qquad\qquad\qquad\qquad\qquad\qquad \forall s \tag{5}$$

$$f_i(s) \geq 0 \qquad\qquad\qquad\qquad\qquad\qquad\qquad\quad \forall s, i \tag{6}$$

In the above linear program, (2) is the objective of the observer. The $f_i(s)$'s represent the probability of the observer taking an action $i \in B$ given the state s. To ensure a well defined probability distribution for each state of the games, (5) and (6) impose the standard sum-equal-to-one and non-negative conditions on the probability of playing each action $i \in B$. The Bellman-like inequality (3) bounds the expected value for any state using expected values of next states plus the expected current reward, assuming the adversary will choose the state transition that minimizes the observer's expected utility. Finally, (4) specifies the base condition when the adversary has reached their destination and the game ends. The size of the linear program is polynomial in the size of the graph.

The solution of this linear program prescribes a randomized stationary strategy $f_i(s)$ for the observer and, from the dual solutions, one can compute a stationary strategy for the adversary. In more detail, the dual linear program is

$$\min \sum_s t_s \tag{7}$$

$$t_s \geq \sum_{\theta,j} \lambda^\theta_{s,j} r(s,i,j,\theta) \qquad \forall s,i \tag{8}$$

$$I_{s=s_0} P(\theta) + \sum_{s' \neq \theta : s \in \nu(s')} \lambda^\theta_{s',s} = \sum_{j \in \nu(s)} \lambda^\theta_{s,j} \qquad \forall \theta \in B, \forall s \neq \theta \tag{9}$$

$$\lambda^\theta_{s,j} \geq 0 \qquad \forall \theta, s, j \tag{10}$$

where $I_{s=s_0}$ is the indicator that equals 1 when $s = s_0$ and 0 otherwise. The dual variables $\lambda^\theta_{s,j}$ can be interpreted as the probability that adversary type θ takes the edge from s to j. These probabilities satisfies the flow conservation constraints (9): given θ, the total flow into s (the left hand side) is equal to the probability that type θ visits s, which should equal the total flow out of s (the right hand side). The variables t_s can be interpreted as the contribution to defender's utility from state s, assuming that the defender is choosing an optimal action at each state (ensured by constraint (8)).

Given the dual solutions $\lambda^\theta_{s,j}$, we can compute a stationary strategy for the adversary: let $\pi(j|\theta,s)$ be the probability that the adversary type θ chooses j at state s. Then for all $\theta \in B$ and $s \neq \theta$, $\pi(j|\theta,s) = \frac{\lambda^\theta_{s,j}}{\sum_{j' \in \nu(s)} \lambda^\theta_{s,j'}}$. It is straightforward to verify that by playing the stationary strategy π, the adversary type θ will visit each edge (s,j) with probability $\lambda^\theta_{s,j}$.

Lemma 1. *Given a stationary strategy for the defender, there exists a best response strategy for the adversary that is also a stationary strategy.*

Proof (Sketch). Given a stationary defender strategy $f_i(s)$, each adversary type θ now faces a Markov Decision Process (MDP) problem, which admits a stationary strategy as its optimal solution.

More specifically, since the state transitions are deterministic and fully controlled by the adversary, each type θ faces a problem of determining the shortest path from s_0 to θ, with the cost of each edge (s,j) as $\sum_{i \in B} f_i(s) r(s,i,j,\theta)$. Looking into the components of $r(s,i,j,\theta)$, since the adversary reward u^θ for reaching target θ occurs exactly once at the target θ, it can be canceled out and the problem is equivalent to the shortest path problem from s_0 to θ with edge cost $d + f_\theta(s)q$. Since edge costs are nonnegative the shortest paths will not involve cycles.

What this lemma implies is that if the defender plays the stationary strategy prescribed by the LP (2), the adversary cannot do better than the value of the LP by deviating to a non-stationary strategy.

Corollary 1. *If the defender plays the stationary strategy $f_i(s)$ given by the solutions of LP (2), the adversary's stationary strategy π as prescribed by LP (7) is a best response, i.e., no non-stationary strategies can achieve a better outcome for the adversary.*

While it is still an open question whether the defender has an optimal stationary strategy, we have shown that if we restrict to stationary strategies for the defender, it is in the best interest of the adversary to also stick to stationary strategies and our LP (2) does not overestimate the value of the game.

4.2   Game-Theoretic Goal Recognition Design Model

One can solve this GTGRD problem by brute-force, i.e., try every subset of edges to block and then for each case solve the resulting LP. The time complexity of this approach grows exponentially in K. Instead, we can encode the choice of edge removal as integer variables added to the LP formulation, resulting in a mixed-integer program (MIP). For example, we could replace (3) with

$$V(\theta, s) \leq \sum_{i \in B} r(s, i, j, \theta) f_i(s) + V(\theta, j) + Mz(s, j) \tag{11}$$

where M is a positive number, and $z(s, j)$ is a 0–1 integer variable indicating whether the action/edge from s to j is blocked. M thus represents the penalty that the attacker incurs if he nevertheless chooses to take the edge from s to j while it is blocked. By making M sufficiently large, we can make the actions of crossing a blocked edge dominated and therefore effectively removing the edges that we block. We also add the constraint $\sum_{s,j} z(s, j) \leq K$.

Dual-Based Greedy Heuristic. The MIP approach scales exponentially in the worst case as the size of the graph and K grows. We propose a heuristic method for selecting edges to block. We first solve the LP for goal recognition and its dual. In particular, we look at the dual variable $\lambda_{s,j}^{\theta}$ for the constraint (3). This dual has the standard interpretation as the *shadow price*: it is the rate of change to the objective if we infinitesimally relax constraint (3).

Looking at the MIP, in particular constraint (11), we see that by blocking off an action from s to j we are effectively relaxing the corresponding LP constraints (3) indexed by θ, s, j for all $\theta \in B$. These are the adversary's incentive constraints for going from s to j, for all adversary types θ.

Utilizing the shadow price interpretation of the duals, the sum of the duals corresponding to the edge from s to j: $\sum_{\theta \in B} \lambda_{s,j}^{\theta}$ gives the rate of change to the objective (i.e. defender's expected utility) if the edge (s, j) is blocked by an infinitesimal amount. Choosing the edge that maximizes this, $\arg\max_{s,j} \sum_{\theta \in B} \lambda_{s,j}^{\theta}$ we get the maximum rate of increase of our utility. These rates of changes hold only when the amount of relaxation (i.e., M) is infinitesimal. However, in practice we can still use this as a heuristic for choosing edges to block.[1]

---

[1] Another perspective: from the previous section we see that $\lambda_{s,j}^{\theta}$ is the probability that adversary type θ traverses the edge s, j. Then if the adversary and defender do not change their strategies after the edge (s, j) is blocked, the defender would receive an additional utility of $M \sum_{\theta \in B} \lambda_{s,j}^{\theta}$ from the adversary's penalty for crossing that edge.

When $K > 1$, we could choose the K edges with the highest dual sums. Alternatively, we can use a greedy approach: pick one edge with the maximum dual sum, place a block on the edge and solve the updated LP for goal recognition, and pick the next edge using the updated duals, and repeat. In our experiments, the latter greedy approach consistently achieved significantly higher expected utilities than the former. Intuitively, by re-solving the LP after adding each edge, we get a more accurate picture of the adversary's adaptations to the blocked edges. Whereas the rates of changes used by the former approach are only accurate when the adversary do not adapt at all to the blocked edges (see Footnote 1). Our greedy heuristic is summarized as follows.

- for $i = 1 \ldots K$:
    - Solve LP (2), updated with the current blocked edges. If edge (s, j) blocked, the corresponding constraint (3) indexed s, j, θ for all θ are modified so that M is added to the right hand side. Get the primal and dual solutions.
    - Take an edge $(s^*, j^*) \in \arg\max_{s,j} \sum_{\theta \in B} \lambda^\theta_{s,j}$, and add it to the set of blocked edges.
- return the set of blocked edges, and the primal solution of the final LP as the defender's stationary strategy.

5   Experiments

Experiments were run on a machine using OSX Yosemite version 10.10.5, with 16 GB of ram and a 2.3 GHz Intel Core i7 processor, and were conducted on grid environments such as the one seen in Fig. 2. In these environments, the adversary is allowed to move to adjacent nodes connected by an edge. S denotes the starting location of the adversary while $T1$ and $T2$ denote the locations of two potential targets.

In Fig. 2, targets $T1$ and $T2$ each have a equal likelihood of being the adversary's intended target. The adversary's timestep penalty d and completion

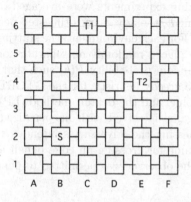

Fig. 2. An instance of GTGR/GTGRD games used in experiments.

reward u^θ are both set to 0. The defender's reward for correctly guessing the adversary's intended target q is set to 10. The attacker penalty value for crossing an edge penalized by the observer is set to 10. The observer is permitted to penalize 3 edges.

5.1 A Comparison of MIP and Greedy Solutions

As seen in Figs. 3 and 4, the mixed integer program and greedy heuristic can yield different results. The mixed integer program yields an expected outcome of 43.3 for the observer, while utilizing the greedy heuristic yields an outcome of 40.0 for the observer. The default expected outcome for the observer (in which no edges are penalized) is 30.0. The following experiments averaged the results of similar grid problems.

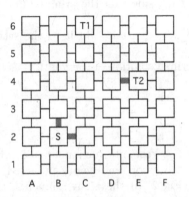
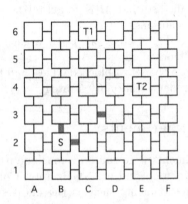

Fig. 3. MIP solution   **Fig. 4.** Greedy solution

5.2 Running Time and Solution Quality

Results from the following experiments were averaged over 1000 grid environments. For each experiment, the adversary's timestep penalty d and completion reward u^θ were set to 0. For each environment, the starting location of the adversary and all targets are placed randomly on separate nodes. Additionally, each target θ is assigned a random probability $P(\theta)$ such that $\sum_{\theta \in B} P(\theta) = 1$. In all of our figures below, the greedy heuristic for the GTGRD is graphed in orange, the MIP is graphed in blue, and the default method (LP) for GTGR is graphed in grey, in which the game is solved with no penalized edges. The defenders reward for correctly guessing the adversary's intended target q was set to 10. The attacker penalty value for crossing an edge penalized by the observer was set to 10. Each game, the observer was permitted to penalize 2 edges.

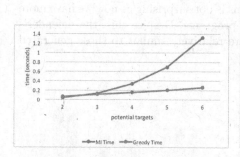

Fig. 5. Average time given targets.   **Fig. 6.** Average outcome given targets.

Various Potential Target Sizes. In this set of experiments, we want to investigate the effect of different potential target sizes (i.e., $|B|$) to the running time (Fig. 5) and solution quality (Fig. 6) of our algorithms. The results are averaged over 1000 simulations of 6 by 6 grids. Each game, the observer was permitted to penalize 2 edges.

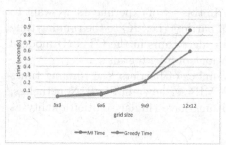

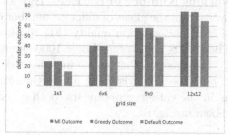

Fig. 7. Average time given size.   **Fig. 8.** Average outcome given size.

As indicated in Fig. 5, the MIP running time increases exponentially while the greedy heuristic running time remains sublinear as we increase the number of potential targets. Moreover, the solution quality (measured by defender's utility) as seen in Fig. 6 suggests that MIP's solution is closely aligned with our greedy heuristics. This gives evidence that our greedy heuristic provides good solution quality while achieving high efficiency. It is no surprise that the defender's utility is higher in the GTGRD setting compared to those of GTGR.

Various Instance Sizes. In this set of experiments, we investigate the effect of different instance sizes (i.e., grids) to the running time (Fig. 7) and solution quality (Fig. 8) of our algorithms.

Unlike our earlier observations on various target sizes, the average running times for both the MIP and our greedy heuristic increase significantly as we increase the instance sizes (see Fig. 7). This is not surprising as now we have more variables and constraints in the integer programs. Despite this, the defender's utilities generated by greedy heuristic are relatively similar to those generated using MIP (see Fig. 8).

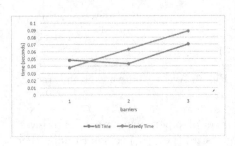
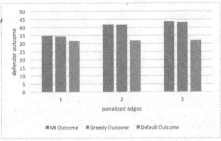

Fig. 9. Average time given penalized edges.

Fig. 10. Average outcome given penalized edges.

Various Number of Barriers/Blocks. In this set of experiments, we want to investigate the effect of different number of barriers (i.e., K) to the running time (Fig. 5) and solution quality (Fig. 6) of our algorithms in the GTGRD models. The results are averaged over 1000 simulations of 6 by 6 grids.

It turns out that as we increase the number of barriers, the running times of our greedy heuristic are longer than the MIP as shown in Fig. 9. Nonetheless, as in the earlier experiments, both algorithms have similar solution quality as shown in Fig. 10.

Various Edge Penalties. Finally, consider the effect of different edge penalties to the solution quality of our greedy heuristic. The results are averaged over 1000 simulations of 6 by 6 grids. As indicated in Fig. 11, the solution gap between the MIP and greedy heuristic as we increase the edge penalty.

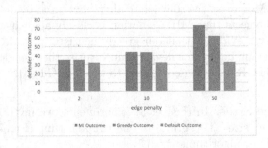

Fig. 11. Average outcome given penalty value.

References

1. Braynov, S.: Adversarial planning and plan recognition: two sides of the same coin. In: Proceedings of the Secure Knowledge Management Workshop (2006)
2. Fudenberg, D., Tirole, J.: Game Theory. The MIT Press, Cambridge (1991)
3. Geib, C., Steedman, M.: On natural language processing and plan recognition. In: Proceedings of the International Joint Conference on Artificial Intelligence (IJCAI), pp. 1612–1617 (2007)
4. Guillarme, N.L., Mouaddib, A.I., Lerouvreur, X., Gatepaille, S.: A generative game-theoretic framework for adversarial plan recognition. In: Proceedings of the Workshop on Distributed and Multi-Agent Planning (2015)
5. Jarvis, P., Lunt, T., Myers, K.: Identifying terrorist activity with AI plan recognition technology. AI Mag. 26(3), 73–81 (2005)
6. Johnson, W.L.: Serious use of a serious game for language learning. Int. J. Artif. Intell. Educ. 20(2), 175–195 (2010)
7. Kautz, H.A.: A formal theory of plan recognition. Ph.D. thesis, Bell Laboratories (1987)
8. Kelley, R., Wigand, L., Hamilton, B., Browne, K., Nicolescu, M., Nicolescu, M.: Deep networks for predicting human intent with respect to objects. In: Proceedings of the International Conference on Human-Robot Interaction (HRI), pp. 171–172 (2012)
9. Keren, S., Gal, A., Karpas, E.: Goal recognition design. In: Proceedings of the International Conference on Automated Planning and Scheduling (ICAPS), pp. 154–162 (2014)
10. Keren, S., Gal, A., Karpas, E.: Goal recognition design for non-optimal agents. In: Proceedings of the AAAI Conference on Artificial Intelligence (AAAI), pp. 3298–3304 (2015)
11. Keren, S., Gal, A., Karpas, E.: Goal recognition design with non-observable actions. In: Proceedings of the AAAI Conference on Artificial Intelligence (AAAI), pp. 3152–3158 (2016)
12. Lee, S.Y., Mott, B.W., Lester, J.C.: Real-time narrative-centered tutorial planning for story-based learning. In: Cerri, S.A., Clancey, W.J., Papadourakis, G., Panourgia, K. (eds.) ITS 2012. LNCS, vol. 7315, pp. 476–481. Springer, Heidelberg (2012). doi:10.1007/978-3-642-30950-2_61
13. Lisý, V., Píbil, R., Stiborek, J., Bosanský, B., Pechoucek, M.: Game-theoretic approach to adversarial plan recognition. In: Proceedings of the European Conference on Artificial Intelligence (ECAI), pp. 546–551 (2012)
14. McQuiggan, S.W., Rowe, J.P., Lee, S., Lester, J.C.: Story-based learning: the impact of narrative on learning experiences and outcomes. In: Woolf, B.P., Aïmeur, E., Nkambou, R., Lajoie, S. (eds.) ITS 2008. LNCS, vol. 5091, pp. 530–539. Springer, Heidelberg (2008). doi:10.1007/978-3-540-69132-7_56
15. Min, W., Ha, E., Rowe, J., Mott, B., Lester, J.: Deep learning-based goal recognition in open-ended digital games. In: Proceedings of the AAAI Conference on Artificial Intelligence and Interactive Digital Entertainment (AIIDE) (2014)
16. Oh, J., Meneguzzi, F., Sycara, K., Norman, T.: ANTIPA: an agent architecture for intelligent information assistance. In: Proceedings of the European Conference on Artificial Intelligence (ECAI), pp. 1055–1056 (2010)
17. Oh, J., Meneguzzi, F., Sycara, K., Norman, T.: An agent architecture for prognostic reasoning assistance. In: Proceedings of the International Joint Conference on Artificial Intelligence (IJCAI), pp. 2513–2518 (2011)

18. Oh, J., Meneguzzi, F., Sycara, K., Norman, T.: Probabilistic plan recognition for intelligent information agents: towards proactive software assistant agents. In: Proceedings of the International Conference on Agents and Artificial Intelligence (ICAART), pp. 281–287 (2011)

19. Raghavan, T.E.S.: Finite-step algorithms for single-controller and perfect information stochastic games. In: Neyman, A., Sorin, S. (eds.) Stochastic Games and Applications. NATO Science Series (Series C: Mathematical and Physical Sciences), vol. 570, pp. 227–251. Springer, Netherlands (2003). doi:10.1007/978-94-010-0189-2_15

20. Ramírez, M., Geffner, H.: Plan recognition as planning. In: Proceedings of the International Joint Conference on Artificial Intelligence (IJCAI), pp. 1778–1783 (2009)

21. Ramírez, M., Geffner, H.: Probabilistic plan recognition using off-the-shelf classical planners. In: Proceedings of the AAAI Conference on Artificial Intelligence (AAAI), pp. 1121–1126 (2010)

22. Ramírez, M., Geffner, H.: Goal recognition over POMDPs: inferring the intention of a POMDP agent. In: Proceedings of the International Joint Conference on Artificial Intelligence (IJCAI), pp. 2009–2014 (2011)

23. Schmidt, C., Sridharan, N., Goodson, J.: The plan recognition problem: an intersection of psychology and artificial intelligence. Artif. Intell. 11(1–2), 45–83 (1978)

24. Son, T.C., Sabuncu, O., Schulz-Hanke, C., Schaub, T., Yeoh, W.: Solving goal recognition design using ASP. In: Proceedings of the AAAI Conference on Artificial Intelligence (AAAI) (2016)

25. Sukthankar, G., Geib, C., Bui, H.H., Pynadath, D., Goldman, R.P.: Plan, Activity, and Intent Recognition: Theory and Practice. Newnes, Boston (2014)

26. Tavakkoli, A., Kelley, R., King, C., Nicolescu, M., Nicolescu, M., Bebis, G.: A vision-based architecture for intent recognition. In: Proceedings of the International Symposium on Advances in Visual Computing, pp. 173–182 (2007)

27. Vilain, M.: Getting serious about parsing plans: a grammatical analysis of plan recognition. In: Proceedings of the National Conference on Artificial Intelligence (AAAI), pp. 190–197 (1990)

28. Wayllace, C., Hou, P., Yeoh, W., Son, T.C.: Goal recognition design with stochastic agent action outcomes. In: Proceedings of the International Joint Conference on Artificial Intelligence (IJCAI) (2016)

Manipulating Adversary's Belief: A Dynamic Game Approach to Deception by Design for Proactive Network Security

Karel Horák[1]([⊠]), Quanyan Zhu[2], and Branislav Bošanský[1]

[1] Department of Computer Science, Faculty of Electrical Engineering,
Czech Technical University in Prague, Prague, Czech Republic
{horak,bosansky}@agents.fel.cvut.cz
[2] Department of Electrical and Computer Engineering, New York University,
New York, USA
quanyan.zhu@nyu.edu

Abstract. Due to the sophisticated nature of current computer systems, traditional defense measures, such as firewalls, malware scanners, and intrusion detection/prevention systems, have been found inadequate. These technological systems suffer from the fact that a sophisticated attacker can study them, identify their weaknesses and thus get an advantage over the defender. To prevent this from happening a proactive cyber defense is a new defense mechanism in which we strategically engage the attacker by using cyber deception techniques, and we influence his actions by creating and reinforcing his view of the computer system. We apply the cyber deception techniques in the field of network security and study the impact of the deception on attacker's beliefs using the quantitative framework of the game theory. We account for the sequential nature of an attack and investigate how attacker's belief evolves and influences his actions. We show how the defender should manipulate this belief to prevent the attacker from achieving his goals and thus minimize the damage inflicted to the network. To design a successful defense based on cyber deception, it is crucial to employ strategic thinking and account explicitly for attacker's belief that he is being exposed to deceptive attempts. By doing so, we can make the deception more believable from the perspective of the attacker.

1 Introduction

As computer systems and devices are becoming increasingly connected and complex in their functionalities, traditional cyber defense technologies (e.g. firewalls, malware scanners, and intrusion detection/prevention systems) have been found inadequate to defend critical cyber infrastructures [23]. Moreover, sophisticated adversaries such as the advanced persistent threats (APTs), can use a combination of social engineering and software exploits to infiltrate the network and inflict cyber and/or physical damages of the defended systems. Therefore, to defend against a sophisticated adversary, we have to accept that the adversary

S. Rass et al. (Eds.): GameSec 2017, LNCS 10575, pp. 273–294, 2017.
DOI: 10.1007/978-3-319-68711-7_15

can study and evade technology-based defenses [20,25]. To move away from the defense paradigm where the attacker has the advantage to the one of defender's advantage, proactive cyber defense is a new defense mechanism in which systems strategically engage the attacker and learn and influence his behaviors.

Cyber deception is a key component of the proactive cyber defense that can create and reinforce attacker's view of the network by revealing or concealing artifacts to the attacker. The attacker needs to pay attention to identifying deceptive artifacts in order to devise the right attack sequence. This becomes challenging in an adversarial environment and the attacker's progress thus becomes slower and less effective. Deception mechanisms, such as honeypots [22,32], honeytokens [3,17], camouflaging [21,28] and moving target defense [12,13,29] are methods that have been used to manipulate the attacker's belief on system parameters and increase their cost of information acquisition.

Understanding deception in a quantitative framework is pivotal to provide rigor, predictability, and design principles. To this end, we analyze deception through a game-theoretic framework [2,16,19,30]. This framework allows making quantitative, credible predictions, and enables the study of situations involving free choice (the option to deceive or not to deceive) and well-defined incentives. Specifically, the class of dynamic games of incomplete information allows modeling the multi-round interactions between an attacker and a defender as well as the information asymmetry that forms the essential part of deception.

In this work, we focus on the applications of cyber deception techniques in the field of network security. Strong proactive incident response strategies can only be devised if we understand the impact of deceptive operations on the attacker's beliefs. To this end, we employ the framework of competitive Markov models with imperfect information, or *partially observable stochastic games* [10,11], to reason about the uncertainties of the two sides of the cyber warfare—the defender of the network and the attacker—and understand how this uncertainty influences their behavior. This framework provides a mathematical formalism of the attacker's belief state to capture his level of engagement and allows the defender to take defensive actions based upon attacker's state of mind.

When the presence of the attacker in the network environment is detected by the sensing systems, the defender can attempt to engage the attacker and start *actively* deceiving him by taking proactive deceptive (and defensive) actions aimed to combat the upcoming attack scenario. He can use the sensing systems to track attacker's further progress and often, by inspecting the log records and/or analyzing the past communication with attacker's command and control servers [5,9], he can also reconstruct a significant part of the history of the attack – thus getting a near-perfect information about the attacker's point of view. We assume that the defender can reconstruct this view *perfectly* which allows us to apply the framework of *one-sided partially observable stochastic games* [11].

To make the deception effective in the long run, we need to make it difficult for the attacker to identify that the deception is employed. An attacker will try to reason about and recognize our deceptive attempts and will adapt his attack plan accordingly—and thus mitigate the impact of the deception. We provide a model which explicitly reasons about attacker's belief about the deception state and we show how important it is for the defender to carefully manipulate this belief to maximize the defensive impact of the cyber deception. We conduct a case study to illustrate the consequences of strategic deception on the security level of the network. Namely, we make the following important observations about cyber deception. First, we observe that the standard incident-response approach which relies on excluding the attacker from the network immediately is inefficient from the perspective of the deception. In fact, it may render the network more vulnerable as it does not take attacker's beliefs into account (we term this phenomenon as the *curse of exclusion*). Second, we observe that it is easier to deceive the attacker when he had already dedicated significant effort to accomplish his goals as he is more greedy about realizing his intents (we term this phenomenon as the *demise of the greedy*).

The rest of the paper is organized as follows. In Sect. 2, we introduce related work on cyber deception and introduce the game-theoretic framework we use. In Sect. 3, we provide a generic approach for reasoning about the deception which accounts for the necessary aspects of the deception, i.e. the informational asymmetry, sequential nature of deception problems and which accounts for the strategic nature of the deception. In Sect. 4, we state the problem from the perspective of cyber deception in network security. Next, we provide a case study illustrating the impact of cyber deception on attacker's beliefs and his ability to inflict damage in Sect. 5. Finally, in Sect. 6 we summarize our main results.

2   Related Work

Typical attacks conducted by advanced attackers consist of multiple stages [24] that can be broadly summarized as *reconnaissance* and realization of attacker's primary goals, e.g. data exfiltration. Underbrink [26] classifies deception techniques into two broad categories – *passive* and *active deception*. The passive deception is targeted against attacker's reconnaissance efforts and relies on a proactively deployed *static* infrastructure of decoy systems, e.g. honeypots [14, 22] or fake documents [4]. Unlike the legitimate users, the attacker does not know about their deceptive nature and may thus reveal his presence by inadvertently interacting with them. The active deception, on the other hand, attempts to *interactively* engage the attacker who has been already detected by the sensing systems. The defender attempts to anticipate probable future actions of the attacker and takes proactive countermeasures against them to prevent the attacker from achieving his goals.

A lot of work has been dedicated to understanding both technological [1, 27] and strategical [6, 18, 31, 32] aspects of passive deception techniques and decoy infrastructures. Considerably less attention has been, however, paid to the active deception. To the best of our knowledge, very few works have focused on the strategical aspects of active deception. [26] has introduced the concept of active deception and the *Legerdemain* approach to active deception was described. The Legerdemain approach secretly manipulates critical assets in the network (such as data files or access credentials) to confuse the attacker and prevent him from getting access to critical resources. A dynamic game model, based on two coupled Markov decision processes, is used to assist the defender in designing the actively deceptive strategy. The model, however, assumes that the attacker will never realize that mechanisms of active deception are applied against him – which simplifies solving the game but makes the model not realistic. In fact, we show that accounting for attacker's belief about the deception is critical for designing strong deceptive strategies.

Our approach reasons explicitly about the belief the attacker has and thus avoids the drawback of the Legerdemain approach. To this end, we use the framework of *one-sided partially observable stochastic games* (one-sided POSGs) [11]. In this class of games, one of the players is assumed to be perfectly informed about the course of the game, which is not the case for the other player. This game-theoretic model has been originally devised to reason about robust defensive strategies by assuming that attacker is able to get a perfect picture of the game. In this work, we provide a novel application of this model to reason about the active deception by assuming that the defender (or deceiver) has already detected the attacker (and thus is able to track his progress) while the attacker (or deceivee) lacks some information about the game (and thus is vulnerable to defender's deceptive attempts). We discuss the way we use this class of games to reason about deception in Sects. 3 and 4 in greater detail.

3   Deception Game Framework

The asymmetry of information plays a major role in many conflicts seen in the real world, starting from the warfare and ending with conflicts as innocent as card games. The success in these operations typically depends on the way we handle the information and in particular on the way we protect our informational advantage. Deception has even evolved to be vital for the survival of many wildlife species, such as chameleons, and has been adopted by armies worldwide.

We cannot, however, expect that a simple presence and naïve use of the deceptive techniques is sufficient to guarantee success – the way we employ them is important to explore. As an example, consider that we have two colored balls, red and blue, and we do not want others to know which one of them we are carrying. To this end, it may seem reasonable to paint each of these balls to the opposite color beforehand and pretend that the red one is, in fact, blue (and vice versa). In such a case, however, other actors will soon discover the principle we use to manipulate the truth and realize that the ball we are carrying is in fact of the opposite color – and hence our attempt to disguise others becomes unsuccessful.

When deciding on the use of deceptive techniques we have to think in a strategic way. We need to understand what impact our *deception strategy* σ_D has on the beliefs of other actors as they will learn and eventually understand the way we misrepresent the truth. The deceived players will derive a *counter-deception strategy* σ_A with the aim to understand the signals they receive and reconstruct the truth (or at least reconstruct how likely each possibility is to be true). Both of these strategies have to account for the beliefs of the players and are thus essentially functions of these beliefs.

We focus on the deception problems where there are two sides of the conflict (or two players). We assume that one side of the conflict, the *deceiver*, knows the truth (i.e. state s of the system), while the other side, the *deceivee*, aims to recognize that. This type of knowledge is often seen in reality. For example, in security problems, the defender usually knows the parameters of the system he is about to defend, e.g., he knows the plans of the facility or the topology of the computer network, and he knows where the important assets are located. In addition, he is equipped with monitoring facilities which allow him to monitor attacker's actions (or, at least, allow him to analyze these actions retrospectively). On the other hand, the attacker is uninformed about the true system parameters and he has to recognize these parameters to plan his activities properly. This setting underlies the need for reasoning about the information and beliefs of the uninformed player as the information is the only advantage we have.

3.1   Deception in a Sequential Setting

We study the deception in a sequential setting, where both the players take sequences of actions to either deceive the adversary, or attempt to recognize the truth, respectively. In each step $t \geq 1$, both the deceiver and the deceivee take an action (a_D and a_A). As a matter of result, the deceivee gets an observation about the true state of the system (e.g. that the ball is painted red) and the state of the system may change (which is then known only to the deceiver again). Moreover the deceiver has to pay a cost associated with his deceptive action and possibly other costs associated with the choice of actions a_D and a_A, denoted $l^{(t)}$. We characterize these costs using a loss function $\mathcal{L}_D$.

The goal of the deceiver is to keep the losses $l^{(t)}$ as low as possible – or at least mitigate them by delaying them in time. This is characterized by the *discounted-sum* objective when the aggregated loss of the deceiver is

$$L = \sum_{t=1}^{\infty} \gamma^{t-1} \cdot l^{(t)}, \tag{1}$$

where $0 < \gamma < 1$ is a constant termed the *discount factor*. In our case, the deceiver is the defender of the system and we aim to devise robust deceptive strategies that account for the worst case scenario, hence we assume that the goal of the deceivee is to maximize the loss L. We also term such games as *zero-sum*.

We aim to understand the *value of deception* $\overline{V}$ and the *value of counter-deception* $\underline{V}$ – and the strategies that induce these values. We define $\overline{V}$ as the expected loss of the deceiver when he is forced to commit himself to a deception strategy σ_D which is then observed by the deceivee who tries to identify the weaknesses of σ_D, i.e.

$$\overline{V} = \inf_{\sigma_D} \sup_{\sigma_A} L(\sigma_D, \sigma_A) \tag{2}$$

where $L(\sigma_D, \sigma_A)$ stands for the expected discounted loss when strategies σ_D and σ_A are followed by the players. Similarly, we define the value of counter-deception $\underline{V}$ as the value where the deceivee is forced to commit himself first to a counter-deceptive strategy σ_A he uses to combat the deception and then the deceiver decides what deceptive techniques he uses, i.e.

$$\underline{V} = \sup_{\sigma_A} \inf_{\sigma_D} L(\sigma_D, \sigma_A). \tag{3}$$

Note that the deceiver can guarantee that the loss will be no higher than $\overline{V}$, while $\underline{V}$ is the minimum loss the deceivee can enforce.

3.2   Game-Theoretic Model

We propose to formulate deception as a partially observable stochastic game with one-sided information (one-sided POSG) [11]. This model has been originally devised to reason about robust strategies of the defender by assuming that the adversary is perfectly informed. The asymmetric nature of the information present in the model, however, makes it convenient to reason about the deception. A deception game based on the model of one-sided POSGs is a tuple $\langle S, \mathcal{A}_A, \mathcal{A}_D, \mathcal{T}, \mathcal{L}_D, O_A, b^0 \rangle$, where

- S is a finite set of states of the system (recall that the true state of the system is known to the deceiver, while the deceivee does not know it). A state may for example represent where both the players have deployed their units in a warfare.
- $\mathcal{A}_D$ is a finite set of actions the deceiver can use to deceive the adversary.
- $\mathcal{A}_A$ is a finite set of actions the adversary, the deceivee, can use to learn more about the system, or potentially in security problems to inflict damage.
- $\mathcal{T} : (S \times \mathcal{A}_A \times \mathcal{A}_D) \to \Delta(O \times S)$ is a transition function representing possible changes to the system (e.g. movements of the units) and observations the deceivee can receive in a probabilistic way.
- $\mathcal{L}_D : (S \times \mathcal{A}_A \times \mathcal{A}_D) \to \mathbb{R}$ is defender's loss function and describes how much the defender loses in each step of the deception game.
- O_A is a finite set of observations the attacker can get about the state of the system.
- $b^0 \in \Delta(S)$ (where $\Delta(S)$ is a probability distribution over S) is the initial belief of the deceivee, where $b_0(s)$ denotes the probability that the initial state of the deception game is s. As an example, the deceivee may know where his units are located, but he may lack the information about the position of deceiver's units. Thus he forms a belief over possible positions of the deceiver in the form of a probability distribution over states that match the current (known) position of units of the deceivee.

A play in the deception game proceeds as follows. First, an initial state of the game s^0 is drawn from b^0. Then, in each step t, players decide simultaneously their actions $(a_D^t, a_A^t) \in \mathcal{A}_D \times \mathcal{A}_A$. Based on their choice, the deceiver loses $l^{(t)} = \mathcal{L}_D(s^{t-1}, a_A^t, a_D^t)$. Then the deceivee receives an observation o^t and the game state changes to s^t with probability $\mathcal{T}(s^{t-1}, a_A^t, a_D^t)(o^t, s^t)$.

Deceiver observes the course of the deception game perfectly, hence he knows what the past states, actions and observations were. He can use all this information to make an informed decision about his next action. He makes this decision based on his deception strategy $\sigma_D : (\mathcal{S}\mathcal{A}_D\mathcal{A}_A\mathcal{O}_A)^*\mathcal{S} \to \Delta(\mathcal{A})$, where $\sigma_D(\omega, a_D)$ denotes the probability that the deceiver chooses an action $a_D \in \mathcal{A}_D$ when the current history is ω.

The deceivee only observes the observations o^t and remembers the actions a_A^t he made. He cannot thus make use of the complete information available to the deceiver. The attacker thus proceeds according to a counter-deception strategy $\sigma_A : (\mathcal{A}_A\mathcal{O})^* \to \Delta(\mathcal{A}_A)$, when $\sigma_A(\omega, a_A)$ stands for the probability that the deceivee uses action a_A given that $\omega \in (\mathcal{A}_A\mathcal{O})^*$ are the actions and observations he has used and seen previously.

The results in [11] show that the players need not remember the histories of the play to make decisions. Instead, they can just keep track of the *belief* $b \in \Delta(\mathcal{S})$ over the states $\mathcal{S}$ of the deception game and play according to one-step strategies $\pi_D^{(b)} : \mathcal{S} \to \Delta(\mathcal{A}_D)$ and $\pi_A^{(b)} \in \Delta(\mathcal{A}_A)$ which are directly functions of beliefs. This emphasizes the fact that the deceivee forms a belief which then directly drives his decisions. The players keep track of the belief using a Bayesian update rule characterized by the following equation:

$$\tau(b, a_A, o, \pi_D)(s') = \frac{1}{K} \sum_{sa_D \in \mathcal{S}\mathcal{A}_D} b(s) \cdot \pi_D(s, a_D) \cdot \mathcal{T}(s, a_A, a_D, o, s') \quad (4)$$

where $\tau(b, a_A, o, \pi_D)$ stands for the updated belief of the deceivee given that the previous belief was b, he played action a_A and received observation o, and the deceiver followed a deception strategy π_D. K stands for the normalization constant.

In the case of zero-sum deception game, the *value of deception* $\overline{V}$ and the *value of counter-deception* $\underline{V}$ have been shown to be equal [11], i.e.

$$\inf_{\sigma_D} \sup_{\sigma_A'} L(\sigma_A', \sigma_D) = \sup_{\sigma_A} \inf_{\sigma_D'} L(\sigma_A, \sigma_D'). \quad (5)$$

We represent the values of deception (or counter-deception) using a convex value function $v^* : \Delta(\mathcal{S}) \to \mathbb{R}$ which maps beliefs over the system states to the expected value of deception for that belief. This value function satisfies the following fixpoint equation

$$v^*(b) = \min_{\pi_D:\mathcal{S}\to\Delta(\mathcal{A}_D)} \max_{\pi_A\in\Delta(\mathcal{A}_A)} \Big[\sum_{sa_Aa_D} b(s) \cdot \pi_A(a_A) \cdot \pi_D(s, a_D) \cdot \mathcal{L}_D(s, a_A, a_D)$$
$$+ \gamma \sum_{a_Ao} \Pr[a_Ao \,|\, b, \pi_A, \pi_D] \cdot v^*(\tau(b, a_A, o, \pi_D)) \Big]. \quad (6)$$

One of the ways to reason about the value of the deception and the associated optimal strategies of the players is to approximate the value function v^* using an approximate value iteration algorithm presented in [11]. We can then derive the optimal strategy for the deceiver by considering the maximizing π_D of Eq. (6) in each step of the interaction.

Remark 1. The convexity of the value function v^* supports our intuition that the deceivee never gets satisfied with being deceived. The value of his counter-deception would never get lower, had he got additional information. For example, assume that the deceivee recognized the true state of the system before he is about to act (i.e. his belief changes from b to b_s, where b_s is a belief where the attacker *knows* the true state). Then, since $b = \sum_{s \in S} b(s) \cdot b_s$ and due to the convexity of v^*, we get

$$\sum_{s \in S} b(s) \cdot v^*(b_s) \geq v^* \left(\sum_{s \in S} b(s) \cdot b_s \right), \tag{7}$$

i.e. if the attacker recognizes the true state (i.e. with probability $b(s)$ he recognizes that the true state is s) and plays accordingly, the loss he is able to cause is greater or equal than in the situation where he has to reason about the state he is in (i.e. his belief is b).

4   Game-Theoretic Approach to Cyber Deception

The ideas we have presented so far are general enough to be applied to reason about the deception in a wide range of scenarios. We are going, however, to focus on the use of the deception in the context of computer networks to improve the security of networked systems. The deception over the networks possesses certain features which allow us to make the model of deception game more specific. Namely, the attacker who is going to be deceived does not know two key properties of the networked system. First, he does not know the topology of the network which he needs to understand to target his attack properly. Second, he does not know whether the defender, the deceiver, already knows about his presence in the network. Understanding both of these aspect is critical from attacker's perspective – and concealing this information from attacker's view is important for the defender to devise strong defensive strategies.

In this section, we describe a general idea how we can use one-sided partially observable stochastic games to reason about active deception in network security, where the defender interactively decides about the actions to mislead the attacker in the course of an attack and mitigate the possible damage to the network. Our model accounts for the uncertainties of the attacker about the topology of the network, whether he has been detected and about defender's actions – both past and upcoming ones. To this end, we represent the states of the game as $S = N \times X_A \times X_D \times D$ where

– N is a set of possible network topologies the defender can choose from based on a fixed distribution $\xi \in \Delta(N)$

- $\mathcal{X}_A$ is a set of possible attack vectors representing the state of an attack (e.g. privileges the attacker has already acquired); $\emptyset \in \mathcal{X}_A$ denotes that the attack has not started yet
- $\mathcal{X}_D$ is a set of possible defense vectors representing the state of defense resources (e.g. dynamic decoy systems deployed in the network); $\emptyset \in \mathcal{X}_D$ denotes that the defender has not deployed any dynamic resources yet
- $\mathcal{D}$ is a set of possible detection states; we assume $\mathcal{D} = \{\texttt{true}, \texttt{false}\}$ denoting whether the attacker has been detected or not by the sensing systems

We denote a state of the game as (n, x_A, x_D, d).

The defender initially chooses a network topology he is going to defend according to a probability distribution $\xi \in \Delta(\mathcal{N})$. We then derive the initial belief of the underlying one-sided POSG $b_0 \in \Delta(\mathcal{S})$ as $b(n, \emptyset, \emptyset, \texttt{false}) = \xi(n)$ and $b_0(\cdot) = 0$ otherwise. This means that we draw the initial network topology from $\xi(n)$ and make both the attack and defense vectors empty, and the attacker is initially undetected.

Once the attacker gets detected by the sensing systems (i.e. $d = \texttt{true}$), the defender may start taking actively actions $a_D \in \mathcal{A}_D$ to combat the attacker's presence in the network. His actions may manipulate the defense vector (e.g. by deploying new defense resources), interfere with actions of the attacker, or they may restrict attacker's access to the network (defense action $\texttt{block} \in \mathcal{A}_D$). We assume that in such case, the attacker is able to change his identity and attack the network again (therefore x_A is set to $\emptyset$ and d to $\texttt{false}$ as we lost track of the attacker when he changed his identity, and the game continues). If the attacker has not been detected yet, however, the defender cannot take any active counteraction (i.e. active deception techniques are not available to him) and he is forced to use action $a_D = \texttt{noop}$. The fact that the defender cannot use any action other than $\texttt{noop}$ when the attacker has not been detected yet allows us to assume a perfect information of the defender, i.e. make the defender be the perfectly informed player in the one-sided partially observable stochastic game. The defender cannot leverage the extra information about the attacker (he would not have in reality) up to the point when the attacker gets detected.

The attacker can choose from attempting to acquire new privileges (and thus manipulating the attack vector x_A), changing his identity (i.e. making $x_A = \emptyset$ and $d = \texttt{false}$) and leveraging his current privileges to cause damage — or combination of any of these. Each action of the attacker is associated with the risk of alerting the defender, we denote the probability of triggering an alert when using action a_A in network n by $p_{trig}(n, a_a)$.

The transition function $\mathcal{T}$ respects the actions the players have taken, i.e. describes possible changes to vectors x_A, x_D and the detection state d in a probabilistic way. Furthermore the attacker receives an observation $(x'_A, o) \in O_A$. The attacker is always aware of his current attack vector, i.e. for any $x'_A \neq x''_A$ the following holds

$$\mathcal{T}((n, x_A, x_D, d), a_A, a_D)((x'_A, o), (n, x''_A, x'_D, d')) = 0. \tag{8}$$

Moreover, once the network topology is chosen it never be changes, i.e. for any $n \neq n'$

$$\mathcal{T}((n, x_A, x_D, d), a_A, a_D)((x'_A, o), (n', x'_A, x'_D, d')) = 0. \tag{9}$$

The detection probabilities (i.e. the probability of transitioning from $d = \mathtt{false}$ to $d' = \mathtt{true}$) are independent of action effects, i.e.

$$\sum_{(x'_A, o)} \mathcal{T}((n, x_A, \emptyset, \mathtt{false}), a_A, \mathtt{noop})((x'_A, o), (n, x'_A, \emptyset, \mathtt{true})) = p_{trig}(n, a_A) \tag{10}$$

The losses $\mathcal{L}_D$ for individual transitions can be set arbitrarily to match the costs (and eventually possible gains if we succeed in exploiting attacker's actions) in the real network and the costs of the deception. We only require that $\mathcal{L}_D((n, x_A, \emptyset, \mathtt{false}), a_A, a_D) = M$ for every $a_D \neq \mathtt{noop}$ where M is a large constant to ensure that the defender does not use active deception techniques when the attacker has not yet been detected.

5   Manipulating Attacker's Belief Using Active Deception

The use of the active deception can significantly improve the security level of the network. In this section, we provide a case study based on a simple game with sets $\mathcal{N}$ and $\mathcal{X}_D$ containing only one element (i.e., $\mathcal{N} = \{n\}$ and $\mathcal{X}_D = \{\emptyset\}$) to illustrate the concept of active deception. In the case of this game, we use the deception only to manipulate attacker's belief over being detected (i.e. the $\mathcal{D}$ part of the state) and we try to make him uncertain about the progress of the attack and eventually take a wrong action. We show that we cannot, however, rely solely on the deceptive actions if we want to maximize the effectiveness of the deceptive operation. The deception is the most effective if it is *stealthy* and the attacker remains unaware that we are trying to deceive him, or, at least, if we make him uncertain about the state of deception.

As soon as the attacker realizes that we are trying to deceive him, his behavior changes significantly. He will attempt to take evasive actions in attempt to lose defender's attention (e.g. by changing his network identity), or, as a matter of last resort, he may opt to inflict severe damage based only on the information he collected so far. These decisions of the attacker make the defender's attempts to contain the attack substantially harder and should be averted (if possible).

To preserve the stealthy nature of the deception, it is crucial that the attacker thinks that the signals he receives are not *too good to be true*. The defender has to manipulate attacker's belief about the deception state carefully if he wants to make the attacker believe that no deceptive operation is taking place and keep him *engaged* in the network.

5.1   Network Topology and the Anatomy of an Attack

We illustrate the concept of active deception using a network topology $n \in \mathcal{N}$ depicted in Fig. 1. We use it as an abstraction of a multilayer network which is commonly adopted in critical network operations, such as power plants or production facilities [15]. Our example network consists of three layers. The outermost layer of the network (Layer 1) is directly exposed to the Internet via demilitarized zones (DMZs) and provides less sensitive services that are used to communicate with the customers and business partners, such as web or email servers.

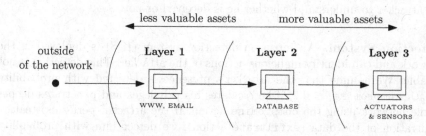

Fig. 1. Network topology (attacker starts outside of the network and attempts to gain access to the most valuable assets in the network)

More critical assets are located in the deeper layers of the network. In our case, the second layer consists of data stores containing confidential data the loss of which may have a severe impact on the company. The third layer is the most critical one since it provides an access to physical devices, such as actuators and sensors, the integrity of which is absolutely essential for the secure operation of the facility. Breach of assets in the Layer 3 of the network may even pose a risk of physical damage, such as in the case of the Stuxnet attack [7,8].

Attack Options. We assume that an attack is initiated from a computer outside of the network ($x_A = \emptyset$). In this section we describe attacker's actions (set of actions $\mathcal{A}_A$) which he can use to acquire new privileges and penetrate deep into the network and to cause damage to it. The attacker attempts to take control of a system in Layer 1 ($x_A = \mathtt{layer}_1$) and then escalates his privileges to take control of the computers located deeper in the network (i.e. acquiring $x_A = \mathtt{layer}_i$) by compromising them (hence we refer to this action of the attacker as $\mathtt{compromise}$). At any point, the attacker can either $\mathtt{wait}$ or leverage the current access. Apart from attempting to compromise a host in the next layer, he has two options:

The first option is to cause significant immediate damage, such as eliminating a physical device in Layer 3 (having the attacker had access to it) – we refer to this action as $\mathtt{takedown}$. Such an action surely attracts the attention of the defender and will lead to the detection of the attacker's presence. Therefore, the attacker is forced to quit the network and possibly repeat his attack later (hence $x_A = \emptyset$ and $d = \mathtt{false}$ as a result).

The second option is to cause smaller amount of damage while attempting not to attract defender's attention. The actions the attacker can use to this purpose include, e.g., a stealthy exfiltration of data or a manipulation of the records in the database – for simplicity we refer to them collectively using the `exfiltrate` action. Nevertheless, even these careful options run into a small risk of being detected. Moreover these options run into the risk that the defender will avert the damage resulting from them by means of active deception and possibly even use the fact that the attacker uses the `exfiltrate` action for his benefit (e.g., to collect evidence; see discussion in Sect. 5.2). This makes it critical for the attacker to understand whether he is deceived or not.

Detection System. An intrusion detection system (IDS) is deployed in the network and can identify malicious actions of the attacker. This detection is not reliable. We assume that the attacker's presence is detected with probability $p_{trig}(n, \texttt{compromise}) = 0.2$, if he escalates his privileges and penetrates deeper in the network using the `compromise` action. If the attacker performs stealthy exfiltration of the data (`exfiltrate` action), we detect him with probability $p_{trig}(n, \texttt{exfiltrate}) = 0.1$. We have chosen these probabilities based on a discussion with an expert, however, the model is general enough to account for any choice of these parameters.

Active Deception. We assume that the passive defensive systems, such as IDS and honeypots, are already in place and we focus on the way the defender can *actively* deceive the attacker when his presence has been detected. We take an abstracted view on defender's actions (set $\mathcal{A}_D$) to focus on the main idea of deception, however, our model is general and these actions can be refined to account for *any* actions the defender can use. In our example, he can either use a stealthy deceptive action and attempt to `engage` the attacker in the network, or he can attempt to exclude the attacker from the network (non-deceptive `block` action). We assume that the `block` action really achieves its goal and all the privileges of the attacker get revoked, and the attacker thus has to start his attack from scratch (i.e. x_A becomes $\emptyset$, and $d = \texttt{false}$). If it were not the case and the `block` action was less powerful, blocking the attacker would have been less tempting and hence the use of deception we are advocating would have been even more desirable. By engaging the attacker we attempt to anticipate the action of the attacker and minimize (or even eliminate) the damage caused by his stealthy damaging action of `exfiltrate`. We cannot, however, contain the more damaging `takedown` action by engaging the attacker – the only way to prevent that kind of damage is to `block` the attacker in time. Note that both of these actions of the defender can only be used once the attacker got detected – otherwise, the defender has to rely on the infrastructure of passive defensive systems (i.e., use `noop` action) as the attacker has to be detected first.

5.2   Game Model

We analyze the active deception in the context of the network presented in Sect. 5.1 using a game-theoretic model of one-sided partially observable stochastic games (see Sect. 3) and we capture the interaction between the defender and the attacker using a transition system depicted in Fig. 2. The state space is divided into two parts. In the upper half, the presence of the attacker in the network has not yet been revealed by the IDS ($d = $ false), therefore, the defender cannot take active countermeasures yet. Triggering an IDS alert switches the game states into the bottom part ($d = $ true) and thus gives the defender an opportunity to decide between engage and block actions.

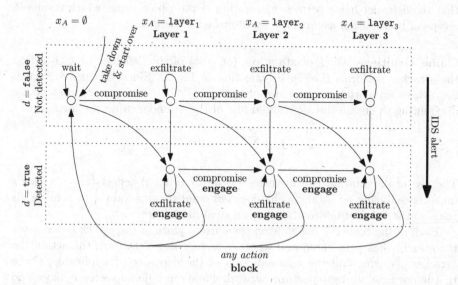

Fig. 2. Transition system of a partially observable stochastic game representing attack on the network from Fig. 1. The attacker can use the takedown action in every layer. The wait action of the attacker has been omitted for clarity and is always applicable.

The arrows in the diagram represent individual transitions in the game (i.e. represent the transition function $\mathcal{T}$). We assume that the transitions in the game are deterministic, except for the transitions between $d = $ false and $d = $ true that are defined using p_{trig}. The attacker never receives an observation that would reveal him some information about the detection state d (i.e. he only gets to know the new attack vector x_A).

If the attacker uses compromise action, he penetrates deeper in the network. If he opts for exfiltrate, he stays in the current layer of the network while possibly gaining access to confidential information. And finally, he can decide to do the immediate damage by the takedown action at any time. In such a case he gets detected and thus returns to the initial state, outside of the network. The defender can stop all this from happening by taking the block action (had he detected the attacker) when the defender is pushed out of the network as well.

The attacker knows his current attack vector x_A and can identify the layer he has penetrated (i.e. he knows the "column" of the transition system where he is located), but he does not know whether he has been detected or not (i.e. whether the game is in the upper or lower half). The defender also does not have perfect information about the state of the attack in reality – namely, he does not know anything about the attacker until the IDS generates an alert. After the alert is generated, however, we assume that he can get a close to perfect information about the attacker by studying the traces he has created in the system. Since the defender cannot make use of the information about the attacker in states where $d = \texttt{false}$ (he cannot take any active countermeasures), we can safely assume that the defender has a *perfect* information in the whole game, which results in a type of information asymmetry we discussed in Sect. 3.

Game Utilities. We associate a loss (or cost) of the defender to each action the attacker performs (i.e. each transition in Fig. 2). Since the attacker takes his actions sequentially, a sequence of costs $l^{(1)}, l^{(2)}, \ldots$ is generated, and we use discounting to obtain the aggregated loss of the defender using the formula

$$L = \sum_{t=1}^{\infty} \gamma^{t-1} \cdot l^{(t)}. \tag{11}$$

The use of discounting (in our case, we use $\gamma = 0.95$) reflects the attacker's impatience during an attack as he does not want to wait forever to achieve his goals as the value of information he can steal diminishes.

Each of the costs $l^{(t)}$ depends on the current state of the attack (what layer the attacker has penetrated and whether he has been detected), the action the attacker performs and the counteraction of the defender (if applicable). Note that in our case, we have just one network n and one defense vector x_D so we do not account for these explicitly. This utility model is general enough to capture any kind of preferences of the defender. The costs we use in our case study are based on a discussion with an expert and are summarized in Table 1. Recall that the players take their action simultaneously and the costs thus depend on their joint action.

The `compromise` action does not cause any immediate harm to the defender and only leaks information to the defender (e.g. about an exploit used) so the loss of the defender is negative ($L_1 = L_7 = -2$). Note that a negative loss is in fact a gain. Moreover, if the defender is already aware of attacker's presence and engages him in the network, he can better understand the techniques used by the attacker and thus his loss is ($L_4 = -4$).

The `exfiltrate` action is already harmful to the defender. If the defender does not take any active countermeasures, the attacker accesses confidential data which implies a significant damage to the defender. Since the assets located deeper in the network are more valuable, we account for this by defining the cost for the defender of $L_2^i = 15i$ for losing data located in the i-th layer.

If the defender realizes that he is dealing with a malicious user, he can minimize or eliminate the risk of losing sensitive data, e.g. by presenting (partly) falsified data to the attacker, using the `engage` action. The attacker then receives useless data and only provides the defender with time to collect the forensic evidence. The loss of the defender is, therefore, negative ($L_5 = -2$) if the attacker

Table 1. Game costs for the game represented in Fig. 2. In each time step, the players take their actions simultaneously and the loss of the defender in the current time step is determined according to their joint action.

State (s^{t-1})		Action		Defender's loss
Attacker's position (x_A)	Detected (d)	Attacker (a_A^t)	Defender (a_D^t)	$\mathcal{L}_D(s^{t-1}, a_A^t, a_D^t)$
any	no	compromise	—	-2 ($= L_1$)
layer$_i$	no	exfiltrate	—	$15i$ ($= L_2^i$)
layer$_i$	no	takedown	—	$25i$ ($= L_3^i$)
any	yes	compromise	engage	-4 ($= L_4$)
layer$_i$	yes	exfiltrate	engage	-2 ($= L_5$)
layer$_i$	yes	takedown	engage	$25i$ ($= L_6^i$)
any	yes	compromise	block	-2 ($= L_7$)
any	yes	exfiltrate	block	0 ($= L_8$)
any	yes	takedown	block	0 ($= L_9$)

exfiltrates data while being engaged. The defender can also prevent the data exfiltration by restricting attacker's access to the network (action block), however, by doing so, he loses the option to collect the evidence and hence the reward is $L_8 = 0$.

If the attacker decides to cause significant immediate damage by the takedown action, the only option of the defender to prevent this from happening is to block the attacker (if applicable) when the loss is $L_9 = 0$. Otherwise, the cost for the defender is $L_3^i = L_6^i = 25i$ (when i represents the layer the attacker is in).

5.3 Optimal Defense Strategy

Once the defender succeeds in detecting the presence of the attacker, he can investigate log records to analyze past attacker's actions and estimate his belief about being detected. The defender can make use of this belief to reason about the defensive measures he should apply and to design an optimal defense strategy. We are aware that in real world deployments, accurate tracking of attacker's belief need not be possible and we discuss this in Sect. 5.5.

The optimal defense strategy incurs expected long-term discounted loss of the defender of 282.154. This is a significant improvement over the common practice nowadays of attempting to block the attacker immediately after he is detected. The *always-block* strategy where the defender is restricted to play only block action once he detects the attacker leads to an expected loss of 429.375. It is also, however, not good to keep the attacker engaged in the network forever (and try to deceive him by never blocking him, and always use the engage action – we refer to this strategy as *always-engage*). Such an approach would not make the deception believable, and the attacker would rather cause the damage and forfeit his current attack attempt, than battle the deception.

We represent the optimal defensive strategy as a mapping from the current position of the attacker (i.e. the layer of the network he penetrated) and his *belief* about being detected (and thus being deceived). Since the defender has only two actions available, we express the probability of playing the engage action only (had he succeeded in detecting the attacker), $\sigma_D(i, b)$, where $i \in \{1, 2, 3\}$ is the current layer and $b \in [0, 1]$ is the attacker's belief about the detection state. Note that $\sigma_D(i, b)$ corresponds to $\pi_D((n, \texttt{layer}_i, x_D, \texttt{true}), \texttt{engage})$, where π_D is the minimax solution of Eq. (6) evaluated for $v^*(\hat{b})$, $\hat{b}(n, \texttt{layer}_i, x_D, \texttt{true}) = b$, $\hat{b}(n, \texttt{layer}_i, x_D, \texttt{false}) = 1 - b$. The optimal defense strategy $\sigma_D(i, b)$ for each of the layers is depicted in Fig. 3.

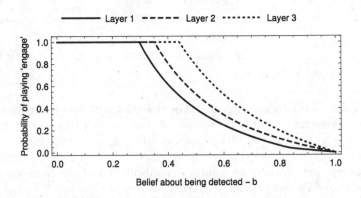

Fig. 3. Optimal defense strategy σ_D for the network from Fig. 1. The optimal strategy of the defender is randomized and depends on the current position of the attacker (the layer he penetrated) and his belief about the detection state.

The optimal defense strategy prescribes the defender to always keep the attacker in the network when the attacker is highly confident that he has not been detected yet. In such a situation, the attacker will opt for data exfiltration, which we can prevent, e.g. by providing him with fake data. At a certain point, however, the attacker starts being worried about being detected and starts considering to cause immediate damage, incur a high loss to the defender and leave the network (i.e. use the takedown action). The defender has to react to this development and think about blocking the attacker by decreasing the probability of keeping the attacker in the network.

Remark 2 (Demise of the greedy). We can observe that the closer the attacker is to his primary goals (or at least the closer he thinks to be), the less concerned he is about the fact that he might be detected and the more greedy he is about realizing his intents. It is thus easier for the defender to deceive the attacker in such a situation. This is caused by the fact that the attacker must have put more effort to get into deeper layers of the network and the damage he can possibly cause now is more significant—thus he is willing to take a greater risk of being detected. This in turn allows the defender to deceive him more efficiently. While in the Layer 1, the attacker starts considering the `takedown` action when he thinks that he is detected with probability 0.298 (and the defender has to react accordingly), in the Layer 3 he delays this decision up to the point when his belief about the detection state is 0.442. We conjecture that this type of behavior of the deceivee can be seen in a wide range of deception problems and the deceiver can capitalize on that.

To better understand the implications of the optimal defense strategy and the need for precise randomization between `engage` and `block` actions, we simulate an attack on the network and depict attacker's belief about being detected when applying the optimal, *always-block* and *always-engage* strategies.

After performing an action and getting feedback from the network, the attacker updates his belief about the detection state from b to b'. Assume that the attacker was in Layer i and he used action $a_A \in \{$ `compromise`, `exfiltrate` $\}$ in the last step and he didn't get blocked. In order to be detected at the current time step, the attacker could have either triggered an alert using his last action (which happens with probability $(1 - b)p_{trig}(n, a_A)$), or he must have been already detected and the defender must have decided not to block the attacker (the probability of which is $b\sigma(i, b)$). The probability of not getting blocked equals to $1 - b(1 - \sigma(i, b))$. We can thus derive a belief update formula (see Eq. (4)) specifically for this game when the updated belief of the attacker b' is the probability of being detected in the next time step:

$$b' = \frac{(1 - b)p_{trig}(n, a_A) + b\sigma(i, b)}{1 - b \cdot [1 - \sigma(i, b)]}. \tag{12}$$

We assume that the attacker conducts an attack that consists of penetrating to the deepest layer of the network using three consecutive `compromise` actions and then the attacker exfiltrates data forever. The comparison of the evolution of attacker's belief while the defender uses the optimal strategy with the always-block and always-engage strategies is shown in Fig. 4.

First of all, we explain why the current best practice in incident response represented by the *always-block* strategy is inferior. Whenever the attacker realizes that he has not been blocked and his access to the network has not been restricted (or limited), he *knows* that he cannot have been detected in the previous time step (since otherwise, the defender would have blocked him according to his *always-block* strategy). His belief about being detected thus depends solely on the detection rate of IDS – which in our experiments is $p_{trig}(n, $ `compromise` $) = 0.2$ when the attacker uses `compromise` action to penetrate deeper into the network (first 3 steps) and $p_{trig}(n, $ `exfiltrate` $) = 0.1$ afterward. Since the attacker remains highly confident that he is not detected at each time step, he can cause a lot of harm by a long-term data exfiltration.

The *always-engage* strategy suffers from playing `engage` action even at times when the attacker becomes highly confident that he has been detected and thus realizes that the data he exfiltrates may be useless. At that point, the attacker

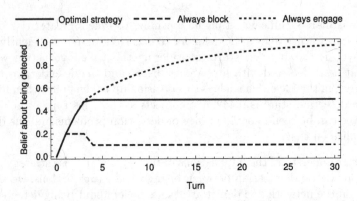

Fig. 4. Evolution of attacker's belief over time. If we block the attacker immediately after detection, he remains highly confident that we cannot employ deceptive actions which allows him to perform long-term data exfiltration. If we always attempt to deceive the attacker by engaging him, he realizes that he likely faces a deception and decides to cause immediate damage – which cannot be prevented by the deceptive **engage** action.

deviates from the assumed attack plan and opts for causing immediate damage and leaving the network temporarily (before launching a new attack).

The optimal defense strategy, on the other hand, stabilizes attacker's belief about being detected at the value of $b = 0.4968$. This is the right belief where the attacker still thinks that it is worth attempting to cause a long term damage by data exfiltration, despite being vulnerable to defender's deceptive attempts.

Remark 3 (Curse of exclusion). This result draws one important conclusion about the use of deception to manipulate attacker's belief. The decision to exclude the attacker from the network (or even more importantly the decision *not* to block him) leaks a valuable piece of information to the attacker. If we do not think about blocking the attacker in a strategic way, the attacker can capitalize on getting this information to devise a powerful attack plan. We have to weigh the use of stealthy and non-stealthy defensive actions carefully not to alert the attacker to the use of deception. The optimal defensive strategy (unlike the *always-block* and *always-engage* strategies) achieves a belief point where no further information leaks to the attacker and the malicious effects of attacker's actions are minimized.

5.4   Engaging the Attacker

In Sect. 5.3 we have shown that the common practice in incident response deployments of blocking the attacker immediately after detection is susceptible to severe drawbacks. We proposed an alternative strategy, based on a game-theoretic model, that postpones the decision to block the attacker to minimize the long-term damage to the network. The key motivation for using this strategy is that by anticipating malicious actions of the attacker, we can minimize negative impacts of his actions and delay his progress. On the other hand, excluding the attacker from the network is only temporary. The attacker is potentially able to reenter the network and cause significant damage before we manage to detect him again.

Our strategy has, however, one more significant advantage since it can be leveraged to decrease false positive rates of the IDS. False detections can have a considerable negative impact on the network operations. By engaging a suspicious user in the network, we can make use of the extra time given by our deceptive strategy to identify the user, infer their objectives and take proper defense actions to reduce the impact of the network defense system on legitimate users. To this end, we can use various types of deceptive signals that do not influence legitimate users considerably, but make the progress of an attacker difficult. These signals are not explicitly captured in our example, but the model is general enough to account for them.

We conducted an experimental evaluation of our game-theoretic strategy to determine the average time between the first IDS alert and the time we decide to block the user. We evaluated our strategy against an advanced attacker who plays a best response to strategy σ_D and we considered only the attacks where the attacker does not decide to quit the network himself. We found out that the average time between detection and the time we decide to restrict attacker's access is in our case 4.577 time steps. In this time window, the defender gets additional alerts from the IDS which may help him to decide about the credibility of the alert better and thus assure that he is about to block a malicious user.

5.5   Robustness of the Model

In real world setups, it need not be possible for the defender to keep track of attacker's belief accurately as a result of failing to reconstruct the exact history of the attacker and/or deficiencies in the model of the network. In this section, we focus on the impact of not knowing the exact IDS detection

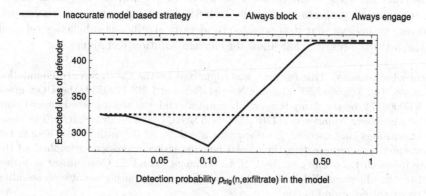

Fig. 5. Expected loss of the defender when using a strategy originating from an inaccurate model. Strategy is computed while assuming detection probability $p'_{trig}(n, \texttt{exfiltrate})$ and this strategy is evaluated in a network with the detection probability of $p_{trig}(n, \texttt{exfiltrate}) = 0.1$.

probabilities. We compute the optimal strategy of the defender based on a model where $p'_{trig}(n, \texttt{exfiltrate})$ does not match the detection probability in the real network. We then evaluate the resulting strategy in the network where the detection probability is $p_{trig}(n, \texttt{exfiltrate}) = 0.1$. Since the model is no longer accurate, the resulting strategies need not be optimal. The experimental evaluation of these strategies is shown in Fig. 5. The experimental results show that our strategy provides significant room for the error in the design of the model, especially if we are pessimistic about the detection rates.

6   Conclusions

We have provided a principled analysis of cyber deception in network security based on game-theoretic foundations. First, we have introduced a generic game-theoretic model for strategic reasoning about deception, then we applied this model to the network security, and we illustrated the impact of active deception on the security level of the network in a case study. Our results have shown that the use of cyber deception techniques can reduce the risks associated with network operations and minimize the damage a sophisticated attacker can inflict to the network. The deceptive operation, however, achieves the maximum efficiency if the attacker is unaware of being deceived. While this result is not surprising, our analysis provides theory supporting this result.

Our work serves as a proof of concept to motivate the interest in thinking about active cyber deception in a strategic way. We used a simplified example to introduce main ideas and discuss the need for reasoning about the belief and adaptation process of the adversary. In the future work, however, we plan to address computational challenges introduced by large networks by leveraging the structure and symmetries found in the problem. An interesting, and also natural, continuation of our work is to relax the assumption that the defender can reconstruct the view of the attacker perfectly. In general, the *two-sided* imperfect information presents significant theoretical and computational challenges, however, we believe that it is possible to identify significant subclasses relevant for the network security that allow for efficient solution techniques.

Acknowledgments. This research was supported by the Czech Science Foundation (grant no. 15-23235S), NSF grants CNS-1544782 and CNS-1720230, the DOE grant DE-NE0008571, by the Army Research Laboratory and was accomplished under Cooperative Agreement Number W911NF-13-2-0045 (ARL Cyber Security CRA). The views and conclusions contained in this document are those of the authors and should not be interpreted as representing the official policies, either expressed or implied, of the Army Research Laboratory or the U.S. Government. The U.S. Government is authorized to reproduce and distribute reprints for Government purposes not with standing any copyright notation here on.

References

1. Achleitner, S., La Porta, T., McDaniel, P., Sugrim, S., Krishnamurthy, S.V., Chadha, R.: Cyber deception: virtual networks to defend insider reconnaissance. In: Proceedings of the 2016 International Workshop on Managing Insider Security Threats, pp. 57–68. ACM (2016)
2. Başar, T., Olsder, G.J.: Dynamic Noncooperative Game Theory. SIAM, Philadelphia (1998)
3. Bercovitch, M., Renford, M., Hasson, L., Shabtai, A., Rokach, L., Elovici, Y.: HoneyGen: an automated honeytokens generator. In: IEEE International Conference on Intelligence and Security Informatics, ISI 2011, pp. 131–136. IEEE (2011)
4. Bowen, B.M., Hershkop, S., Keromytis, A.D., Stolfo, S.J.: Baiting inside attackers using decoy documents. In: Chen, Y., Dimitriou, T.D., Zhou, J. (eds.) SecureComm 2009. LNICSSITE, vol. 19, pp. 51–70. Springer, Heidelberg (2009). doi:10.1007/978-3-642-05284-2_4
5. Dagon, D., Qin, X., Gu, G., Lee, W., Grizzard, J., Levine, J., Owen, H.: HoneyStat: local worm detection using honeypots. In: Jonsson, E., Valdes, A., Almgren, M. (eds.) RAID 2004. LNCS, vol. 3224, pp. 39–58. Springer, Heidelberg (2004). doi:10.1007/978-3-540-30143-1_3
6. Durkota, K., Lisý, V., Bošanský, B., Kiekintveld, C.: Approximate solutions for attack graph games with imperfect information. In: Khouzani, M.H.R., Panaousis, E., Theodorakopoulos, G. (eds.) GameSec 2015. LNCS, vol. 9406, pp. 228–249. Springer, Cham (2015). doi:10.1007/978-3-319-25594-1_13
7. Falliere, N., Murchu, L.O., Chien, E.: W32. stuxnet dossier. White Paper Symantec Corp. Secur. Response 5(6), 2–3 (2011). https://www.symantec.com/content/en/us/enterprise/media/security_response/whitepapers/w32_stuxnet_dossier.pdf
8. Gostev, A., Soumenkov, I.: Stuxnet/Duqu: The evolution of drivers (2011). http://www.securelist.com/en/analysis/204792208/Stuxnet_Duqu
9. Gu, G., Zhang, J., Lee, W.: BotSniffer: detecting botnet command and control channels in network traffic. In: Proceedings of the 15th Annual Network and Distributed System Security Symposium (NDSS 2008) (2008)
10. Hansen, E.A., Bernstein, D.S., Zilberstein, S.: Dynamic programming for partially observable stochastic games. AAAI 4, 709–715 (2004)
11. Horák, K., Bošanský, B., Pěchouček, M.: Heuristic search value iteration for one-sided partially observable stochastic games. In: Proceedings of the Thirty-First AAAI Conference on Artificial Intelligence (AAAI 2017) (2017)
12. Jajodia, S., Ghosh, A.K., Subrahmanian, V., Swarup, V., Wang, C., Wang, X.S. (eds.): Moving Target Defense II - Application of Game Theory and Adversarial Modeling. Advances in Information Security, vol. 100. Springer, New York (2013)
13. Jajodia, S., Ghosh, A.K., Swarup, V., Wang, C., Wang, X.S. (eds.): Moving Target Defense - Creating Asymmetric Uncertainty for Cyber Threats. Advances in Information Security, vol. 54. Springer, New York (2011)
14. Kreibich, C., Crowcroft, J.: Honeycomb: creating intrusion detection signatures using honeypots. ACM SIGCOMM Comput. Commun. Rev. 34(1), 51–56 (2004)
15. Kuipers, D., Fabro, M.: Control systems cyber security: Defense in depth strategies. United States, Department of Energy (2006)
16. Manshaei, M.H., Zhu, Q., Alpcan, T., Bacşar, T., Hubaux, J.P.: Game theory meets network security and privacy. ACM Comput. Surv. (CSUR) 45(3), 25 (2013)

17. McRae, C.M., Vaughn, R.B.: Phighting the phisher: using web bugs and honeytokens to investigate the source of phishing attacks. In: 40th Annual Hawaii International Conference on System Sciences 2007, HICSS 2007, p. 270c. IEEE (2007)

18. Mohammadi, A., Manshaei, M.H., Moghaddam, M.M., Zhu, Q.: A game-theoretic analysis of deception over social networks using fake avatars. In: Zhu, Q., Alpcan, T., Panaousis, E., Tambe, M., Casey, W. (eds.) GameSec 2016. LNCS, vol. 9996, pp. 382–394. Springer, Cham (2016). doi:10.1007/978-3-319-47413-7_22

19. Osborne, M.J., Rubinstein, A.: A Course in Game Theory. MIT Press, Cambridge (1994)

20. Pawlick, J., Farhang, S., Zhu, Q.: Flip the cloud: cyber-physical signaling games in the presence of advanced persistent threats. In: Khouzani, M.H.R., Panaousis, E., Theodorakopoulos, G. (eds.) GameSec 2015. LNCS, vol. 9406, pp. 289–308. Springer, Cham (2015). doi:10.1007/978-3-319-25594-1_16

21. Rostami, M., Koushanfar, F., Rajendran, J., Karri, R.: Hardware security: threat models and metrics. In: Proceedings of the International Conference on Computer-Aided Design, pp. 819–823. IEEE Press (2013)

22. Spitzner, L.: Honeypots: Tracking Hackers, vol. 1. Addison-Wesley Reading, Boston (2003)

23. Stech, F.J., Heckman, K.E., Strom, B.E.: Integrating cyber-D&D into adversary modeling for active cyber defense. In: Jajodia, S., Subrahmanian, V., Swarup, V., Wang, C. (eds.) Cyber Deception, pp. 1–22. Springer, Cham (2016). doi:10.1007/978-3-319-32699-3_1

24. Symantec: Preparing for a cyber attack (2013). http://www.symantec.com/content/en/us/enterprise/other_resources/b-preparing-for-a-cyber-attack-interactive-SYM285k_050913.pdf. Accessed 17 Apr 2017

25. Tankard, C.: Advanced persistent threats and how to monitor and deter them. Netw. Secur. **2011**(8), 16–19 (2011)

26. Underbrink, A.: Effective cyber deception. In: Jajodia, S., Subrahmanian, V., Swarup, V., Wang, C. (eds.) Cyber Deception, pp. 115–147. Springer, Cham (2016). doi:10.1007/978-3-319-32699-3_6

27. Vollmer, T., Manic, M.: Cyber-physical system security with deceptive virtual hosts for industrial control networks. IEEE Trans. Industr. Inf. **10**(2), 1337–1347 (2014)

28. Weinstein, W., Lepanto, J.: Camouflage of network traffic to resist attack (CONTRA). In: DARPA Information Survivability Conference and Exposition 2003. Proceedings, vol. 2, pp. 126–127. IEEE (2003)

29. Zhu, Q., Başar, T.: Game-theoretic approach to feedback-driven multi-stage moving target defense. In: Das, S.K., Nita-Rotaru, C., Kantarcioglu, M. (eds.) GameSec 2013. LNCS, vol. 8252, pp. 246–263. Springer, Cham (2013). doi:10.1007/978-3-319-02786-9_15

30. Zhu, Q., Basar, T.: Game-theoretic methods for robustness, security, and resilience of cyberphysical control systems: games-in-games principle for optimal cross-layer resilient control systems. IEEE Control Syst. **35**(1), 46–65 (2015)

31. Zhu, Q., Clark, A., Poovendran, R., Başar, T.: Deceptive routing games. In: IEEE 52nd Annual Conference on Decision and Control (CDC), pp. 2704–2711. IEEE (2012)

32. Zhu, Q., Clark, A., Poovendran, R., Basar, T.: Deployment and exploitation of deceptive honeybots in social networks. In: IEEE 52nd Annual Conference on Decision and Control (CDC), pp. 212–219. IEEE (2013)

A Stochastic Game-Theoretic Model for Smart Grid Communication Networks

Xiaobing He[✉] and Hermann de Meer

Department of Informatics and Mathematics, University of Passau,
Innstr. 43, 94032 Passau, Germany
{Xiaobing.He,Hermann.DeMeer}@uni-passau.de

Abstract. The increasing adoption of new information and communication technology assets in smart grids is making smart grids vulnerable to cyber threats, as well as raising numerous concerns about the adequacy of current security approaches. As a single act of penetration is often not sufficient for an attacker to achieve his/her goal, multistage cyber attacks may occur. This paper looks at the stochastic and dynamic nature of multistage cyber attacks in smart grid use cases and develops a stochastic game-theoretic model to capture the interactions between the attacker and the defender in multistage cyber attack scenarios. Due to the information asymmetry of the interactions between the attacker and the defender, neither of both players knows the exact current game state. This paper proposes a belief-updating mechanism for both players to form a common belief about the current game state. In order to assess threats of multistage cyber attacks, it further discusses the computation of Nash equilibria for the designed game model.

Keywords: Asymmetric information · Positive stop probability · Stochastic game · Multistage cyber attacks · Smart grid · Threat assessment

1 Introduction

Network security is a critical concern with regard to cyber-physical systems. For a long time, security operators have been interested in knowing what an attacker can do to a cyber-physical system and what can be done to prevent or counteract cyber attacks [3,14]. It is suggested that risk assessment must be integral to the overall life cycle of the smart grid systems. A cyber threat assessment helps the system administrator to better understand the effectiveness of the current network security solution and determine the best approach to secure the system against a particular threat, or a class of threats. By offering a deep analysis of existing or potential threats, system administrators are given a clear assessment of the risks to their systems, while possessing a clear vision about the kind of security countermeasures that the respective utility should invest in.

Attack scenarios are dynamically changing in smart grid communication networks, for example, because of existing of legacy and new systems in smart

© Springer International Publishing AG 2017
S. Rass et al. (Eds.): GameSec 2017, LNCS 10575, pp. 295–314, 2017.
DOI: 10.1007/978-3-319-68711-7_16

grid communication networks. Multistage cyber attacks, as important threats in smart grid communication networks, make use of a variety of different exploits, propagation methods, and payloads, resulting in the emergence of many more sophisticated cyber attacks. Current protection mechanisms, which rely on isolation techniques, such as firewalls, data diodes, and zoning concepts, are not sufficiently applicable in cyber-physical systems. For more than a decade, game-theoretic approaches have been recognized as useful tools to handle network attacks [2,7,13,15]. Significant results from game theory concerning cyber situation awareness and network security risk assessment in conventional information and communication technology (ICT) systems have been reported [14,30]. But the application of game theory for the assessment of threats from multistage cyber attacks and the prediction of an attacker's actions in smart grid communication networks are still in their infancy nowadays.

Threat assessment for multistage cyber attacks is not straightforward, given that, at any stage of a cyber attack, the attacker may decide not to proceed or change his/her attack actions. Since the attacker has motivations (costs versus benefits) and finite resources to launch a further attack at any stage, the stage at which the multistage attack stops is not necessary predetermined (stochastic). This paper accounts for this by adding a stopping time to the stochastic model. It is to be noted that an attacker who doesnot have any resource limitations (from an economic point of view) is beyond the scope of this paper. The stop of the attack or the change of attack actions at any stage makes a threat assessment extremely challenging, as it is difficult to know what the attacker will do or to assess possible cyber or physical impacts resulting from his/her attack actions in the next stage.

Cyber attacks on smart grid communication networks can cause physical damage to the power grid. Many existing stochastic game-theoretic threat assessment methods assume symmetric information among the players, which implies that all the players share the same information, i.e., the same signal observed and the same knowledge about states/payoffs in a game. However, in many situations, this assumption is unrealistic. There are many games arising out of communication networks, electronic commerce systems, and society's critical infrastructures involving players with different kinds of information about the game state and action processes over time [11,23,29]. For instance, in cyber-security systems, the attacker knows his/her own skill set, while the defender knows the current and planned resource characteristics of the system. In short, the attacker and the defender do not share their available information with each other.

This paper attempts to design a stochastic game-theoretic model with asymmetric information and positive stop probabilities in order to assess the threat of multistage cyber attacks in smart grid communication networks. The positive stop probability means that the probability of the game to end at any state is positive. Unlike random failures, attackers have motivations and capabilities to launch further attacks. Both the attacker and the defender will act in consideration of the consequences of their corresponding actions, with such consequences including satisfactions, risk versus effort, and effectiveness. In each state of the game, if launching a further attack would have limited benefits, and take months of time and huge amount of computers and memory, the attacker will most probably stop his/her attack. Once the defender observed these phenomena regarding the attacker, he/she will not deploy any corresponding countermeasures. Therefore, this situation will be accounted for by adding a stop probability to the stochastic model; and such a stop probability is positive. The designed stochastic game-theoretic model extends an existing stochastic game-theoretic model with specific characteristics of attacker-defender interactions in smart grid communication networks. The objectives of this attacker-defender stochastic game-theoretic model is to assess cyber attack scenarios at an early stage of the attack, where the defender makes correct optimal proactive defence decisions. Therefore, a defence system can be prewarned, security resources can be better allocated to defeat or mitigate future attacks, and security incidents can be avoided. This paper considers the worst-case scenario where the attacker has complete knowledge of the architecture/infrastructure of the system and hosts' vulnerabilities in the system, and the attacker has full knowledge of the target smart grid defense configurations. Section 2 provides a non-exhaustive overview of existing game-theoretic approaches for cyber attacks, while Sect. 3 presents an attacker-defender stochastic game-theoretic model to represent the attacker-defender interactions. Section 4 analyses the belief-updating mechanisms and presents the feasible computation of Nash equilibria. Finally, Sect. 5 concludes the paper ans discusses future works.

2   Related Work

A game consists of players (in this paper, the attacker and the defender), strategies (i.e., actions of players) available to each player, and utilities depending on the joint decisions of all players. Game theory depicts dynamic interactions between players, involving a complementary methodology of attack trees and/or attack graphs in face of changing attack patterns.

Ismail et al. [10] modelled the problem of optimizing the distribution of defence resources on communication equipment as a one-shot game [22] between the attacker and the defender. That game took into account the interdependency between the cyber and physical components in the power grid. It was assumed that the initial risk, the immediate risk on a node before any incidents or failure propagations is a positive real number and evaluated using other risk assessment methods. The immediate risk and the future cascading risk from interdependent electrical and communication infrastructures were balanced in [10]. The interdependency between the electrical and communication infrastructures were modelled as a weighted directed interdependency graph. Each communication equipment was associated with a load. The worst-case scenario, where both the attacker and the defender have complete knowledge of the architecture of the system, was considered in [10]. The utility functions of both players are composed of three parts: the reward for an attack, the cost of attacking/defending, and the impact of redundant communication equipment. The impact of attacks in the electric and communication infrastructures was evaluated by solving power flow equations and using attack graphs, in conjunction with other risk assessment methods. The dataset of the Polish electric transmission system, provided in the MATPOWER computational packages, was taken as a case study to validate the proposed game-theoretic model, while Nash equilibria for the attacker and the defender for each type of communication equipment in the case study were presented.

Jiang et al. [30] proposed a two-player non-cooperative, zero-sum, and finite stochastic game for the attacker and the defender in computer networks. A Markov chain for a privilege model and a privilege-escalating attack taxonomy were presented. By making use of the developed stochastic game model, a Markov chain for the privilege model, and a cost-sensitive model, the attacker's behaviour and the optimal defence strategy for the defender were predicted. He et al. [8] studied a network security risk assessment-oriented game-theoretic attack-defence model to quantify the probability of threats. The payoff matrix was formulated from a cost-benefit analysis, where the cost to the defender when taking actions was made up of the operational cost, the response cost, and the response negative cost. Combined with the vulnerability associated with the nodes, risks of the system were computed as the sum of the threat value of all nodes.

Guillarme et al. [6] presented an attack stochastic game model for adversarial intention recognition for situations featuring strategic interactions between an attacker and a defender. The attack stochastic game model is a coupling of discounted stochastic games and probabilistic attack graphs, although it suffers from zero-sum constraints. In the attack stochastic game model, it was assumed that both the attacker's action and the defender's action, as well as the states experienced by players, were fully observable to both players. This model was inverted to infer the intention of an attacker from observations of his/her (sub-)optimal actions. However, this model does not have the ability to detect intention changes, while the scalability is the principal limitation of this attack stochastic game model.

Nguyen et al. [21] studied a two-player zero-sum stochastic game-theoretic approach to provide the defender with guidelines to allocate his/her resources to secure his/her communication and computer networks. Linear influence networks [19] were used to present the interdependency of nodes in terms of security assets and vulnerabilities. He et al. [9] investigated game-theoretic risk assessment in smart grid communication networks and noticed that the data acquisition and data interpretation for risk assessment and prediction had not been intensively explored. Therefore, [9] established a surveillance architecture to monitor message transactions in communication networks, while surveillance observations were further interpreted as Dirichlet-distributed security events with certain probabilities. By taking the interactions between possible suspicious nodes and the security operators as a repeated zero-sum transmitting-monitoring game, a game-theoretic risk assessment framework was established to compute and forecast the risk of network security impairment. Rass and Zhu [25] presented a sequence of nested finite two-player zero-sum games for developing effective protective layers and designing defence-in-depth strategies against advanced persistent attacks (APTs). In the game-theoretical model, nodes in an infrastructure were equidistantly separated into different levels according to their layers in the infrastructure. Within each level, the game structure was determined by the nodes' vulnerabilities and their distances from the target node. The authors of [25] discussed some closed form solutions for their APTs games and analytically formulated infrastructure design problems to optimize the quality of security across several layers. Under the framework of the HyRiM project, Rass et al. [24] investigated an extensive form game as a risk mitigation tool for defending against APTs. An APT was modelled as a zero-sum one-shot game with complete information, but uncertainty was observed in the game payoffs. Based on a topological vulnerability analysis and an established attack graph, all the attack vectors covered in enumerated attack paths (from the root node to the target node in the attack graph) made up the attacker's action space. By defining players' payoffs as probability-distributed values, instead of real numbers, [24] provided a relative new approach to tackling ambiguous and inconsistent expert opinions in risk management.

The proposed game-theoretic model in this paper differs from the aforementioned approaches in the sense that the model captures the key characteristics (e.g., information asymmetry) of the interactions between the attacker and the defender in smart grid communication networks. None of theses precursor works has looked at the stochastic and dynamic nature of attacks in smart grid use cases (modelled as stochastic games). Both decision makers, the attacker and the defender, have asymmetric information about the underlying system state, while they both maintain a belief (i.e., a probability distribution) about the current system state. This paper provides a common belief-updating mechanism for the attacker and the defender to refresh such a belief. The objectives of this research include contributing towards safety improvements for relevant stakeholders (e.g., smart grid equipment manufacturers, utility companies) in power distributed grids and making recommendations about allocating security resources to reduce cyber security incidents or even safety-related events.

3 Attacker-Defender Stochastic Game-Theoretic Model

To assess threats of multistage attacks, the strategic interactions between the attacker and the defender are modelled as a stochastic game (which covers the step occurrence dependency in multistage attacks). In such a game, the possible actions of the players are restricted, such that there exists an equilibrium point in which the attacker has no chance to successfully obtain his/her ultimate goal. This section introduces action spaces and state transition probabilities of the game between the attacker and the defender. This work designs the attacker-defender stochastic game-theoretic model by a description of an existing stochastic game model and an extension of this model according to the characteristics of the interactions of the attacker and the defender in smart grid communication networks.

3.1 Players

An attacker and a defender are the key "players" in the designed stochastic game-theoretic model. There could be many attackers who are trying to launching attacks and many defenders in the network to protect the system, but this work abstracts those attackers and defenders as one attacker and one defender, respectively. The attacker attains his/her ultimate target via multiple stages. The concept of the defender denotes the security operator (security operator and system administrator are used interchangeably in this paper) who has the task of deploying available defence countermeasures to protect the underlying system, while the attacker attempts to reach the target or the most critical assets located at the centre of the smart grid. This model considers that each of the players has some finite resources to perform actions at each stage of the game. The attacker is considered to be a resource-constrained, determined and rational player. In this way, the attacker will give up when he/she finds it is out of his/her capability to launch any further attacks. Furthermore, it is assumed that once an attack is initiated, the attacker him/herself will never revert the system to any of the previous state (for example, to recover the system from a malfunctioning state to a normal operational state). In this work, the attacker is only able to perform a single action in his/her turn. It is also assumed that the defender does not know whether or not there is an attacker, as that in real systems. Furthermore, the attacker is assumed to be always aware of the active defence mechanisms. Moreover, the defender does not know the objectives and strategies of an attacker. A successful attack may or may not be observable to the defender. The attacker strategically and dynamically chooses his/her targets and attack methods in order to achieve his/her goals, while the defender defines security policies and implements security measures (including email filtering, detection software, patches to prevent and detect attacks, and repairing the system after disruption).

3.2 State Transition Probabilities

A multistage attack, by exploiting vulnerabilities, makes the network system transition from one state to another. However, such a transition also depends on the active defence mechanisms. Therefore, the probability that the state will transition from one to another depends on the joint actions of both players. Unlike accidental failures, an attacker will consider the consequences of his/her actions and compare the reward versus the cost of each elementary attack action [27]. Therefore, the transition probabilities from one state to another depend not only on the decisions of both players to take action, but also the success probability of an attacker going through with his/her action. The probability of success for the attacker at state s is denoted as $p_{suc}(y_{s,b})$ (this work assumes the second player to be the attacker and $y_{s,b}$ (which will be defined later in Eq. (4)) to be the probability that his/her action b is taken at state $s \in S$, S is the state space and $k = |S|$ is the number of game states). Obviously, whether an action by an attacker succeeds depends on the available exploitable vulnerabilities of an asset in the smart grid communication network. For example, attacking an asset with no exploitable vulnerability has zero probability of success. In the attacker-defender stochastic game-theoretic model, success probabilities of an attacker's actions are assigned, based on the intuition and experience (e.g., case studies, common vulnerability scoring system (CVSS), knowledge engineering). Principally, the action of the defender also involves a success probability (e.g., IDSs have detection rates); to simplify the underlying problem, however, such a success probability of the defender with his/her actions is always assumed to be one.

The probability for player 1 (player 1 is the defender) to take action $a \in AS_1$ at state s is denotes as $x_{s,a}$ (which will be defined later in Eq. (3)), while the probability for player 2 (player 2 is the attacker) to take action $b \in AS_2$ at state s is denoted as $y_{s,a}$. Both players take actions simultaneously, meaning that both players take action independently of one another. Thus, when actions $a \in AS_1$ and $b \in AS_2$ are taken from both players, the state transition probability from game state $s \in S$ to state $s' \in S$ can be calculated as

$$q(s'|s,a,b) = x_{s,a} \cdot y_{s,b} \cdot p_{suc}(y_{s,b}).$$

For example, if the probability for player 1 to take action "IDS deployment" is 0.5, the probability for player 2 to take action "Exploit" is 0.4, and the probability that the attacker will successful obtain his/her (sub)goal is 0.2, the game will move from state "normal" to state "malfunctioning" with a state transition probability of

$$q(\text{malfunctioning}|\text{normal}, \text{IDS deployment}, \text{Exploit}) = 0.5 \cdot 0.4 \cdot 0.2 = 0.04.$$

Depending on the exploitable vulnerabilities, it may be that there is no transition between certain game states. For example, it may not be possible for the network to transition from a normal functioning state to a totally failed state without going through any intermediate states. In this work, infeasible state transitions are assigned with a transition probability of zero and hence ignored. Both players make their moves simultaneously, with state transition probabilities being common knowledge to them.

3.3 Game Formalization

In the previous subsections, this paper elaborates players in a game play. At each stage of the game for multistage attacks, the play is in a given state, with every player choosing an action from his/her available action space. With a state transition probability (which is jointly controlled by both players), the current state of the game, and the collection of actions that the players choose, the game will go to another state with an immediate payoff received by each player. Each player has his/her own costs of executing actions, thus the payoff of the game cannot only be described by rewards. Although there may be a dependence of rewards and losses among player's payoffs, because of players' own action execution costs, the payoffs of the attacker and the defender do not sum up to zero. Therefore, the interaction between the attacker and the defender is non-zero-sum. The game is also played with positive stop probabilities in each game state, since the game will end when the attacker decides to stop his/her attacks (completely inactive) and the defender keeps his/her defence countermeasures unchanged. Besides, this paper notices that none of the players knows the exact state of the system, while both players have different kinds of private information about the state and action processes over time. Therefore, in order to apply game theory to assess multistage attacks in smart grid communication networks, the asymmetric information, non-zero-sum, and positive stop probability characteristics of the interaction between the attacker and the defender should be taken into account.

The next concern is on the game type that appropriately captures the players' interactions in the case of multistage cyber attacks. Both players do not know the exact state of the game, but maintain a belief about the current state of the game (where a belief is a probability distribution over the possible states of the game). Taking a two-player non-zero-sum two-stage game for instance, suppose the game has two states and both players do not know the current state of the game (either in state s_1 or state s_2), but they have a belief $\rho_1 = (\rho_1(s_1), \rho_1(s_2)) = (0.8, 0.2)$ about the current state, that is, there is a 80% likelihood that the current game at stage 1 is in state s_1, while there is a 20% likelihood that the current game at stage 1 is in state s_2. The most relevant existing game model that can partially solve this problem is the stochastic game with lack of information on one side (SGLIOS) with positive stop probabilities. Thus, this paper considers SGLIOS with positive stop probabilities as a basic game model and extends it to include the non-zero-sum and information asymmetry of the interactions between the attacker and the defender in smart grid communication networks.

This work starts with the definition of SGLIOS with positive stop probabilities described in [18]. The model of SGLIOS with positive stop probabilities is a two-person zero-sum game and states are a finite set $S = \{s_1, s_2, \cdots, s_\ell, \cdots, s_k\}$ ($k = |S|$ denotes the number of states). Associated with each state s_ℓ ($\ell \in \{1, 2, \cdots, k\}$) is a matrix game $\mathbf{G}_{\{s_\ell\}}$ of size $m_1 \times m_2$, where $m_1 = |AS_1|$ (the number of actions of player 1), $m_2 = |AS_2|$ (the number of actions of player 2), and $\mathbf{G}_{\{s_\ell\}} = \{g_{\{s_\ell\}}(a, b) : AS_1 \times AS_2 \to \mathbb{R} | a = 1, 2, \cdots, m_1; b = 1, 2, \cdots, m_2; \ell = 1, 2, \cdots, k\}$. Additionally, $\emptyset$ is adjoined to S, where $\emptyset$ represents the end of the game. In SGLIOS with positive stop probabilities, at any stage N, there is a probability distribution over states in S. throughout this paper, N takes values from $\mathbb{N}$ and $\mathbb{N}$ is the set of natural number. Player 1 is informed about such a probability distribution at every game stage, but player 2 is never informed about that. There is a probability $\rho_1 \in \Delta(S)$ about the initial state, where $\Delta(S)$ is the set of all probability distributions on S. State transition probabilities are denoted as $q(\cdot|s, a, b)$, which depends on the current state s and actions a and b taken by the defender and the attacker, respectively. Because of the positive stop probability assumption, the sum of transition probabilities from state s to all possible next game state s' is less than one, i.e., $\sum_{s' \in \{S - \emptyset\}} q(s'|s, a, b) < 1$, $\forall a \in AS_1, b \in AS_2$. Both players make their moves simultaneously and both of them are informed of their choices (a, b). The game will either end with a probability of $q(\emptyset|s, a, b) > 0$ or transition to a new state s' with a probability of $q(s'|s, a, b) > 0$. Although both players remember actions taken by them, player 2 is not informed of the received immediate payoff $g_{\{s\}}(a, b)$ (which only player 1 knows) of the game. SGLIOS with positive stop probabilities is played with perfect recall (i.e., at each stage each player remembers all past actions chosen by all players and player 1 knows all *past* states that have occurred). There is a common knowledge among both players before they move at stage N and such a common knowledge is a sequence of the form $h_N = \{(a_1, b_1), (a_2, b_2), \cdots, (a_{N-1}, b_{N-1})\}$ (where $a_\ell \in AS_1$ is the action chosen from player 1 at the ℓ stage, $b_\ell \in AS_2$ is the action chosen from player 2 at the ℓ stage, and $\ell \in \{1, 2, \cdots, N-1\}$). The common knowledge h_N is also called *history* and it represents the choices of actions (i.e., pure strategies) of the two players up to (and excluding) stage N. SGLIOS with positive stop probabilities restricts its attention to behavioural strategies [12].

When the game is in state s at stage N, the action chosen by the players can be deterministic or randomized. A mixed strategy corresponds to a distribution over actions (i.e., pure strategies). Let $\mathbf{x}_s$ ($s \in S$) denote the mixed strategy of player 1 in state s and $\mathbf{y}_s$ ($s \in S$) denote the mixed strategy of player 2 at state s. The strategies $\mathbf{x}_s$ and $\mathbf{y}_s$ in state s are used to assign probabilities over the action set AS_1 and AS_2 with cardinality m_1 and m_2, respectively. And the mixed strategies $\mathbf{x}_s$ and $\mathbf{y}_s$ are defined as

$$\mathbf{x}_s := \{(x_{s,1}, \cdots, x_{s,a}, \cdots, x_{s,m_1}) \in \mathbb{R}_+^{m_1} | \sum_{a=1}^{m_1} x_{s,a} = 1, 0 \leq x_{s,a} \leq 1\}, \quad (1)$$

$$\mathbf{y}_s := \{(y_{s,1}, \cdots, y_{s,b}, \cdots, y_{s,m_2}) \in \mathbb{R}_+^{m_2} | \sum_{b=1}^{m_2} y_{s,b} = 1, 0 \leq y_{s,b} \leq 1\}, \quad (2)$$

where

$$x_{s,a} := \mathbb{P}(a|s, h_N), \quad\quad\quad\quad (3)$$
$$y_{s,b} := \mathbb{P}(b|s, h_N), \quad\quad\quad\quad (4)$$

and $x_{s,a}$ and $y_{s,b}$ represent the probability that player 1 takes action a and player 2 takes action b, respectively. It is to be noted that actions of players are independently chosen among each other, since both players are playing simultaneously. Let $\mathbf{x} = (\mathbf{x}_{s_1}, \mathbf{x}_{s_2}, \cdots, \mathbf{x}_{s_\ell}, \cdots, \mathbf{x}_{s_k})$ be a vector of mixed strategies for player 1 and $\mathbf{x} \in \Omega^{m_1}$ (Ω^{m_1} is the set of all probability vectors of length m_1). Correspondingly, let $\mathbf{y} = (\mathbf{y}_{s_1}, \mathbf{y}_{s_2}, \cdots, \mathbf{y}_{s_\ell}, \cdots, \mathbf{y}_{s_k})$ be a vector of mixed strategies for player 2 and $\mathbf{x} \in \Omega^{m_2}$ (Ω^{m_2} is the set of all probability vectors of length m_2). Let E be a random variable representing the stage the game ends and h_N be the common knowledge among players up to (and excluding) stage N. At each stage N, if player 1 takes action a and player 2 took action b, player 1 receives an immediate payoff $g_{\{s_N\}}(a, b)$, The total payoff function $\mathcal{H}(\cdot)$ (with strategies from both players as parameters) in SGLIOS with positive stop probabilities is given as

$$\mathcal{H}(\mathbf{x}, \mathbf{y}) = \sum_{N=1}^{\infty} \mathcal{R}_N(\mathbf{x}, \mathbf{y}) \tag{5}$$

$$= \sum_{N=1}^{\infty} \mathbb{E}_{\mathbf{x}, \mathbf{y}}\big(\rho_N(s)\mathbf{G}_{\{s\}}|E > N\big) \cdot \mathbb{P}(E > N),$$

where $\mathbb{P}(E > N)$ means that the game does not end at stage N and the stage E where game ends is longer than N. The expectation operator $\mathbb{E}_{\mathbf{x}, \mathbf{y}}(\cdot | E > N)$ is used to mean that player 1 plays strategy $\mathbf{x}$ and player 2 plays strategy $\mathbf{y}$, under the condition that the game does not end at stage N. Equation (5) assumes that the game stage can go to infinite (∞). However, because of the positive stop probability assumption, the game will end after a finite number of stages [28]. Therefore, the game of SGLIOS with positive stop probabilities is a finite game. The fundamental tool in SGLIOS with positive stop probabilities is an updating mechanism which gives at each stage N the belief ρ_N, the posterior distribution on the state space given the history h_N up to stage N. Player 1 is informed about the belief ρ_N but player 2 does not. The updating mechanism for the belief ρ_N is working in this way: initially both players choose strategies $\mathbf{x}$ and $\mathbf{y}$ and give them to chance (chance is a special player, who can be the environment of the system) who then at stage 1 chooses s_1 according to ρ_1. Then the action pair (a_1, b_1) is chosen according to $(\mathbf{x}_{s_1}, \mathbf{y}_{s_1})$ and an immediate payoff $g_{\{s_1\}}(a_1, b_1)$ is received by player 1. Provided that the game does not end, chance chooses another state s_2 according to $\rho_2(s_2) := \mathbb{P}(s_2 | a_1, b_1, E > 2)$ or decides to end the game according to $\mathbb{P}(E = 2 | a_1, b_1)$. At stage N, chance decides the game to go to state s_N according to $\rho_N(s_N) := \mathbb{P}(s_N | h_N, E > N)$ or ends the game according to $\mathbb{P}(E = N | E > N - 1, h_N)$. The value $\rho_N(s)$ represents that the chance believes that the current game state is $s \in S$. It is proved in [18] that the value of the game of SGLIOS with positive stop probabilities exists and is a continuous function on the state space; and there exists also a stationary optimal strategy for the informed player, i.e., player 1. The optimal strategy of player 1 depends only on the updated probability of the current state which he/she independently knows.

Since the interaction between the attacker and the defender in smart grid use cases is non-zero-sum, it is needed to extend SGLIOS with positive stop probabilities (which is zero-sum) to non-zero-sum cases. The game matrices should be first identified. Each player (player 1 or player 2) has his/her own game matrix, which is composed of two parts: his/her reward/loss as the result of an attack and the cost of carrying out his/her action. Essentially, both two players are with contradictory objectives and they are competing with each other. The objective of each player is to maximize his/her own total payoff with strategies $\mathbf{x}$ and $\mathbf{y}$

$$\mathcal{H}_1(\mathbf{x}, \mathbf{y}) = \sum_{N=1}^{\infty} \mathcal{R}_{1,N}(\mathbf{x}, \mathbf{y}) = \sum_{N=1}^{\infty} \mathbb{E}_{\mathbf{x}, \mathbf{y}}(\rho_N(s)\mathbf{G}_{\{1,s\}}|E > N) \cdot \mathbb{P}(E > N), \quad (6)$$

$$\mathcal{H}_2(\mathbf{x}, \mathbf{y}) = \sum_{N=1}^{\infty} \mathcal{R}_{2,N}(\mathbf{x}, \mathbf{y}) = \sum_{N=1}^{\infty} \mathbb{E}_{\mathbf{x}, \mathbf{y}}(\rho_N(s)\mathbf{G}_{\{2,s\}}|E > N) \cdot \mathbb{P}(E > N). \quad (7)$$

The reason why both the attacker and the defender share the same belief value $\rho_N(s)$ will be given out in Sect. 4.1. Another characteristic of the interaction between the defender and the attacker is the information asymmetry, where each player has private information about the state of the network system, while such private information among players is asymmetric. The asymmetry stems from the fact that the attacker has knowledge of a particular vulnerability which can be exploited; while the defender knows how to use resources to defend against all possible attacks. In other words, one player either deliberately distorts or does not disclose all the relevant information to another player, during their interaction phases. Since no player completely knows the exact state s of the game, it is assumed that each player (player 1 or player 2) observes a private local state $s_{\{1\}}$ or $s_{\{2\}}$ of the game and the state of the game is composed of both private local states $s = \{s_{\{1\}}, s_{\{2\}}\}$. Each player has to form a belief about the exact state s up to stage N. It is assumed that each player knows all *past* states that have occurred, which means when the game goes to next state, the previous one state will be publicly known to all players. Provided that the game has not ended, the history h_N is common information available to both players whereas private information is only available to that specific player.

According to [18], players can forget the sequence of previous states. So without loss of generality, it is assumed that the state of the two-player game at $N+1$ stage (assuming that the game does not end at N stage) evolves according to the current state s_N and all previous strategies from both players. Similarly, the private local state of each player is evolving according to the current local state $s_{\{1,N\}}$ for player 1 or $s_{\{2,N\}}$ for player 2 and all previous strategies from both players. It is obviously that, at any stage N, the local state $s_{\{1,N\}}$ for player 1 is independent of the local state $s_{\{2,N\}}$ for player 2. Therefore, when both players have taken actions $a \in AS_1$ and $b \in AS_2$, the state transition probability in the case of information asymmetry among players is defined as

$$q(s_N|s_{N-1}.a, b) := \mathbb{P}(s_N|s_{N-1}, a, b)$$
$$= \mathbb{P}(s_{\{1,N\}}|s_{\{1,N-1\}}, a, b) \cdot \mathbb{P}(s_{\{2,N\}}|s_{\{2,N-1\}}, a, b). \quad (8)$$

The choice of actions for each player at stage N may depend on all past strategies from both players and the player's current local state (the local state is one part of the game state $s_N = \{s_{\{1,N\}}, s_{\{2,N\}}\}$), which is consistent with Eqs. (3) and (4). Given the fact that no player can observe the current game state s_N ($s_N \in S$) at stage N and each player observes only a private local current game state $s_{\{1,N\}}$ or $s_{\{2,N\}}$, the probability for player 1 to choose action a and the probability for player 2 to choose action b at stage N are defined as

$$x_{s_{\{1,N\}},a} := \mathbb{P}(a|s_{\{1,N\}}, h_N) \tag{9}$$

and

$$y_{s_{\{2,N\}},b} := \mathbb{P}(b|s_{\{2,N\}}, h_N), \tag{10}$$

respectively.

It is to noteworthy that by knowing the strategy of the other player, one player can make inference about the other player's private information $s_{\{1,N\}}$ (if this player is player 2) or $s_{\{2,N\}}$ (if this player is player 1) from observing their actions. If a player knows the local private state of the other player, he/she can further predict the action of the other player in next stage. Provided that the game continues, state s_N is chosen according to $\rho_N(s_N) = \mathbb{P}(s_N|h_N, E > N)$, the immediate payoff $g_{\{1,s_N\}}(a_N, b_N)$ is received at player 1(correspondingly, $g_{\{2,s_N\}}(a_N, b_N)$ is received at player 2), and both two players computes his/her belief $\rho_{N+1}(s_{N+1})$ on next game state s_{N+1}.

4   Game Analysis

This section analyses the previously specified game model and finds Nash equilibria to construct an attack scenario in which the adversary cannot succeed in performing multistage cyber attacks and arriving at his/her ultimate target. In the previously specified game model, players have asymmetric information about the current state of the game, therefore, each player has to form a belief about the current state of the game. In SGLIOS with positive stop probabilities, player 1 (who can be assumed to be the defender) is informed about the belief value on the current game state but player 2 (who can be assumed to be the attacker) does not. Under the assumption that the true state of the game is independent of the action taken by player 2, the belief value in SGLIOS with positive stop probabilities is not conditional on the strategy taken by player 2 [18]. However, this assumption is not applicable in attacker-defender games where strategies from both player decide the state and the process of the game. Therefore, new belief system updating mechanisms should be described and belief system updates account for a central technical contribution in this paper. To assist equilibria computation for the designed attacker-defender stochastic game-theoretic model, this section first provides the belief update mechanism and then elaborates an easy-to-follow method for Nash equilibria computation.

4.1   Belief System Updates

Actions taken by both players can be summarized through a belief ρ_N of game states. For example, in SGLIOS with positive stop probabilities, under the assumption that the current state of the game is independent of player 2's actions, the belief ρ_N summarizes actions taken by player 1 [18]. In the game of asymmetric information, at stage N, the current game state is unknown to both players; player 1 privately observes a local state $s_{\{1,N\}}$ and player 2 privately observes another local state $s_{\{2,N\}}$. To consist with [18] and the recent work on stochastic game with asymmetric information [23,29], in this work, belief ρ_N on the current state s_N of the game is defined as $\rho_N(s_N) := \mathbb{P}(s_N | h_N, E > N)$.

Provided that the game does not end at N stage, which means the condition $\mathbb{P}(E > N)$ satisfies, for any history $h_N = \{(a_1, b_1), (a_2, b_2), \cdots, (a_{N-1}, b_{N-1})\}$, it can be observed that player's belief about the current game state s_N is

$$\rho_N(s_N) := \mathbb{P}(s_N | h_N)$$
$$= \mathbb{P}(s_{\{1,N\}}, s_{\{2,N\}} | h_N). \tag{11}$$

Because of the independence of private local states $s_{\{1,N\}}$ and $s_{\{2,N\}}$, Eq. (11) can be further written as

$$\rho_N(s_N) = \mathbb{P}(s_{\{1,N\}}, s_{\{2,N\}} | h_N) \tag{12}$$
$$= \mathbb{P}(s_{\{1,N\}} | h_N) \cdot \mathbb{P}(s_{\{2,N\}} | h_N).$$

The probability $\mathbb{P}(s_{\{1,N\}} | h_N)$ can be viewed as the probability that player 2 believes that player 1 will be in state $s_{\{1,N\}}$ based on the history h_N of past actions taken from both players. Player 2 might also derive this probability $\mathbb{P}(s_{\{1,N\}} | h_N)$ at N stage based on his/her private local states, however, since the private local states $s_{\{1,N\}}$ and $s_{\{2,N\}}$ ($N \in \mathbb{N}$) are independent, the probability $\mathbb{P}(s_{\{1,N\}} | h_N, s_{\{2,1\}}, s_{\{2,2\}}, \cdots, s_{\{2,N-1\}})$ would be the same as the probability $\mathbb{P}(s_{\{1,N\}} | h_N)$. Therefore, knowledge of private state information $(s_{\{2,1\}}, s_{\{2,2\}}, \cdots, s_{\{2,N-1\}})$ from player 2 does not affect the probability $\mathbb{P}(s_{\{1,N\}} | h_N)$. For player 2, the probability $\mathbb{P}(s_{\{2,N\}} | h_N)$ can be viewed as the probability that player 2 believes that his/her private local state at stage N is $s_{\{2,N\}}$ based on the history of actions from both players. It is to be noted that player 2 knows his current private local state $s_{\{2,N\}}$. However, this paper assumes that after taking any action and before arriving in state $s_{\{2,N\}}$, player 2 can also has a probability $\mathbb{P}(s_{\{2,N\}} | h_N)$ about his/her private local state $s_{\{2,N\}}$. Based on probabilities that player 1 will in state $s_{\{1,N\}}$ and he/she him/herself will be in state $s_{\{2,N\}}$ at stage N, player 2 can derive the probability $\rho_N(s_N)$ that the current game state is s_N at stage N. Similarly, player 1 can also derive the probability that player 2 will be in state $s_{\{2,N\}}$ at stage N with probability $\mathbb{P}(s_{\{2,N\}} | h_N)$ and the probability that he/she him/herself will be in state $s_{\{1,N\}}$ with probability $\mathbb{P}(s_{\{1,N\}} | h_N)$. Therefore, both players can obtain the same belief value that the game play is in state s_N at stage N.

4.2 Finding Nash Equilibria

When dealing with strategic players with inter-dependent payoffs (for example, the attacker's rewards might somehow be losses of the defender), investigating equilibria, mostly notably Nash equilibria, is a method of predicting players' decisions. If we restrict our attention to pure strategies (i.e., actions), a Nash equilibrium may not exists, this is the reason that this work considers only behaviour strategies and the probability used by both players to choose among pure strategies. The attacker-defender game with asymmetric information has finite states and the action spaces AS_1 and AS_2 are finite. The major differences between this attacker-defender game and the SGLIOS with positive stop probabilities are that this attacker-defender game is a non-zero-sum one and the belief system updates in this attacker-defender game are jointly conditioned on strategies from both players. In the SGLIOS with positive stop probabilities, the belief is conditioned only on the strategy of the informed player; while in the attacker-defender game, the belief is conditioned on strategies of both players. If the probability that taking action b_{N-1} is zero, the history h_N will not be observed, which will not happen under the assumption that the game does not end at $N-1$ stage. It was said that the belief in the SGLIOS with positive stop probabilities is continuous [17]. The same continuity property extends to the belief in the proposed attacker-defender game. In the designed attacker-defender game, both players are informed about the belief of game states. Hence, each player can be taken as the informed player in the SGLIOS with positive stop probabilities. It is proved in [18] that the informed player has a stationary optimal strategy. However, [18] does not provide a systematic way to find such optimal strategies.

The designed attacker-defender game is non-zero-sum. It is stated in [20] that every non-zero-sum stochastic game has at least one (not necessary unique) Nash equilibrium in stationary strategies and finding these equilibria is non-trivial. The attacker-defender game with uncertainty about current game state for both players makes it extremely challenging. Given the strategies of both players, players continue to accumulate the immediate payoffs. Once the end state of the game is reached, the game is over and no more accumulations are possible. Each player wishes to maximize his/her expected payoff at state s_N. This maximization, in turn, yields player's value of the game. Hence, if the value of the game Γ_N exists, let the vector of values for player 1 be $\mathbf{v}_1$, where $\mathbf{v}_1 = (v_{1,s_1}, v_{1,s_2}, \cdots, v_{1,s_\ell}, \cdots, v_{1,s_k})$ (v_{1,s_ℓ} is player 1's value of the game in state $s_{1,\ell}$ and $v_{1,s_\ell} \in \mathbb{R}$ ($\ell \in \{1, 2, \cdots, k\}$)) and the vector of values for player 2 be $\mathbf{v}_2$, where $\mathbf{v}_2 = (v_{2,s_1}, v_{2,s_2}, \cdots, v_{2,s_\ell}, \cdots, v_{2,s_k})$ (v_{2,s_ℓ} is player 2's value of the game in state s_ℓ and $v_{2,s_\ell} \in \mathbb{R}$ ($\ell \in \{1, 2, \cdots, k\}$)). The value of each player (either the attacker or the defender) includes both short-term (i.e., immediate) payoff and long-term payoff (which is given by the expected value of the sum of state payoffs from the current state) [4]. Taking the value for player 1 for instance, his/her value can be recursively defined as (that for player 2 can be defined in the same way)

$$v_{1,s_N}(\rho_N(s_N)) := \max_{\mathbf{x}_N} \min_{\mathbf{y}_N} \sum_{\mathbf{x}_N, \mathbf{y}_N} \left(\rho_N(s_N) \mathbf{G}_{\{1,s_N\}} + \mathbf{T}_1(s_N, \mathbf{v}) \right), \qquad (13)$$

where matrix $\mathbf{T}_1(s_N, \mathbf{v})$ is used to represent the long-term payoff (i.e., the future payoff) in a matrix form. The vector $\mathbf{v}$ is a value vector (a sub-vector of the game value vector that is defined above) for player 1 and it depends on the states that the current state s_N can transition to.

A pair of strategy sequence $(\mathbf{x}^*, \mathbf{y}^*)$ forms (Nash) equilibria with strategy pair $(\mathbf{x}_{s_N}^*, \mathbf{y}_{s_N}^*)$ if

$$\mathcal{H}_1(\mathbf{x}^*, \mathbf{y}^*) \geq \mathcal{H}_1(\mathbf{x}, \mathbf{y}^*), \forall \mathbf{x} \in \Omega^{m_1},$$
$$\mathcal{H}_2(\mathbf{x}^*, \mathbf{y}^*) \geq \mathcal{H}_2(\mathbf{x}^*, \mathbf{y}), \forall \mathbf{y} \in \Omega^{m_2},$$

where $\geq$ is used to mean at every stage N, the left-hand-side with strategy profile $(\mathbf{x}_{s_N}^*, \mathbf{y}_{s_N}^*)$ is greater than the right-hand-side with strategy $(\mathbf{x}_{s_N}, \mathbf{y}_{s_N}^*)$ or strategy $(\mathbf{x}_{s_N}^*, \mathbf{y}_{s_N})$. Therefore, the pair of strategy profile $(\mathbf{x}_{s_N}^*, \mathbf{y}_{s_N}^*)$ $(N \in \mathbb{N})$ is said to be a Nash equilibrium strategy. At this equilibrium, there is no incentive for either player to deviate from his/her equilibrium strategy $\mathbf{x}_{s_N}^*$ or $\mathbf{y}_{s_N}^*$ at any stage N of the game. In each pair of equilibrium strategies, a strategy for one player is a best-response to the other player and vice versa. A deviation means that one or both of them may have a lower expected payoff, i.e., $\mathcal{H}_1(\mathbf{x}, \mathbf{y}^*)$ or $\mathcal{H}_2(\mathbf{x}^*, \mathbf{y})$.

In order to find Nash equilibria for the designed attacker-defender non-zero-sum game in smart grid communication networks, based on the formed work [5,26], this paper studies nonlinear programming (NLP) formulation of the attacker-defender non-zero-sum stochastic game with finite number of strategies and asymmetric information. The theorem and proof of a global minimum to be a (Nash) equilibrium with equilibrium payoff can be found in [1,5], this work is not going to repeat them here again, whereas it provides here an easy-to-follow method to find such (Nash) equilibria in the designed attacker-defender game.

Assuming the game has M stage, where the game ends after the M stage (i.e., $E > M$, $M \geq 1$ and $M \in \mathbb{N}$). The equilibrium solution $(\mathbf{x}^*, \mathbf{y}^*)$ for M-stages games can be obtained by solving the following nonlinear programming problem:

$$\text{minimize} \quad \sum_{N=1}^{M-1} \left(v_{1,s_M} - \mathbf{x}_{s_M} \cdot \rho_M(s_M) \cdot \mathbf{G}_{1,s_M} \cdot \mathbf{y}_{s_M}^T + v_{2,s_M} - \mathbf{x}_{s_M} \cdot \rho_M(s_M) \cdot \right.$$
$$\mathbf{G}_{2,s_M} \cdot \mathbf{y}_{s_M}^T + v_{1,s_N} - \mathbf{x}_{s_N} \cdot (\rho_N(s_N) \mathbf{G}_{1,s_N} + \mathbf{T}_1(s_N, \mathbf{v})) \cdot \mathbf{y}_{s_N}^T$$
$$\left. + v_{2,s_N} - \mathbf{x}_{s_N} \cdot (\rho_N(s_N) \mathbf{G}_{2,s_N} + \mathbf{T}_2(s_N, \mathbf{v})) \cdot \mathbf{y}_{s_N}^T \right),$$

subject to

(i) $\rho_M(s_M)\mathbf{G}_{1,s_M}\mathbf{y}_{s_M}^T \le v_{1,s_M}\mathbf{J}_{m_1}^T$,

(ii) $\rho_M(s_M)\mathbf{G}_{2,s_M}^T\mathbf{x}_{s_M}^T \le v_{2,s_M}\mathbf{J}_{m_2}^T$,

(iii) $\rho_N(s_N)\mathbf{G}_{1,s_N}\mathbf{y}_{s_N}^T + \mathbf{T}_1(s_N,\mathbf{v})\mathbf{y}_{s_N}^T \le v_{1,s_N}\mathbf{J}_{m_1}^T, \forall N \in \{1,2,\cdots,M-1\}$,

(iv) $\rho_N(s_N)\mathbf{G}_{2,s_N}^T\mathbf{x}_{s_N}^T + \mathbf{T}_2(s_N,\mathbf{v})^T\mathbf{x}_{s_N}^T \le v_{2,s_N}\mathbf{J}_{m_2}^T, \forall N \in \{1,2,\cdots,M-1\}$,

(v) $\sum_{a=1}^{m_1} x_{s_N,a} = 1 \quad \forall a \in AS_1, N \in \{1,2,\cdots,M\}$,

(vi) $x_{x_N,a} \ge 0 \quad \forall a \in AS_1, N \in \{1,2,\cdots,M\}$,

(vii) $\sum_{b=1}^{m_2} y_{s_N,b} = 1 \quad \forall b \in AS_2, N \in \{1,2,\cdots,M\}$,

(viii) $y_{s_N,b} \ge 0 \quad \forall b \in AS_2, N \in \{1,2,\cdots,M\}$,

(ix) $\rho_N(s_N) = \mathbb{P}(s_N|h_N, E > M), N \in \{1,2,\cdots,M\}$.

Constraints (i) and (iv) are the value bounds for the attacker-defender game, which are satisfied for any pair of strategy profile. The mixed strategies $\mathbf{x}_{s_N}$ and $\mathbf{y}_{s_N}$ ($N = \{1,2,\cdots,M\}$) are defined in Eqs. (1) and (2), respectively. Constraints (v)–(viii) are conditions that the probability $x_{s_N,a}$ to select action a for player 1 in state s_N and the probability $y_{s_N,b}$ to select action b for player 2 in state s_N is greater than zero and the sum of all such probabilities for each player is one. Any pair of strategy profile satisfies constraints (v)–(viii). The constraint (ix) is a prior belief constraint and the belief ρ_1 for the first stage, which is presumed to be known to both players, is a probability distribution over state space S, i.e., $\rho_1 \in \Delta(S)$. Because of the recursion definition of belief values of constraint (ix) and the recursive optimization involved in the long-term payoff (i.e., $\mathbf{T}_1(s_N,\mathbf{v})$ or $\mathbf{T}_2(s_N,\mathbf{v})$) of constraints (iii) and (iv), it is non-trivial to find global minima.

In an one-stage game, each player (either the attacker or the defender) would play with the stationary strategy that maximizes his/her expected immediate payoff at the current game stage. Hence $(\mathbf{x}_{s_1}^*, \mathbf{y}_{s_1}^*)$ will be one optimal strategy profile. There can be mutiple stationary Nash equilibria in each game state and hence there will be multiple global minima. For example, for a stochastic game with one stage and the payoff matrix for player 1 (who has three actions: A, B and C) and player 2 (who has two actions: D and E) is $\mathbf{G}_{\{1,s_1\}}$ and $\mathbf{G}_{\{2,s_1\}}$, respectively (to be noted that those values in payoff matrices are artificial numbers for illustration)

$$
\mathbf{G}_{\{1,s_1\}} = \begin{array}{c|cc} & D & E \\ \hline A & 6 & 2 \\ B & 1 & 3 \\ C & 5 & 4 \end{array}, \text{ and } \quad \mathbf{G}_{\{2,s_1\}} = \begin{array}{c|cc} & D & E \\ \hline A & 4 & 3 \\ B & 1 & 5 \\ C & 2 & 2 \end{array}.
$$

Presuming that each player knows that the probability distribution $\rho_1(s_1)$ is 1, and the game value for player 1 (the row player) and player 2 (the column player) are denoted as v_{1,s_1} and v_{2,s_1}, respectively. Therefore, the nonlinear programming formulation of this one-stage game can be expressed as

$$\text{minimize} \left(v_{1,s_1} - \mathbf{x}_{s_1} \cdot \begin{bmatrix} 6 & 2 \\ 1 & 3 \\ 5 & 4 \end{bmatrix} \cdot \mathbf{y}_{s_1}^T + v_{2,s_1} - \mathbf{x}_{s_1} \cdot \begin{bmatrix} 4 & 3 \\ 1 & 5 \\ 2 & 2 \end{bmatrix} \cdot \mathbf{y}_{s_1}^T \right),$$

subject to

$$\text{(i)} \quad \begin{bmatrix} 6 & 2 \\ 1 & 3 \\ 5 & 4 \end{bmatrix} \mathbf{y}_{s_1}^T \leq v_{1,s_1} \begin{bmatrix} 1 & 1 & 1 \end{bmatrix}^T,$$

$$\text{(ii)} \quad \begin{bmatrix} 4 & 3 \\ 1 & 5 \\ 2 & 2 \end{bmatrix}^T \mathbf{x}_{s_1}^T \leq v_{2,s_1} \begin{bmatrix} 1 & 1 \end{bmatrix}^T,$$

$$\text{(iii)} \quad \sum_{a=1}^{3} x_{s_1,a} = 1 \quad \forall a \in \{A, B, C\},$$

$$\text{(iv)} \quad x_{s_1,a} \geq 0 \quad \forall a \in \{A, B, C\},$$

$$\text{(v)} \quad \sum_{b=1}^{2} x_{s_1,b} = 1 \quad \forall b \in \{D, E\},$$

$$\text{(vi)} \quad x_{s_1,b} \geq 0 \quad \forall b \in \{D, E\}.$$

There are three stationary mixed equilibria available for this one-stage game (by solving a constrained minimization problem), which are shown in Table 1 with their corresponding values for each player. All Nash equilibria and game values in Table 1 are further verified by the Gambit software tool [16]. Suppose that the first player is the defender of a system and the second player is the attacker. For the first Nash equilibrium in Table 1, to obtain maximum payoffs ("6" for the defender and "4" for the attacker, as shown in Table 1), the defender is suggested play the pure strategy "A" with a probability of 1 (i.e., play the action "A" in all game repetitions) and the attacker play the pure strategy "D" with a probability of 1. The same interpretation can be applied to the third Nash equilibrium, i.e., the defender plays the pure strategy "C" with a probability of 1 and the attacker plays the pure strategy "E" with a probability of 1 to maximise their payoffs. Regarding the second Nash equilibrium, the game suggests that the defender play his/her pure strategy "C" with a probability of 1, while it suggests that the the attacker play his/her pure strategy "D" with a probability of approximately 0.67 and his/her pure strategy "E" with a probability of approximately 0.33. If actions (i.e., pure strategies) are continuously and taking daily (24 h), the mixed Nash equilibrium strategy $\left(\dfrac{2}{3}, \dfrac{1}{3} \right)$ for the attacker can

Table 1. Nash equilibria and their corresponding game values in the sampled game.

# of Nash equilibrium	Player 1			Player 2		Game value	
	A	B	C	D	E	Player 1	Player 2
1	1	0	0	1	0	6	4
2	0	0	1	2/3	1/3	14/3	2
3	0	0	1	0	1	4	2

also be interpreted that the attacker temporarily runs the pure strategy "D" for approximately 16 h and runs the pure strategy "E" for the remainder of the day. If the actions "D" and "E" are instantaneous actions (which are taken at discrete time instants), the mixed Nash equilibrium strategy $\left(\frac{2}{3}, \frac{1}{3}\right)$ for the attacker can be interpreted as the (asymptotic) frequency with which the strategies "D" and "E" are chosen in the game. After obtaining the mixed Nash equilibrium, the defender and the attacker can subsequently use it in the following way: when the game begins, both players (the defender and the attacker) randomly choose actions (i.e., pure strategies) from their corresponding action spaces, a game payoff from the chosen action pair will be received at each player. When the game is played again, both players again randomly choose actions from their corresponding action spaces in this round. It is to be noted that the actions from both players in this round may be different from that taken in the previous one. A game payoff will again be received at each player. The actions in each round are chosen randomly, however, the player should be aware of that the (asymptotic) frequency of chosen actions must be that suggested from the mixed Nash equilibrium. Therefore, when averaging payoffs in all repetitions of the game, the average payoff is optimal for each player only if the actions are chosen with their frequencies that are prescribed by the equilibrium strategy. For example, for the attacker, in any game round, he/she should always aware of that the (asymptotic) frequency of choosing actions "D" and "E" in all game repetitions should be $\frac{2}{3}$ and $\frac{1}{3}$, respectively.

5   Conclusion and Future Work

To assess the threat of multistage cyber attacks in smart grid communication networks, this paper designs a stochastic game-theoretic model according to the characteristics of the interactions between the attacker and the defender in smart grid use cases. Firstly, the majority of the existing game-theoretic threat and risk assessment models are reviewed. Then, this paper elaborates players and state transition probabilities of the designed stochastic game-theoretic model. Since each player has partial knowledge of the game state, a belief-updating mechanism for both players to form a common belief about the current state of the game is proposed. Moreover, this paper discusses the use of nonlinear programming for Nash equilibria computation. One important aim of future work is the application of the proposed stochastic game-theoretic model to evaluate a multistage cyber attack scenario. Additionally, cyber attacks can also introduce disruptive events in power grids. Therefore, further studies of payoff formulation with an understanding of cascading effects of multistage cyber attacks would be of great significance.

Acknowledgments. The research leading to the results presented in this paper was supported by the European Commission's Project No. 608090, HyRiM (Hybrid Risk Management for Utility Networks) under the 7th Framework Programme (FP7-SEC-2013-1). The authors acknowledge Stefan Rass from Alpen-Adria-Universität Klagenfurt for his invaluable discussions and comments.

References

1. Barron, E.N.: Game Theory: An Introduction. Wiley, Hoboken (2007)
2. Chen, L., Leneutre, J.: Fight jamming with jamming – a game theoretic analysis of jamming attack in wireless networks and defense strategy. J. Comput. Netw.: Int. J. Comput. Telecommun. Network. **55**(9), 2259–2270 (2011)
3. European Union Agency for Network and Information Security. ENISA smart grid security recommendations. Technical report, European Union Agency for Network and Information Security (2012)
4. Feinberg, E.A., Shwartz, A. (eds.): Handbook of Markov Decision Processes: Methods and Applications, vol. 40. Springer US, New York (2002). doi:10.1007/978-1-4615-0805-2
5. Filar, J.A., Schultz, T.A., Thuijsman, F., Vrieze, O.J.: Nonlinear programming and stationary equilibria in stochastic games. Math. Program. **50**(1), 227–237 (1991)
6. Le Guillarme, N., Mouaddib, A-I., Gatepaille, S., Bellenger, A.: Adversarial intention recognition as inverse game-theoretic plan for threat assessment. In: IEEE 28th International Conference on Tools with Artificial Intelligence, January 2017
7. Hamman, S.T., Hopkinson, K.M., McCarty, L.A.: Applying Behavioral game theory to cyber-physical systems protection planning. In: Cyber-Physcial Systems: Foundations, Principles and Applications, pp. 251–264. Elsevier (2017)
8. He, W., Xia, C., Wang, H., Zhang, C., Ji, Y.: A game theoretical attack-defense model oriented to network security risk assessment. In: 2008 International Conference on Computer Science and Software Engineering. IEEE (2008)
9. He, X., Sui, Z., de Meer, H.: Game-theoretic risk assessment in communication networks. In: IEEE 16th International Conference on Environment and Electrical Engieering (EEEIC). IEEE, June 2016
10. Ismail, Z., Leneutre, J., Bateman, D., Chen, L.: A methodology to apply a game theoretic model of security risks interdependencies between ICT and electric infrastructures. In: Zhu, Q., Alpcan, T., Panaousis, E., Tambe, M., Casey, W. (eds.) GameSec 2016. LNCS, vol. 9996, pp. 159–171. Springer, Cham (2016). doi:10.1007/978-3-319-47413-7_10
11. Jones, M.G.: Asymmetric information games and cyber security. Ph.D. Thesis, Georgia Institute of Technology (2013)
12. Kuhn, H.W.: Extensive games and the problem of information. Ann. Math. Stud. **28**(28), 193–216 (1953)

13. Liang, X., Xiao, Y.: Game theory for network security. IEEE Commun. Surv. Tutorials **15**(1), 472–486 (2013)
14. Lye, K.-W., Wing, J.M.: Game strategies in network security. Int. J. Inf. Secur. **4**(1–2), 71–86 (2005)
15. Manshaei, M.H., Zhu, Q., Alpcan, T., Başar, T., Hubaux, J.-P.: Game theory meets network security and privacy. J. ACM Comput. Surv. (CSUR) **45**(3), 25:1–25:39 (2013)
16. Mckelvey, R.D., McLennan, A.M., Turocy, T.L.: Gambit: software tools for game theory, Version 14.1.0 (2014). http://www.gambit-project.org. Accessed 04 June 2017
17. Melolidakis, C.: On stochastic games with lack of information on one side. Int. J. Game Theory **18**(1), 1–29 (1989)
18. Melolidakis, C.: Stochastic games with lack of information on one side and positive stop probabilities. In: Raghavan, T.E.S., Ferguson, T.S., Parthasarathy, T., Vrieze, O.J. (eds.) Stochastic Games and Related Topics. TDLC, vol. 7, pp. 113–126. Springer, Netherlands (1991). doi:10.1007/978-94-011-3760-7_10
19. Miura-Ko, R.A., Yolken, B., Bambos, N., Mitchell, J.: Security investment games of interdependent organizations. In: 46th Annual Allerton Conference on Communication, Control, and Computing. IEEE (2009)
20. Nash, J.: Non-cooperative Games. Ph.D. Thesis, Princeton University (1950)
21. Nguyen, K.C., Alpcan, T., Başar, T.: Stochastic games for security in networks with interdependent nodes. In: International Conference on Game Theory for Networks. IEEE, June 2009
22. Osborne, M.J., Rubinstein, A.: A Course in Game Theory. MIT Press, Cambridge (1994)
23. Ouyang, Y.: On the interaction of information and decision in dynamic network systems. Ph.D. Thesis, University of Michigan (2016)
24. Rass, S., König, S., Schauer, S.: Defending against advanced persistent threats using game-theory. PLoS ONE **12**(1), 1–43 (2017)
25. Rass, S., Zhu, Q.: GADAPT: a sequential game-theoretic framework for designing defense-in-depth strategies against advanced persistent threats. In: Zhu, Q., Alpcan, T., Panaousis, E., Tambe, M., Casey, W. (eds.) GameSec 2016. LNCS, vol. 9996, pp. 314–326. Springer, Cham (2016). doi:10.1007/978-3-319-47413-7_18
26. Rothblum, U.G.: Solving stopping stochastic games by maximizing a linear function subject to quadratic constraints. In: Game Theory and Related Topics, pp. 103–105 (1979)
27. Sallhammar, K., Knapskog, S.J.: Using game theory in stochastic models for quantifying security. In: The 9th Nordic Workshop on Secure IT-Systems (2004)
28. Shapley, S.L.: Stochastic games. Proc. Natl. Acad. Sci. U.S.A. **39**(10), 1095–1100 (1953)
29. Vasal, D.: Dynamic decision problems with cooperative and strategic agents and asymmetric information. Ph.D. Thesis, University of Michigan (2016)
30. Zhang, H., Jiang, W., Tian, Z., Song, X.: A stochastic game theoretic approach to attack prediction and optimal active defense strategy decision. In: International Conference on Networking, Sensing and Control (ICNSC), pp. 648–653. IEEE (2008)

A Stackelberg Game and Markov Modeling of Moving Target Defense

Xiaotao Feng[1]([✉]), Zizhan Zheng[2], Prasant Mohapatra[3], and Derya Cansever[4]

[1] Department of Electrical and Computer Engineering, University of California, Davis, Davis, USA
xtfeng@ucdavis.edu
[2] Department of Computer Science, Tulane University, New Orleans, USA
zzheng3@tulane.edu
[3] Department of Computer Science, University of California, Davis, Davis, USA
pmohapatra@ucdavis.edu
[4] U.S. Army Research Laboratory, Adelphi, USA
derya.h.cansever.civ@mail.mil

Abstract. We propose a Stackelberg game model for Moving Target Defense (MTD) where the defender periodically switches the state of a security sensitive resource to make it difficult for the attacker to identify the real configurations of the resource. Our model can incorporate various information structures. In this work, we focus on the worst-case scenario from the defender's perspective where the attacker can observe the previous configurations used by the defender. This is a reasonable assumption especially when the attacker is sophisticated and persistent. By formulating the defender's problem as a Markov Decision Process (MDP), we prove that the optimal switching strategy has a simple structure and derive an efficient value iteration algorithm to solve the MDP. We further study the case where the set of feasible switches can be modeled as a regular graph, where we solve the optimal strategy in an explicit way and derive various insights about how the node degree, graph size, and switching cost affect the MTD strategy. These observations are further verified on random graphs empirically.

1 Introduction

In cybersecurity, it is often the case that an attacker knows more about a defender than the defender knows the attacker, which is one of the major obstacles to achieve effective defense. Such information asymmetry is a consequence of time asymmetry, as the attacker often has abundant time to observe the defender's behavior while remaining stealthy. This is especially the case for incentive-driven targeted attacks, such as Advanced Persistent Threats (APT). These attacks are highly motivated and persistent in achieving their goals. To this end, they may intentionally act in a "low-and-slow" fashion to avoid immediate detection [1].

Recognizing the shortage of traditional cyber-defense techniques in the face of advanced attacks, Moving Target Defense (MTD) has been recently proposed

© Springer International Publishing AG 2017
S. Rass et al. (Eds.): GameSec 2017, LNCS 10575, pp. 315–335, 2017.
DOI: 10.1007/978-3-319-68711-7_17

as a promising approach to reverse the asymmetry in information or time in cybersecurity [2]. MTD is built upon the key observation that to achieve a successful compromise, an attacker requires knowledge about the system configuration to identify vulnerabilities that he is able to exploit. However, the system configuration is under the control of the defender, and multiple configurations may serve the system's goal, albeit with different performance security tradeoffs. Thus, the defender can periodically switch between configurations to increase the attacker's uncertainty, which in turn increases attack cost/complexity and reduces the chance of a successful exploit in a given amount of time. This high level idea has been applied to exploit the diversity and randomness in various domains, including computer networks [3], system platforms [4], runtime environment, software code, and data representation [5].

Early work on MTD mainly focus on empirically studies of domain specific dynamic configuration techniques. More recently, decision and game theoretic approaches have been proposed to reason about the incentives and strategic behavior in cybersecurity to help derive more efficient MTD strategies. In particular, a stochastic game model for MTD is proposed in [6], where in each round, each player takes an action and receives a payoff depending on their joint actions and the current system state, and the latter evolves according to the joint actions and a Markov model. Although this model is general enough to capture various types of configurations and information structures and can be used to derive adaptive MTD strategies, solutions obtained are often complicated, making it difficult to derive useful insights for practical deployment of MTD. Moreover, existing stochastic game models for MTD focus on Nash Equilibrium based solutions and do not exploit the power of commitment for the defender. To this end, Bayesian Stackelberg games (BSG) has been adapted to MTD recently [7]. In this model, before the game starts, the defender commits to a mixed strategy – a probability distribution over configurations – and declare it to the attacker, assuming the latter will adopt a best response to this randomized strategy. Note that, the defender's mixed strategy is independent of real time system states, so does the attacker's response. Thus, a BSG can be considered as a repeated game without dynamic feedback. Due to its simplicity, efficient algorithms have been developed to solve BSG in various settings, with broad applications in both physical and cyber security scenarios [8]. However, a direct application of BSG to MTD as in [8] ignores the fact that both the attacker and the defender can adapt their strategies according to the observations obtained during the game.

In this paper, we propose a non-zero-sum Stackelberg game model for MTD that incorporates real time states and observations. Specifically, we model the defender's strategy as a set of transition probabilities between configurations. Before the game starts, the defender declares its strategy to the attacker. Both players take rounds to make decisions and moves. In the beginning of each round, the defender moves from the current configuration to a new one (or stay on the current one) according to the transition probabilities. Note that this is more general than [8], where the defender picks the next configuration independently of the current one. Our approach also allows us to model the long-term switching cost in a more accurate way. Moreover, we assume that the attacker can get some feedback during the game. This is especially true for advanced attacks. In this paper, we consider the extreme case where the attacker knows the previous configuration used by the defender in the beginning of each round (even if it fails in the previous round). This is the worst-case scenario from the defender's perspective. However, our model can be readily extended to settings where the attacker gets partial feedback or no feedback.

To derive the optimal MTD strategy for the defender, we model the defender's problem as a Markov decision process (MDP). Under the assumptions that all the configurations have the same value to the defender and require the same amount of effort to compromise for the attacker, we prove that the optimal stationary strategy has a simple structure. Based on this observation, we derive efficient value iteration algorithm to solve the MDP. We further study the case where the switching cost between any pair of configurations is either a unit or infinite. In this case, the configuration space can be modeled as a directed graph. When the graph is regular, we derive the optimal strategy in an explicit way and prove that it is always better to have a higher degree in the graph, but the marginal improvement decreases when the diversity increases. This observation is further verified on random graphs empirically.

We have made the following contributions in this paper

- We propose a Stackelberg game model for moving target defense that combines Markovian defense strategies and realtime feedback.
- We model the defender's problem as a Markov decision process and derive efficient algorithms based on some unique structural properties of the game.
- We derive various insights on efficient MTD strategies using our models. In particular, we study how the diversity of the configuration space affects the effectiveness of MTD, both analytically and empirically.

The remainder of the paper is organized as follows. We introduce the related work in Sect. 2 and propose the game model in Sect. 3. Detailed solutions for optimal strategies and a special case study are presented in Sect. 4. The performance of optimal strategies under different scenarios are evaluated via numerical study in Sect. 5. Finally, we conclude the paper in Sect. 6.

2   Related Work

As a promising approach to achieve proactive defense, MTD techniques have been investigated in various cybersecurity scenarios [2–5]. A fundamental challenge of large scale deployment of MTD, however, is to strike a balance between the risk of being attacked and the extra cost introduced by MTD including the extra resource added, the migration costs and the time overhead. To this end, game theory provides a proper framework to analyze and evaluate the key tradeoffs involved in MTD [9].

In this paper, we propose a non-zero-sum Stackelberg game model for MTD where the defender plays as the leader and the attacker plays as the follower and both players make their decision sequentially. Sequential decision making with limited feedback naturally models many security scenarios. Recently, inspired by poker games, an efficient sub-optimal solution for a class of normal-form games with sequential strategies is proposed in [10]. However, the solution is only applicable to zero-sum games, while the MTD game is typically non-zero-sum as the defender usually has a non-zero migration cost.

Stackelberg game models have been extensively studied in cybersecurity as they capture the fact that a targeted attacker may observe a finite number of defender's actions and then estimate the defender's strategy [11]. This is especially true for an APT attacker. By exercising the power of commitment, the defender (leader) can take advantages of being observed to alert the attacker.

In the context of MTD, several Stackelberg game models have been proposed [7,8,12]. In particular, a Stackelberg game is proposed for dynamic platform defense against uncertain threat types [7]. However, this work does not consider the moving cost for platform transitions, which should be taken into consideration on strategy design. A Stackelberg game for MTD against stealthy attacks is proposed in [12], where it is shown that MTD can be further improved through strategic information disclosure. One limitation of this work is that the authors only consider a one-round game.

More recently, a Bayesian Stackelberg Game (BSG) model is proposed for MTD in Web applications [8], where multiple types of attackers with different expertise and preferences are considered. Both theoretical analysis and experimental studies are given in [8]. However, to adapt the classic BSG model to MTD, the defender's strategy is defined as a probability distribution over states and is *i.i.d.* over rounds, which is a strong limitation. In contrast, we defined the defender's strategy as the set of transition probabilities between states. Such a Markovian strategy is not only more natural in the context of MTD, but also allows us to incorporate real time feedback available to the players.

Our model is similar in spirit to stochastic game models [6] and recent Markov models for MTD [13,14]. However, existing stochastic game models for MTD focus on Nash Equilibria instead of Stackelberg Equilibria. Moreover, solutions to stochastic games are often complicated and hard to interpret. More recently, several Markov models for MTD have been proposed [13,14]. Due to the complexity of these models, only preliminary analytic results for some special cases are provided. In particular, these work focus on analyzing the expected time needed for the attacker to compromise the resource under some simple defense strategies.

3   Game Model

In this section, we formally present our MTD game model. There are two players in the game who fight for a security sensitive resource. The one who protects the resource is called the defender while the one who tries to comprise the resource is called the attacker. Below we discuss each element of the game model in details.

Resource: We consider a single resource with N features, where for the i-th feature, there are m_i possible configurations that can be chosen by the defender, denoted by $\mathbf{c}_i$ with $|\mathbf{c}_i| = m_i$. We define the *state* of the resource at any time as the set of configurations of all the features, $s = \{c_i \in \mathbf{c}_i, i = 1, 2, \cdots, N\}$. For example, the resource can represent a critical cyber system with features such as its processor architecture, operating system, storage system, virtual machine instances, network address space, and communication channels, etc. Each feature has several possible configurations such as Windows/Linux for operating system, a range of IP addresses for network address space and so on. Moreover, the concept of resource is not limited to the cyber world. It can also represent physical entities such as military units, vulnerable species, and antiques.

We define a state as *valid* if it is achievable by the defender and the resource can function properly under that state. Although the maximum possible states of the resource can be $\prod_{i=1}^{N} m_i$, typically only a small number of them are valid. For instance, consider a mobile app that with two features: {program language $\in$ {Objective-C, Java, JavaScript}, operating system $\in$ {iOS, Android}}. The maximum number of states for the app is 6. However, since a Java based app is incompatible with iOS, and an Objective-C based app is incompatible with Android, there are only 4 valid states. We denote the set of valid states as $V = \{1, 2, \cdots, |V|\}$.

Defender: To protect the resource, the defender periodically switches the state to make it difficult for the attacker to identify the real state of the resource. A switch is achieved by changing the configurations of one or more features and is subject to a cost. Note that not all the switches between valid states are feasible as it can be extremely difficult or even impossible to switch between two valid states in some cases.

Attacker: We assume that the attacker can potentially attack all the valid states of the resource. Note that if the defender knows that the attacker does not have technical expertise to attack certain states, then the defender should always keep the resource in those states. We leave the case where the defender is uncertain about the attacker's capability in the future work.

Before each attack, the attacker selects an attack scheme that targets at a specific configuration combination (state) of the resource. We assume that the attacker can compromise the resource successfully if and only if the selected attack scheme matches the real state of the resource. Due to this 1-1 correspondence, we simply define the attacker's action space as the set of valid states V. We further assume that the attacker can only observe and exploit the state of the resource but cannot modify it through successful attacks. That is, the state of the resource is completely under the control of the defender.

The rules of the MTD game are introduced below.

1. The game is a turn based Stackelberg Game in which the defender plays as the leader and the attacker plays as the follower.
2. The game starts at turn $t = 0$ with the resource initially in state $s_0 \in V$ (chosen by the defender), and lasts for a possibly infinite number of turns T.

3. Each turn begins when the defender takes action. We assume that the defender moves periodically and normalize the length of each turn to a unit.
4. At the beginning of turn t, the defender switches the resource from s_t to s_{t+1} with a switching cost $c_{s_t s_{t+1}}$, and the attacker selects one state $a_t \in V$ to attack. We assume that the attacker attacks once each turn. Moreover, both switching and attacking are effective instantly.
5. If the attacker is successful at turn t (that is, if $a_t = s_{t+1}$), he obtains a reward of 1, while the defender incurs a loss of 1 (not including the switching cost). Otherwise, there is no reward obtained or loss incurred.

A Graphical View: We can model the set of states and state switches as a directed graph. For example, Fig. 1a shows a fully connected graph with the set of states as nodes and state switches as links. We then eliminate some invalid states and invalid switches to get Fig. 1b. The defender chooses one node as initial state s_0 at the beginning of the game. The attacker selects one node a_t as the target in each turn. Every valid state has a self loop meaning that no switch is always one option for the defender. We define the *outdegree* (or *degree* for short) of a node as the number of outgoing links from the node, or equivalently, the number of states that can be switched to from the state. We define the *neighbor* of state s as a set $N(s) = \{s' \in V | c_{ss'} \neq \infty\}$, $\forall s \in V$. The degree of node s is equal to $|N(s)|$.

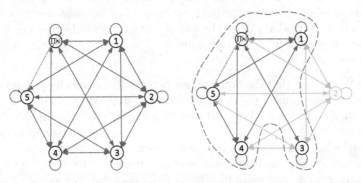

(a) A fully connected graph   (b) A subgraph after elimination of invalid states and valid links

Fig. 1. All the possible switch pairs modeled by a graph

The graph can be uniquely determined by V and a matrix $C = \{c_{ss'}\}_{|V| \times |V|}$, where $c_{ss'}$ represents the switching cost between two states s and s'. There is no link between s and s' if $c_{ss'} = \infty$, and $c_{ss'} = 0$ if $s' = s$. We expect that the switching costs can be learned from history data and domain knowledge [8].

Consider again the example given above. There are four valid states corresponding to four nodes. Let nodes $1, 2, 3$ and 4 represent {Objective-C, iOS},

{JavaScript, iOS}, {JavaScript, Android} and {Java, Android}, respectively. An example of the cost matrix $\mathcal{C}$ and the corresponding graph are given in Fig. 2. In this example, if the current state of the resource is at node 1, the defender may keep the state at node 1 without any expense, or switch the state from node 1 to node 2 or node 3 with a switching cost 0.8 and 1.5, respectively. However, the defender cannot switch the resource from node 1 to node 4 in one step as there is no direct link between them.

$$
\mathcal{C} = \begin{pmatrix}
0 & 0.8 & 1.5 & \infty \\
0.7 & 0 & 0.6 & 1.6 \\
1.3 & 0.5 & 0 & 0.4 \\
\infty & 1.2 & 0.4 & 0
\end{pmatrix}
$$

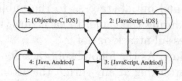

Fig. 2. A resource with 4 states and 14 switch pairs

3.1 Attacker's Strategy

We define the attacker's strategy and payoff in this subsection. In order to decide a_t, the attacker forms a prior belief $\mathbf{q}_t = \{q_s \mid s \in V\}$ regarding the probability distribution of states according to the feedback obtained during the game and the previous actions (to be discussed). For the sake of simplicity, we assume that the attacking cost is identical for all the states and it is always beneficial to attack. Thus, the attacker always selects $a_t = \arg\max_{s \in V} q_s$ at turn t.

3.2 Defender's Strategy and Cost

The defender's objective is to strike a balance between the loss from attacks and the cost of switching states. To this end, the defender commits to a strategy and declares it to the attacker before the game starts. As in Bayesian Stackelberg Games, the defender should adopt a randomized strategy taking into account the possible response of the attacker. In this work, we define the defender's strategy as a set of transition probabilities $P = \{p_{ss'}\}_{|V| \times |V|}$, where $p_{ss'}$ is the probability of switching the resource to s' given that the current state is s. The defender commits to an optimal P in the beginning and then samples the state in each turn according to P. We require that $p_{ss'} = 0$ if $c_{ss'} = \infty$ and $\sum_{s' \in V} p_{ss'} = 1$, $\forall s \in V$. Given a pair of states s_t, s_{t+1}, the defender's cost at turn t can be then defined as follows:

$$
c(s_t, s_{t+1}) = 1_{\{a_t = s_{t+1}\}} + c_{s_t s_{t+1}} \tag{1}
$$

The first term in (1) represents the loss from being attacked where $1_{\{a_t = s_{t+1}\}} = 1$ if $a_t = s_{t+1}$ and is 0 otherwise. The second term depicts the switching cost.

3.3 Feedback During the Game

The main purpose of MTD is to reverse information asymmetry. Thus, it is critical to define the information structure of the game. We assume that both players know the defender's strategy and all the information about the resource such as V and C before the game starts. However, the players have different feedback during the game:

- **Defender:** As the leader of Stackelberg game, the defender declares her strategy P and initial state s_0 to the public. The defender would not change P and C during the game. In each turn, the defender knows if the attacker has a successful attack or not.
- **Attacker:** As the follower of Stackelberg game, the attacker knows P and s_0. After attacking at any turn t, the attacker knows if the attack is successful or not. If the attack is successful, the attacker knows s_t immediately. Otherwise, we assume that the attacker spends this turn to learn s_t and will know s_t at the end of this turn. In both cases, $\mathbf{q}_t = \mathbf{p}_{s_t}$, where $\mathbf{p}_{s_t}$ represents the s_t-th row in P. This is the worst-case scenario from the defender's perspective. We will leave the case where attacker only gets partial feedback or no feedback to the future work.

3.4 Defender's Problem as a Markov Decision Process

Given the feedback structure defined above, we have $a_t = \mathrm{argmax}_{s \in V} p_{s_t s}$ for any t. Hence, the defender's expected loss at turn t is:

$$E\left[1_{\{a_t = s_{t+1}\}}\right] = E\left[1_{\{s_{t+1} = \mathrm{argmax}_{s \in V} p_{s_t s}\}}\right] = \max \mathbf{p}_{s_t} \tag{2}$$

Therefore, given P and s_t, the defender's expected cost at turn t is

$$c_P(s_t) \triangleq E_{s_{t+1}}\left[c(s_t, s_{t+1})\right]$$
$$= \max \mathbf{p}_{s_t} + \sum_{s_{t+1} \in N(s_t)} p_{s_t s_{t+1}} c_{s_t s_{t+1}} \tag{3}$$

In this work, we consider the defender's objective to be minimizing its long-term discounted cost defined as $\sum_{t=0}^{\infty} \alpha^t c(s_t)$ where $\alpha \in (0, 1)$ is the discounted factor. One interpretation of α is that the defender would prefer to minimize the cost at current turn rather than future turns because she is not sure if the attacker will attack at the next turn. A higher discount factor indicates that the defender is more patient.

For a given P and an initial state s_0, the state of the resource involves according to a Markov chain with V as its state space and P as the transition probabilities. Thus, the defender's problem can be considered as a discounted

Markov decision problem where the defender's strategy and the transition probabilities coincide. We can rewrite the defender's long-term cost with the initial state $s_0 = s$ as follows:

$$
\begin{aligned}
C_P(s) &= \sum_{t=0}^{\infty} c_P(s_t) \\
&= c_P(s) + \alpha \sum_{s' \in N(s)} p_{ss'} E\left[\sum_{t=0}^{\infty} \alpha^t c(s_{t+1}, s_{t+2}) \mid s_1 = s' \right] \\
&= c_P(s) + \alpha \sum_{s' \in N(s)} p_{ss'} C_P(s')
\end{aligned}
\tag{4}
$$

3.5 Discussion About the MTD Model

In the BSG model for MTD in [8], the defender's strategy is defined as a probability distribution $\mathbf{x} = \{x_s \mid \forall s \in V\}$ over states, and the expected switching cost is defined as $\sum_{s,s' \in V} c_{ss'} x_s x_{s'}$. This model implies that at each turn, the defender samples the next state independent of the current state of the resource. In contrast, we define the defender's strategy as a set of transition probabilities between states. Our choice is not only more natural for MTD, but also considers a richer set of defense strategies. Note that different transition probability matrices may lead to the same stationary distribution of states, but with different switching costs, which cannot be distinguished using the formulation in [8]. Our approach provides a more accurate definition of the defender's real cost. We show that by modeling the problem as a MDP, we can still find the optimal defense strategy in this more general setting. Moreover, the MDP can be solved in an explicit way under certain system settings, which provides useful insights to the design of MTD strategies, as we discuss below.

4 Defender's Optimal Strategy and Cost

In this section, we solve the defender's optimal strategy as well as the optimal cost under different scenarios. Recall that the defender's problem is to find a strategy such that the cost in (4) is minimized from any initial state. Let $C^*(s)$ denote the defender's optimal cost with an initial state s, where

$$
C^*(s) = \min_P C_P(s)
\tag{5}
$$

According to the theory of MDP, it is possible to find an optimal strategy P^* that simultaneously optimizes the cost for any initial state $s \in V$; that is,

$$
P^* = \mathrm{argmin}_P C_P(s), \forall s \in V
\tag{6}
$$

4.1 Algorithms for Solving the MDP

According to (3) and (4), we expand $C_P(s)$ in (5) and rewrite $C^*(s)$ in the following form,

$$C^*(s) = \min_P \left[\max \mathbf{p}_s + \sum_{s' \in N(s)} (c_{ss'} + \alpha C_P(s')) p_{ss'} \right] \tag{7}$$

In order to solve (7), we employ the standard value iteration algorithm to find the defender's optimal cost as well as the optimal strategy. Algorithm 1 shows the value iteration algorithm, where $C^\tau(s)$ is the cost at state s in the τ−th iteration. Initially, the value of $C^\tau(s)$ is set to 0 for all s. In each iteration, the algorithm updates $C^\tau(s)$ by finding the optimal strategy that solves (7) using the costs in the previous iteration (step 1), which involves solving a Min-Max problem.

Although the value iteration algorithm is standard, solving the Min-Max problem in step 1 of Algorithm 1 directly is computationally expensive. Note that the decision variables $p_{ss'}$ can take any real value in $[0, 1]$. One way to solve the problem is to approximate the search space $[0, 1]$ by a discrete set $\{0, \frac{1}{M}, \frac{2}{M}, ..., \frac{M-1}{M}, 1\}$ where M is a parameter. The search space over all the neighbors of s has a size of $O(M^{|V|})$. A suboptimal solution can be obtained by searching over this space, which is expensive when M and $|V|$ are large. Rather than solving it directly, we first derive some properties of the MDP, which helps reduce the computational complexity significantly.

Algorithm 1. Value Iteration Algorithm for the MTD game

Input: V, $\mathcal{C}$, α, ϵ.
Output: P^*, $C^*(s)$.
1: Set $\tau = 0$, $C^\tau(s) = 0$, $\forall s \in V$; $\{C^\tau(s)$ is the cost at state s in the τ−th iteration$\}$
2: **repeat**
3:     $\tau = \tau + 1$;
4:     $\mathbf{p}_s^* = \text{argmin}_{\mathbf{p}_s} \left[\max \mathbf{p}_s + \sum_{s' \in N(s)} p_{ss'} \left(c_{ss'} + \alpha C^{\tau-1}(s') \right) \right]$, $\forall s \in V$;
5:     $C^\tau(s) = C_{P^*}(s)$, $\forall s \in V$;
6: **until** $\sum_{s \in V} |C^\tau(s) - C^{\tau-1}(s)| \leq \epsilon$
7: $C^*(s) = C^\tau(s)$, $\forall s \in V$

Before presenting the results, we first give some definitions. Fix a state s. For any $s' \in N(s)$, let $\theta_{s'} = c_{ss'} + \alpha C^{\tau-1}(s')$ denote the coefficient of $p_{ss'}$ in the second term of the Min-Max problem in the τ-th iteration. Let $s^1, s^2, ..., s^{N(s)}$ denote the set of neighbors of s sorted according to their θ values nondecreasingly. We abuse the notation a little bit and let $\theta_i = \theta_{s^i}$.

The following lemma shows that the Min-Max problem can be simplified as a minimization problem.

Lemma 1. *Let P be the optimal solution to the Min-Max problem in the τ-th iteration of Algorithm 1. We have $p_{ss^1} = \max \mathbf{p}_s$.*

Proof. Assume $p_{ss^1} < \mathbf{p}_s$. Let $p_{ss^i} = \max \mathbf{p}_s$ for some $s^i \in N(s)$ and $p_{ss^i} = p_{ss^1} + \epsilon_1$ for some $\epsilon_1 > 0$. By the definition of s^1, there is $\epsilon_2 \geq 0$ such that $\theta_i = \theta_1 + \epsilon_2$. From the definition of P and s^i, we have

$$G^\tau(s) = p_{ss^i} + \sum_{s^j \in N(s)} p_{ss^j} \theta_j$$

$$= p_{ss^1} \theta_1 + p_{ss^i} (1 + \theta_i) + \sum_{s^j \in N(s) \setminus \{s^1, s^i\}} p_{ss^j} \theta_j$$

$$= p_{ss^1} \theta_1 + (p_{ss^1} + \epsilon_1)(1 + \theta_1 + \epsilon_2) + \sum_{s^j \in N(s) \setminus \{s^1, s^i\}} p_{ss^j} \theta_j$$

$$> (p_{ss^1} + \epsilon_1)(1 + \theta_1) + p_{ss^1}(\theta_1 + \epsilon_2) + \sum_{s^j \in N(s) \setminus \{s^1, s^i\}} p_{ss^j} \theta_j$$

$$= p_{ss^i}(1 + \theta_1) + p_{ss^1} \theta_i + \sum_{s^j \in N(s) \setminus \{s^1, s^i\}} p_{ss^j} \theta_j \tag{8}$$

The value in (8) can be obtained by a strategy P' that switches the values of p_{ss^1} and p_{ss^i} while keeping everything else in P unchanged. This contradicts the optimality of P.

According to Lemma 1, the Min-Max problem in the τ-th iteration can be simplified as follows:

$$C^\tau(s) = \min_P \left[p_{ss^1} + \sum_{s^j \in N(s)} \theta_j p_{ss^j} \right]$$

$$= \min_P \left[(1 + \theta_1) p_{ss^1} + \sum_{s^j \in N(s) \setminus \{s^1\}} \theta_j p_{ss^j} \right] \tag{9}$$

The following lemma gives a further relation among the elements in the optimal solution to the Min-Max problem.

Lemma 2. *Let P be the optimal solution to the Min-Max problem in the τ-th iteration of Algorithm 1. If $i < j$, then $p_{ss^i} \geq p_{ss^j} \ \forall s^i, s^j \in N(s)$.*

Proof. Assume $p_{ss^i} < p_{ss^j}$ for some $i < j$. Then we have $p_{ss^j} = p_{ss^i} + \epsilon$ for some $\epsilon > 0$. It follows that

$$C^\tau(s) = \max \mathbf{p}_s + \sum_{s^k \in N(s)} \theta_k p_{ss^k}$$

$$= \max \mathbf{p}_s + \theta_i p_{ss^i} + \theta_j(p_{ss^i} + \epsilon) + \sum_{s^k \in N(s) \setminus \{s^i, s^j\}} \theta_k p_{ss^k}$$

$$> \max \mathbf{p}_s + \theta_i(p_{ss^i} + \epsilon) + \theta_j p_{ss^i} + \sum_{s^k \in N(s) \setminus \{s^i, s^j\}} \theta_k p_{ss^k} \tag{10}$$

The value in (10) can be obtained by a strategy P' that switches p_{ss^i} and p_{ss^j} while keeping everything else in P unchanged. This contradicts the optimality of P.

From Lemmas 1 and 2, we can obtain a complete characterization of the optimal solution to the Min-Max problem, as stated in the following proposition.

Proposition 1. *Let P be the optimal solution to the Min-Max problem in the τ-th iteration of Algorithm 1. Let $k < |N(s)|$ be the smallest positive integer such that $\theta_{k+1} > \frac{1+\sum_{i=1}^{k+1}\theta_i}{k+1}$, then we have $p_{ss^i} = \frac{1}{k}$, $\forall i \leq k$ and $p_{ss^i} = 0$, $\forall i > k$. If no such k exists, $p_{ss^i} = \frac{1}{|N(s)|}$, $\forall i \in N(s)$.*

Proof. First note that since $\theta_1 < 1 + \theta_1$, we must have $k \geq 1$ (if it exists). We first show that $p_{ss^i} = 0$ $\forall i > k$. Assume $p_{ss^j} = \epsilon > 0$ for some $j > k$. From Lemma 1, we have

$$C^\tau(s) = p_{ss^1} + \sum_{s^j \in N(s)} \theta_j p_{ss^j}$$

$$\geq p_{ss^1} + \sum_{i=1}^{k} \theta_i p_{ss^i} + \theta_j \epsilon$$

$$> p_{ss^1} + \sum_{i=1}^{k} \theta_i p_{ss^i} + \frac{1 + \sum_{i=1}^{k}\theta_i}{k+1}\epsilon$$

$$= (p_{ss^1} + \frac{\epsilon}{k+1}) + \sum_{i=1}^{k} \theta_i(p_{ss^i} + \frac{\epsilon}{k+1}) \qquad (11)$$

Consider another strategy P' where $p'_{ss^i} = p_{ss^1} + \frac{\epsilon}{k+1}$ for all $i \leq k$ and $p'_{ss^i} = 0$ for all $i > k$. According to (11), a smaller cost $(p_{ss^1} + \frac{\epsilon}{k+1}) + \sum_{i=1}^{k} \theta_i(p_{ss^i} + \frac{\epsilon}{k+1})$ can be obtained by adopting P'. This contradicts the optimality of $C^\tau(s)$.

We then show that $p_{ss^i} = \frac{1}{k}$ for all $i \leq k$. To this end, we first prove the following claim: $\theta_i \leq 1 + \theta_1$ for all $i \leq k$. We prove the claim by induction. For $i = 1$, it is clear that $\theta_1 \leq 1 + \theta_1$. Assume the claim is true for all $i \leq m - 1 < k$. We need to show that $\theta_m \leq 1 + \theta_1$. Since $\theta_m \leq \frac{1+\sum_{i=1}^{m}\theta_i}{m}$, we have $(m-1)\theta_m \leq 1 + \theta_1 + \sum_{i=2}^{m-1}\theta_i \leq 1 + \theta_1 + (m-2)(1+\theta_1) = (m-1)(1+\theta_1)$, which implies $\theta_m \leq 1 + \theta_1$.

To show that $p_{ss^i} = \frac{1}{k}$ for all $i \leq k$, it suffices to show that $p_{ss^1} = \frac{1}{k}$. Assume $C_P(s)$ obtains the minimum value at P^* where $p_{ss^1} > \frac{1}{k}$. Without loss of generality, assume $p_{ss^1} > p_{ss^2}$. Then there exists an $\epsilon > 0$ such that $p_{ss^1} - \epsilon \geq \frac{1}{k}$ and $p_{ss^1} - \epsilon \geq p_{ss^2} + \epsilon$. Consider another strategy P'' where $p''_{ss^1} = p_{ss^1} - \epsilon$, $p''_{ss^2} = p_{ss^2} + \epsilon$, $p''_{ss^i} = p_{ss^i}$ for $i \geq 3$. We have

$$C_{P''}(s) = p_{ss^1} - \epsilon + \theta_1(p_{ss^1} - \epsilon) + \theta_2(p_{ss^2} + \epsilon) + \sum_{i=3}^{k} \theta_i p_{ss^i}$$

$$= C_{P^*}(s) - (1 + \theta_1 - \theta_2)\epsilon$$

$$< C_{P^*}(s) \qquad (12)$$

where the last inequality follows from the claim above. This contradicts the optimality of P. Therefore, $p_{ss^1} = \frac{1}{k}$, which implies that $p_{ss^i} = \frac{1}{k}$ for all $i \leq k$.

If $\theta_k \leq \frac{1+\sum_{i=1}^{k}\theta_i}{k}$ for all $k \leq |N(s)|$, we can use a similar argument as above to show that $C_P(s) \geq p_{ss^1} + \theta_1$, where the equality can be achieved by setting $p_{ss^1} = \frac{1}{|N(s)|}$, which implies that $p_{ss^i} = \frac{1}{k}$ for all i.

Proposition 1 has several important implications. First, each row of the optimal P has at most two different values 0 and $\frac{1}{k}$, where k is bounded by the degree of the corresponding node. This implies that the defender may move the resource to several states with the same switching probability even if their switching costs are different. Second, depending on the structure of the state graph, the defender may prefer switching to a state with larger cost or never switch the resource from one state to another even if there is a link between them. Third, for any state s, the value of k in the $(\tau+1)$-th iteration only depends on the s-th row of C and $\{C^\tau(s)|s \in V\}$ from the τ-th iteration. Thus, the minimization problem in (9) can be easily solved. Forth, according to the proof of Proposition 1, if $\theta_k \leq 1+\theta_1$ for $\forall k \in [1, |N(s)|]$, then $p_{ss^1} = \frac{1}{|N(s)|}$. Otherwise, $p_{ss^1} = \frac{1}{k}$.

According to the above observations, we can derive an efficient solution to the step 4 in Algorithm 1, as shown in Algorithm 2.

Algorithm 2. Solving the Min-Max problem in the τ-th iteration of Algorithm 1

Input: $V, C, C^{\tau-1}(\cdot), \alpha$.
Output: P^*.
1: **for** $s \in V$ **do**
2:    $\{s^1, s^2, ..., s^{|N|}\} \leftarrow$ a nondecreasing ordering of $s' \in N(S)$ in terms of $c_{ss'} + \alpha C^{\tau-1}(s')$;
3:    $\theta_i \leftarrow c_{ss^i} + \alpha C^{\tau-1}(s^i), \forall s^i \in N(s)$;
4:    $k \leftarrow 1$;
5:    **while** $\theta_{k+1} \leq \frac{1+\sum_{i=1}^{k+1}\theta_i}{k+1}$ and $k < |N(s)|$ **do**
6:       $k \leftarrow k+1$
7:    **end while**
8:    $p^*_{ss^i} = \frac{1}{k}$, for all $i \leq k$, $p^*_{ss^i} = 0$, for all $i \geq k+1$;
9: **end for**

The running time of Algorithm 2 is dominated by sorting the neighbors of a node according to their θ values. Thus, the complexity of the algorithm is bounded by $O(|V|^2 \log |V|)$. This is much faster than the searching approach with complexity of $O(M^{|V|})$.

4.2   Solving the MDP in Regular Graphs

In this section, we consider a special case of the MTD game where each state has $K+1$ neighbors (including itself) and the switching costs between two distinct switchable states have the same value $c > 0$ as the beginning step. In this case, the state switching graph becomes a regular graph (with self loops on all the nodes). Intuitively, the regular graphs are hard to attack since all the vertices (states) look the same. It will be beneficial for the defender to construct regular or approximately regular graphs to protect the resource if this hypothesis is true. We will show that explicit formulas can be obtained for the MDP under this scenario.

Due to the symmetric nature of the regular graph, it is easy to see that the defender has the same optimal cost at every state. Let $C^{(K)}$ denote the optimal cost when each state has $K+1$ neighbors. We have

$$C^{(K)} = \max \mathbf{p}_s + \sum_{s' \in N(s)} p_{ss'}(c_{ss'} + \alpha C^{(K)})$$

$$\overset{(a)}{=} p_{ss}(1 + \alpha C^{(K)}) + \sum_{s' \in N(s) \backslash s} p_{ss'}(c + \alpha C^{(K)}) \tag{13}$$

where (a) is due to the fact that $c_{ss} + \alpha C^{(K)} = \alpha C^{(K)} < c_{ss'} + \alpha C^{(K)}$ for any $s' \neq s$, which implies that p_{ss} is the maximum element in $\mathbf{p}_s$ according to Lemma 1. If $c > 1$, then $\theta_2 = c + \alpha C^{(K)} > \frac{1 + \alpha C^{(K)} + c + \alpha C^{(K)}}{2} = \frac{1 + \theta_1 + \theta_2}{2}$. We have $p_{ss} = 1$ and $p_{ss'} = 0$ for all $s' \neq s$ according to Proposition 1, and $C^{(K)} = \frac{1}{1-\alpha}$. In this case, the defender will keep the resource at the original state all the time. If $c \leq 1$, then $\theta_k \leq \frac{1 + \sum_{i=1}^{k} \theta_k}{k}$ for all $k \leq K + 1$. We have $p_{ss'} = \frac{1}{K+1}$ for all $s' \in N(s)$ according to Proposition 1. In this case, we can solve the value of $C^{(K)}$ as

$$C^{(K)} = \frac{1}{K+1}(1 + \alpha C^{(K)}) + \frac{K}{1+K}(c + \alpha C^{(K)})$$

$$\Rightarrow C^{(K)} = \frac{1 + Kc}{(1-\alpha)(1+K)} \tag{14}$$

Putting the two cases together, we have

$$C^{(K)} = \begin{cases} \frac{1}{1-\alpha} & \text{if } c > 1, \\ \frac{1+Kc}{(1-\alpha)(1+K)} & \text{if } c \leq 1. \end{cases}$$

Assume $c \leq 1$ in the rest of this section. It is clearly that $C^{(K)}$ is increasing with c. Taking the partial derivative of $C^{(K)}$ w.r.t. K, we have

$$\frac{\partial C^{(K)}}{\partial K} = -\frac{1-c}{(1-\alpha)(1+K)^2} < 0 \tag{15}$$

Therefore, $C^{(K)}$ is strictly decreasing with K. Further, we find that $C^{(K)}$ is a convex function of K by taking the second partial derivative of $C^{(K)}$ w.r.t. K,

$$\frac{\partial^2 C^{(K)}}{\partial K^2} = \frac{1-c}{(1-\alpha)(1+K)^3} > 0 \tag{16}$$

which implies that for larger K, the marginal decrease of $C^{(K)}$ is smaller. We further notice that $C^{(K)}$ is independent of the number of valid states $|V|$ and total links in the graph. Hence, adding more states and switching pairs is not always helpful. For example, in a 8-node regular graph with $K = 2$, the defender has an optimal cost of $\frac{1+2c}{3(1-\alpha)}$. However, given the same switching cost and discount factor, the defender has a smaller cost of $\frac{1+3c}{4(1-\alpha)}$ in a 4-node regular graph with $K = 3$.

5   Numerical Results

In this section, we examine our proposed model with numerical study under different system scenarios and configurations.

5.1   Warm-up Example

We first use a simple example to illustrate the defender's optimal strategy P^* and optimal cost C^*. We consider a resource with $n = |V|$ valid states and model the valid state switches as an Erdős - Rényi $G(n, p)$ random graph [15], where every possible link between two distinct states occurs independently with a probability $p \in (0, 1)$.

Figure 3a shows a small state switching graph sampled from $G(10, 0.6)$ (we also add self links to all the nodes). The switching costs between any two distinct connected states follow the uniform distribution $U(0, 2)$ as shown in Fig. 4, and the discount factor is set to 0.5. Figure 5 gives the defender's optimal strategy P^* and optimal cost $C^*(s)$. The s-th row of C^* represents the optimal cost with an initial state s. Figure 3b highlights the optimal strategy P^*, where from a current state s, the resource may switch to any of the neighboring states connected by red links with an equal probability. From the optimal P^* given in

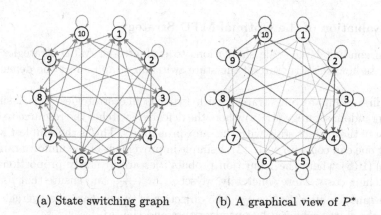

(a) State switching graph   (b) A graphical view of P^*

Fig. 3. An example of the MTD game where the state switching graph is sampled from the Erdős - Rényi random graph $G(10, 0.6)$.

$$C = \begin{bmatrix} 0.00 & 1.48 & 0.91 & 0.95 & 1.64 & 1.12 & \infty & 1.82 & 0.23 & \infty \\ 0.60 & 0.00 & \infty & 0.07 & 0.28 & 0.60 & \infty & 0.16 & \infty & 1.43 \\ 0.50 & \infty & 0.00 & 0.08 & 1.28 & 0.90 & \infty & 1.52 & 0.64 & 1.05 \\ 1.91 & 0.61 & 0.69 & 0.00 & \infty & 0.42 & 1.58 & 0.56 & 1.59 & \infty \\ 1.37 & 0.39 & 1.63 & \infty & 0.00 & 0.58 & 0.30 & 0.07 & \infty & \infty \\ 1.75 & 1.56 & 0.55 & 1.34 & 1.29 & 0.00 & 1.22 & 0.15 & \infty & 0.98 \\ \infty & \infty & \infty & 1.94 & 1.47 & 1.23 & 0.00 & 1.75 & 1.82 & 0.44 \\ 1.39 & 1.69 & 1.24 & 0.01 & 0.28 & 1.48 & 0.99 & 0.00 & \infty & 0.33 \\ 1.31 & \infty & 1.67 & 0.58 & \infty & \infty & 1.93 & \infty & 0.00 & 0.17 \\ \infty & 1.87 & 1.06 & \infty & \infty & 1.30 & 0.88 & 1.95 & 1.02 & 0.00 \end{bmatrix}$$

Fig. 4. Switching cost matrix

$$P^* = \begin{bmatrix} 0.50 & 0.00 & 0.00 & 0.00 & 0.00 & 0.00 & 0.00 & 0.00 & 0.50 & 0.00 \\ 0.00 & 0.25 & 0.00 & 0.25 & 0.25 & 0.00 & 0.00 & 0.25 & 0.00 & 0.00 \\ 0.00 & 0.00 & 0.50 & 0.50 & 0.00 & 0.00 & 0.00 & 0.00 & 0.00 & 0.00 \\ 0.00 & 0.25 & 0.00 & 0.25 & 0.00 & 0.25 & 0.00 & 0.25 & 0.00 & 0.00 \\ 0.00 & 0.33 & 0.00 & 0.00 & 0.33 & 0.00 & 0.00 & 0.33 & 0.00 & 0.00 \\ 0.00 & 0.00 & 0.00 & 0.00 & 0.00 & 0.50 & 0.00 & 0.50 & 0.00 & 0.00 \\ 0.00 & 0.00 & 0.00 & 0.00 & 0.00 & 0.00 & 0.50 & 0.00 & 0.00 & 0.50 \\ 0.00 & 0.00 & 0.00 & 0.33 & 0.33 & 0.00 & 0.00 & 0.33 & 0.00 & 0.00 \\ 0.00 & 0.00 & 0.00 & 0.33 & 0.00 & 0.00 & 0.00 & 0.00 & 0.33 & 0.33 \\ 0.00 & 0.00 & 0.25 & 0.00 & 0.00 & 0.00 & 0.25 & 0.00 & 0.25 & 0.25 \end{bmatrix} \quad C^* = \begin{bmatrix} 1.2401 \\ 0.8639 \\ 1.1009 \\ 1.1511 \\ 0.9463 \\ 1.0798 \\ 1.5231 \\ 0.9358 \\ 1.2668 \\ 1.6877 \end{bmatrix}$$

Fig. 5. The defender's optimal strategy P^* and the corresponding optimal cost $C^*(s)$

Fig. 5, we can make some interesting observations. First, the defender abandons some switching pairs and only switches the resource to the rest of states with equal probability. Second, the defender may prefer switching to a state with larger switching cost. For example, when the resource is currently at state 5, the probability of switching to state 2 is higher than the probability of switching state 7, even though $c_{52} > c_{57}$ ($c_{52} = 0.39, c_{57} = 0.30$). Third, a state s with more neighbors does not necessarily has smaller $C^*(s)$. For instance, state 2 has 7 neighbors and state 6 has 9 neighbors, but $C^*(2) = 0.8639 < C^*(6) = 1.0798$.

5.2 Evaluation of the Optimal MTD Strategy

We then conduct large scale simulations to evaluate our MTD strategies and investigate how the structure of the state switching graph affect the defender's cost.

We first compare our strategy with two baseline strategies: (1) A simple uniform random strategy (URS) where the defender switches the resource to each neighbor of the current state with the same probability. This is the simplest MTD strategy one can come up with. (2) A simple improvement of the uniform random strategy (IRS) where the transition probabilities are inversely proportional to the switching costs. More concretely, we set $p_{ss} = \frac{1}{|N(s)|}$ and ensure that $p_{ss'}c_{ss'}$ is a constant for all $s' \in N(s)\backslash s$. The objective is to compare the average cost over all the states achieved by our algorithm and the two baselines.

The state switching graph is sampled from $G(50, 0.1)$. 100 samples are generated. We set the discount factor $\alpha = 0.5$. The switching costs between two distinct connected nodes follow an uniform distribution $U[0, 2a]$ where a varies between 0.2 and 1.

Figure 6 shows the mean average cost over all the random graphs generated. As we expected, the optimal strategy (OS) has significant better performance than the two baselines, especially when the mean switching cost becomes larger. One thing to highlight is that, although URS is the simplest strategy that one can think of, it may actually perform better than a more complicated strategy such as IRS in certain scenarios. Hence, one has to be careful when adapting a heuristic based strategy to MTD. This observation also indicates the importance of developing optimal strategies for MTD.

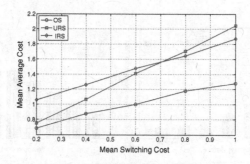

Fig. 6. Mean average cost vs. mean switching cost

5.3   Impact of Switching Graph Structures

In Sect. 4.2, we have derived explicit relations between the optimal defense cost and the structure of the switching graph when the graph is regular. It is interesting to know if such relations hold in more general settings. In this section, we conduct simulations to answer this question for random graphs. To have a fair comparison between regular graphs and random graphs, we set the switching costs between distinct connected nodes to a constant c in this section. We consider two scenarios.

We first fix $|V| = 128$ and the switching cost $c = 0.5$, and vary the average degree K of the switching graph, by using different values of p in the $G(128, p)$ model. We compare this case with a regular graph with the same K. Figure 7a gives the mean average costs for the two models. We observe that when the average degree increases, the defender's optimal cost follows a similar trend in both models. In particular, the cost reduces sharply in the small degree regime, which is consistent with our analysis in Sect. 4.2. In addition, the defender's performance in regular graphs is always better than that in random graphs, especially when the average degree is small. This can be explained by the convexity of $C^{(K)}$ over K shown in Sect. 4.2. More specifically, the degree distribution of a random graph is more diverse than that of a regular graph with the same average degree. Due to the convexity of $C^{(K)}$, we have $C^{(K+\epsilon)} + C^{(K-\epsilon)} > 2C^{(K)}$

(ϵ is a small positive integer), which implies that a graph where the degree distribution is more concentrated has better performance. In addition, the gap between $C^{(K+\epsilon)} + C^{(K-\epsilon)}$ and $2C^{(K)}$ is bigger for smaller K. Hence, regular graphs perform much better than random graphs when the average degree is small.

We then fixe the average degree $K = 8$ and vary $|V|$ and the switching cost c. From Fig. 7b, we observe that the defender's optimal costs in different $|V|$ are almost the same when both the average degree and the switching cost are fixed. Moreover, by increasing the switching cost, the defender's optimal cost in the random graph model increases linearly. Both observations are consistent with our analysis for the regular model in Sect. 4.2.

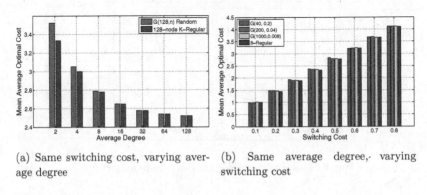

(a) Same switching cost, varying average degree

(b) Same average degree, varying switching cost

Fig. 7. Mean average optimal cost under different settings

5.4 Rate of Convergence

Previous studies have analyzed the convergence rate of discounted MDP [16]. We will examine the convergence speed of proposed Algorithm 1 using simulations with a similar setup as in Sect. 5.3. In Fig. 8a, we vary both $|V|$ and the mean

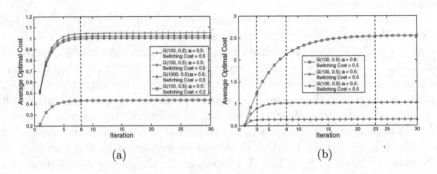

(a)

(b)

Fig. 8. Rate of Convergence with different parameters

switching cost c, while fixing the discount factor $\alpha = 0.5$. We observe that each curve converges to a relative stable value after 8 iterations. We then fix $|V|$, p, and mean switching cost c, while varying the discount factor α. From Fig. 8b, we observe that the convergence speed gets slower with larger α, which is expected. We draw the conclusion that the main factor that affects the convergence rate of Algorithm 1 is the discount factor.

5.5 Suggestions to the Defender

Based on the results and observations above, we make the following suggestions to the defender for holding a more secured resource:

- Due to the fact that the defender's cost is largely determined by the average degree of the switching graph, adding more switching pairs can help reduce the cost. In particular, for a given number of states, the average degree can be maximized adopting a complete graph where the resource can switch between any two states.
- Since the defender's cost is approximately convex with the average degree and linear with the switching cost, the defender should pay more attention to increasing the number of states rather than reducing the switching cost if the average degree is small. While if the average degree is already large enough, reducing switching cost is more useful.
- Introducing a large number of states is not always helpful. The main reason is that the attacker could obtain full feedback about the previous configuration used by the defender in our model. Under this assumption, adding more states does not necessarily means that the defender has more choice to switch. Instead of increasing the number of states, adding more switching pairs is more beneficial to the defender.

6 Conclusion

In this paper, we propose a Stackelberg game model for Moving Target Defense (MTD) between a defender and an attacker. After fully characterizing the player's strategies, payoffs and feedback structures, we model the defender's problem on optimizing the switching strategy as a Markov Decision Process (MDP) and further derive an efficient value iteration algorithm to solve the MDP. By employing a directed graph to illustrate the pattern of switching states, we obtain the relation between defender's performance and the properties of the graph in an explicit way when the graph is regular. Similar results are further verified on random graphs empirically. Through theoretical analysis and numerical study of the proposed model, we have derived several insights and made suggestions to the defender towards more efficient MTD.

Acknowledgement. The effort described in this article was partially sponsored by the U.S. Army Research Laboratory Cyber Security Collaborative Research Alliance under Contract Number W911NF-13-2-0045. The views and conclusions contained in this document are those of the authors, and should not be interpreted as representing the official policies, either expressed or implied, of the Army Research Laboratory or the U.S. Government. The U.S. Government is authorized to reproduce and distribute reprints for Government purposes, notwithstanding any copyright notation hereon. This research was also supported in part by a grant from the Board of Regents of the State of Louisiana LEQSF(2017-19)-RD-A-15.

References

1. Tankard, C.: Advanced persistent threats and how to monitor and deter them. Netw. Secur. **2011**(8), 16–19 (2011)
2. Jajodia, S., Ghosh, A.K., Swarup, V., Wang, C., Wang, X.S.: Moving Target Defense: Creating Asymmetric Uncertainty for Cyber Threats, vol. 54. Springer, New York (2011). doi:10.1007/978-1-4614-0977-9
3. Jafarian, J.H., Al-Shaer, E., Duan, Q.: Openflow random host mutation: transparent moving target defense using software defined networking. In: Proceedings of the First Workshop on Hot topics in Software Defined Networks, pp. 127–132 (2012)
4. Salamat, B., Jackson, T., Wagner, G., Wimmer, C., Franz, M.: Runtime defense against code injection attacks using replicated execution. IEEE Trans. Depend. Secur. Comput. **8**(4), 588–601 (2011)
5. Nguyen-Tuong, A., Evans, D., Knight, J.C., Cox, B., Davidson, J.W.: Security through redundant data diversity. In: IEEE International Conference on Dependable Systems and Networks, pp. 187–196 (2008)
6. Zhu, Q., Başar, T.: Game-theoretic approach to feedback-driven multi-stage moving target defense. In: Das, S.K., Nita-Rotaru, C., Kantarcioglu, M. (eds.) GameSec 2013. LNCS, vol. 8252, pp. 246–263. Springer, Cham (2013). doi:10.1007/978-3-319-02786-9_15
7. Carter, K.M., Riordan, J.F., Okhravi, H.: A game theoretic approach to strategy determination for dynamic platform defenses. In: Proceedings of the First ACM Workshop on Moving Target Defense, pp. 21–30 (2014)
8. Sengupta, S., Vadlamudi, S.G., Kambhampati, S., Doupé, A., Zhao, Z., Taguinod, M., Ahn, G.-J.: A game theoretic approach to strategy generation for moving target defense in web applications. In: International Conference on Autonomous Agents and MultiAgent Systems (AAMAS), pp. 178–186 (2017)
9. Nochenson, A., Heimann, C.F.L.: Simulation and game-theoretic analysis of an attacker-defender game. In: Grossklags, J., Walrand, J. (eds.) GameSec 2012. LNCS, vol. 7638, pp. 138–151. Springer, Heidelberg (2012). doi:10.1007/978-3-642-34266-0_8
10. Lisỳ, V., Davis, T., Bowling, M.H.: Counterfactual regret minimization in sequential security games. In: Association for the Advancement of Artificial Intelligence (AAAI), pp. 544–550 (2016)
11. Yin, Z., Korzhyk, D., Kiekintveld, C., Conitzer, V., Tambe, M.: Stackelberg vs. nash in security games: interchangeability, equivalence, and uniqueness. In: International Conference on Autonomous Agents and Multiagent Systems (AAMAS), pp. 1139–1146 (2010)

12. Feng, X., Zheng, Z., Mohapatra, P., Cansever, D., Swami, A.: A signaling game model for moving target defense. In: IEEE Conference on Computer Communications (INFOCOM) (2017)
13. Zhuang, R., DeLoach, S.A., Ou, X.: A model for analyzing the effect of moving target defenses on enterprise networks. In: Proceedings of the 9th Annual Cyber and Information Security Research Conference, pp. 73–76 (2014)
14. Maleki, H., Valizadeh, S., Koch, W., Bestavros, A., van Dijk, M.: Markov modeling of moving target defense games. In: ACM Workshop on Moving Target Defense, pp. 81–92 (2016)
15. Erdős, P., Rényi, A.: On the evolution of random graphs. Publ. Math. Inst. Hung. Acad. Sci. 5(1), 17–60 (1960)
16. Puterman, M.L.: Markov Decision Processes: Discrete Stochastic Dynamic Programming. Wiley, New York (2014)

Proactive Defense Against Physical Denial of Service Attacks Using Poisson Signaling Games

Jeffrey Pawlick[✉] and Quanyan Zhu

Department of Electrical and Computer Engineering, New York University Tandon School of Engineering, 6 MetroTech Center, Brooklyn, NY 11201, USA
{jpawlick,quanyan.zhu}@nyu.edu

Abstract. While the Internet of things (IoT) promises to improve areas such as energy efficiency, health care, and transportation, it is highly vulnerable to cyberattacks. In particular, distributed denial-of-service (DDoS) attacks overload the bandwidth of a server. But many IoT devices form part of cyber-physical systems (CPS). Therefore, they can be used to launch "physical" denial-of-service attacks (PDoS) in which IoT devices overflow the "physical bandwidth" of a CPS. In this paper, we quantify the population-based risk to a group of IoT devices targeted by malware for a PDoS attack. In order to model the recruitment of bots, we develop a "Poisson signaling game," a signaling game with an unknown number of receivers, which have varying abilities to detect deception. Then we use a version of this game to analyze two mechanisms (legal and economic) to deter botnet recruitment. Equilibrium results indicate that (1) defenders can bound botnet activity, and (2) legislating a minimum level of security has only a limited effect, while incentivizing active defense can decrease botnet activity arbitrarily. This work provides a quantitative foundation for proactive PDoS defense.

1 Introduction to the IoT and PDoS Attacks

The Internet of things (IoT) is a "dynamic global network infrastructure with self-configuring capabilities based on standard and interoperable communication protocols where physical and virtual 'things' have identities, physical attributes, and virtual personalities" [2]. The IoT is (1) decentralized, (2) heterogeneous, and (3) connected to the physical world. It is *decentralized* because nodes have "self-configuring capabilities," some amount of local intelligence, and incentives which are not aligned with the other nodes. The IoT is *heterogeneous* because diverse "things" constantly enter and leave the IoT, facilitated by "standard and interoperable communication protocols." Finally, IoT devices are *connected*

Q. Zhu—This work is partially supported by an NSF IGERT grant through the Center for Interdisciplinary Studies in Security and Privacy (CRISSP) at New York University, by the grant CNS-1544782, EFRI-1441140, and SES-1541164 from National Science Foundation (NSF) and DE-NE0008571 from the Department of Energy.

S. Rass et al. (Eds.): GameSec 2017, LNCS 10575, pp. 336–356, 2017.
DOI: 10.1007/978-3-319-68711-7_18

Fig. 1. Conceptual diagram of a PDoS attack. (1) Attack sponsor hires botnet herder. (2) Botnet herder uses server to manage recruitment. (3) Malware scans for vulnerable IoT devices and begins cascading infection. (4) Botnet herder uses devices (*e.g.*, HVAC controllers) to deplete bandwidth of a cyber-physical service (*e.g.*, electrical power).

to the physical world, *i.e.*, they are part of cyber-physical systems (CPS). For instance, they may influence behavior, control the flow of traffic, and optimize home lighting.

1.1 Difficulties in Securing the Internet of Things

While the IoT promises gains in efficiency, customization, and communication ability, it also raises new challenges. One of these challenges is security. The social aspect of IoT devices makes them vulnerable to attack through social engineering. Moreover, the dynamic and heterogeneous attributes of the IoT create a large attack surface. Once compromised, these "things" serve as vectors for attack. The most notable example has been the Mirai botnet attack on Dyn in 2016. Approximately 100,000 bots—largely belonging to the (IoT)—attacked the domain name server (DNS) for Twitter, Reddit, Github, and the New York Times [15]. A massive flow of traffic overwhelmed the bandwidth of the DNS.

1.2 Denial of Cyber-Physical Service Attacks

Since IoT devices are part of CPS, they also require physical "bandwidth." As an example, consider the navigation app Waze [1]. Waze uses real-time traffic information to find optimal navigation routes. Due to its large number of users, the app also influences traffic. If too many users are directed to one road, they can consume the physical bandwidth of that road and cause unexpected congestion. An attacker with insider access to Waze could use this mechanism to manipulate transportation networks.

Another example can be found in healthcare. Smart lighting systems (which deploy, *e.g.*, *time-of-flight* sensors) detect falls of room occupants [22]. These systems alert emergency responders about a medical situation in an assisted living center or the home of someone who is aging. But an attacker could potentially trigger many of these alerts at the same time, depleting the response bandwidth of emergency personnel.

Such a threat could be called a denial of *cyber-physical* service attack. To distinguish it from a cyber-layer DDoS, we also use the acronym *PDoS* (*Physical Denial of Service*). Figure 1 gives a conceptual diagram of a PDoS attack. In the rest of the paper, we will consider one specific instance of a PDoS attack, although our analysis is not limited to this example. We consider the infection and manipulation of a population of IoT-based heating, ventilation, and air conditioning (HVAC) controllers in order to cause a sudden load shock to the power grid. Attackers either disable demand response switches used for reducing peak load [6], or they unexpectedly activate inactive loads. This imposes risks ranging from frequency droop to load shedding and cascading failures.

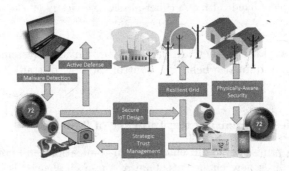

Fig. 2. PDoS defense can be designed at multiple layers. Malware detection and active defense can combat initial infection, secure IoT design and strategic trust can reduce the spread of the malware, and CPS can be resilient and physically-aware. We focus on detection and active defense.

1.3   Modeling the PDoS Recruitment Stage

Defenses against PDoS can be designed at multiple layers (Fig. 2). The scope of this paper is limited to defense at the stage of botnet recruitment, in which the attacker scans a wide range of IP addresses, searching for devices with weak security settings. Mirai, for example, does this by attempting logins with a dictionary of factory-default usernames and passwords (*e.g.* `root/admin`, `admin/admin`, `root/123456`) [12]. Devices in our mechanism identify these suspicious login attempts and use active defense to learn about the attacker or report his activity.

In order to quantify the risk of malware infection, we combine two game-theoretic models known as signaling games [7,14] and Poisson games [18]. Signaling games model interactions between two parties, one of which possesses information unknown to the other party. While signaling games consider only two players, we extend this model by allowing the number of target IoT devices to be a random variable (r.v.) that follows a Poisson distribution. This captures the fact that the malware scans a large number of targets. Moreover, we allow the targets to have heterogeneous abilities to detect malicious login attempts.

1.4   Contributions and Related Work

We make the following principle contributions:

1. We describe an IoT attack called a *denial of cyber-physical service* (*PDoS*).
2. We develop a general model called *Poisson signaling games* (*PSG*) which quantifies one-to-many signaling interactions.
3. We find the pure strategy equilibria of a version of the PSG model for PDoS.
4. We analyze legal and economic mechanisms to deter botnet recruitment, and find that (1) defenders can bound botnet activity, and (2) legislating a minimum level of security has only a limited effect, while incentivizing active defense, in principle, can decrease botnet activity arbitrarily.

Signaling games are often used to model deception and trust in cybersecurity [16,19,21]. Poisson games have also been used to model malware epidemics in large populations [11]. Wu *et al.* use game theory to design defense mechanisms against DDoS attacks [24]. But the defense mechanisms mitigate the actual the flood of traffic against a target system, while we focus on botnet recruitment. Bensoussan *et al.* use a susceptible-infected-susceptible (SIS) model to study the growth of a botnet [5]. But IoT devices in our model maintain beliefs about the reliability of incoming messages. In this way, our paper considers the need to trust legitimate messages. Finally, *load altering attacks* [4,17] to the power grid are an example of PDoS attacks. But PDoS attacks can also deal with other resources.

In Sect. 2, we review signaling games and Poisson games. In Sect. 3, we combine them to create Poisson signaling games (PSG). In Sect. 4, we apply PSG to quantify the population risk due to PDoS attacks. Section 5 obtains the perfect Bayesian Nash equilibria of the model. Some of these equilibria are harmful for power companies and IoT users. Therefore, we design proactive mechanisms to improve the equilibria in Sect. 6. We underline the key contributions in Sect. 7.

2   Signaling Games and Poisson Games

This section reviews two game-theoretic models: signaling games and Poisson games. In Sect. 3, we combine them to create PSG. PSG can be used to model many one-to-many signaling interactions in addition to PDoS.

2.1   Signaling Games with Evidence

Signaling games are a class of dynamic, two-player, information-asymmetric games between a sender S and a receiver R (*c.f.* [7,14]). Signaling games *with evidence* extend the typical definition by giving receivers some exogenous ability to detect deception[1] [20]. They are characterized by the tuple

$$\Phi_{\mathbf{SG}} = \left(X, M, E, A, q^S, \delta, u^S, u^R\right).$$

---

[1] This is based on the idea that deceptive senders have a harder time communicating some messages than truthful senders. In interpersonal deception, for instance, lying requires high cognitive load, which may manifest itself in external gestures [23].

First, S posses some private information unknown to R. This private information is called a *type*. The type could represent, *e.g.*, a preference, a technological capability, or a malicious intent. Let the finite set X denote the set of possible types, and let $x \in X$ denote one particular type. Each type occurs with a probability $q^S(x)$, where $q^S : X \to [0,1]$ such that (s.t.) $\sum_{x \in X} q^S(x) = 1$ and $\forall x \in X, q^S(x) \geq 0$.

Based on his private information, S communicates a *message* to the receiver. The message could be, *e.g.*, a pull request, the presentation of a certificate, or the execution of an action which partly reveals the type. Let the finite set M denote the set of possible messages, and let $m \in M$ denote one particular type. In general, S can use a strategy in which he chooses various m with different probabilities. We will introduce notation for these *mixed strategies* later.

In typical signaling games (*e.g.* Lewis signaling games [7,14] and signaling games discussed by Crawford and Sobel [7]), R only knows about x through m. But this suggests that deception is undetectable. Instead, signaling games with evidence include a *detector*[2] which emits evidence $e \in E$ about the sender's type [20]. Let $\delta : E \to [0,1]$ s.t. for all $x \in X$ and $m \in M$, we have $\sum_{e \in E} \delta(e \mid x, m) = 1$ and $\delta(e \mid x, m) \geq 0$. Then $\delta(e \mid x, m)$ gives the probability with which the detector emits evidence e given type x and message m. This probability is fixed, not a decision variable. Finally A be a finite set of *actions*. Based on m and e, R chooses some $a \in A$. For instance, R may choose to accept or reject a request represented by the message. These can also be chosen using a mixed-strategy.

In general, x, m, and a can impact the utility of S and R. Therefore, let $u^S : M \times A \to \mathbb{R}^{|X|}$ be a vector-valued function such that $u^S(m, a) = \left[u_x^S(m, a) \right]_{x \in X}$. This is a column vector with entries $u_x^S(m, a)$. These entries give the utility that S of each receiver of type $x \in X$ obtains for sending a message m when the receiver plays action a. Next, define the utility function for R by $u^R : X \times M \times A \to \mathbb{R}$, such that $u^R(x, m, a)$ gives the utility that R receives when a sender of type x transmits message m and R plays action a.

2.2   Poisson Games

Poisson games were introduced by Roger Myerson in 1998 [18]. This class of games models interactions between an unknown number of players, each of which belongs to one type in a finite set of types. Modeling the population uncertainty using a Poisson r.v. is convenient because merging or splitting Poisson r.v. results in r.v. which also follow Poisson distributions.

Section 3 will combine signaling games with Poisson games by considering a sender which issues a command to a pool of an unknown number of receivers, which all respond at once. Therefore, let us call the players of the Poisson game

---

[2] This could literally be a hardware or software detector, such as email filters which attempt to tag phishing emails. But it could also be an abstract notion meant to signify the innate ability of a person to recognize deception.

"receivers," although this is not the nomenclature used in the original game. Poisson games are characterized by the tuple

$$\Phi^{\mathbf{PG}} = \left(\lambda, Y, q^R, A, \tilde{u}^R \right).$$

First, the population parameter $\lambda > 0$ gives the mean and variance of the Poisson distribution. For example, λ may represent the expected number of mobile phone users within range of a base station. Let the finite set Y denote the possible types of each receiver, and let $y \in Y$ denote one of these types. Each receiver has type y with probability $q^R(y)$, where $\sum_{y \in Y} q^R(y) = 1$ and $\forall y \in Y$, $q^R(y) > 0$.

Because of the decomposition property of the Poisson r.v., the number of receivers of each type $y \in Y$ also follows a Poisson distribution. Based on her type, each receiver chooses an action a in the finite set A. We have deliberately used the same notation as the action for the signaling game, because these two actions will coincide in the combined model.

Utility functions in Poisson games are defined as follows. For $a \in A$, let $c_a \in \mathbb{Z}_+$ (the set of non-negative integers) denote the count of receivers which play action a. Then let c be a column vector which contains entries c_a for each $a \in A$. Then c falls within the set $\mathbb{Z}(A)$, the set of all possible integer counts of the number of players which take each action.

Poisson games assume that all receivers of the same type receive the same utility. Therefore, let $\tilde{u}^R : A \times \mathbb{Z}(C) \to \mathbb{R}^{|Y|}$ be a vector-valued function such that $\tilde{u}^R(a, c) = \left[\tilde{u}_y^R(a, c) \right]_{y \in Y}$. The entries $\tilde{u}_y^R(a, c)$ give the utility that receivers of each type $y \in Y$ obtain for playing an action a while the vector of the total count of receivers that play each action is given by c. Note that this is different from the utility function of receivers in the signaling game. Given the strategies of the receivers, c is also distributed according to a Poisson r.v.

3   Poisson Signaling Games

Figure 3 depicts Poisson signaling games (PSG). PSG are characterized by combining $\Phi_{\mathbf{SG}}$ and $\Phi^{\mathbf{PG}}$ to obtain the tuple

$$\Phi_{\mathbf{SG}}^{\mathbf{PG}} = \left(X, Y, M, E, A, \lambda, q, \delta, U^S, U^R \right).$$

3.1   Types, Actions, and Evidence, and Utility

As with signaling games and Poisson games, X denotes the set of types of S, and Y denotes the set of types of R. M, E, and A denote the set of messages, evidence, and actions, respectively. The Poisson parameter is λ.

The remaining elements of $\Phi_{\mathbf{SG}}^{\mathbf{PG}}$ are slightly modified from the signaling game or Poisson game. First, $q : X \times Y \to [0, 1]^2$ is a vector-valued function such that $q(x, y)$ gives the probabilities $q^S(x)$, $x \in X$, and $q^R(y)$, $y \in Y$, of each type of sender and receiver, respectively.

As in the signaling game, δ characterizes the quality of the deception detector. But receivers differ in their ability to detect deception. Various email clients, for example, may have different abilities to identify phishing attempts. Therefore, in PSG, we define the mapping by $\delta : E \to [0,1]^{|Y|}$, s.t. the vector $\delta(e \mid x, m) = [\delta_y(e \mid x, m)]_{y \in Y}$ gives the probabilities $\delta_y(e \mid x, m)$ with which each receiver type y observes evidence e given sender type x and message m. This allows each receiver type to observe evidence with different likelihoods[3].

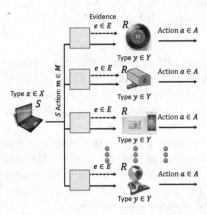

Fig. 3. PSG model the third stage of a PDoS attack. A sender of type x chooses an action m which is observed by an unknown number of receivers. The receivers have multiple types $y \in Y$. Each type may observe different evidence $e \in E$. Based on m and e, each type of receiver chooses an action a.

The utility functions U^S and U^R are also adjusted for PSG. Let $U^S : M \times \mathbb{Z}(A) \to \mathbb{R}^{|X|}$ be a vector-valued function s.t. the vector $U^S(m, c) = [U_x^S(m, c)]_{x \in X}$ gives the utility of senders of each type x for sending message m if the count of receivers which choose each action is given by c. Similarly, let $U^R : X \times M \times A \times \mathbb{Z}(A) \to \mathbb{R}^{|Y|}$ be a vector-valued function s.t. $U^R(x, m, a, c) = [U_y^R(x, m, a, c)]_{y \in Y}$ gives the utility of receivers of each type $y \in Y$. As earlier, x is the type of the sender, and m is the message. But note that a denotes the action of *this particular receiver*, while c denotes the count of overall receivers which choose each action.

3.2 Mixed-Strategies and Expected Utility

Next, we define the nomenclature for mixed-strategies and expected utility functions. For senders of each type $x \in X$, let $\sigma_x^S : M \to [0,1]$ be a mixed strategy

---

[3] In fact, although all receivers with the same type y have the same likelihood $\delta_y(e \mid x, m)$ of observing evidence e given sender type x and message m, our formulation allows the receivers to observe different actual realizations e of the evidence.

such that $\sigma_x^S(m)$ gives the probability with which he plays each message $m \in M$. For each $x \in X$, let Σ_x^S denote the set of possible σ_x^S. We have

$$\Sigma_x^R = \left\{ \bar{\sigma} \mid \sum_{m \in M} \bar{\sigma}(m) = 1 \text{ and } \forall m \in M, \, \bar{\sigma}(m) \geq 0 \right\}.$$

For receivers of each type $y \in Y$, let $\sigma_y^R : A \to [0,1]$ denote a mixed strategy such that $\sigma_y^R(a \mid m, e)$ gives the probability with which she plays action a after observing message m and action e. For each $y \in Y$, the function σ_y^R belongs to the set

$$\Sigma_y^R = \left\{ \bar{\sigma} \mid \sum_{a \in A} \bar{\sigma}(a) = 1 \text{ and } \forall a \in A, \, \bar{\sigma}(a) \geq 0 \right\}.$$

In order to choose her actions, R forms a belief about the sender type x. Let $\mu_y^R(x \mid m, e)$ denote the likelihood with which each R of type y who observes message m and evidence e believes that S has type x. In equilibrium, we will require this belief to be consistent with the strategy of S.

Now we define the expected utilities that S and each R receive for playing mixed strategies. Denote the expected utility of a sender of type $x \in X$ by $\bar{U}_x^S : \Sigma_x^S \times \Sigma^R \to \mathbb{R}$. Notice that all receiver strategies must be taken into account. This expected utility is given by

$$\bar{U}_x^S(\sigma_x^S, \sigma^R) = \sum_{m \in M} \sum_{c \in \mathbb{Z}(A)} \sigma_x^S(m) \, \mathbb{P}\{c \mid \sigma^R, x, m\} \, U_x^S(m, c).$$

Here, $\mathbb{P}\{c \mid \sigma^R, x, m\}$ is the probability with which the vector c gives the count of receivers that play each action. Myerson shows that, due to the aggregation and decomposition properties of the Poisson r.v., the entries of c are also Poisson r.v. [18]. Therefore, $\mathbb{P}\{c \mid \sigma^R, x, m\}$ is given by

$$\mathbb{P}\{c \mid \sigma^R, x, m\} = \prod_{a \in A} e^{\lambda_a} \frac{\lambda_a^{c_a}}{c_a!}, \quad \lambda_a = \lambda \sum_{y \in Y} \sum_{e \in E} q^R(y) \, \delta_y(e \mid x, m) \, \sigma_y^R(a \mid m, e).$$

(1)

Next, denote the expected utility of each receiver of type $y \in Y$ by $\bar{U}_y^R :$ $\Sigma_y^R \times \dot{\Sigma}^R \to \mathbb{R}$. Here, $\bar{U}_y^R(\theta, \sigma^R \mid m, e, \mu_y^R)$ gives the expected utility when this particular receiver plays mixed strategy $\theta \in \Sigma_y^R$ and the population of all types of receivers plays the mixed-strategy vector σ^R. The expected utility is given by

$$\bar{U}_y^R(\theta, \sigma^R \mid m, e, \mu_y^R) = \sum_{x \in X} \sum_{a \in A} \sum_{c \in \mathbb{Z}(A)}$$

$$\mu_y^R(x \mid m, e) \, \theta(a \mid m, e) \, \mathbb{P}\{c \mid \sigma^R, x, m\} \, U_y^R(x, m, a, c),$$

(2)

where again $\mathbb{P}\{c \mid \sigma^R, x, m\}$ is given by Eq. (1).

3.3   Perfect Bayesian Nash Equilibrium

First, since PSG are dynamic, we use an equilibrium concept which involves *perfection*. Strategies at each information set of the game must be optimal for the remaining subgame [8]. Second, since PSG involve incomplete information, we use a *Bayesian* concept. Third, since each receiver chooses her action without knowing the actions of the other receivers, the Poisson stage of the game involves a *fixed point*. All receivers choose strategies which best respond to the optimal strategies of the other receivers. Perfect Bayesian Nash equilibrium (PBNE) is the appropriate concept for games with these criteria [8].

Consider the two chronological stages of PSG. The second stage takes place among the receivers. This stage is played with a given m, e, and μ^R determined by the sender (and detector) in the first stage of the game. When m, e, and μ^R are fixed, the interaction between all receivers becomes a standard Poisson game. Define $BR_y^R : \Sigma^R \to \mathcal{P}(\Sigma_y^R)$ (where $\mathcal{P}(\mathbb{S})$ denotes the power set of $\mathbb{S}$) such that the best response of a receiver of type y to a strategy profile σ^R of the other receivers is given by the strategy or set of strategies

$$BR_y^R \left(\sigma^R \,|\, m, e, \mu_y^R \right) \triangleq \arg\max_{\theta \in \Sigma_y^R} \bar{U}_y^R \left(\theta, \sigma^R \,|\, m, e, \mu_y^R \right). \tag{3}$$

The first stage takes place between the sender and the set of receivers. If we fix the set of receiver strategies σ^R, then the problem of a sender of type $x \in X$ is to choose σ_x^S to maximize his expected utility given σ^R. The last criteria is that the receiver beliefs μ^R must be consistent with the sender strategies according to Bayes' Law. Definition 1 applies PBNE to PSG.

Definition 1. *(PBNE) Strategy and belief profile $(\sigma^{S*}, \sigma^{R*}, \mu^R)$ is a PBNE of a PSG if all of the following hold [8]:*

$$\forall x \in X, \ \sigma_x^{S*} \in \arg\max_{\sigma_x^S \in \Sigma_x^S} \bar{U}_x^S(\sigma_x^S, \sigma^{R*}), \tag{4}$$

$$\forall y \in Y, \forall m \in M, \forall e \in E, \ \sigma_y^{R*} \in BR_y^R \left(\sigma^{R*} \,|\, m, e, \mu_y^R \right), \tag{5}$$

$$\forall y \in Y, \forall m \in M, \forall e \in E, \ \mu_y^R (d \,|\, m, e) \in \frac{\delta_y (e \,|\, d, m) \, \sigma_d^S (m) \, q^S (d)}{\sum_{\tilde{x} \in X} \delta_y (e \,|\, \tilde{x}, m) \, \sigma_{\tilde{x}}^S (m) \, q^S (\tilde{x})}, \tag{6}$$

if $\sum_{\tilde{x} \in X} \delta_y (e \,|\, \tilde{x}, m) \, \sigma_{\tilde{x}}^S (m) \, q^S (\tilde{x}) > 0$, and $\mu_y^R (d \,|\, m, e) \in [0, 1]$, otherwise. We also always have $\mu_y^R (l \,|\, m, e) = 1 - \mu_y^R (d \,|\, m, e)$.

Equation (4) requires the sender to choose an optimal strategy given the strategies of the receivers. Based on the message and evidence that each receiver observes, Eq. (5) requires each receiver to respond optimally to the profile of the strategies of the other receivers. Equation (6) uses Bayes' law (when possible) to obtain the posterior beliefs μ^R using the prior probabilities q^S, the sender strategies σ^S, and the characteristics δ_y, $y \in Y$ of the detectors [20].

4   Application of PSG to PDoS

Section 3 defined PSG in general, without specifying the members of the type, message, evidence, or action sets. In this section, we apply PSG to the recruitment stage of PDoS attacks. Table 1 summarizes the nomenclature.

S refers to the agent which attempts a login attempt, while R refers to the device. Let the set of sender types be given by $X = \{l, d\}$, where l represents a legitimate login attempt, while d represents a malicious attempt. Malicious S attempt to login to many devices through a wide IP scan. This number is drawn from a Poisson r.v. with parameter λ. Legitimate S only attempt to login to one device at a time. Let the receiver types be $Y = \{k, o, v\}$. Type k represents weak receivers which have no ability to detect deception and do not use active defense. Type o represents strong receivers which can detect deception, but do not use active defense. Finally, type v represents active receivers which can both detect deception and use active defense.

Table 1. Application of PSG to PDoS recruitment

Set	Elements
Type $x \in X$ of S	l : legitimate, d : malicious
Type $y \in Y$ of R	k : no detection; o : detection; v : detection & active defense
Message $m \in M$ of S	$m = \{m^1, m^2, \ldots\}$, a set of $|m|$ password strings
Evidence $e \in E$	b : suspicious, n : not suspicious
Action $a \in A$ of R	t : trust, g : lockout, f : active defense

4.1   Messages, Evidence Thresholds, and Actions

Messages consist of sets of consecutive unsuccessful login attempts. They are denoted by $m = \{m^1, m^2, \ldots\}$, where each $m^1, m^2, \ldots$ is a string entered as an attempted password[4]. For instance, botnets similar to Mirai choose a list of default passwords such as [12]

$$m = \{\texttt{admin}, \texttt{888888}, \texttt{123456}, \texttt{default}, \texttt{support}\}.$$

Of course, devices can lockout after a certain number of unsuccessful login attempts. Microsoft Server 2012 recommends choosing a threshold at 5 to 9 [3]. Denote the lower end of this range by $\tau_L = 5$. Let us allow all attempts with $|m| < \tau_L$. In other words, if a user successfully logs in before τ_L, then the PSG does not take place. (See Fig. 5).

The PSG takes place for $|m| \geq \tau_L$. Let $\tau_H = 9$ denote the upper end of the Microsoft range. After τ_L, S may persist with up to τ_H login attempts, or he may not persist. Let p denote persist, and w denote not persist. Our goal is to force malicious S to play w with high probability.

---

[4] A second string can also be considered for the username.

For R of types o and v, if S persists and does not successfully log in with $|m| \leq \tau_H$ login attempts, then $e = b$. This signifies a suspicious login attempt. If S persists and does successfully login with $|m| \leq \tau_H$ attempts, then $e = n$, *i.e.*, the attempt is not suspicious[5].

If a user persists, then the device R must choose an action a. Let $a = t$ denote trusting the user, *i.e.*, allowing login attempts to continue. Let $a = g$ denote locking the device to future login attempts. Finally, let $a = f$ denote using an active defense such as reporting the suspicious login attempt to an Internet service provider (ISP), recording the attempts in order to gather information about the possible attacker, or attempting to block the offending IP address.

4.2   Characteristics of PDoS Utility Functions

The nature of PDoS attacks implies several features of the utility functions U^S and U^R. These are listed in Table 2. Characteristic 1 (C1) states that if S does not persist, both players receive zero utility. C2 says that R also receives zero utility if S persists and R locks down future logins. Next, C3 states that receivers of all types receive positive utility for trusting a benign login attempt, but negative utility for trusting a malicious login attempt. We have assumed that only type v receivers use active defense; this is captured by C4. Finally, C5 says that type v receivers obtain positive utility for using active defense against a malicious login attempt, but negative utility for using active defense against a legitimate login attempt. Clearly, C1-C5 are all natural characteristics of PDoS recruitment.

Table 2. Characteristics of PDoS utility functions

#	Notation
C1	$\forall x \in X, y \in Y, a \in A, c \in \mathbb{Z}(A),$ $U_x^S(w,c) = U_y^R(x,w,a,c) = 0.$
C2	$\forall x \in X, y \in Y, c \in \mathbb{Z}(A),$ $U_y^R(x,p,g,c) = 0.$
C3	$\forall y \in Y, c \in \mathbb{Z}(A),$ $U_y^R(d,p,t,c) < 0 < U_y^R(l,p,t,c).$
C4	$\forall x \in X, c \in \mathbb{Z}(A),$ $U_k^R(x,p,f,c) = U_o^R(x,p,f,c) = -\infty.$
C5	$\forall c \in \mathbb{Z}(A),$ $U_v^R(l,p,f,c) < 0 < U_v^R(d,p,f,c).$

---

[5] For strong and active receivers, $\delta_y(b \mid d, p) > \delta_y(b \mid l, p)$, $y \in \{o, v\}$. That is, these receivers are more likely to observe suspicious evidence if they are interacting with a malicious sender than if they are interacting with a legitimate sender. Mathematically, $\delta_k(b \mid d, p) = \delta_k(b \mid l, p)$ signifies that type k receivers do not implement a detector.

4.3 Modeling the Physical Impact of PDoS Attacks

The quantities c_t, c_g, and c_f denote, respectively, the number of devices that trust, lock down, and use active defense. Define the function $Z : \mathbb{Z}(A) \to \mathbb{R}$ such that $Z(c)$ denotes the load shock that malicious S cause based on the count c. $Z(c)$ is clearly non-decreasing in c_t, because each device that trusts the malicious sender becomes infected and can impose some load shock to the power grid.

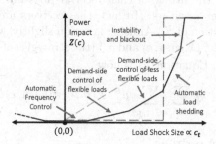

Fig. 4. Conceptual relationship between load shock size and damage to the power grid. Small shocks are mitigated through automatic frequency control or demand-side control of flexible loads. Large shocks can force load shedding or blackouts. (Color figure online)

The red (solid) curve in Fig. 4 conceptually represents the mapping from load shock size to damage caused to the power grid based on the mechanisms available for regulation. Small disturbances are regulated using automatic frequency control. Larger disturbances can significantly decrease frequency and should be mitigated. Grid operators have recently offered customers *load control switches*, which automatically deactivate appliances in response to a threshold frequency decrease [10]. But the size of this voluntary demand-side control is limited. Eventually, operators impose involuntary load shedding (*i.e.*, rolling blackouts). This causes higher inconvenience. In the worst case, transient instability leads to cascading failures and blackout [9].

The yellow and orange dashed curves in Fig. 4 provide two approximations to $Z(c)$. The yellow curve, $\tilde{Z}_{\text{lin}}(c)$, is linear in c^t. We have $\tilde{Z}_{\text{lin}}(c) = \omega_d^t c^t$, where ω_d^t is a positive real number. The orange curve, $\tilde{Z}_{\text{step}}(c)$, varies according to a step function, *i.e.*, $Z(c) = \Omega_d^t \mathbf{1}_{\{c_t > \tau_t\}}$, where Ω_d^t is a positive real number and $\mathbf{1}_{\{\bullet\}}$ is the indicator function. In this paper, we derive solutions for the linear approximation. Under this approximation, the utility of malicious S is given by

$$U_d^S(m, c) = \tilde{Z}_{\text{lin}}(c) + \overset{g}{\underset{d}{\omega}} c_g + \overset{f}{\underset{d}{\omega}} c_f = \overset{t}{\underset{d}{\omega}} c_t + \overset{g}{\underset{d}{\omega}} c_g + \overset{f}{\underset{d}{\omega}} c_f.$$

where $\omega_d^g < 0$ and $\omega_d^f < 0$ represent the utility to malicious S for each device that locks down or uses active defense, respectively.

Using $\tilde{Z}_{\text{lin}}(c)$, the decomposition property of the Poisson r.v. simplifies $\bar{U}_x^S(\sigma_x^S, \sigma^R)$. We show in Appendix A that the sender's expected utility depends on the expected values of each of the Poisson r.v. that represent the number of receivers who choose each action c_a, $a \in A$. The result is that

$$\bar{U}_x^S(\sigma_x^S, \sigma^R) = \lambda \sigma_x^S(p) \sum_{y \in Y} \sum_{e \in E} \sum_{a \in A} q^R(y) \, \delta_y(e \mid x, p) \, \sigma_y^R(a \mid p, e) \, \overset{a}{\underset{x}{\omega}}. \quad (7)$$

Next, assume that the utility of each receiver does not depend directly on the actions of the other receivers. (In fact, the receivers are still endogenously coupled through the action of S.) Abusing notation slightly, we drop c (the count of receiver actions) in $U_y^R(x, m, a, c)$ and σ^R (the strategies of the other receivers) in $\bar{U}_y^R(\theta, \sigma^R \mid m, e, \mu_y^R)$. Equation (2) is now

$$\bar{U}_y^R\left(\theta \mid m, e, \mu_y^R\right) = \sum_{x \in X} \sum_{a \in \{t, f\}} \mu_y^R(x \mid m, e) \, \theta(a \mid m, e) \, \mathcal{U}_y^R(x, m, a).$$

5   Equilibrium Analysis

In this section, we obtain the equilibrium results by parameter region. In order to simplify analysis, without loss of generality, let the utility functions be the same for all receiver types (except when $a = f$), i.e., $\forall x \in X$, $U_k^R(x, p, t) = U_o^R(x, p, t) = U_v^R(x, p, t)$. Also without loss of generality, let the quality of the detectors for types $y \in \{o, v\}$ be the same: $\forall e \in E$, $x \in X$, $\delta_o(e \mid x, p) = \delta_v(e \mid x, p)$.

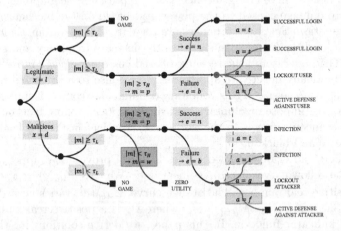

Fig. 5. Model of a PSG under Lemma 1. Only one of many R is depicted. After the types x and y, of S and R, respectively, are drawn, S chooses whether to persist beyond τ_L attempts. Then R chooses to trust, lockout, or use active defense against S based on whether S is successful. Lemma 1 determines all equilibrium strategies except $\sigma_d^{S*}(\bullet)$, $\sigma_o^{R*}(\bullet \mid p, b)$, and $\sigma_v^{R*}(\bullet \mid p, b)$, marked by the blue and red items. (Color figure online)

5.1 PSG Parameter Regime

We now obtain equilibria for a natural regime of the PSG parameters. First, assume that legitimate senders always persist: $\sigma_l^S(p) = 1$. This is natural for our application, because IoT HVAC users will always attempt to login. Second, assume that R of all types trust login attempts which appear to be legitimate (*i.e.*, give evidence $e = n$). This is satisfied for

$$q^S(d) < \frac{U_k^R(l,p,t)}{U_k^R(l,p,t) - U_k^R(d,p,t)}. \tag{8}$$

Third, we consider the likely behavior of R of type o when a login attempt is suspicious. Assume that she will lock down rather than trust the login. This occurs under the parameter regime

$$q^S(d) > \frac{\tilde{U}_o^R(l,p,t)}{\tilde{U}_o^R(l,p,t) - \tilde{U}_o^R(d,p,t)}, \tag{9}$$

using the shorthand notation

$$\tilde{U}_o^R(l,p,t) = U_o^R(l,p,t)\,\delta_0(b\,|\,l,p), \quad \tilde{U}_o^R(d,p,t) = U_o^R(d,p,t)\,\delta_0(b\,|\,d,p).$$

The fourth assumption addresses the action of R of type v when a login attempt is suspicious. The optimal action depends on her belief $\mu_o^R(d\,|\,p,b)$ that S is malicious. The belief, in turn, depends on the mixed-strategy probability with which malicious S persist. We assume that there is some $\sigma_d^S(p)$ for which R should lock down ($a = g$). This is satisfied if there exists a real number $\phi \in [0,1]$ such that, given[6] $\sigma_d^S(p) = \phi$,

$$\bar{U}_v^R(t\,|\,p,b,\mu_v^R) > 0, \quad \bar{U}_v^R(f\,|\,p,b,\mu_v^R) > 0. \tag{10}$$

This simplifies analysis, but can be removed if necessary.

Lemma 1 summarizes the equilibrium results under these assumptions. Legitimate S persist, and R of type o lock down under suspicious login attempts. All receiver types trust login attempts which appear legitimate. R of type k, since she cannot differentiate between login attempts, trusts all of them. The proof follows from the optimality conditions in Eqs. (4–6) and the assumptions in Eqs. (8–10).

Lemma 1 (*Constant PBNE Strategies*). *If $\sigma_d^S(p) = 1$ and Eqs. (8–10) hold, then the following equilibrium strategies are implied:*

$$\sigma_l^{S*}(p) = 1, \ \sigma_o^{R*}(g\,|\,p,b) = 1, \ \sigma_k^{R*}(t\,|\,p,b) = 1,$$

$$\sigma_o^{R*}(t\,|\,p,n) = \sigma_v^{R*}(t\,|\,p,n) = \sigma_k^{R*}(t\,|\,p,n) = 1.$$

Figure 5 depicts the results of Lemma 1. The remaining equilibrium strategies to be obtained are denoted by the red items for S and the blue items for R. These strategies are $\sigma_o^{R*}(\bullet\,|\,p,b)$, $\sigma_v^{R*}(\bullet\,|\,p,b)$, and $\sigma_d^{S*}(p)$. Intuitively, $\sigma_d^{S*}(p)$ depends on whether R of type o and type v will lock down and/or use active defense to oppose suspicious login attempts.

---

[6] We abuse notation slightly to write $\bar{U}_v^R(a\,|\,m,e,\mu_y^R)$ for the expected utility that R of type v obtains by playing action a.

5.2   Equilibrium Strategies

The remaining equilibrium strategies fall into four parameter regions. In order to delineate these regions, we define two quantities.

Let $TD_v^R(U_v^R, \delta_v)$ denote a threshold which determines the optimal action of R of type v if $\sigma_d^S(p) = 1$. If $q^S(d) > TD_v^R(U_v^R, \delta_v)$, then the receiver uses active defense with some probability. Equation (3) can be used to show that

$$TD_v^R\left(U_v^R, \delta_v\right) = \frac{\tilde{U}_v^R\left(l, p, f\right)}{\tilde{U}_v^R\left(l, p, f\right) - \tilde{U}_v^R\left(d, p, f\right)},$$

where we have used the shorthand notation:

$$\tilde{U}_v^R\left(l, p, f\right) := U_v^R\left(l, p, f\right)\delta_v\left(b \,|\, l, p\right), \quad \tilde{U}_v^R\left(d, p, f\right) := U_v^R\left(d, p, f\right)\delta_v\left(b \,|\, d, p\right).$$

Next, let $BP_d^S\left(\omega_d, q^R, \delta\right)$ denote the benefit which S of type d receives for choosing $m = p$, $i.e.$, for persisting. We have

$$BP_d^S\left(\omega_d, q^R, \delta\right) := \sum_{y \in Y}\sum_{e \in E}\sum_{a \in A} q^R\left(y\right)\delta_y\left(e \,|\, d, p\right)\sigma_y^R\left(a \,|\, p, e\right)\omega_d^a.$$

If this benefit is negative, then S will not persist. Let $BP_d^S\left(\omega_d, q^R, \delta \,|\, a_k, a_o, a_v\right)$ denote the benefit of persisting when receivers use the pure strategies:

$$\sigma_k^R\left(a_k \,|\, p, b\right) = \sigma_o^R\left(a_o \,|\, p, b\right) = \sigma_v^R\left(a_v \,|\, p, b\right) = 1.$$

We now have Theorem 1, which predicts the risk of malware infection in the remaining parameter regions. The proof is in Appendix B.

Theorem 1 *(PBNE within Regions)*. *If $\sigma_d^S(p) = 1$ and Eqs. (8–10) hold, then $\sigma_o^{R*}(\bullet \,|\, p, b)$, $\sigma_v^{R*}(\bullet \,|\, p, b)$, and $\sigma_d^{S*}(p)$ vary within the four regions listed in Table 3.*

In the *status quo* equilibrium, strong and active receivers lock down under suspicious login attempts. But this is not enough to deter malicious senders from persisting. We call this the status quo because it represents current scenarios in which botnets infect vulnerable devices but incur little damage from being locked out of secure devices. This is a poor equilibrium, because $\sigma_d^{S*}(p) = 1$.

In the *active deterrence* equilibrium, lockouts are not sufficient to deter malicious S from fully persisting. But since $q^S(d) > TD_v^R$, R of type v use active defense. This is enough to deter malicious S: $\sigma_d^{S*}(p) < 1$. In this equilibrium, R of type o always locks down: $\sigma_o^{R*}(g \,|\, p, b) = 1$. R of type v uses active defense with probability

$$\sigma_v^{R*}\left(f \,|\, p, b\right) = \frac{\omega_d^t\, q^R\left(k\right) + \omega_d^g\left(q^R\left(o\right) + q^R\left(v\right)\right)}{\left(\omega_d^g - \omega_d^f\right) q^R\left(v\right)\delta_v\left(v \,|\, d, p\right)}, \tag{11}$$

and otherwise locks down: $\sigma_v^{R*}(g \mid p, b) = 1 - \sigma_v^{R*}(f \mid p, b)$. Deceptive S persist with reduced probability

$$\sigma_d^{S*}(p) = \frac{1}{q^S(d)} \left(\frac{\tilde{U}_v^R(l, p, f)}{\tilde{U}_v^R(l, p, f) - \tilde{U}_v^R(d, p, f)} \right). \tag{12}$$

In the *resistant attacker* equilibrium, $q^S(d) > TD_v^R$. Therefore, R of type v use active defense. But $BP_d^S(\bullet \mid t, g, f) > 0$, which means that the active defense is not enough to deter malicious senders. This is a "hopeless" situation for defenders, since all available means are not able to deter malicious senders. We still have $\sigma_d^{S*}(p) = 1$.

In the *vulnerable attacker* equilibrium, there is no active defense. But R of type o and type v lock down under suspicious login attempts, and this is enough to deter malicious S, because $BP_d^S(\bullet \mid t, g, g) < 0$. R of types o and v lock down with probability

$$\sigma_o^{R*}(g \mid p, b) = \sigma_v^{R*}(g \mid p, b) = \frac{\omega_d^t}{(q^R(0) + q^R(v)) \delta_o(b \mid d, p)(\omega_d^t - \omega_d^g)}, \tag{13}$$

and trust with probability $\sigma_o^{R*}(t \mid p, b) = \sigma_v^{R*}(t \mid p, b) = 1 - \sigma_o^{R*}(g \mid p, b)$. Deceptive S persist with reduced probability

$$\sigma_d^{S*}(p) = \frac{1}{q^S(d)} \left(\frac{\tilde{U}_o^R(l, p, t)}{\tilde{U}_o^R(l, p, t) - \tilde{U}_o^R(d, p, t)} \right). \tag{14}$$

The *status quo* and *resistant attacker* equilibria are poor results because infection of devices is not deterred at all. The focus of Sect. 6 will be to shift the PBNE to the other equilibrium regions, in which infection of devices is deterred to some degree.

Table 3. Equilibrium regions of the PSG for PDoS

	$q^S(d) < TD_v^R(\bullet)$	$q^S(d) > TD_v^R(\bullet)$
$BP_d^S(\bullet \mid t, g, g) < 0,$	Vulnerable Attacker	
	$\sigma^{S*}(p) < 1$	
$BP_d^S(\bullet \mid t, g, f) < 0$	$0 < \sigma_o^{R*}(t \mid p, b), \sigma_o^{R*}(g \mid p, b) < 1$	
	$0 < \sigma_v^{R*}(t \mid p, b), \sigma_v^{R*}(g \mid p, b) < 1$	
$BP_d^S(\bullet \mid t, g, g) > 0,$		Active Deterrence
		$\sigma^{S*}(p) < 1$
$BP_d^S(\bullet \mid t, g, f) < 0$	Status Quo	$\sigma_o^{R*}(g \mid p, b) = 1$
	$\sigma^{S*}(p) = 1$	$0 < \sigma_v^{R*}(g \mid p, b),$
	$\sigma_o^{R*}(g \mid p, b) = 1$	$\sigma_v^{R*}(f \mid p, b) < 1$
$BP_d^S(\bullet \mid t, g, g) > 0,$	$\sigma_v^{R*}(g \mid p, b) = 1$	Resistant Attacker
		$\sigma^{S*}(p) = 1$
$BP_d^S(\bullet \mid t, g, f) > 0$		$\sigma_o^{R*}(g \mid p, b) = 1$
		$\sigma_v^{R*}(f \mid p, b) = 1$

6 Mechanism Design

The equilibrium results are delineated by the quantities q^S, $TD_v^R(U_v^R, \delta_v)$ and $BP_d^S(\omega_d, q^R, \delta)$. These quantities are functions of the parameters q^S, q^R, δ_o, δ_v, ω_d, and U_v^R. Mechanism design manipulates these parameters in order to obtain a desired equilibrium. We discuss two possible mechanisms.

6.1 Legislating Basic Security

Malware which infects IoT devices is successful because many IoT devices are poorly secured. Therefore, one mechanism design idea is to legally require better authentication methods, in order to decrease $q^R(k)$ and increase $q^R(o)$.

The left-hand sides of Figs. 6, 7 and 8 depict the results. Figure 6(a) shows that decreasing $q^R(k)$ and increasing $q^R(o)$ moves the game from the *status quo* equilibrium to the *vulnerable attacker* equilibrium. But Fig. 7(a) shows that this only causes a fixed decrease in $\sigma_d^{S*}(p)$, regardless of the amount of decrease in $q^R(k)$. The reason, as shown in Fig. 8(a), is that as $q^R(o)$ increases, it is incentive-compatible for receivers to lock down with progressively lower probability $\sigma_y^{R*}(g \mid p, b)$, $y \in \{o, v\}$. Rather than forcing malicious S to not persist,

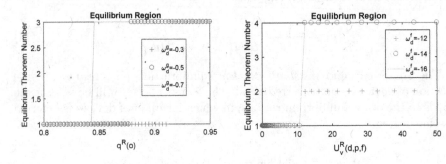

Fig. 6. Equilibrium transitions for (a) legal and (b) active defense mechanisms. The equilibrium numbers signify: 1-*status quo*, 2-*resistant attacker*, 3-*vulnerable attacker*, 4-*active deterrence*.

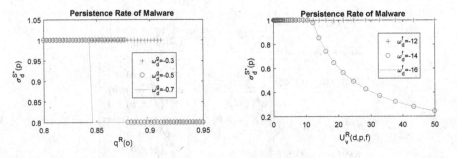

Fig. 7. Malware persistence rate for (a) legal and (b) active defense mechanisms.

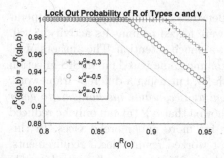 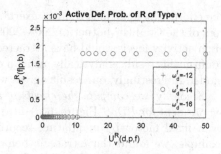

Fig. 8. Probabilities of opposing malicious S. Plot (a): probability that R lock down with the legal mechanism. Plot (b): probability that R use active defense.

increasing $q^R(o)$ only decreases the incentive for receivers to lock down under suspicious login attempts.

6.2 Incentivizing Active Defense

One reason for the proliferation of IoT malware is that most devices which are secure (*i.e.*, R of type $y = o$) do not take any actions against malicious login attempts except to lock down (*i.e.*, to play $a = g$). But there is almost no cost to malware scanners for making a large number of login attempts under which devices simply lock down. There is a lack of economic pressure which would force $\sigma_d^{S*}(p) < 1$, unless $q^R(0) \approx 1$.

This is the motivation for using active defense such as reporting the activity to an ISP or recording the attempts in order to gather information about the attacker. The right hand sides of Figs. 6, 7 and 8 show the effects of providing an incentive $U_v^R(d, p, f)$ for active defense. This incentive moves the game from the *status quo* equilibrium to either the *resistant attacker* equilibrium or the *vulnerable attacker* equilibrium, depending on whether $BP_d^S(\bullet \mid t, g, f)$ is positive (Fig. 6(b)). In the *vulnerable attacker* equilibrium, the persistence rate of malicious S is decreased (Fig. 7(b)). Finally, Fig. 8(b) shows that only a small amount of active defense $\sigma_v^{R*}(f \mid p, b)$ is necessary, particularly for high values of[7] ω_d^f.

7 Discussion of Results

The first result is that *the defender can bound the activity level of the botnet*. Recall that the *vulnerable attacker* and *active deterrence* equilibria force $\sigma_d^{S*}(p) < 1$. That is, they decrease the persistence rate of the malware scanner. But another interpretation is possible. In Eqs. (14) and (12), the product $\sigma_d^{S*}(p) q^S(d)$ is bounded. This product can be understood as the *total activity* of botnet scanners: a combination of prior probability of malicious senders and

---

[7] In Fig. 8(b), $\sigma_v^{R*}(f \mid p, b) = 1$ for $\omega_d^f = -12$.

the effort that malicious senders exert[8]. Bensoussan *et al.* note that the operators of the Confiker botnet of 2008–2009 were forced to limit its activity [5,13]. High activity levels would have attracted too much attention. The authors of [5] confirm this result analytically, using a dynamic game based on an SIS infection model. Interestingly, our result agrees with [5], but using a different framework.

Secondly, we *compare the effects of legal and economic mechanisms to deter recruitment for PDoS.* Figures 6(a)–8(a) showed that $\sigma_d^{S*}(p)$ can only be reduced by a fixed factor by mandating security for more and more devices. In this example, we found that strategic behavior worked against legal requirements. By comparison, Figs. 6(b)–8(b) showed that $\sigma_d^{S*}(p)$ can be driven arbitrarily low by providing an economic incentive $U_v^R(d, p, f)$ to use active defense.

Future work can evaluate technical aspects of mechanism design such as improving malware detection quality. This would involve a non-trivial trade-off between a high true-positive rate and a low false-positive rate. Note that the model of Poisson signaling games is not restricted PDoS attacks. PSG apply to any scenario in which one sender communicates a possibly malicious or misleading message to an unknown number of receivers. In the IoT, the model could capture the communication of a roadside location-based service to a set of autonomous vehicles, or spoofing of a GPS signal used by multiple ships with automatic navigation control, for example. Online, the model could apply to deceptive opinion spam in product reviews. In interpersonal interactions, PSG could apply to advertising or political messaging.

A   Simplification of Sender Expected Utility

Each each component of c is distributed according to a Poisson r.v. The components are independent, so $\mathbb{P}\{c \mid \sigma^R, x, m\} = \prod_{a \in A} \mathbb{P}\{c_a \mid \sigma^R, x, m\}$. Recall that S receives zero utility when he plays $m = w$. So we can choose $m = p$:

$$\bar{U}_x^S(\sigma_x^S, \sigma^R) = \sigma_x^S(p) \sum_{c \in \mathbb{Z}(A)} \prod_{a \in A} \mathbb{P}\{c_a \mid \sigma^R, x, p\} \left(\overset{t}{\underset{x}{\omega}} c_t + \overset{g}{\underset{x}{\omega}} c_g + \overset{f}{\underset{x}{\omega}} c_f \right).$$

Some of the probability terms can be summed over their support. We are left with

$$\bar{U}_x^S(\sigma_x^S, \sigma^R) = \sigma_x^S(p) \sum_{a \in A} \overset{a}{\underset{x}{\omega}} \sum_{c_a \in \mathbb{Z}_+} c_a \mathbb{P}\{c_a \mid \sigma^R, x, p\}. \tag{15}$$

The last summation is the expected value of c_a, which is λ_a. This yields Eq. (7).

---

[8] A natural interpretation in an evolutionary game framework would be that $\sigma_d^{S*}(p) = 1$, and $q^S(d)$ decreases when the total activity is bounded. In other words, malicious senders continue recruiting, but some malicious senders drop out since not all of them are supported in equilibrium.

B   Proof of Theorem 1

The proofs for the *status quo* and *resistant attacker* equilibria are similar to the proof for Lemma 1. The *vulnerable attacker* equilibrium is a partially-separating PBNE. Strategies $\sigma_o^{R*}(g \mid p, b)$ and $\sigma_v^{R*}(g \mid p, b)$ which satisfy Eq. (13) make malicious senders exactly indifferent between $m = p$ and $m = w$. Thus, they can play the mixed-strategy in Eq. (14), which makes strong and active receivers exactly indifferent between $a = g$ and $a = t$. The proof of the *vulnerable attacker* equilibrium follows a similar logic.

References

1. Free community-based mapping, traffic and navigation app. Waze Mobile. https://www.waze.com/
2. Visions and challenges for realising the internet of things. Technical report, CERP IoT Cluster, European Commission (2010)
3. Account lockout threshold. Microsoft TechNet (2014). https://technet.microsoft.com/en-us/library/hh994574(v=ws.11).aspx
4. Amini, S., Mohsenian-Rad, H., Pasqualetti, F.: Dynamic load altering attacks in smart grid. In: Innovative Smart Grid Technologies Conference, pp. 1–5. IEEE (2015)
5. Bensoussan, A., Kantarcioglu, M., Hoe, S.R.C.: A game-theoretical approach for finding optimal strategies in a botnet defense model. In: Alpcan, T., Buttyán, L., Baras, J.S. (eds.) GameSec 2010. LNCS, vol. 6442, pp. 135–148. Springer, Heidelberg (2010). doi:10.1007/978-3-642-17197-0_9
6. Byers, T.: Demand response and the IoT: using data to maximize customer benefit. Comverge Blog (2017). http://www.comverge.com/blog/february-2017/demand-response-and-iot-using-data-to-maximize-cus/
7. Crawford, V.P., Sobel, J.: Strategic information transmission. Econom. J. Econom. Soc. **50**(6), 1431–1451 (1982)
8. Fudenberg, D., Tirole, J.: Game Theory, vol. 393. MIT Press, Cambridge (1991)
9. Glover, J.D., Sarma, M.S., Overbye, T.: Power System Analysis & Design, SI Version. Cengage Learning, Boston (2012)
10. Hammerstrom, D.J.: Part II. Grid friendly appliance project. In: GridWise Testbed Demonstration Projects. Pacific Northwest National Laboratory (2007)
11. Hayel, Y., Zhu, Q.: Epidemic protection over heterogeneous networks using evolutionary poisson games. IEEE Trans. Inf. Forensics Secur. **12**(8), 1786–1800 (2017)
12. Herzberg, B., Bekerman, D., Zeifman, I.: Breaking down mirai: An IoT DDoS botnet analysis. Incapsula Blog, Bots and DDoS, Security (2016). https://www.incapsula.com/blog/malware-analysis-mirai-ddos-botnet.html
13. Higgins, K.J.: Conficker botnet 'dead in the water', researcher says, 2010. Dark Reading. http://www.darkreading.com/attacks-breaches/conficker-botnet-dead-in-the-water-researcher-says/d/d-id/1133327?
14. Lewis, D.: Convention: A Philosophical Study. Wiley, New York (2008)
15. Meyer, R.: How a Bunch of Hacked DVR Machines Took Down Twitter and Reddit. The Atlantic, Darya Ganj (2016)
16. Mohammadi, A., Manshaei, M.H., Moghaddam, M.M., Zhu, Q.: A game-theoretic analysis of deception over social networks using fake avatars. In: Zhu, Q., Alpcan, T., Panaousis, E., Tambe, M., Casey, W. (eds.) GameSec 2016. LNCS, vol. 9996, pp. 382–394. Springer, Cham (2016). doi:10.1007/978-3-319-47413-7_22

17. Mohsenian-Rad, A.-H., Leon-Garcia, A.: Distributed internet-based load altering attacks against smart power grids. IEEE Trans. Smart Grid **2**(4), 667–674 (2011)
18. Myerson, R.B.: Population uncertainty and poisson games. Int. J. Game Theor. **27**(3), 375–392 (1998)
19. Pawlick, J., Farhang, S., Zhu, Q.: Flip the cloud: cyber-physical signaling games in the presence of advanced persistent threats. In: Khouzani, M.H.R., Panaousis, E., Theodorakopoulos, G. (eds.) GameSec 2015. LNCS, vol. 9406, pp. 289–308. Springer, Cham (2015). doi:10.1007/978-3-319-25594-1_16
20. Pawlick, J., Zhu, Q.: Deception by design: evidence-based signaling games for network defense. In: Workshop on the Economics of Information Security and Privacy, Delft, The Netherlands (2015)
21. Pawlick, J., Zhu, Q.: Strategic trust in cloud-enabled cyber-physical systems with an application to glucose control. IEEE Trans. Inf. Forensics and Secur. (2017, to appear)
22. Radke, R.J., Woodstock, T-K., Imam, M.H., Sanderson, A.C., Mishra, S.: Advanced sensing and control in the smart conference room at the center for lighting enabled systems and applications. In: SID Symposium Digest of Technical Papers, vol. 47, pp. 193–196. Wiley Online Library (2016)
23. Vrij, A., Mann, S.A., Fisher, R.P., Leal, S., Milne, R., Bull, R.: Increasing cognitive load to facilitate lie detection: the benefit of recalling an event in reverse order. Law Hum. Behav. **32**(3), 253–265 (2008)
24. Wu, Q., Shiva, S., Roy, S., Ellis, C., Datla, C.: On modeling and simulation of game theory-based defense mechanisms against DoS and DDoS attacks. In: Proceedings of Spring Simulation Multiconference, p. 159. Society for Computer Simulation International (2010)

A Large-Scale Markov Game Approach to Dynamic Protection of Interdependent Infrastructure Networks

Linan Huang[(✉)], Juntao Chen, and Quanyan Zhu

Department of Electrical and Computer Engineering, New York University,
2 Metrotech Center, Brooklyn 11201, USA
{lh2328,jc6412,qz494}@nyu.edu

Abstract. The integration of modern information and communication technologies (ICTs) into critical infrastructures (CIs) improves its connectivity and functionalities yet also brings cyber threats. It is thus essential to understand the risk of ICTs on CIs holistically as a cyber-physical system and design efficient security hardening mechanisms. To this end, we capture the system behaviors of the CIs under malicious attacks and the protection strategies by a zero-sum game. We further propose a computationally tractable approximation for large-scale networks which builds on the factored graph that exploits the dependency structure of the nodes of CIs and the approximate dynamic programming tools for stochastic Markov games. This work focuses on a localized information structure and the single-controller game solvable by linear programming. Numerical results illustrate the proper tradeoff of the approximation accuracy and computation complexity in the new design paradigm and show the proactive security at the time of unanticipated attacks.

1 Introduction

Recent advances in information and communication technologies (ICTs), such as 5G networks and the Internet of Things (IoTs), have witnessed a tight integration of critical infrastructures to improve the quality of facility services and the operational efficiency. However, the direct application of current ICT systems exposes infrastructures to cyber vulnerabilities, which can compromise the functionalities of the infrastructures and inflict a significant economic loss. For example, the cyber attacks on Ukrainian power systems have left 230,000 people without electricity. The WannaCry ransomware attacks have infected thousands of computers worldwide and invalidated critical services such as hospitals and manufacturing plants, causing an estimated loss of $4 billion.

As shown in Fig. 1, the cyber-physical nature of the interdependent infrastructure systems increases the probability of the attacks and the failure

This research is partially supported by NSF grants EFRI-1441140, SES-1541164, CNS-1544782, DOE grant DE-NE0008571, and a DHS CIRI grant.

S. Rass et al. (Eds.): GameSec 2017, LNCS 10575, pp. 357–376, 2017.
DOI: 10.1007/978-3-319-68711-7_19

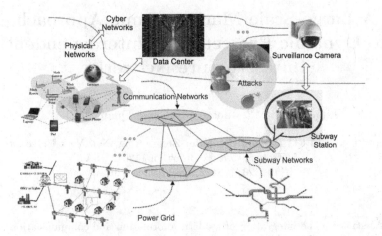

Fig. 1. Cyber networks on the top are interdependent with physical systems on the bottom which consists of critical infrastructures such as the power grid, subway, and communication networks. The healthy functioning of components of the physical system, e.g., subway stations depends on the well-being of other subway stations and cross-layer nodes (e.g. power substations and surveillance cameras). This interdependency allows adversaries to attack different types of nodes to compromise the entire cyber-physical infrastructure.

rates for both systems. For example, a terrorist can use cyber attacks to compromise the surveillance camera of an airport, government building, or public area, and stealthily plant a bomb without being physically detected. The physical damage of infrastructure systems can also assist attackers to intrude into cyber systems such as data centers and control rooms. Moreover, the cyber, physical, and logical connectivity among infrastructures creates dependencies and inter-dependencies between nodes or components within an infrastructure and across the infrastructures. As a result, the failure of one component in either cyber or physical system can lead to a cascading failure over multiple infrastructures. It is thus essential to design effective defense mechanisms to harden both the cyber and physical security at the nodes of the infrastructure.

To this end, we first develop a zero-sum game framework to capture the adversarial interactions between the attack and the defense of the interdependent critical infrastructures (ICIs). The attacker aims to compromise the cyber and physical components of ICIs that are under his control and inflict maximum loss on the system. The defense of the ICIs seeks to invest resources to minimize the loss by implementing cost-effective defense mechanisms. To capture the dynamics of the ICIs, we use a binary state variable to describe the state of each node. The attacker's strategy can affect the transition probability of a node's state from a normal operation mode to a failure mode. The saddle-point equilibrium analysis of the zero-sum dynamic game provides a systematic way to design defense strategies for the worst-case attack schemes.

In our work, we focus on the class of Markov games whose transition kernel is controlled by the attacker yet the defender can choose state-dependent actions to mitigate the economic loss or increase attacking costs at each state. The single-controller assumption reduces the computation of the saddle-point equilibrium strategies into a linear program. One challenge in computing security strategies arises from the large-scale nature of the infrastructure systems together with an exponentially growing number of global states. To address it, we use linear function approximation techniques for the value function and exploit the sparse network structure to formulate the factored Markov game to reduce the computational complexity. Another challenge is the implementability of the security strategies. The global stationary policies of both players are difficult to implement since the knowledge of the global state of each infrastructure is not often accessible. Hence we restrict the security strategies to a decentralized information structure and use the factored Markov game framework to compute approximately distributed policies.

Numerical results illustrate the implementable distributed policies, significant computation reductions, reasonable accuracy losses, and the impacts of different information structures and the interdependent networks. Firstly, we observe that fewer attacks happen when defenders are present in the system because attacks tend to avoid attacking nodes equipped with safeguard procedures. The security strategy is proactive, e.g., nodes choose to protect themselves in advance at a normal state when their neighbors are observed to be attacked. Secondly, the approximation scheme yields a significant reduction of the complexity while maintaining a reasonable level of accuracy. Thirdly, we observe that a node can improve its security performance with more information about the global state of the multi-layer infrastructures. Besides, when strengthening every node is too costly, we choose to consolidate every other connecting node in the network to mitigate cascading failures.

2 Literature Review

A lot of works have been devoted to understand the interdependent networks by conceptual frameworks [24], dependency classification including physical, cyber, geographic, and logical types [17], and input-output or agent-based model construction [16]. The authors in [1,3] have proposed a game-theoretic framework to capture the decentralized decision-making nature of interdependent CIs. To analyze and manage the risks of CIs due to the interdependencies, various models have been proposed, e.g., based on network flows [9], numerical simulations [8], dynamic coupling [18], and the ones summarized in [14]. Game-theoretic methods have been extensively used to model the cyber security with applications to infrastructures [12,20,21,23]. Zhu et al. have proposed a proactive feedback-driven moving target defense mechanism to secure the computer networks [19]. In [15], a *FlipIt* game framework has been used to model the security in cloud-enabled cyber-physical systems. The authors in [2,4] have addressed the multi-layer cyber risks management induced by attacks in Internet of things through a contract-theoretic approach. In [22,23], Markov games model have been used to deal with network security. Our paper differs from the previous works by proposing a factored Markov game framework and developing computational methods for the dynamic protection policies of large-scale interdependent CIs. The computation limitation caused by the curse of dimension urges researchers to find scalable methods. A number of works have focused on the linear programming formulation of Markov Decision Processes (MDP) and complexity reduction of the objective and constraints of the linear programming [6,11]. In [5], the authors have reduced the number of constraints by proper sampling and derived its error bound. Li and Shamma [10] has formulated a linear program of the asymmetric zero-sum game and reduced its computational complexity to polynomial time. Further, [13] extends to take into account of the incomplete information leveraging low-rank features.

3   Mathematical Model

This section introduces in Subsect. 3.1 a zero-sum Markov game model over interdependent infrastructure networks to understand the interactions between an attacker and a defender at the nodes of infrastructures. The solution concept of the saddle-point equilibrium strategies is presented in Subsect. 3.2 and the computational issues of the equilibrium is discussed in Subsect. 3.3.

3.1   Network Game Model

The dynamic and complex infrastructure networks can be represented by nodes and links. For example, in an electric power system, a node can be a load bus or a generator and the links represent the transmission lines. Similarly, in a water distribution system, a node represents a source of water supply, storage or users, and the links can represent pipes for water delivery. Consider a system of I interdependent infrastructures. Let $\mathcal{G}^i = (\mathcal{N}^i, \mathcal{E}^i)$ be the graph representation of infrastructure $i, i \in \mathcal{I} := \{1, 2, \cdots, I\}$, where $\mathcal{N}^i = \{n_1^i, n_2^i, \cdots, n_{m_i}^i\}$ is the set of m^i nodes in the infrastructure and $\mathcal{E}^i = \{e_{j,k}^i\}$ is the set of directed links connecting nodes n_j^i and n_k^i. The directed link between two nodes indicates either physical, cyber or logical influences from one node to the other. For example, the state of node n_j^i in the electric power system can influence the state of node n_k^i through the physical connection or the market pricing. The dependencies across the infrastructures can be captured by adding interlinks. Let $\mathcal{E}^{i,j}$ be the set of directed interlinks between nodes in infrastructure i and infrastructure j. In particular, let $\varepsilon_{n_k^i, n_l^j} \in \mathcal{E}^{i,j}$ denote the interlink between n_k^i and n_l^j. Hence, the composed network can be represented by the graph $\mathcal{G} = (\mathcal{N}, \mathcal{E})$, where $\mathcal{N} = \cup_{i=1}^I \mathcal{N}^i$ and $\mathcal{E} = \left(\cup_{i=1}^I \mathcal{E}^i \right) \bigcup \left(\cup_{i \neq j} \mathcal{E}^{i,j} \right)$.

Denote by $X_j^i \in \mathcal{X}_j^i$ the state of node n_j^i that can take values in the state space $\mathcal{X}_j^i$. We let $\mathcal{X}_j^i = \{0, 1\}$ be binary random variables for all $i = 1, 2, \cdots, I$ and $j \in \mathcal{N}^i$. Here, $X_j^i = 1$ means that node n_j^i is functional in a normal mode;

$X_j^i = 0$ indicates that node n_j^i is in a failure mode. The state of infrastructure i can be thus denoted by $X^i = [X_1^i, X_2^i, \cdots, X_{m_i}^i] \in \mathcal{X}^i := \prod_{j=1}^{m_i} \mathcal{X}_j^i$ and the state of the whole system is denoted by $X = [X^1, X^2, \cdots, X^I] \in \prod_{i=1}^{I} \mathcal{X}^i$. The state transition of a node n_j^i from state $x_j^{i'} \in \mathcal{X}_j^i$ to state $x_j^i \in \mathcal{X}_j^i$ is governed by a stochastic kernel $p_{i,j}(x_j^{i'}|x, d_j^i, a_j^i) := Pr(X_j^i = x_j^{i'}|X = x, d_j^i, a_j^i)$, which depends on the protection policy $d_j^i \in \mathcal{D}_j^i$ adopted at node n_j^i as well as the adversarial behavior $a_j^i \in \mathcal{A}_j^i$, where $\mathcal{D}_j^i, \mathcal{A}_j^i$ are feasible sets for the infrastructure protection and the adversary, respectively. The state transition of a node depends on the entire system state of the interdependent infrastructure. It, in fact, captures the interdependencies between nodes in one infrastructure and across infrastructures. The infrastructure protection team or defender determine the protection policy with the goal of hardening the security and improving the resilience of the interdependent infrastructure. On the other hand, an adversary aims to create damage on the nodes that he can compromise and inflict maximum damage on the infrastructure in a stealthy manner, e.g., creating cascading and wide-area failures. Let $\mathcal{M}_a^i \subseteq \mathcal{N}^i$ and $\mathcal{M}_d^i \subseteq \mathcal{N}^i$ be the set of nodes that an adversary can control and the system action vector of the adversary is $\mathbf{a} = [a_j^i]_{j \in \mathcal{M}_a^i, i \in \mathcal{I}} \in \mathcal{A} := \prod_{i \in \mathcal{I}} \prod_{j \in \mathcal{N}^i} A_j^i$ with $|\mathcal{M}_a^i| = \bar{m}_{a,i}$. The system action vector for infrastructure protection is $\mathbf{d} = [d_j^i]_{j \in \mathcal{M}_d^i, i \in \mathcal{I}} \in \mathcal{D} := \prod_{i \in \mathcal{I}} \prod_{j \in \mathcal{N}^i} D_j^i$ with $|\mathcal{M}_d^i| = \bar{m}_{d,i}$. At every time $t = 1, 2, \cdots$, the pair of action profiles $(\mathbf{d}_t, \mathbf{a}_t)$ taken at t and the kernel $\mathbb{P}$ defined later determine the evolution of the system state trajectory. Here, we use add subscript t to denote the action taken time t. The conflicting objective of both players can be captured by a long-term cost J over an infinite horizon:

$$J := \sum_{i \in \mathcal{I}, j \in \mathcal{N}^i} \sum_{t=1}^{\infty} \gamma^t c_j^i(X_t, d_{j,t}^i, a_{j,t}^i), \tag{1}$$

where $\gamma \in (0, 1)$ is a discount factor; $X_t \in \mathcal{X}$ is the system state at time t; $c_j^i : \mathcal{X} \times \mathcal{D}_j^i \times \mathcal{A}_j^i \to \mathbb{R}_+$ is the stage cost function of the node n_j^i. Let $\mathcal{U}_j^i, \mathcal{V}_j^i$ be the sets of admissible strategies for the defender and the adversary, respectively. Here, we consider a feedback protection policy $\mu_j^i \in \mathcal{U}_j^i$ as a function of the information structure $F_{j,t}^i$, i.e., $d_{j,t}^i = \mu_j^i(F_{j,t}^i)$. Likewise, we consider the same class of policies for the adversary, i.e., $a_{j,t}^i = \nu_j^i(F_{j,t}^i), \nu_j^i \in \mathcal{V}_j^i$.

The policy can take different forms depending on the information structure. For example, if $F_{j,t}^i = X_t$, i.e., each node can observe the whole state across infrastructures, then the policy is a global stationary policy, denoted by $\mu_j^{i,\mathrm{GS}} \in \mathcal{U}_j^{i,\mathrm{GS}}$, where $\mathcal{U}_j^{i,\mathrm{GS}}$ is the set of all admissible global stationary policies.. If $F_{j,t}^i = X_{j,t}^i$, i.e., each node can only observe its local state, then the policy is a local stationary policy, denoted by $\mu_j^{i,\mathrm{LS}} \in \mathcal{U}_j^{i,\mathrm{LS}}$, where $\mathcal{U}_j^{i,\mathrm{LS}}$ is the set of all admissible local stationary policies. If $F_{j,t}^i = X_t^i$, i.e., each node can observe the infrastructure-wide state, then the policy is an infrastructure-dependent stationary policy, denoted by $\mu_j^{i,\mathrm{ID}} \in \mathcal{U}_j^{i,\mathrm{ID}}$, where $\mathcal{U}_j^{i,\mathrm{ID}}$ is the set of all admissible infrastructure-dependent stationary policies.

Similarly, an adversary chooses a policy $\nu^i_{j,t}$, i.e., $a^i_{j,t} = \nu^i_j(F^i_{j,t})$. Denote by $\mu^i = [\mu^i_1, \mu^i_2, \cdots, \mu^i_{m_i}]$, $\nu^i = [\nu^i_1, \nu^i_2, \cdots, \nu^i_{m_i}]$ the protection and attack policies for infrastructure i, respectively, and let $\boldsymbol{\mu} = [\mu^1, \mu^2, \cdots, \mu^I]$ and $\boldsymbol{\nu} = [\nu^1, \nu^2, \cdots, \nu^I]$. Note that although both policies are determined only by the information structure and are independent of each other, the total cost function J depends on them both because of the coupling of the system stage cost $c(X_t, \mathbf{d}, \mathbf{a}) := \sum_{i,j} c^i_j(X_t, d^i_{j,t}, a^i_{j,t})$ and the system transition probability $\mathbb{P}(X' = x'|X = x, \mathbf{d}, \mathbf{a}) := \prod_{i \in \mathcal{I}, j \in \mathcal{N}_i} p_{i,j}(x^{i'}_j|x, d^i_j, a^i_j)$. Therefore, with $\mathcal{U} = \prod_{i \in \mathcal{I}, j \in \mathcal{N}_i} \mathcal{U}^i_j$ and $\mathcal{V} = \prod_{i \in \mathcal{I}, j \in \mathcal{N}_i} \mathcal{V}^i_j$, the total cost function $J : \mathcal{X} \times \mathcal{U} \times \mathcal{V} \to \mathbb{R}_+$ starting at initial state x^0 can be written as the expectation of the system stage cost regarding the system state transition probability, i.e., $J(x^0, \boldsymbol{\mu}, \boldsymbol{\nu}) := \sum_{t=0}^{\infty} \gamma^t E_{\mathbf{d}, \mathbf{a}|x^0}[c(X_t, \mathbf{d}, \mathbf{a})]$. Hence a security strategy for the infrastructure protection achieves the optimal solution $J^*(x^0)$ to the following minimax problem, which endeavors to minimize the system cost under the worst attacking situation $\max_{\boldsymbol{\nu} \in \mathcal{V}} J(x^0, \boldsymbol{\mu}, \boldsymbol{\nu})$, i.e.,

$$J^*(x^0) = \min_{\boldsymbol{\mu} \in \mathcal{U}} \max_{\boldsymbol{\nu} \in \mathcal{V}} J(x^0, \boldsymbol{\mu}, \boldsymbol{\nu}). \tag{2}$$

3.2   Zero-Sum Markov Games

The non-cooperative objective function (2) leads to the solution concept of *Saddle-Point Equilibrium* in game theory. Let J^d and J^a be the total cost function regarding the defender and the attacker respectively.

Definition 1. *A Saddle-Point Equilibrium (SPE) $(\boldsymbol{\mu}^*, \boldsymbol{\nu}^*) \in \mathcal{U} \times \mathcal{V}$ of the discounted zero-sum Markov games with two players satisfies the following inequalities:*

$$J(x^0, \boldsymbol{\mu}, \boldsymbol{\nu}^*) \geq J(x^0, \boldsymbol{\mu}^*, \boldsymbol{\nu}^*) \geq J(x^0, \boldsymbol{\mu}^*, \boldsymbol{\nu}), \forall \boldsymbol{\nu} \in \mathcal{V}, \boldsymbol{\mu} \in \mathcal{U}, \forall x^0 \in \prod_{i=1}^{I} \mathcal{X}^i. \tag{3}$$

The value $J^*(x^0)$ achieved under the saddle-point equilibrium of the game (2) for a given initial condition x^0 is called the value function of a two-player zero-sum game, i.e.,

$$J^*(x^0) := J(x^0, \boldsymbol{\mu}^*, \boldsymbol{\nu}^*) = \min_{\boldsymbol{\mu} \in \mathcal{U}} \max_{\boldsymbol{\nu} \in \mathcal{V}} J(x^0, \boldsymbol{\mu}, \boldsymbol{\nu}) = \max_{\boldsymbol{\nu} \in \mathcal{V}} \min_{\boldsymbol{\mu} \in \mathcal{U}} J(x^0, \boldsymbol{\mu}, \boldsymbol{\nu}). \tag{4}$$

By focusing on the class of global stationary policies, i.e., $\mu^{i,\text{GS}}_j \in \mathcal{U}^{i,\text{GS}}_j$ and $\nu^{i,\text{GS}}_j \in \mathcal{V}^{i,\text{GS}}_j$, we can characterize the value function $J^*(x^0)$ using dynamic programming principles. The action pair $(\mathbf{d}^*, \mathbf{a}^*)$ with $d^{i*}_j = \mu^{i*,\text{GS}}_j(x)$ and $a^{i*}_j = \nu^{i*,\text{GS}}_j(x)$ satisfies the following Bellman equation:

$$J^*(x) = c(x, \mathbf{d}^*, \mathbf{a}^*) + \gamma \sum_{x' \in \prod_{i=1}^{I} \mathcal{X}^i} \mathbb{P}(x'|x, \mathbf{a}^*, \mathbf{d}^*) J^*(x'), \forall x. \tag{5}$$

The first term is the reward of current stage x and the second term is the expectation of the value function over all the possible next stage x'. The optimal action pairs $(\mathbf{d}^*, \mathbf{a}^*)$ guarantee that the value function starting from x equals the current stage cost plus the expectation starting at the next stage x'. By solving the Bellman equation (5) for every state x, we can obtain the saddle-point equilibrium strategy pairs $(\boldsymbol{\mu}^*, \boldsymbol{\nu}^*)$ in global stationary policies.

The global stationary saddle-point policies in pure strategies may not always exist. The Bellman equation (6) can be solved under mixed-strategy action spaces. Let the mixed-strategy actions for the attacker and the defender be $\phi^a(x, \mathbf{a})$ and $\phi^d(x, \mathbf{d})$, where $\phi^d(x, \mathbf{d})$ (or $\phi^a(x, \mathbf{a})$) denotes the probability of taking action $\mathbf{d}$ (or $\mathbf{a}$) at the global state x for a defender (or an attacker). The saddle-point mixed-strategy action pair $(\phi^{a*}(x, \mathbf{a}), \phi^{d*}(x, \mathbf{d}))$ satisfy the following generalized Bellman equation:

$$J^*(x) = \sum_{\mathbf{a} \in \mathcal{A}} \phi^{a*}(x, \mathbf{a}) \sum_{\mathbf{d} \in \mathcal{D}} \left[c(x, \mathbf{d}, \mathbf{a}) + \gamma \sum_{x'} \mathbb{P}(x'|x, \mathbf{a}, \mathbf{d}) J^*(x') \right] \phi^{d*}(x, \mathbf{d}), \forall x.$$

(6)

The existence of the mixed-strategy action pair is guaranteed when the action spaces $\mathcal{A}$ and $\mathcal{D}$ are finite. Hence solving (6) for every state x, we can obtain the mixed-strategy saddle-point equilibrium strategy pairs $(\hat{\boldsymbol{\mu}}^*, \hat{\boldsymbol{\nu}}^*)$ in global stationary policies, where $\hat{\boldsymbol{\mu}}, \hat{\boldsymbol{\nu}}$ are the mixed strategy extension of $\boldsymbol{\mu}, \boldsymbol{\nu}$, respectively.

3.3   Mathematical Programming Perspective

One way to compute the mixed-strategy equilibrium solutions for zero-sum games is to use the mathematical programming as follows:

$$\min_{J^*(x), \phi^d(x, \mathbf{d})} \sum_{x \in \prod_{i=1}^I \mathcal{X}^i} \alpha(x) J^*(x)$$

subject to :

(a) $J^*(x) \geq \sum_{\mathbf{d} \in \mathcal{D}} \left[c(x, \mathbf{d}, \mathbf{a}) + \gamma \sum_{x' \in \prod_{i=1}^I \mathcal{X}^i} \mathbb{P}(x'|x, \mathbf{a}, \mathbf{d}) J^*(x') \right] \phi^d(x, \mathbf{d}), \forall x, \mathbf{a}$

(b) $\sum_{\mathbf{d} \in \mathcal{D}} \phi^d(x, \mathbf{d}) = 1,$ $\hspace{4cm} \forall x$

(c) $\phi^d(x, \mathbf{d}) \geq 0,$ $\hspace{4.5cm} \forall x, \mathbf{d}$

(7)

Constraints (b) and (c) reflect $\phi^d(x, \mathbf{d})$ as a probability measure. State-dependent weights $\alpha(x)$ are positive and satisfy $\sum_x \alpha(x) = 1$. Solutions of this problem provide us the value function $J^*(x)$ and the optimal defender policy $\phi^{d*}(x, \mathbf{d})$.

3.4   Single-Controller Markov Game

In the single-controller game, one player's action entirely determines transition probabilities. This structure captures the fact that the failure probability of a node in the infrastructure depends on the action taken by the attacker once the node is attacked. We focus on an attacker-controlled game Γ^a where the stochastic kernel for each node possesses $p_{i,j}(x_j^{i'}|x; d_j^i, a_j^i) = p_{i,j}(x_j^{i'}|x, a_j^i), \forall x_j^{i'}, x, d_j^i, a_j^i$ and the system transition probability $\mathbb{P}(X' = x'|X = x, \mathbf{d}, \mathbf{a}) = \mathbb{P}(X' = x'|X = x, \mathbf{a})$. Because the transition probability is independent of $\mathbf{d}$ and $\sum_{\mathbf{d}} \phi^{d*}(x, \mathbf{d}) \equiv 1$, bi-linear programming (7) can be reduced into a linear program (LP) where the prime LP is described as follows:

$$\min_{J^*(x), \phi^d(x,\mathbf{d})} \sum_{x' \in \prod_{i=1}^I \mathcal{X}^i} \alpha(x') J^*(x')$$

subject to :

$$(a) \ J^*(x) \geq \sum_{\mathbf{d} \in \mathcal{D}} c(x, \mathbf{d}, \mathbf{a}) \phi^d(x, \mathbf{d}) + \gamma \sum_{x' \in \prod_{i=1}^I \mathcal{X}^i} \mathbb{P}(x'|x, \mathbf{a}) J^*(x'), \forall x, \mathbf{a}$$

$$\tag{8}$$

$$(b) \ \sum_{\mathbf{d} \in \mathcal{D}} \phi^d(x, \mathbf{d}) = 1, \qquad\qquad\qquad\qquad\qquad \forall x$$

$$(c) \ \phi^d(x, \mathbf{d}) \geq 0, \qquad\qquad\qquad\qquad\qquad\qquad \forall x, \mathbf{d}$$

The major challenge to solve the LP is the large-scale nature of the infrastructure networks, which is known as the curse of dimension. Take (8) for an instance, we have $|\prod_{i=1}^I \mathcal{X}^i|$ variables in the LP objective and the number of constraints is $|\prod_{i=1}^I \mathcal{X}^i| \times |\mathcal{A}| + |\prod_{i=1}^I \mathcal{X}^i| + |\prod_{i=1}^I \mathcal{X}^i| \times |\mathcal{D}|$. If we have n nodes in the network of CIs and all nodes can be attacked and defended, then we will have $N := 2^n$ variables and $N^2 + N + N^2$ constraints, which both grow exponentially with the number of nodes. The high computation cost restrains the direct computation using the LP with a large number of nodes.

4   Factored Markov Game

To address the issue of the combinatorial explosion of the state size or the curse of dimensionality, we develop a factored Markov game framework in this section. The transition kernel is assumed to be sparse based on the fact that compromising a node usually has local influence within sequential stages. We first use factor graphs to represent the sparsity of the probability transition matrix. Next, we introduce an approximation method for the value function and then reorganize terms and eliminate variables by exploiting the factored structure. Finally, we refer our reader to an overall structure diagram of this work in Fig. 2.

4.1   Factored Structure

Define Ω_l as the set that contains all the parent nodes of node l. Parent nodes refer to the nodes that affect node l through physical, cyber or logic interactions. The network example in Fig. 3 is a bi-directed graph that represent a 3-layer

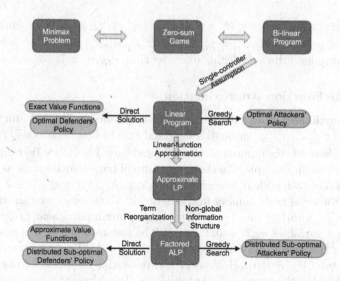

Fig. 2. In this overall structural diagram, blue squares show a sequence of techniques used in the problem formulation. The linear programming technique yields the exact value functions and the optimal defender's policy. The factored approximate linear program yields an approximate value function and distributed sub-optimal defender's policy. The greedy search method solves for the attacker's policy. (Color figure online)

interdependent critical infrastructures. Then, Ω_l contains node l's neighbors, e.g., $\Omega_{1,1} = \{n_1^1, n_2^1, n_1^2, n_7^3\}$. Note that we do not distinguish the dependence inside (links in black) and across (links in blue) layers when considering the stochastic kernel. We use a global index l to unify the 2D index of $\{i, j\}$, e.g., $l := \sum_{i'=1}^{i} i' m^{i'} + j$, which transforms the multi-layer network into a larger single network with $n = \sum_{i \in \mathcal{I}} m^i$ nodes. In this way, we can write $\Omega_{1,1} = \{n_1^1, n_2^1, n_1^2, n_7^3\}$ as $\Omega_1 = \{n_1, n_2, n_6, n_{19}\}$ and $p_{i,j}(x_j^{i'}|x, d_j^i, a_j^i), \forall i \in \mathcal{I}, j \in \mathcal{N}^i$ equivalently as $p_l(x_{l'}|x, d_l, a_l), \forall l = 1, 2, \cdots, n$. Define $x_{\Omega_l} := [x_l]_{l \in \Omega_l}$ as the state

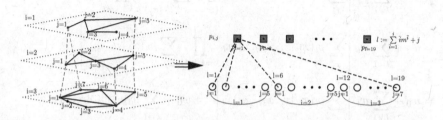

Fig. 3. The left network shows a 3-layer example of CIs with blue lines representing the interdependencies across layers. The right bipartite graph shows a factor graph representation of the sparse transition probability. The total node number $n = \sum_{i=1,2,3} m^i = 5 + 5 + 7 = 19$. (Color figure online)

vector of the nodes inside set Ω_l, e.g., $x_{\Omega_1} = [x_1, x_2, x_6, x_{19}]$. Then, each node's kernel will be $p_{i,j}(x_j^{i\,'}|x, d_j^i, a_j^i) = p_{i,j}(x_j^{i\,'}|x_j^i, x_{\Omega_{i,j}}, d_j^i, a_j^i)$ due to the sparsity, or in the global index form $p_l(x_l{}'|x, d_l, a_l) = p_l(x_l'|x_l, x_{\Omega_l}, d_l, a_l)$.

4.2   Linear Function Approximation

We first approximate the high dimensional space spanned by the cost function vector $\mathbf{J} = [J^*(x')]_{x' \in \prod_{i=1}^{l} \mathcal{X}^i}$ through a weighted sum of basis functions $h_l(x'), l = 0, 1, \cdots, k$, where k is the number of 'features' and $h_0(x') \equiv 1, \forall x'$. Take infrastructure networks as an example. We choose a group of basis which serves as an indicator function of each node n_j^i's working state, e.g., $h_{i,j}(x') = x_j^{i\,'}, \forall i \in \mathcal{I}, j \in \mathcal{N}_j^i$. Let $k = n$, the total node number in the network. To this end, we can substitute $J^*(x') = \sum_{l=0}^{k} w_l h_l(x')$ into (8) to obtain an approximate linear programming (ALP) with k variables $w_l, l = 0, 1, \cdots, k$. The feature number k is often much smaller than the system state number 2^n. Hence the ALP reduces the involving variables in the LP objective. However, the exponentially growing number of constraints still makes the computation prohibitive. To address this issue, we further reduce the computational complexity in the following sections with similar techniques in [7].

4.3   Term Re-organization

The system transition matrix $\mathbb{P}(x'|x, \mathbf{a})$ has the dimension of $N \times N \times |\mathcal{A}|$ in constraint (a) of (8). Here, we choose indicator functions of each node $h_l(x') = x_l, \forall x', l = \{1, 2, \cdots, n\}$ as the set of basis functions, which turns out to yield a good trade off between the accuracy and computation complexity as shown in Sect. 5. We observe that the right-most term of constraint (a) of (8) can be rewritten as follows:

$$\sum_{x' \in \prod_{i=1}^{l} \mathcal{X}^i} \mathbb{P}(x'|x, \mathbf{a}) \sum_{l=0}^{n} w_l h_l(x') = w_0 + \sum_{l=1}^{n} w_l \left[\sum_{x_1', \cdots, x_n'} \prod_{k=1}^{n} p_k(x_k'|x_k, a_k) x_l \right]$$

$$= w_0 + \sum_{l=1}^{n} w_l \left[\sum_{x_l'} p_l(x_l'|x_l, x_{\Omega_l}, a_l) x_l \sum_{\{x_1', \cdots, x_n'\} \backslash \{x_l'\}} \prod_{k=1, k \neq l}^{n} p_k(x_k'|x_k, a_k) \right]$$

$$= w_0 + \sum_{l=1}^{n} w_l \left[\sum_{x_l'} p_l(x_l'|x_l, x_{\Omega_l}, a_l) x_l \prod_{k=1, k \neq l}^{n} \sum_{x_k'} p_k(x_k'|x_k, a_k) \right]$$

$$= w_0 + \sum_{l=1}^{n} w_l \left[\sum_{x_l'} p_l(x_l'|x_l, x_{\Omega_l}, a_l) x_l \right] := w_0 + \sum_{l=1}^{n} w_l g_l(x_l, x_{\Omega_l}, a_l),$$

where the symbol $\sum_{\{x_1, \cdots, x_n\} \backslash \{x_l\}}$ means that a summation over all variables except x_l, and $g_l(x_l, x_{\Omega_l}, a_l) := p_l(x_l' = 1|x_l, x_{\Omega_l}, a_l)$. To this end, we reduce $N = 2^n$ summations over the huge dimension system transition matrix into $n + 1$ summations over the local stochastic kernel.

4.4   Restricted Information Structure

The second step is to deal with $\sum_{\mathbf{d}} c(x, \mathbf{d}, \mathbf{a})\phi^d(x, \mathbf{d})$ in the constraint (a) of (8). The saddle-point strategies studied in Sect. 3.2 belong to a class of global stationary policies in which the actions taken by the players are dependent on the global state information. The implementation of the policies is often restricted to the local information that is specific to the type of the infrastructure. For example, the Metropolitan Transportation Authority (MTA) may not be able to know the state of nodes in the power grid operated by Con Edison. Thus, MTA cannot make its policy based on the states of power nodes. Therefore, one way to approximate the optimal solution is to restrict the class of policies to stationary policies with local observations. We consider a time-invariant information structure of the defender $F_{j,t}^i \equiv F_j^i$. By unifying with the global index, we let $F_l := F_j^i$. Define $\phi_l^d(x, d_l)$ as the probability of node l choosing d_l at state x. Therefore, $\phi^d(x, \mathbf{d}) = \prod_{l=1}^n \phi_l^d(x, d_l) = \prod_{l=1}^n \phi_l^d(F_l, d_l)$ and $F_l = [x_{\bar{\Omega}_l}]$, where $\bar{\Omega}_l$ is the set of nodes which node l can observe. Note that not all nodes can be protected, i.e., $|\mathcal{D}| \leq N$. We let $d_l \equiv 0$ if node l cannot be defended.

$$
\sum_{\mathbf{d} \in \mathcal{D}} c(x, \mathbf{d}, \mathbf{a})\phi^d(x, \mathbf{d}) = \sum_{\mathbf{d} \in \mathcal{D}} \sum_{k=1}^n c_k(x_k, d_k, a_k) \prod_{l=1}^n \phi_l^d(F_l, d_l)
$$

$$
= \sum_{k=1}^n \left[\sum_{d_w, w=1,\cdots,|D|} c_k(x_k, d_k, a_k)\phi_k^d(F_k, d_k) \prod_{l=1, l\neq k}^n \phi_l^d(F_l, d_l) \right]
$$

$$
= \sum_{k=1}^n \left[\sum_{d_k} c_k(x_k, d_k, a_k)\phi_k^d(F_k, d_k) \prod_{l=1, l\neq k}^n \sum_{d_l} \phi_l^d(F_l, d_l) \right] \tag{9}
$$

$$
= \sum_{k=1}^n \left[\sum_{d_k \in \{0,1\}} c_k(x_k, d_k, a_k)\phi_k^d(F_k, d_k) \right].
$$

Therefore, the ALP with the restricted information structure can be further rewritten as follows to form the factored ALP:

$$
\min_{\mathbf{w},\phi_l^d(F_l,d_l)} \sum_{l=0}^n \alpha(w_l)w_l h_l(x)
$$

subject to :

$$
(a)\ 0 \geq \sum_{k=1}^n \sum_{d_k} c_k(x_k, d_k, a_k)\phi_k^d(F_k, d_k) + \sum_{l=0}^n w_l[\gamma g_l(x_l, x_{\Omega_l}, a_l) - h_l(x)], \ \forall x, a_k \tag{10}
$$

$$
(b)\ \sum_{d_i \in \{0,1\}} \phi_l^d(F_l, d_l) = 1, \hspace{4cm} \forall l, F_l
$$

$$
(c)\ 0 \leq \phi_l^d(F_l, d_l) \leq 1, \hspace{4cm} \forall l, F_l, d_l
$$

To this end, the number of constraints (b) $n \times |F_l|$ and (c) $n \times |F_l| \times 2$ relates only to the node number n and the domain of each node's information structure.

Remark: For a general zero-sum game with bi-linear formulation (7), we can extend constraint (a) with the same factored technique for all x, a_l,

$$0 \geq \sum_{l=1}^{n} \sum_{d_l} c_l(x_l, d_l, a_l)\phi_l^d(F_l, d_l) + \sum_{l=0}^{n} w_l\left[\gamma \sum_{d_l} g_l(x_l, x_{\Omega_l}, a_l, d_l)\phi_l^d(F_l, d_l) - h_l(x)\right],$$

where the second term is bi-linear in the variable of w_l and $\phi_l^d(F_l, d_l)$.

4.5 Variable Elimination

Constraint (a) of (10) can be further rewritten as one nonlinear constraint using the variable elimination method (see Sect. 4.2.2 of [6]) as follows:

$$0 \geq \max_{a_1,\cdots,a_n} \max_{x_1,\cdots,x_n} \sum_{k=1}^{n} \sum_{d_k} c_k(x_k, d_k, a_k)\phi_k^d(F_k, d_k) + \sum_{l=0}^{n} w_l[\gamma g_l(x_l, x_{\Omega_l}, a_l) - h_l(x)]. \quad (11)$$

For simplicity, we have provided above an inequality for the case of a local information structure $\phi_l^d(F_l, d_l) = \phi_l^d(x_l, x_{\Omega_l}, d_l)$ and $|F_l| = 2^{|\Omega_l|+1}$.

First, we eliminate variables of the attackers' action. Define $f_l(x_l, x_{\Omega_l}, a_l) := w_l[\gamma g_l(x_l, x_{\Omega_l}, a_l) - h_l(x_l)] + \sum_{d_l} c_l(x_l, d_l, a_l)\phi_l^d(x_l, d_l), l = 1, 2, \cdots, n$. We separate w_0, the weight of the constant basis, to the left-hand side and (11) becomes

$$(1 - \gamma)w_0 \geq \max_{x_1,\cdots,x_n} \max_{a_1,\cdots,a_n} \sum_{l=1}^{n} f_l(x_l, x_{\Omega_l}, a_l)$$

$$= \max_{x_1,\cdots,x_n} \sum_{l=1}^{n} \max_{a_l} f_l(x_l, x_{\Omega_l}, a_l) := \max_{x_1,\cdots,x_n} \sum_{l=1}^{n} e_l(x_l, x_{\Omega_l}). \quad (12)$$

To achieve the global optimal solution of (10), we require the following constraints for each l:

$$e_l(x_l, x_{\Omega_l}) \geq f_l(x_l, x_{\Omega_l}, a_l), \forall x_l, x_{\Omega_l}, a_l. \quad (13)$$

Note that if node n_l cannot be attacked, we take $a_l \equiv 0$ and arrive at a simplified form:

$$e_l(x_l, x_{\Omega_l}) = f_l(x_l, x_{\Omega_l}, 0), \forall x_l, x_{\Omega_l}. \quad (14)$$

The second step is to eliminate the variable of each node's state following a given order of $\mathcal{O} = \{p_1, p_2, \cdots, p_n\}$, where $\mathcal{O}$ is a perturbation of $\{1, 2, \cdots, n\}$. The RHS of (12) is rewritten as:

$$\max_{x_{p_2},\cdots,x_{p_n}} \sum_{l=\{1,\cdots,n\}\backslash\mathcal{K}} e_k(x_k, x_{\Omega_k}) + \max_{x_{p_1}} \sum_{k\in\mathcal{K}} e_k(x_k, x_{\Omega_k})$$

$$= \max_{p_2,\cdots,p_n} \sum_{l=\{1,\cdots,n\}\backslash\mathcal{K}} e_k(x_k, x_{\Omega_k}) + E_1(\mathcal{E}), \quad (15)$$

where the set $\mathcal{K} := \{k : P_1 \in \{\Omega_k \cup \{k\}\}\}$ and E_1's domain $\mathcal{E} := \{x_j : j \in \cup_{k \in \mathcal{K}} \Omega_k \cup \{k\} \setminus \{P_1\}\}$. The variable x_{p_1} is eliminated and similar new constrains are generated to form the new LP, i.e., $E_1(\mathcal{E}) \geq \sum_{k \in \mathcal{K}} e_k(x_k, x_{\Omega_k})$ for all variables included in $\mathcal{E}$.

We repeat the above procedure of variable eliminations and constraints generation for n times following the order $\mathcal{O}$ and finally reach the equation $(1 - \gamma)w_0 \geq E_n$, where E_n is a parameter independent of state and action variables. This method is suitable for a sparse network where each e_l has a domain involving a small set of node variables.

Example 1. Let us take a four node example in Fig. 4 to show how the variable elimination works. With node 2 immune to attacks, we have (14) to be $e_2(x_1, x_2) = f_1(x_1, x_2, 0), \forall x_1, x_2$. For node 1, (13) leads to four new inequality constraints $e_1(x_1) \geq f_1(x_1, a_1), \forall x_1, a_1$. Similarly, we have $2^4 = 16$ inequalities for node 3, i.e., $e_3(x_2, x_3, x_4) \geq f_3(x_2, x_3, x_4, a_3), \forall x_2, x_3, x_4, a_3$ and $2^3 = 8$ for node 4, i.e., $e_4(x_3, x_4) \geq f_3(x_3, x_4, a_4), \forall x_3, x_4, a_4$. After that, we eliminate all action variables and (12) becomes

$$(1 - \gamma)w_0 \geq \max_{x_1, x_2, x_3, x_4} e_1(x_1) + e_2(x_1, x_2) + e_3(x_2, x_3, x_4) + e_4(x_3, x_4). \quad (16)$$

Suppose an elimination order $\mathcal{O} = \{3, 2, 4, 1\}$, the RHS of (16) is rewritten as

$$\max_{x_1, x_2, x_4} e_1(x_1) + e_2(x_1, x_2) + \max_{x_3} e_3(x_2, x_3, x_4) + e_4(x_3, x_4)$$
$$= \max_{x_1, x_2, x_4} e_1(x_1) + e_2(x_1, x_2) + E_1(x_2, x_4).$$

New constraints are generated: $E_1(x_2, x_4) \geq e_3(x_2, x_3, x_4) + e_4(x_3, x_4), \forall x_2, x_3, x_4$. Then, we can repeat the above process and eliminate x_2, x_4, x_1 in sequence, i.e.,

$$\max_{x_1, x_2, x_4} e_1(x_1) + e_2(x_1, x_2) + E_1(x_2, x_4)$$
$$= \max_{x_1, x_4} e_1(x_1) + \max_{x_2} E_1(x_2, x_4) + e_2(x_1, x_2)$$
$$= \max_{x_1, x_4} e_1(x_1) + E_2(x_1, x_4) = \max_{x_1} \max_{x_4} e_1(x_1) + E_2(x_1, x_4) = \max_{x_1} E_3(x_1) = E_4.$$

Along with the above process, new constraints appear $E_2(x_1, x_4) \geq E_1(x_2, x_4) + e_2(x_1, x_2), \forall x_1, x_2, x_4$; $E_3(x_1) \geq e_1(x_1) + E_2(x_1, x_4), \forall x_1, x_4$ and $E_4 \geq E_3(x_1), \forall x_1$. Finally, (16) becomes $(1 - \gamma)w_0 \geq E_4$.

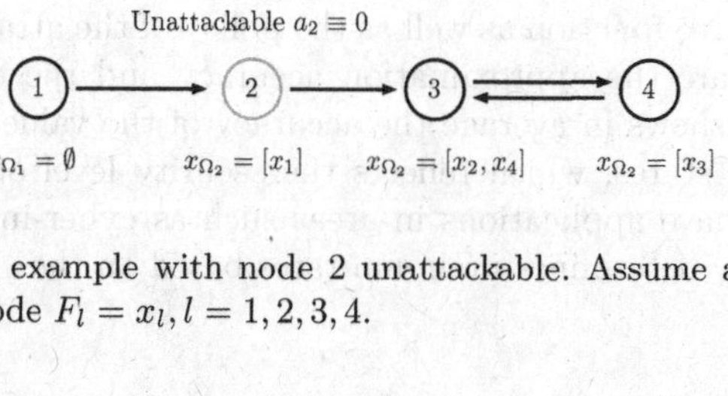

Fig. 4. A four node example with node 2 unattackable. Assume a local information structure for each node $F_l = x_l, l = 1, 2, 3, 4$.

The new LP in this example contains 51 constraints while the original constraint (a) possess $2^{(4+3)} = 128$ constraints. With the increase of the node number and a sparse topology, our factored framework greatly reduces the exponential computation complexity.

4.6 Distributed Policy of Attacker

We use the greedy search for the approximate saddle-point policy of the attacker, i.e., for all $x_1, \cdots, x_n$,

$$\mathbf{a}^* \in arg \max_{a_1, \cdots, a_n} \sum_{k=1}^{n} \sum_{d_k \in \{0,1\}} c_k(x_k, d_k, a_k) \phi_k^{d*}(F_k, d_k) + \sum_{l=0}^{n} w_l \gamma g_l(x_l, x_{\Omega_l}, a_l).$$

Separate w_0 in the second term and obtain:

$$\mathbf{a}^* \in arg \max_{a_1, \cdots, a_n} \sum_{k=1}^{n} \sum_{d_k \in \{0,1\}} c_k(x_k, d_k, a_k) \phi_k^{d*}(F_k, d_k) + w_k \gamma g_k(x_k, x_{\Omega_k}, a_k).$$

Exchange the argmax and the summation, and we arrive at

$$\mathbf{a}^* \in \sum_{k=1}^{n} arg \max_{a_k} \sum_{d_k \in \{0,1\}} c_k(x_k, d_k, a_k) \phi_k^{d*}(F_k, d_k) + w_k \gamma g_k(x_k, x_{\Omega_k}, a_k).$$

Therefore, we can obtain a distributed attack's policy of node k which is fully determined by the state of itself and its parent nodes x_k, x_{Ω_k} and the state of nodes observable for the defender F_k, i.e.,

$$a_k \in arg \max_{a_k} \sum_{d_k \in \{0,1\}} c_k(x_k, d_k, a_k) \phi_k^{d*}(F_k, d_k) + w_k \gamma g_k(x_k, x_{\Omega_k}, a_k), \forall x_k, x_{\Omega_k}, F_k.$$

Remark: Under a local information structure with $F_l = x_l$, the defender decides its action at node l based on x_l and yet the attacker requires state information of x_l and x_{Ω_l}. The difference in the structures of the policies is caused by the distinct factored structures of the cost function and the attacker-controlled transition probability matrix. The former $c_k(x_k, d_k, a_k)$ contains only x_k and the latter $g_l(x_l, x_{\Omega_l}, a_l)$ contains both x_l and x_{Ω_l}.

5 Numerical Experiments

We implement our framework of the factored single-controller game and investigate the LP objective function as well as the policy of the attacker and defender. Besides, we compare the approximation accuracy and the computation time. The LP objective shows in average the accuracy of the value functions starting at different initial states, which reflects the security level of the system. This risk analysis can have applications in areas such as cyber-insurance where the insurance company determines premium rates based on the risk level of insured systems.

5.1 Transition Probability and Cost

To illustrate the algorithm, we take one node's failure probability proportional to the failure number of its neighboring nodes. After one node is attacked, it can infect the connecting nodes and increase their failing risks. Besides, a node has a larger failure probability if it is targeted directly by attackers. In an attacker-controlled game, the defender cannot change the failure probability yet can positively affect the cost function.

The system stage cost is the sum of the local stage cost of each node $c(x, \mathbf{a}, \mathbf{d}) = \sum_{l=1}^{n} c_l(x_l, a_l, d_l)$, where $c_l(x_l, a_l, d_l) = \xi_1(1-x_l) - \xi_2 a_l + \xi_3 d_l - \xi_4 a_l d_l$ with positive weights $\xi_i > 0, i = 1, 2, 3, 4$. The explicit form consists of four terms: the loss for faulty nodes, a cost of applying attacks, protection costs, and a reward of protecting a node which is being attacked. The second and fourth terms are negative because more costs from attackers means more rewards to the defender with the zero-sum assumption. The ordering of $\xi_1 > \xi_4 > \xi_3 > \xi_2$ is assumed because the functionality of nodes serves as our primary goal. Protections are more costly than attacks, however, once an adversary attacks the node that possesses defensive strategies, e.g., a honeypot, which will create a significant loss for the attacker.

5.2 Approximation Accuracy

We use a directed ring topology to show the accuracy of the linear function approximation under the local information structure assumption. The comparison is limited to a network with 7 or fewer number of nodes due to the state explosion of the exact LP as shown in Table 1. The computational time indicates the increasing efficiency of the approximate algorithm as the number of nodes

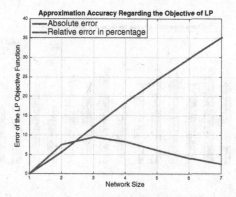

Fig. 5. Approximation accuracy for a directed ring topology. We use $obj(exact)$ and $obj(ALP)$ as the value of the objective function for the exact and approximate LP respectively. The blue and red lines are the absolute error $obj(ALP) - obj(exact)$ and the relative error $(obj(exact) - obj(ALP))/obj(exact)$ respectively. (Color figure online)

Table 1. Time cost (units: seconds) for the directed ring with a increasing node number.

Network size	2	3	4	5	6	7
Exact LP	0.214229	0.629684	3.329771	34.51808	178.6801	1549.016
ALP	2.664791	2.755704	2.749961	2.769759	3.238778	3.534943

increases. Figure 5 illustrates the fact that the growth of the network size causes an increase of the absolute error, which is inevitable due to the growth of difference $2^n - n$ as n grows. In particular, the linear growth of the ALP variables $w_i, i \in \{0, 1, ..., n\}$, may not catch up with the exponential growth of the exact LP variables $v(\mathbf{x}), x \in \mathcal{X}$. Fortunately, the relative error decreases when the number of nodes in the network is larger than 3. Therefore, the error becomes negligible with a large number of nodes.

5.3   Various Information Structure

In Fig. 6, we compare the influence of global and local information structure of the defender to the exact LP. Recall that the y-axis shows the optimal cost of the system and a smaller value introduces a more secure system. Then, a local information structure in red brings a higher system cost than a global information structure in green for all initial states. It shows that more knowledge can help defender better correspond to the threat from the attacker. We can understand this with an example of the information structure of its neighboring nodes. Since the failure of its neighboring nodes increases its risk of being attacked, it tends to defend itself even when it is still working yet all its neighbors fail. Apparently, a defender with local information structure cannot achieve that.

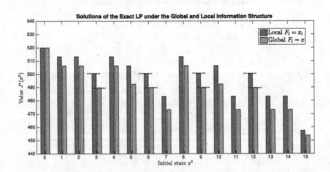

Fig. 6. Value functions of different initial states in a four node directed ring topology. State $0, 1, \cdots, 15$ is a decimalization of 2^4 different states from 0000 to 1111. Because the topology is symmetric, the number of working nodes determines the value. For example, states $3, 6, 9, 12$ share the same value in either global or local information structure because they all have two working nodes. Besides, a better initial state 1111 with all nodes working causes less loss of the system.

5.4   Network Effect

We reorganize the value-initial state pair $(J^*(x^0), x^0)$ of a 6-node ring topology in the top of Fig. 7 in an increasing order. Then, we see that the number of faulty nodes dominates the order of value. However, when the number of failures is the same, the location of the failure has an impact on $J^*(x^0)$, and a high degree of the failure aggregation results in a less secure system. For example, $J^*(111000) > J^*(110010) > J^*(101010)$ because the dense pattern of the state vector 111000 is more likely to cause a cascading failure in the network than a sparse one 101010. These results suggest protecting a node by strengthening all the connected nodes if we cannot consolidate every node due to a limited budget.

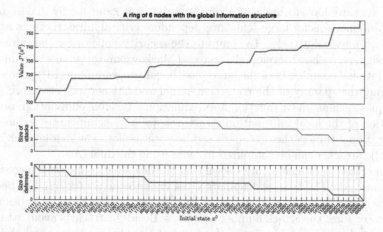

Fig. 7. Value function $J^*(x^0)$ and the number of defending nodes at the optimal policy for different initial states x^0 in a 6-node ring example. From the value function (the blue line), the size of failures (the number of failure nodes) as well as the location of the failures affect the security level of the system. At the equilibrium policy, the number of defending nodes (the red line) is proportional to the number of the working nodes in the network. The attacker (the green line) decreases the number of nodes to attack as more nodes have been taken down. (Color figure online)

5.5   Optimal Policy

The global stationary policies of defenders and attackers for a 6-node ring topology is shown in Fig. 7 in red and green respectively. We observe that the size of defense is proportional to the number of working nodes in the network while the attacker compromises less nodes with the size of failure. Since the defender can only affect the system through the reward function, the defense's policy follows an opposite pattern of the value function. The attacker, on the other hand, has a more irregular pattern, because it can also influence the transition of the system.

Other results of the approximated policy are summarized below. The local stationary defender policy is to defend a normal node with a higher probability.

The defender does not defend the faulty nodes, because the recovery of a failed node cannot mitigate the loss. Furthermore, if we reduce the cost of state failure ξ_1 or increase the defense cost ξ_3, we observe that the defender is less likely to defend. The sub-optimal distributed attacker policy avoids attacking node l when nodes in Ω_l are working. With an increase in ξ_4, the total number of attacks decreases to avoid attacking protected nodes. Thus, the presence of the defender results in fewer attacks. Besides, when node k cannot be attacked, then, naturally node k will not be defended, and attacker tends to decrease attack levels on the parent nodes of k.

6   Conclusion

In this work, we have formulated a zero-sum dynamic game model to design protection mechanisms for large-scale interdependent critical infrastructures against cyber and physical attacks. To compute the security policies for the infrastructure designers, we have developed a factored Markov game approach to reduce the computational complexity of the large-scale linear programming (LP) problem by leveraging the sparsity of the transition kernel and the network structure. With techniques such as linear function approximations, variable elimination, and the restriction of the local information structures, we have significantly reduced the computational time of the defender's saddle-point policy. The saddle-point strategy of the attacker can be computed likewise using the dual LP.

Numerical experiments have shown that the defender's policy can successfully thwart attacks. The lack of defenders gives rises to the attack number because the attack cost is negligible comparing to the system loss. As more nodes equip with protections, the attack number decreases. Besides, attackers avoid attacking nodes with healthy neighboring nodes because they have a larger probability of survival and are also more likely to be protected. The global stationary policy of defender of each state depends on the security level at that state because of the single-controller assumption. Moreover, with more information or observations of the system states available to the defender, the infrastructure is shown to be more secure under the saddle-point equilibrium security policy. Finally, a ring topology example has illustrated an increasing approximation accuracy when the number of nodes grows. It also shows that the localized information structure introduces an acceptable approximation error.

References

1. Chen, J., Zhu, Q.: Interdependent network formation games with an application to critical infrastructures. In: American Control Conference (ACC), pp. 2870–2875 (2016)
2. Chen, J., Zhu, Q.: Optimal contract design under asymmetric information for cloud-enabled internet of controlled things. In: Zhu, Q., Alpcan, T., Panaousis, E., Tambe, M., Casey, W. (eds.) GameSec 2016. LNCS, vol. 9996, pp. 329–348. Springer, Cham (2016). doi:10.1007/978-3-319-47413-7_19

3. Chen, J., Zhu, Q.: Resilient and decentralized control of multi-level cooperative mobile networks to maintain connectivity under adversarial environment. In: IEEE Conference on Decision and Control (CDC), pp. 5183–5188 (2016)
4. Chen, J., Zhu, Q.: Security as a service for cloud-enabled internet of controlled things under advanced persistent threats: a contract design approach. IEEE Trans. Inf. Forensics Secur. **12**(11), 2736–2750 (2017)
5. De Farias, D.P., Van Roy, B.: On constraint sampling in the linear programming approach to approximate dynamic programming. Math. Oper. Res. **29**(3), 462–478 (2004)
6. Guestrin, C., Koller, D., Parr, R., Venkataraman, S.: Efficient solution algorithms for factored MDPs. J. Artif. Intell. Res. **19**, 399–468 (2003)
7. Huang, L., Chen, J., Zhu, Q.: A factored MDP approach to optimal mechanism design for resilient large-scale interdependent critical infrastructures. In: Proceedings of 2017 Workshop on Modeling and Simulation of Cyber-Physical Energy Systems (MSCPES), CPS Week, 18–21 April 2017, Pittsburgh, PA, USA (2017)
8. Korkali, M., Veneman, J.G., Tivnan, B.F., Hines, P.D.: Reducing cascading failure risk by increasing infrastructure network interdependency. arXiv preprint arXiv:1410.6836 (2014)
9. Lee II, E.E., Mitchell, J.E., Wallace, W.A.: Restoration of services in interdependent infrastructure systems: a network flows approach. IEEE Trans. Syst. Man Cybern. Part C (Appl. Rev.) **37**(6), 1303–1317 (2007)
10. Li, L., Shamma, J.: LP formulation of asymmetric zero-sum stochastic games. In: 2014 IEEE 53rd Annual Conference on Decision and Control (CDC), pp. 1930–1935. IEEE (2014)
11. Malek, A., Abbasi-Yadkori, Y., Bartlett, P.: Linear programming for large-scale Markov decision problems. In: International Conference on Machine Learning, pp. 496–504 (2014)
12. Manshaei, M.H., Zhu, Q., Alpcan, T., Başar, T., Hubaux, J.P.: Game theory meets network security and privacy. ACM Comput. Surv. (CSUR) **45**(3), 25 (2013)
13. Monga, A., Zhu, Q.: On solving large-scale low-rank zero-sum security games of incomplete information. In: 2016 IEEE International Workshop on Information Forensics and Security (WIFS), pp. 1–6. IEEE (2016)
14. Ouyang, M.: Review on modeling and simulation of interdependent critical infrastructure systems. Reliab. Eng. Syst. Saf. **121**, 43–60 (2014)
15. Pawlick, J., Farhang, S., Zhu, Q.: Flip the cloud: cyber-physical signaling games in the presence of advanced persistent threats. In: Khouzani, M.H.R., Panaousis, E., Theodorakopoulos, G. (eds.) GameSec 2015. LNCS, vol. 9406, pp. 289–308. Springer, Cham (2015). doi:10.1007/978-3-319-25594-1_16
16. Pederson, P., Dudenhoeffer, D., Hartley, S., Permann, M.: Critical infrastructure interdependency modeling: a survey of US and international research. Ida. Nat. Lab. **25**, 27 (2006)
17. Rinaldi, S.M., Peerenboom, J.P., Kelly, T.K.: Identifying, understanding, and analyzing critical infrastructure interdependencies. IEEE Control Syst. **21**(6), 11–25 (2001)
18. Rosato, V., Issacharoff, L., Tiriticco, F., Meloni, S., Porcellinis, S., Setola, R.: Modelling interdependent infrastructures using interacting dynamical models. Int. J. Crit. Infrastruct. **4**(1–2), 63–79 (2008)
19. Zhu, Q., Başar, T.: Game-theoretic approach to feedback-driven multi-stage moving target defense. In: Das, S.K., Nita-Rotaru, C., Kantarcioglu, M. (eds.) GameSec 2013. LNCS, vol. 8252, pp. 246–263. Springer, Cham (2013). doi:10.1007/978-3-319-02786-9_15

20. Zhu, Q., Basar, T.: Game-theoretic methods for robustness, security, and resilience of cyberphysical control systems: games-in-games principle for optimal cross-layer resilient control systems. IEEE Control Syst. **35**(1), 46–65 (2015)
21. Zhu, Q., Fung, C., Boutaba, R., Basar, T.: Guidex: a game-theoretic incentive-based mechanism for intrusion detection networks. IEEE J. Sel. Areas Commun. **30**(11), 2220–2230 (2012)
22. Zhu, Q., Li, H., Han, Z., Basar, T.: A stochastic game model for jamming in multi-channel cognitive radio systems. In: 2010 IEEE International Conference on Communications (ICC), pp. 1–6. IEEE (2010)
23. Zhu, Q., Tembine, H., Başar, T.: Network security configurations: a nonzero-sum stochastic game approach. In: American Control Conference (ACC), pp. 1059–1064. IEEE (2010)
24. Zimmerman, R., Zhu, Q., De Leon, F., Guo, Z.: Conceptual modeling framework to integrate resilient and interdependent infrastructure in extreme weather. J. Infrastruct. Syst. **23**, 04017034 (2017)

VIOLA: Video Labeling Application for Security Domains

Elizabeth Bondi[1]([✉]), Fei Fang[2], Debarun Kar[1], Venil Noronha[1],
Donnabell Dmello[1], Milind Tambe[1], Arvind Iyer[3], and Robert Hannaford[3]

[1] University of Southern California, Los Angeles, CA, USA
bondi@usc.edu
[2] Carnegie Mellon University, Pittsburgh, PA, USA
[3] Air Shepherd, Berkeley Springs, WV, USA

Abstract. Advances in computational game theory have led to several successfully deployed applications in security domains. These game-theoretic approaches and security applications learn game payoff values or adversary behaviors from annotated input data provided by domain experts and practitioners in the field, or collected through experiments with human subjects. Beyond these traditional methods, unmanned aerial vehicles (UAVs) have become an important surveillance tool used in security domains to collect the required annotated data. However, collecting annotated data from videos taken by UAVs efficiently, and using these data to build datasets that can be used for learning payoffs or adversary behaviors in game-theoretic approaches and security applications, is an under-explored research question. This paper presents VIOLA, a novel labeling application that includes (i) a workload distribution framework to efficiently gather human labels from videos in a secured manner; (ii) a software interface with features designed for labeling videos taken by UAVs in the domain of wildlife security. We also present the evolution of VIOLA and analyze how the changes made in the development process relate to the efficiency of labeling, including when seemingly obvious improvements surprisingly did not lead to increased efficiency. VIOLA enables collecting massive amounts of data with detailed information from challenging security videos such as those collected aboard UAVs for wildlife security. VIOLA will lead to the development of a new generation of game-theoretic approaches for security domains, including approaches that integrate deep learning and game theory for real-time detection and response.

Keywords: UAV · Security · Video surveillance · Labeling application

1 Introduction

Security has already widely benefited from the use of game theory to develop better protection strategies. Game-theoretic approaches have led to applications that have been successfully deployed in infrastructure security domains such

© Springer International Publishing AG 2017
S. Rass et al. (Eds.): GameSec 2017, LNCS 10575, pp. 377–396, 2017.
DOI: 10.1007/978-3-319-68711-7_20

as protecting airports, ports and metro systems [28], as well as in green security domains such as protecting wildlife, forests, and fisheries [8,9,11]. In these game-theoretic approaches and security applications, input data are needed to determine the payoff structure of the game, to learn the behavioral models of the players, and to predict where attackers are more likely to attack. In previous efforts, the data were provided by domain experts directly [24], recorded by practitioners in the field over months or years [13,19], or collected through human subject experiments on platforms such as Amazon Mechanical Turk (AMT) [12].

With the recent use of unmanned aerial vehicle (UAV) technology in security domains, videos taken by UAVs have become an emerging source of massive data [10], especially in the domain of wildlife protection (e.g., the PAWS security games application [8]). For example, detecting wildlife from UAV videos can help estimate the animal distribution density, which decides the payoff structure of the game. Detecting poachers and their movement patterns could lead to successful learning of attackers' behavioral models, which is an important topic in security games [12,20]. Data collected from UAVs can not only be used to provide input data to the game-theoretic models, but can also enable the development of a new generation of game-theoretic tools for security. The data can be used to train or fine-tune a deep neural network to automatically detect attackers from the video taken by the UAVs in real-time.

Unfortunately, collecting labeled data from videos taken by UAVs can be a labor-intensive, time-consuming task. To our knowledge, there is no existing application that focuses on assisting in the labeling of videos taken by UAVs in security domains. Existing applications for labeling images [6,7] cannot be directly applied to labeling videos, as treating each frame as a separate image can lead to inefficiency since it does not exploit the correlation between frames. Video labeling applications such as VATIC [29] attempt to choose key frames for labeling, or track objects through the video. However, in UAV videos with camera motion, possibly collected using a different wavelength, these methods may not apply and may lead to inaccurate results or extra work for labelers, since the position of the objects in the video may change abruptly and the lack of color bands makes the tracking much more difficult. Furthermore, these applications are often paired with AMT to get labeled video datasets from online workers. However, in a security domain with sensitive data, meaning data that would provide attackers with some knowledge of defenders' strategies should it be shared, it may be undesirable to use AMT. This would then require finding labelers, and setting up an internal system to keep the process organized.

In this paper, we focus on better collection of labeled data from UAVs to provide input for game-theoretic approaches for security, and in particular to security game applications for wildlife conservation such as PAWS [8]. There has been work on labeling tools in domains such as computer vision and cyber security [5,6], but there exists no work on labeling tools for game-theoretic approaches in security domains. Most previous work on game theory for security ignores where the payoffs and behavioral models come from, and we fill the gap.

In particular, we will focus on labeling videos taken by long wave thermal infrared (hereafter referred to as thermal infrared) cameras installed on UAVs, in the domain of wildlife security. We present VIOLA (VIdeO Labeling Application), a novel application that assists labeling objects of interest such as wildlife and poachers. VIOLA includes a workload distribution framework to efficiently gather human labels from videos in a secured manner. We distribute the work of labeling the videos and reviewing the labels amongst a small group of labelers to ensure efficiency and data security. VIOLA also provides an easy-to-use interface, with a set of features designed for UAV videos in the wildlife security domain, such as allowing for moving multiple bounding boxes simultaneously and tracking bright spots in the video automatically. We will also discuss the various stages of development to create VIOLA, and we will analyze the impact of different labeling procedures and versions of the labeling application on efficiency, with a particular emphasis on the surprising results that showed some changes did not increase the efficiency.

2   Related Work

Game-theoretic approaches have been widely used in infrastructure and green security domains [28]. In green security domains such as protecting wildlife from poaching, multiple research efforts in artificial intelligence and conservation biology have attempted to estimate wildlife distribution and poacher activities [8]; such efforts often rely on months or years of recorded data [13,19]. With the recent advances in unmanned aerial vehicle (UAV) technology, there is an opportunity to provide detailed data about wildlife and poachers for game-theoretic approaches. Since a poacher is rewarded for successfully poaching wildlife, the wildlife distribution determines the payoff structure of the game. Poachers' behavioral models can be inferred from poaching activities and be used to design better patrol strategies with game-theoretic reasoning. In addition, game-theoretic patrolling with alarm systems [1,4] has been studied. UAVs can provide input for such systems in real-time using computer vision, particularly by detecting attackers or suspicious human beings in the UAV videos.

Detecting attackers in the UAV videos is related to object detection. Recently, great progress has been achieved in computer vision by deep learning in object detection and recognition [25,26]. However, state-of-the-art detectors cannot be directly applied to our aerial videos because most methods focus on detection in high resolution, visible spectrum images. An alternative approach to this detection is to track moving objects throughout videos. Tracking of both single and multiple objects in videos has been studied extensively [31]. These methods also rely on high resolution visible spectrum videos. Single object trackers use discriminant features from high resolution videos to establish correspondences [14]. Much of multi-object tracking research is directed towards pedestrians [3, 17,32], and primarily focuses on visible spectrum videos with high resolution, or videos taken from a fixed camera (except [17]).

Simpler and more general tracking algorithms exist that do not necessarily have these dependencies, such as the Lucas-Kanade tracker for optical flow [15], popular in the OpenCV package, and general correlation-based tracking [16]. Small moving objects can also be detected by a background subtraction method after applying video stabilization [22]. Because these methods are more general, they are still applicable to our domain and were explicitly tested, but still did not perform well in many cases. For example, since the video stabilization and background subtraction method assumes a planar surface, in the case of more complex terrain, there were many noisy detections. Instead of using tracking for detection, we therefore decided to focus on deep learning.

In order to use deep learning-based detection methods with aerial, thermal infrared data, hand-labeled training data are required to fine-tune the networks or even train them from scratch. In addition to video labeling applications such as VATIC [29], there has been work on semi-automatic labeling [30] and label propagation [2] which combines the effort of human labelers and algorithms to speed up the labeling process for videos. This work often focuses on how to select the frames for human labelers to label and how to propagate the labels for the remaining frames. This is difficult for our domain because of the motion of UAVs, and because it is often hard for humans to tell which objects are of interest without seeing the object's motion. As a result, we sought to develop our own labeling application, VIOLA. The first key component of the application is a workload distribution framework. A common framework for image and video labeling is a majority voting framework [18,21,23,27]. VIOLA uses a framework based upon [7] to efficiently gather labels from a small group of labelers. We examine the framework further in Sects. 6 and 7.

3   Domain

There has recently been increased use of UAVs for security surveillance. UAVs are able to cover more ground than a stationary camera and can provide the defenders more advanced notice of a potential threat. To detect suspicious human activities at night, the UAVs can be equipped with thermal infrared cameras. This is the type of UAV video we deal with in our domain, since poaching often occurs at night. We will specifically be able to use these types of data to detect poachers and provide advanced notice to park rangers, and use these detections to provide input for patrol generation tools such as PAWS.

In order to accomplish this, we need labeled data from the thermal infrared, UAV videos in the form of rectangular "bounding boxes" for objects of interest (animals and poachers) in each frame, with a color corresponding to their classification. However, the movement of UAVs and the thermal infrared images make it extremely difficult to label videos in this domain. First, thermal infrared cameras are low-resolution, and typically show warmer objects as brighter pixels in the image, although the polarity could be reversed occasionally. Different phenomena could also cause brighter pixels without a warm object. For example, the ground warms during the day, and then emits heat at night, which can be reflected under a tree canopy and lead to an amplified signal that might look like a human or animal. Furthermore, vegetation often looks bright and similar to objects of interest, as in Fig. 1, where there are three humans labeled with bounding boxes, amongst many other bright objects. Second, since the data are captured aboard a moving UAV, these data often vary drastically. For example, the resolution, and therefore size of targets, is very different throughout our dataset because the UAV flies at varying altitudes.

In addition to difficult, variable video data to begin with, some videos may have many objects of interest in them, whereas some videos may not have any objects of interest at all. It sometimes takes a long time to determine if there are any objects of interest, and it also often takes a long time to label when there are many objects of interest. To illustrate the variation in the number of objects of interest, we analyze the historical videos we get from our collaborator. Figure 2 shows a histogram of the average number of labels per frame, meaning that all frames in the video were counted, regardless of whether or not they were labeled, and a histogram of the average number of labels per labeled frame, meaning only frames that had at least one label were counted.

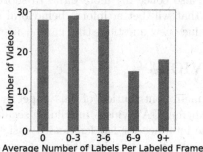

Fig. 1. An example of a thermal infrared frame, where the three humans outlined by the white boxes look very similar to the surrounding vegetation.

Although we focus on UAV videos in wildlife security domains, similar challenges in UAV videos in other security domains can be expected. Therefore, the application VIOLA we introduce in this paper can potentially be applied to other security domains to provide input for game-theoretic approaches.

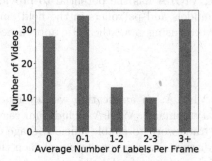

Fig. 2. A histogram with the number of videos for average objects of interest per frame (left), and the average objects of interest per labeled frame (right).

4   Example Game-Theoretic Uses

We now provide two more specific examples of game-theoretic approaches that may be derived from the data acquired using VIOLA. First, we focus on using the labeled data directly for behavioral models. Second, we discuss using the labeled data to train deep learning models for further data analysis.

With the labels provided by VIOLA and information about each frame, such as GPS and camera angle, we can locate poachers exactly throughout labeled videos. As such, we know the exact location of poaching activities and could use this information to learn how the poachers make decisions on where to poach. In particular, we could use an existing behavioral model, such as SUQR [20], and the location of poaching activity derived from the labels to update or improve the behavioral model for poachers, which would better inform patrol strategies. Furthermore, we could analyze the movement of the poachers, and a new behavioral model could be built using these movement patterns, in which poachers could choose a path instead of simply choosing a target to attack. This new behavioral model could be exploited to plan game-theoretic patrols.

In addition to directly using the labels from VIOLA for behavioral models, the labels could be used to train a deep learning model to automatically identify poachers in real-time video streams. Similarly, we could use the output from the deep learning algorithm for behavioral models, and the automated identification would allow us to circumvent the need for human labelers when incorporating data collected in the future into the behavioral models. Moreover, patrollers could make online decisions during patrols without the need for additional personnel to monitor the videos in the field. The ability to make online decisions during patrols could lead to new models of game-theoretic patrolling. Patrols could even be made for the UAVs themselves, which could introduce some behavioral challenges. The UAVs could also potentially be used as a deterrent, so flying UAVs could serve to both detect and deter poaching activities, while also collecting more data. In short, VIOLA has the potential to provide data that will better inform behavioral models and patrollers in the field, and introduce new questions that can be answered using game-theoretic approaches.

5   VIOLA

The main contribution of this paper is VIOLA, an application we developed for labeling UAV videos in wildlife security domains. VIOLA includes an easy-to-use interface for labelers and a basic framework to enable efficient usage of the application. In this section, we first discuss the user interface and then the framework for work distribution and training process for labelers.

5.1   User Interface of VIOLA

The user interface of VIOLA was written in Java and Javascript, and hosted on a server through a cloud computing service so it could be accessed using a URL from anywhere with an internet connection.

Before labeling, labelers were asked to login to ensure data security (Fig. 3a). The first menu that appears after login (Fig. 3b) asks the labeler which mode they would like, whether they would like to label a new video or review a previous submission. Then, after choosing "Label", the second menu (Fig. 3c) asks them to choose a video to label. Figure 4 is an example of the next screen used for labeling, also with sample bounding boxes that might be drawn at this stage. Along the top of the screen is an indication of the mode and the current video name, and along the bottom of the screen is a toolbar. First, in the bottom left corner, is a percentage indicating progress through the video. Then, there are four buttons used to navigate through the video. The two arrows move backwards or forwards, the play button advances frames at a rate of one frame per second, and the square stop button returns to the first frame of the video. The next button is the undo button, which removes the bounding boxes just drawn in the current frame, just in case they are too tiny to easily delete. Also to help with the nuisance of creating tiny boxes by accident while drawing a new bounding box or while moving existing bounding boxes, there is a filter on bounding box size. The trash can button deletes the labeler's progress and takes them back to the first menu after login (Fig. 3b). Otherwise, work is automatically saved after each change and re-loaded each time the browser is closed and re-opened. The application asks for confirmation before deleting the labeler's progress and undoing bounding boxes to prevent accidental loss of work. The check-mark button is used to submit the labeler's work, and is only pressed when the whole video is finished. Again, there is a confirmation screen to avoid accidentally submitting half of a video. The copy button and the slider will be described further in Sect. 6. The eye button allows the labeler to toggle the display of the bounding boxes on the frame, which is often helpful during review to check that the labels are correct. Finally, the question mark button provides a help menu with a similar summary of the controls of the application (Fig. 5). Notice the bounding boxes surrounding the animals in this video are colored red. Humans would be colored blue. This is also included in the help menu.

To draw bounding boxes, the labeler can simply click and drag a box around the object of interest, then click the box until the color reflects the class. Deleting a bounding box is done by pressing SHIFT and click, and selecting multiple bounding boxes is done by pressing CTRL and click, which allows the labeler to move multiple bounding boxes at once. Finally, while advancing frames, bounding boxes drawn in the current frame are moved to the next frame. It only happens the first time a frame is viewed since it could otherwise add redundant bounding boxes or replace the bounding boxes originally added by the labeler.

If "Review" is chosen in the first menu after login, the second menu also asks the labeler to choose a video to review, and then a third menu (Fig. 3d) asks them to choose a labeling submission to review. It finally displays the video with the labels from that particular submission, and they may begin reviewing the submission. The two differences between the labeling and review modes in the application are (i) that the review mode displays an existing set of labels and (ii) that labels are not moved to the next frame in review mode.

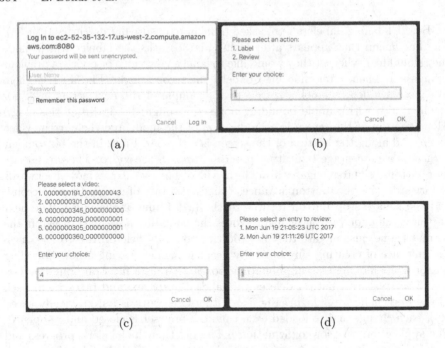

Fig. 3. The menus to begin labeling.

5.2 Use of VIOLA

Our goal in labeling the challenging videos in the wildlife security domain is first to keep the data secure, and second, to collect more usable labels to provide input for game-theoretic tools for security. In addition, we aim for getting exhaustive labels with high accuracy and consistency. To achieve these goals, we distribute the work among a small group of labelers in a secured manner, assign labelers to either provide or review others' labels, and supply guidelines for the labelers.

Distribution of Work. To keep the data (historical videos from our collaborators) secure, instead of using AMT, we recruit a small group of labelers, in this work 13. Labelers are given a username and password to access the labeling interface, and the images on the labeling interface cannot be downloaded.

In order to achieve label accuracy, we use a framework of label and review. The idea is simply that one person labels a video, and another person checks, or reviews, the labels of the first person. By checking the work of the labeler, the reviewer must agree or disagree with the original set of labels instead of creating their own. Upon disagreement, the reviewer can change the original labels. This was primarily chosen because it was clean, leading to one set of final labels.

We use spreadsheets to share both assignments and completion progress with the team of labelers. We ask labelers to include the time it took for them to complete their assignment in order to help make future assignments more reasonable in terms of time commitment, and in order to track the efficiency and success of

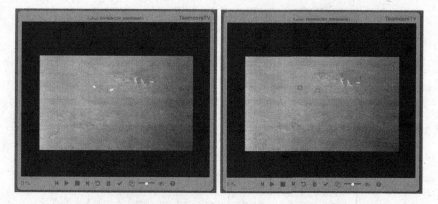

Fig. 4. An example of a frame (left) and labeled frame (right) in a video. This is the next screen displayed after all of the menus, and allows the labeler to navigate through the video and manipulate or draw bounding boxes throughout.

the application itself. In addition, we split long videos into segments to make it easier to respect labelers' time commitments, and to finish extremely long videos quickly. There are also some videos that have long periods of nothingness, which are easier to ignore when the video is split.

Guidelines and Training for Labelers. In order to achieve accuracy and consistency of labels, we provide guidelines and training for the labelers. During the training, we show the labelers several examples of the videos and point out the features of interest. We provide them with general guidelines on how to start labeling a video, as below.

In general, the process for labeling should be:

- Watch the video once all the way through and try to decide what you see.
- Once you have an idea of what is happening in the video by going through it, return to the beginning of the video and start labeling.
- Make and move bounding boxes.
- Send screenshots (including the percentage in the videos) if you need help.

In general, the process for reviewing should be:

- Refer to the guidelines and special circumstances directions.
- Go through the video, and use the eye button to check the original labels.
- Move, create, or delete bounding boxes as necessary, either as you go or after watching the whole video. Try not to resize the bounding boxes unless they are much too big or too small. Only change the classification and add or delete boxes if certain, and please confirm with us if not.
- Send screenshots (including the percentage in the videos) if you need help.

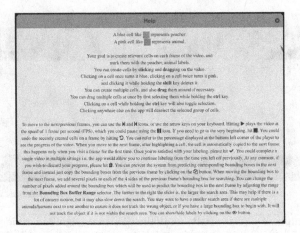

Fig. 5. Help screen detailing the controls of the application (? icon). (Color figure online)

Fig. 6. Three consecutive frames where the middle frame has ghosting. The middle frame is "in between" the left and right frames.

We also provide special instructions for the videos in our domain of interest, including a few key clues. For example, animals tend to be in herds, obviously shaped like animals, and/or significantly brighter than the rest of the scene, and humans tend to be moving. We also provide the following additional guidelines.

Directions for special circumstances:

- Only label when objects are bright since the polarity changes occasionally
- If something is occluded completely: do not label
- If something is occluded but you can still see most features of them: label
- If something is shaped like a human but never moves: do not label
- If something is cutoff halfway in/out of the frame: do not label
- If there are "ghosts" (Fig. 6): do not label
- If you cannot recognize an individual (i.e., distinct poachers and animals): do not label

The final instruction about distinct objects is one of the more difficult instructions to follow in practice because often, the aerial view and small targets make it difficult to tell if there are one or more animals. The movement instruction is also

Table 1. Changes made throughout development.

Version	Change	Date of change	Brief description
1	-	-	Draws and edits boxes, navigates video, copies boxes to next frame
2	Multiple box selection	3/23/17	Moves multiple boxes at once, to increase labeling speed
3	Five majority to review	3/24/17	Requires only two people per video instead of five to improve overall efficiency
4	Labeling days	4/12/17	Has labelers assemble to discuss difficult videos
5	Tracking	6/17/17	Copies and automatically moves boxes to next frame

difficult, since with so few pixels on objects plus camera motion, it sometimes looks like objects are moving that are not. In these ambiguous cases, labelers are encouraged to seek help. In cases of disagreement after discussion, we err on the side of caution and only label certain objects.

6 Development

Thanks in large part to feedback provided by the labelers, we were able to make improvements throughout the development of the application to the current version discussed in Sect. 5.1. In the initial version of the application, we had five people label a single video, and then automatically checked for a majority consensus among these five sets of labels. We used the Intersection over Union (IoU) metric to check for overlap with a threshold of 0.5 [7]. If at least three out of five sets of labels overlapped, it was deemed to be consensus, and we took the bounding box coordinates of the first labeler. Our main motivation for having five opinions per video was to compensate for the difficulty of labeling thermal infrared data, though we also took into account the work of [18,23]. The interface of the initial version allowed the user to draw and manipulate bounding boxes, navigate through the video, save work automatically, and submit the completed video. Boxes were copied to the next frame and could be moved individually. To get where we are today, the changes were as listed in Table 1.

The most significant change made during the development process was the transition from five labelers labeling the same video and using majority voting to get the final labels (referred to as "MajVote") to having one labeler label the video followed by a reviewer reviewing the labels (referred to as "LabelReview"). We realized that having five people label a single video was very time consuming, and the quality of the labels was still not perfect because of the ambiguity of labeling thermal infrared data, which led to little consensus. Furthermore, when there was consensus, there were three to five different sets of coordinates to

---

Algorithm 1. Basic Tracking Algorithm

```
 1: bufferPixels ← userInput
 2: for all boxesPreviousFrame do
 3:     if boxSize > sizeThreshold then
 4:         newBoxCoordinates ← boxCoordinates
 5:     else
 6:         searchArea ← newFrame[boxCoordinates + bufferPixels]
 7:         thresholdedImage ← THRESHOLD(searchArea, threshold)
 8:         components ← CONNECTEDCOMPONENTS(thresholdedImage)
 9:         if numberComponents > 0 then
10:             newBoxCoordinates ← GETLARGESTCOMPONENT(components)
11:         else
12:             newBoxCoordinates ← boxCoordinates
13:         end if
14:     end if
15:     COPYANDMOVEBOX(newFrame, newBoxCoordinates)
16: end for
```

---

consider. Switching to LabelReview eliminated this problem, providing a cleaner and also time-saving solution. Another change, "Labeling Days", consisted of meeting together in one place for several hours per week so labelers were able to discuss ambiguities with us or their peers during labeling. Finally, the tracking algorithm (Algorithm 1) was added to automatically track the bounding boxes when the labeler moves to the new frame. The goal was to improve labeling efficiency, as the labelers would be able to label a single frame, then simply check that the labels were correct.

An example of the tracking process in use is shown in Fig. 7. First, the labeler drew two bounding boxes around the animals (Fig. 7a), then adjusted the search size for the tracking algorithm using the slider in the toolbar (Fig. 7b). The tracking algorithm was applied to produce the new bounding box location (Fig. 7c). In contrast, the copy feature, activated when the copy button was selected on the toolbar, only copied the boxes to the same location (Fig. 7d). In this case, since there was movement, and the animals were large and far from one another, the tracking algorithm correctly identified the animals in consecutive frames. If several bright objects were in the search region, it could track incorrectly and copying could be better. One direction of future work is to improve the tracking algorithm by setting thresholds automatically and accounting for close objects.

7   Analysis

In this section, we analyze how the changes we made during the development of VIOLA affect labeling efficiency. To do this, we examine two questions: (i) how the changes affect the overall efficiency of the data collection process, which is measured by the total person time needed to get a final label – a label confirmed by the five majority voting or the reviewer that can be used for game-theoretic analysis or deep learning algorithms; (ii) how the changes affect the individual

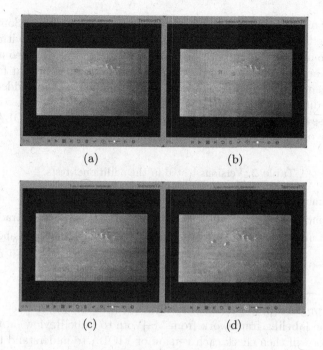

(a)   (b)

(c)   (d)

Fig. 7. A sample labeling process.

efficiency, or the person time needed for an individual labeler or reviewer to provide or check a label. In addition, we examine whether other desired properties of the data collection process, such as exhaustiveness, have been achieved.

To analyze efficiency, we first went through the person time data collected during VIOLA's development. Any changes made were deployed immediately to make faster progress. These person time data came from different videos and labelers. They inherently took different amounts of time to label, since the videos varied in their content. To mitigate the intrinsic heterogeneity, we divide the videos into four groups, $(0, 1)$, $[1, 2)$, $[2, 3)$, and $[3, +\infty)$, based on the average number of labels per frame, since it was an important indicator of the difficulty of labeling a video. There were other factors affecting the difficulty of labeling videos, so videos in the same group may still have had high variation. Because of this, we remove the top and bottom 5% of time per label entries.

Also due to these concerns, we collected additional person time data in a more controlled environment. We gave six unique videos that contained animals but no poachers to the labelers to label. The labelers had not seen these videos previously. We distributed the work among the labelers so as to get one set of final labels for each video under each of the versions of VIOLA (as shown in Table 1). We asked the labelers to label for no more than 15 min on each video. To accommodate the labelers' schedules and coordinate their schedules to set up meetings, which are necessary for LabelDays and Tracking, we gave the labelers 2 to 4 days to label the videos under each version. As such, it was difficult to get multiple sets of labels for each video or get labels for more videos. Some labelers were not able to complete checking all of the frames in the video within 15 min, so we use the minimum checked frame among labelers for each video under each version, and analyze efficiency using person time data up until that frame only. Also, note that since some labelers were asked to label the same video multiple times under different versions, the labelers likely got faster as time went on. To mitigate these effects, we randomly ordered the five versions of VIOLA for them to label. The order is shown in Table 2.

Table 2. Versions tested in the additional tests.

Version number	1	2	3	4	5
Version name	Basic	MultiBox	Review	LabelDays	Tracking
Framework used	MajVote	MajVote	LabelReview	LabelReview	LabelReview
Test order	Fourth	Third	First	Second	Fifth

We will proceed in this section by first focusing on the impact of the key change in the labeling framework from MajVote to LabelReview on the overall efficiency. We will then check each version of VIOLA to understand the impact of other changes. Because of the surprising results, we will particularly examine videos in which these features helped and in which they did not.

7.1   From MajVote to LabelReview

Figure 8a and b show the comparison on overall efficiency between MajVote and LabelReview. The total person time per final label is lower on average when we use LabelReview, based on data collected through both the development process and additional tests. During the development process, there were only seven videos for which we got final labels from five full sets of labels using MajVote, two of which did not produce any consensus labels. There were more than 70 videos for which we got final labels through LabelReview. During the additional tests, we tested two versions using MajVote and three versions using LabelReview, which means the value of each bar is averaged over two or three samples, respectively. We exclude one sample for Video C where no consensus labels were achieved through MajVote. The LabelReview efficiency for Video D is 0.63 with a standard error of 0.09 but it is too small to appear in Fig. 8b.

In addition to having more labelers involved, one reason that MajVote leads to a higher person time per final label is the lack of consensus. Figure 9 shows that there were large discrepancies in the number of labels between individual labelers, which led to fewer consensus labels (zero in Videos I and M).

Figure 10 shows that MajVote leads to many fewer final labels than Label-Review for the videos in the additional tests. This indicates that using Label-Review can get us closer to the goal of exhaustively labeling all of the objects of interest when compared to MajVote.

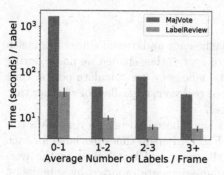

(a) Data from development process.

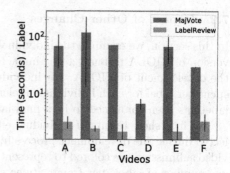

(b) Data from additional tests.

Fig. 8. Comparison of overall efficiency with different labeling frameworks.

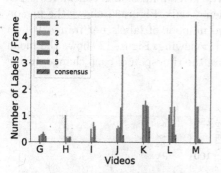

(a) For the seven videos with five sets of labels during development process.

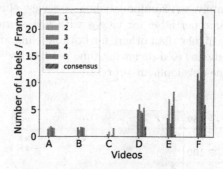

(b) For the six videos used in the additional tests under version Basic.

Fig. 9. Number of labels per frame for individual labelers and for consensus.

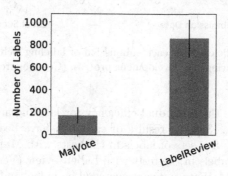

Fig. 10. Number of final labels for MajVote and LabelReview in additional tests.

7.2   Impact of Other Changes

In this section, we examine the individual efficiency and overall efficiency of each version of VIOLA to analyze the impact of every other change we made during the development of VIOLA. For individual efficiency, we calculate person time spent per label for each individual labeler or reviewer, regardless of whether that label has been confirmed to be a final label.

We first show results of individual efficiency based on person time data collected during the development process in Fig. 11. Person times per label for each video submission are colored to represent the group which is decided by the average number of labels per frame. Video submissions are reported by submission date since the date submitted indicates which version of the application was used for the video. The dates on which features were added, given in Table 1, are used to color the background of the plot. Finally, each submission is considered separately, to examine labeling or review efficiency only. Figure 11 shows the person time per label for videos with low average number of labels per frame $(0-1)$ is higher than others for both labeling and reviewing. Figure 12 shows the mean labeling and reviewing time per label within the timespan of each change during the development process.

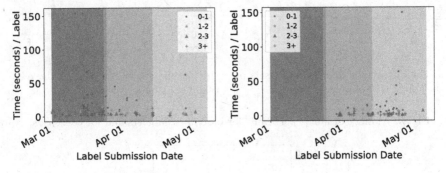

Fig. 11. Individual efficiency for each submission of labeling (left) and review (right) with data collected during the development process. (Color figure online)

We next examine the individual efficiency for labeling and reviewing in the additional tests (Fig. 13). The results of each test have been shown by video, since there were only five sets of labels in the tests with MajVote (Version 1–2) and only one set of labels in the tests with LabelReview (Version 3–5). The five sets of labels in the MajVote tests are averaged by video, and the standard error bars are included. Figure 13 shows that each of the changes we made resulted in an improvement on the individual efficiency for some, but not all, of the videos.

Multiple Box Selection. The feature of multiple box selection was added to improve the individual efficiency of labeling. Checking the first two groups in Figs. 12 and 13, we notice that surprisingly, this feature improves individual

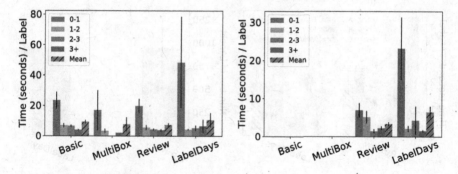

Fig. 12. Average individual efficiency of labeling (left) and review (right) with data collected during the development process.

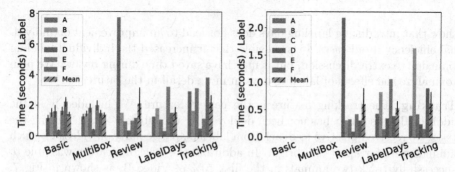

Fig. 13. Individual efficiency for each submission and average efficiency of labeling (left) and review (right) with data collected from the additional tests.

efficiency for some of the videos (e.g., Video F), but not all of the videos. One possible explanation is that in videos where there are many animals that do not move much over time, the changing position of the bounding boxes is mainly due to the movement of the camera. In this case, using multiple box selection and moving all of the bounding boxes in the same direction simultaneously is helpful. However, in other videos where there are only one or two animals in each frame, it may be faster to move the boxes separately, particularly if an animal moves.

Labeling Days. Labeling days were introduced with the aim to increase the overall efficiency. Figure 14 shows the average person time per final label has slightly reduced from Review to LabelDays during the additional tests, and the person time per final label has reduced for Videos A, C, and F. Figure 14 also shows the number of final labels has remained the same on average. The results indicate that introducing labeling days may help improving the efficiency and exhaustiveness of labeling, at least for some more complex videos. Subjective feedback from the labelers also indicated that introducing labeling days made it easier for them to deal with ambiguous cases, when it is difficult to maintain consistency and accuracy despite the guidelines. However, Figs. 12, 13, and 14

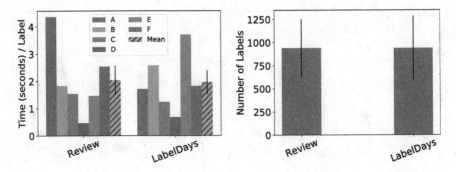

Fig. 14. Overall efficiency (left) and number of final labels (right) with version Review and LabelDays during the additional tests.

show that introducing labeling days does not lead to an improvement on individual efficiency in all cases. It is possible that it increased the individual labeling time due to extra discussion, but it may have saved time during review. We plan to analyze the effects of labeling days in more detail in the future.

Tracking. The tracking feature is the newest feature. We included it in the additional tests but it has not been deployed for the labelers to use. During the tests, we received positive feedback from labelers, particularly on videos in which animals were far apart and bright. In addition, the tracking feature was able to successfully track two animals in the first 10% of Video B, as shown in Fig. 7. Unexpectedly, the initial results from the additional tests do not show a positive effect on time per label or number of labels. We believe this is due to the fact that it does not find a brightness threshold automatically, and is likely to track the wrong object when multiple objects are within the same search region. We plan to continue developing this feature given its promise in the cases where animals are far apart and bright.

Summary. This section thus shows that while some of our proposed improvements actually led to increased efficiency, particularly the switch from MajVote to LabelReview, in other cases (e.g., multiple box selection), surprisingly, it only increased efficiency in some videos. This result indicates that we must not simply add features on the intuition that they are bound to improve performance, as they may only be useful for certain videos.

8   Conclusions

In conclusion, we presented VIOLA, which provides a labeling and reviewing framework to gather labeled data from a small group of people in a secure manner, and a labeling interface with both general features for difficult video data, and specific features for our green security domain to track wildlife and poachers. We analyzed the impact of the framework and the features on labeling efficiency, and found that some changes did not improve efficiency in general, but worked only in particular types of videos.

We plan to utilize the labeled data we acquired in this work to estimate the animal distribution and predict poachers' movement patterns, which are important for game-theoretic approaches such as generating patrol strategies as in PAWS. In addition, we will use the dataset to train deep neural networks to automatically detect wildlife and poachers in real-time, and develop novel game-theoretic approaches that incorporate real-time information to plan UAV and human patrol routes. VIOLA can be adopted to detect objects of interest in other types of surveillance videos, with widespread applications to various security domains.

Acknowledgments. This research was supported by UCAR N00173-16-2-C903, with the primary sponsor being the Naval Research Laboratory (Z17-19598). It was also partially supported by the Harvard Center for Research on Computation and Society Fellowship and the Viterbi School of Engineering Ph.D. Merit Top-Off Fellowship.

References

1. Alpcan, T., Basar, T.: A game theoretic approach to decision and analysis in network intrusion detection. In: Proceedings of 42nd IEEE Conference on Decision and Control, vol. 3, pp. 2595–2600. IEEE (2003)
2. Badrinarayanan, V., Galasso, F., Cipolla, R.: Label propagation in video sequences. In: CVPR, pp. 3265–3272. IEEE (2010)
3. Bae, S.H., Yoon, K.J.: Robust online multi-object tracking based on tracklet confidence and online discriminative appearance learning. In: CVPR (2014)
4. Basilico, N., Nittis, G.D., Gatti, N.: A security game combining patrolling and alarm-triggered responses under spatial and detection uncertainties. In: AAAI, pp. 397–403 (2016)
5. Catania, C.A., Bromberg, F., Garino, C.G.: An autonomous labeling approach to support vector machines algorithms for network traffic anomaly detection. Expert Syst. Appl. **39**(2), 1822–1829 (2012)
6. Deng, J., Dong, W., Socher, R., Li, L.J., Li, K., Fei-Fei, L.: Imagenet: a large-scale hierarchical image database. In: CVPR, pp. 248–255. IEEE (2009)
7. Everingham, M., Van Góol, L., Williams, C.K.I., Winn, J., Zisserman, A.: The pascal visual object classes (VOC) challenge. Int. J. Comput. Vision **88**(2), 303–338 (2010)
8. Fang, F., Nguyen, T.H., Pickles, R., Lam, W.Y., Clements, G.R., An, B., Singh, A., Tambe, M., Lemieux, A.: Deploying PAWS: field optimization of the protection assistant for wildlife security. In: AAAI, pp. 3966–3973 (2016)
9. Haskell, W., Kar, D., Fang, F., Tambe, M., Cheung, S., Denicola, E.: Robust protection of fisheries with COmPASS. In: IAAI, pp. 2978–2983 (2014)
10. Hodgson, J.C., et al.: Precision wildlife monitoring using unmanned aerial vehicles. Sci. Rep. **6**, 22574 (2016). doi:10.1038/srep22574
11. Johnson, M.P., Fang, F., Tambe, M.: Patrol strategies to maximize pristine forest area. In: AAAI (2012)
12. Kar, D., Fang, F., Fave, F.D., Sintov, N., Tambe, M.: "A Game of Thrones": when human behavior models compete in repeated Stackelberg security games. In: AAMAS (2015)

13. Kar, D., Ford, B., Gholami, S., Fang, F., Plumptre, A., Tambe, M., Driciru, M., Wanyama, F., Rwetsiba, A., Nsubaga, M., Mabonga, J.: Cloudy with a chance of poaching: adversary behavior modeling and forecasting with real-world poaching data. In: AAMAS, pp. 159–167 (2017)
14. Kristan, M., Matas, J., Leonardis, A., Felsberg, M., Cehovin, L., Fernández, G., Vojir, T., Hager, G., Nebehay, G., Pflugfelder, R.: The visual object tracking vot2015 challenge results. In: ICCV Workshops, pp. 1–23 (2015)
15. Lucas, B.D., Kanade, T., et al.: An iterative image registration technique with an application to stereo vision (1981)
16. Ma, C., Yang, X., Zhang, C., Yang, M.H.: Long-term correlation tracking. In: CVPR, pp. 5388–5396 (2015)
17. Milan, A., Leal-Taixé, L., Reid, I., Roth, S., Schindler, K.: Mot16: a benchmark for multi-object tracking. arXiv preprint arXiv:1603.00831 (2016)
18. Nguyen, P., Kim, J., Miller, R.C.: Generating annotations for how-to videos using crowdsourcing. In: CHI 2013 Extended Abstracts on Human Factors in Computing Systems, pp. 835–840 (2013)
19. Nguyen, T.H., Sinha, A., Gholami, S., Plumptre, A., Joppa, L., Tambe, M., Driciru, M., Wanyama, F., Rwetsiba, A., Critchlow, R., et al.: Capture: a new predictive anti-poaching tool for wildlife protection. In: AAMAS, pp. 767–775 (2016)
20. Nguyen, T.H., Yang, R., Azaria, A., Kraus, S., Tambe, M.: Analyzing the effectiveness of adversary modeling in security games. In: AAAI, pp. 718–724 (2013)
21. Nguyen-Dinh, L.V., Waldburger, C., Roggen, D., Tröster, G.: Tagging human activities in video by crowdsourcing. In: ICMR, pp. 263–270 (2013)
22. Pai, C.H., Lin, Y.P., Medioni, G.G., Hamza, R.R.: Moving object detection on a runway prior to landing using an onboard infrared camera. In: CVPR, pp. 1–8. IEEE (2007)
23. Park, S., Mohammadi, G., Artstein, R., Morency, L.P.: Crowdsourcing micro-level multimedia annotations: the challenges of evaluation and interface. In: CrowdMM, pp. 29–34 (2012)
24. Pita, J., Jain, M., Western, C., Portway, C., Tambe, M., Ordonez, F., Kraus, S., Paruchuri, P.: Deployed ARMOR protection: the application of a game theroetic model for security at the Los Angeles International Airport. In: AAMAS (2008)
25. Redmon, J., Divvala, S., Girshick, R., Farhadi, A.: You only look once: unified, real-time object detection. In: CVPR, pp. 779–788 (2016)
26. Ren, S., He, K., Girshick, R., Sun, J.: Faster R-CNN: towards real-time object detection with region proposal networks. In: NIPS, pp. 91–99 (2015)
27. Sheng, V.S., Provost, F., Ipeirotis, P.G.: Get another label? Improving data quality and data mining using multiple, noisy labelers. In: KDD, pp. 614–622 (2008)
28. Tambe, M.: Security and Game Theory: Algorithms, Deployed Systems, Lessons Learned. Cambridge University Press, Cambridge (2011)
29. Vondrick, C., Patterson, D., Ramanan, D.: Efficiently scaling up crowdsourced video annotation. Int. J. Comput. Vis. **101**, 184–204 (2013)
30. Yan, R., Yang, J., Hauptmann, A.: Automatically labeling video data using multi-class active learning. In: ICCV (2003)
31. Yilmaz, A., Javed, O., Shah, M.: Object tracking: a survey. ACM Comput. Surv. (CSUR) **38**(4), 13 (2006)
32. Zhang, L., Li, Y., Nevatia, R.: Global data association for multi-object tracking using network flows. In: CVPR, pp. 1–8. IEEE (2008)

On the Economics of Ransomware

Aron Laszka[1]([✉]), Sadegh Farhang[2], and Jens Grossklags[3]

[1] University of Houston, Houston, USA
alaszka@uh.edu
[2] Pennsylvania State University, State College, USA
[3] Technical University of Munich, Munich, Germany

Abstract. While recognized as a theoretical and practical concept for over 20 years, only now ransomware has taken centerstage as one of the most prevalent cybercrimes. Various reports demonstrate the enormous burden placed on companies, which have to grapple with the ongoing attack waves. At the same time, our strategic understanding of the threat and the adversarial interaction between organizations and cybercriminals perpetrating ransomware attacks is lacking.

In this paper, we develop, to the best of our knowledge, the first game-theoretic model of the ransomware ecosystem. Our model captures a multi-stage scenario involving organizations from different industry sectors facing a sophisticated ransomware attacker. We place particular emphasis on the decision of companies to invest in backup technologies as part of a contingency plan, and the economic incentives to pay a ransom if impacted by an attack. We further study to which degree comprehensive industry-wide backup investments can serve as a deterrent for ongoing attacks.

Keywords: Ransomware · Backups · Security economics · Game theory

1 Introduction

Already in 1996, Young and Yung coined the term *cryptovirological attacks* and provided a proof-of-concept implementation of what could now be considered a major building block of ransomware malware [35]. Due to the perceived seriousness of this attack approach, they also suggested that "access to cryptographic tools should be well controlled."

Malware featuring ransomware behavior was at first deployed at modest scale (e.g., variants of PGPCoder/GPCode between approximately 2005–2010), and often suffered from technical weaknesses, which even led a researcher in the field to proclaim that "ransomware as a mass extortion mean is certainly doomed to failure" [11]. However, later versions of GPCode already used 1024-bit RSA key encryption; a serious threat even for well-funded organizations.

Ransomware came to widespread prominence with the CryptoLocker attack in 2013, which utilized Bitcoin as a payment vehicle [21]. Since then, the rise of ransomware has been dramatic, culminating (so far) with the 2017 attack

© Springer International Publishing AG 2017
S. Rass et al. (Eds.): GameSec 2017, LNCS 10575, pp. 397–417, 2017.
DOI: 10.1007/978-3-319-68711-7_21

waves of the many variants of the WannaCrypt/WannaCry and the Petya ransomwares. Targets include all economic sectors and devices ranging from desktop computers, entire business networks, industrial facilities, and also mobile devices. Security industry as well as law enforcement estimates for the amount of money successfully extorted and the (very likely much larger) overall damage caused by ransomware attacks differ widely. However, the figures are significant (see Sect. 5). Observing these developments, in a very recent retrospective article, Young and Yung bemoan the lack of adequate response focused on ransomware attacks by all stakeholders even though the threat was known for over 20 years [36].

As with any security management decision, there is a choice between doing nothing to address a threat, or selecting an appropriate investment level. In the case of responding to a sophisticated ransomware attack, this primarily concerns decisions on how to invest in backup and recovery technologies, and whether to pay a ransom in case of a successful attack. These decisions are interrelated.

The empirical evidence is mixed (and scarce). It is probably fair to say that backup technologies have always been somewhat of a stepchild in the overall portfolio of security technologies. In 2001, a survey showed that only 41% of the respondents did data backups and 69% had not recently facilitated a backup; at the same time, 25% reported to have lost data [7]. In 2009, another survey found backup usage of less than 50%; and 66% reported to have lost files (42% within the last 12 months) [15]. In a backup awareness survey that has been repeated annually since 2008, the figures for individuals who *never* created backups have been slowly improving. Starting at 38% in 2008, in the most recent survey in June 2017 only 21% reported to have never made a backup. Still, only 37% now report to create at least monthly backups [3], despite the heightened media attention given to ransomware.

Regarding ransom payment behavior, IBM surveyed 600 business leaders in the U.S about ransomware, and their data management practices and perceptions. Within their sample, almost 50% of the business representatives reported ransomware attacks in their organizations. Interestingly, 70% of these executives reported that ransom payments were made in order to attempt a resolution of the incident. About 50% paid over $10,000 and around 20% reported money transfers of over $40,000 [14]. In contrast, a different survey of ransomware-response practices found that 96% of those affected (over the last 12 months) did *not* pay a ransom [18]. However, the characteristics of the latter sample are not described [18]. Finally, recent cybercrime measurement studies have tracked the approximate earnings for particular ransomware campaigns (for example, by tracking related Bitcoin wallets). These studies typically do not succeed in pinpointing the percentage of affected individuals or organizations paying the ransom (e.g., [17,21]).

Our work targets two key aspects of a principled response to sophisticated ransomware attacks. First, we develop an economic model to further our strategic understanding of the adversarial interaction between organizations attacked by ransomware and ransomware attackers. As far as we know, our work is the first such game-theoretic model.

Second, we study the aforementioned response approaches to diminish the economic impact of the ransomware threat on organizations. As such our model focuses on organizations' decision-making regarding backup investments (as part of an overall contingency plan), which is an understudied subject area. We further determine how backup security investments interact with an organization's willingness to pay a ransom in case of a ransomware attack.

Further, we numerically show how (coordinated) backup investments by organizations can have a deterrent effect on ransomware attackers. Since backup investments are a private good and are not subject to technical interdependencies, this observation is novel to the security economics literature and relatively specific to ransomware. Note, for example, that in the context of cyberespionage and data breaches to exfiltrate data, such a deterrence effect of backup investments is unobservable.

We proceed as follows. In Sect. 2, we develop our game-theoretic model. We conduct a thorough analysis of the model in Sect. 3. In Sect. 4, we complement our analytic results with a numerical analysis. We discuss additional related work on ransomware as well as security economics in Sect. 5, and offer concluding remarks in Sect. 6.

2   Model

We model ransomware attacks as a multi-stage, multi-defender security game. Table 1 shows a list of the symbols used in our model.

2.1   Players

On the defenders' side, players model organizations that are susceptible to ransomware attacks. Based on their characteristics, we divide these organizations into two groups (e.g., hospitals and universities). We will refer to these two groups as group 1 and group 2, and we let set G_1 and set G_2 denote their members, respectively. On the attacker's side, there is a single player, who models cybercriminals that may develop and deploy ransomware. Note that we model attackers as a single entity since our goal is to understand and improve the behavior of defenders; hence, competition between attackers is not our current focus. Our model—and many of our results—could be extended to multiple attackers in a straightforward manner.

2.2   Strategy Spaces

With our work, we focus on the mitigation of ransomware attacks through backups (as a part of contingency plans), and we will not consider the organizations' decisions on preventative effort (e.g., firewall security policies). The tradeoff between mitigation and preventative efforts has been subject of related work [12].

Table 1. List of symbols

Symbol	Description
G_j	Set of organizations belonging to group j
W_j	Initial wealth of organizations in group j
β	Discounting factor for uncertain future losses
F_j	Cost of data loss due to random failures in group j
L_j	Cost of permanent data loss due to ransomware attacks in group j
T_j	Loss from business interruptions due to ransomware in group j
D	Base difficulty of perpetrating ransomware attacks
C_B	Unit cost of backup effort
C_A	Unit cost of attack effort
C_D	Fixed cost of developing ransomware
b_i	Backup effort of organization i
p_i	Decision of organization i about ransom payment
a_j	Attacker's effort against group j
r	Ransom demanded by the attacker
$V_j(a_1, a_2)$	Probability of an organization $i \in G_j$ becoming compromised

We let $b_i \in \mathbb{R}_+$ denote the backup effort of organization i, which captures the frequency and coverage of backups as well as contingency plans and preparations. Compromised organizations also have to decide whether they pay the ransom or sustain permanent data loss. We let $p_i = 1$ if organization i pays, and $p_i = 0$ if it does not pay.

The attacker first decides whether it wishes to engage in cybercrime using ransomware. If the attacker chooses to engage, then it has to select the amount of effort spent on perpetrating the attacks. We let $a_1 \in \mathbb{R}_{\geq 0}$ and $a_2 \in \mathbb{R}_{\geq 0}$ denote the attacker's effort spent on attacking group 1 and group 2, respectively. If the attacker chooses not to attack group j (or not to engage in cybercrime at all), then $a_j = 0$. We assume that each organization within a group falls victim to the attack with the same probability $V_j(a_1, a_2)$, which depends on the attacker's effort, independently of the other organizations. Since the marginal utility of attack effort is typically decreasing, we assume that the infection probability $V_j(a_1, a_2)$ is

$$V_j(a_1, a_2) = \frac{a_j}{D + (a_1 + a_2)}, \tag{1}$$

where D is the base difficulty of attacks. In the formula above, the numerator expresses that as the attacker increases its effort on group j, more and more organizations fall victim. Meanwhile, the denominator captures the decreasing marginal utility: as the attacker increases its attack effort, compromising additional targets becomes more and more difficult. In practice, this corresponds to the increasing difficulty of finding new targets as organizations are becoming aware of a widespread ransomware attack and are taking precautions, etc.

The attacker also has to choose the amount of ransom r to demand from compromised organizations in exchange for restoring their data and systems.

Stages. The game consists of two stages:

- Stage I: Organizations choose their backup efforts b, while the attacker chooses its attack effort a_1 and a_2, as well as its ransom demand r.
- Stage II: Each organization $i \in G_j$ becomes compromised with probability $V_j(a_1, a_2)$. Then, organizations that have fallen victim to the attack choose whether to pay the ransom or not, which is represented by p.

2.3 Payoffs

Defender's Payoff. If an organization i, which belongs to group $j \in \{1, 2\}$, has not fallen victim to a ransomware attack, then its payoff is

$$\mathcal{U}_{O_i}\big|_{\text{not compromised}} = W_j - C_B \cdot b_i - \beta \frac{F_j}{b_i}, \tag{2}$$

where W_j is the initial wealth of organizations in group j, F_j is their loss resulting from corrupted data due to random failures[1], and C_B is the unit cost of backup effort. The parameter β is a behavioral discount factor, which captures the robust empirical observation that individuals underappreciate the future consequences of their current actions [24]. The magnitude of β is assumed to be related to underinvestment in security and privacy technologies [1,13]; in our case, procrastination of backup investments [4].[2]

Otherwise, we have two cases. If organization i decides to pay the ransom r, then its payoff is

$$W_j - C_B \cdot b_i - \beta \left(\frac{F_j}{b_i} + T_j + r \right),$$

where T_j is the loss resulting from temporary business interruption due to the attack. On the other hand, if organization i does not pay the ransom, then its

---

[1] Since we interpret effort b_i primarily as the frequency of backups, the fraction $\frac{1}{b_i}$ is proportional to the expected time since the last backup. Consequently, we assume that data losses are inversely proportional to b_i. Note that alternative interpretations, such as assuming b_i to be the level of sophistication of backups (e.g., air-gapping), which determines the probability that the backups remain uncompromised, also imply a similar relationship.

[2] We are unaware of any behavioral study that specifically investigates the impact of the *present bias* behavioral discount factor on backup decisions, but industry experts argue strongly for its relevance. For example, in the context of the 2017 WannaCry ransomware attacks a commentary about backups stated: "This may be stating the obvious, but it's still amazing to know the sheer number of companies that keep procrastinating over this important task [32]."

payoff is

$$W_j - C_B \cdot b_i - \beta \left(\frac{F_j + L_j}{b_i} + T_j \right),$$

where L_j is the loss resulting from permanent data loss due to the ransomware attack. Using p_i, we can express a compromised organization's payoff as

$$\mathcal{U}_{O_i} \Big|_{\text{compromised}} = W_j - C_B \cdot b_i - \beta \left(\frac{F_j + (1 - p_i) \cdot L_j}{b_i} + T_j + p_i \cdot r \right). \quad (3)$$

By combining Eqs. (2) and (3) with V_j, we can express the expected utility of an organization $i \in G_j$ as

$$\mathrm{E}\left[\mathcal{U}_{O_i}\right] = (1 - V_j(a_1, a_2)) \left[W_j - C_B \cdot b_i - \beta \frac{F_j}{b_i} \right]$$

$$+ V_j(a_1, a_2) \left[W_j - C_B \cdot b_i - \beta \left(\frac{F_j + (1 - p_i) \cdot L_j}{b_i} + T_j + p_i \cdot r \right) \right]. \quad (4)$$

Attacker's Payoff. For the attacker's payoff, we also have two cases. If the attacker decides not to participate (i.e., if $a_1 = 0$ and $a_2 = 0$), then its payoff is simply zero. Otherwise, its payoff depends on the number of organizations that have fallen victim and decided to pay. We can calculate the expected number of victims who pay the ransom as

$$\mathrm{E}[\text{number of victims who pay the ransom}] = \sum_j \sum_{i \in G_j} V_j(a_1, a_2) \cdot p_i \quad (5)$$

since each organization $i \in G_j$ is compromised with probability V_j, and $p_i = 1$ if organization i chooses to pay (and $p_i = 0$ if it does not pay).

Then, we can express the attacker's expected payoff simply as

$$\mathrm{E}\left[\mathcal{U}_A\right] = \left[\sum_j \sum_{i \in G_j} V_j(a_1, a_2) \cdot p_i \right] \cdot r - C_A \cdot (a_1 + a_2) - C_D, \quad (6)$$

where C_A is the unit cost of attack effort, and C_D is the fixed cost of developing a ransomware, which the attacker must pay if it decides to engage (i.e., if $a_1 > 0$ or $a_2 > 0$).

2.4 Solution Concepts

We assume that every player is interested in maximizing its expected payoff, and we use subgame-perfect Nash equilibrium as our solution concept. We also assume that organizations always break ties (i.e., when both paying and not paying are best responses) by choosing to pay. Note that the latter assumption has no practical implications, it only serves to avoid pathological mathematical cases.

Further, in our numerical analysis in Sect. 4, we will use the *social optimum* concept for comparison with the Nash equilibrium results. In the social optimum, a social planner can coordinate the decisions of organizations, such that it yields the maximum aggregate outcome for the organizations, subject to an optimal response by the attacker (who is not guided by the social planner).

3   Analysis

In this section, we analyze our proposed game-theoretic model of the ransomware ecosystem. Our solution concept, as mentioned in Sect. 2.4, is the subgame perfect Nash equilibrium. Hence, in our analysis, we first calculate each organization's decision in Stage II in Sect. 3.1. In other words, we derive under what conditions a victim organization will pay the requested ransom from the attacker. Then, we calculate the best-response backup strategy for each organization in Stage I of the game in Sect. 3.2. Third, we calculate the attacker's best-response, i.e., demanded ransom and the attacker's effort, in Sect. 3.3. By calculating the attacker's and the organizations' best-responses, we can then derive the Nash equilibrium in Sect. 3.4.

3.1   Compromised Organizations' Ransom Payment Decisions

We begin our analysis by studying the compromised organizations' best-response payment strategies in the second stage of the game.

Lemma 1. *For organization $i \in G_j$, paying the ransom (i.e., $p_i = 1$) is a best response if and only if*

$$r \leq \frac{L_j}{b_i}. \tag{7}$$

Proof of Lemma 1 is provided in Appendix A.1.

Lemma 1 means that an organization will pay the demanded ransom if the demanded value is not higher than the average permanent data loss due to ransomware attack.

3.2   Organizations' Backup Decisions

We next study the organizations' best-response backup strategies in the first stage. We assume that compromised organizations will play their best responses in the second stage (see Lemma 1), but we do not make any assumptions about the attacker's effort or ransom strategies. We first characterize the organizations' best-response backup strategies when they do not face any attacks (Lemma 2) and then in the case when they are threatened by ransomware (Lemma 3).

Lemma 2. *If the attacker chooses not to attack group j (i.e., $a_j = 0$), then the unique best-response backup strategy for organization $i \in G_j$ is*

$$b_i^* = \sqrt{\beta \frac{F_j}{C_B}}. \tag{8}$$

Proof of Lemma 2 is provided in Appendix A.2.

Note that in Lemma 2, an organization chooses its backup strategy by considering data loss due to random failures rather than data loss due to ransomware attack since that organization is not chosen to be attacked by the attacker.

Lemma 3 calculates an organization's best-response backup strategy. Note that an organization chooses its backup strategy at Stage I. In this stage, an organization does not know whether it is the target of a ransomware attack and if that organization is the target of a ransomware attack, whether the attack is successful.

Lemma 3. *If the attacker chooses to attack group j (i.e., $a_j > 0$), then the best-response backup strategy b_i^* for organization $i \in G_j$ is*

- *if $b_j^{low} > \frac{L_j}{r}$, then $b_i^* = b_j^{high}$;*
- *if $b_j^{high} < \frac{L_j}{r}$, then $b_i^* = b_j^{low}$;*
- *otherwise, $b_i^* \in \left\{ b_j^{low}, b_j^{high} \right\}$ (the one that gives the higher payoff or both if the resulting payoffs are equal),*

where $b_j^{low} = \sqrt{\beta \frac{F_j}{C_B}}$ and $b_j^{high} = \sqrt{\beta \frac{(F_j + V_j(a_1, a_2)L_j)}{C_B}}$.

Proof of Lemma 3 is provided online in the extended version of the paper [20].

Lemma 3 shows the best-response backup strategy when an organization is under attack. If the demanded ransom value is high, i.e., $r > \frac{L_j}{b_j^{low}}$, an organization takes into account the data loss due to ransomware attack as well as the data loss due to random failure when choosing the backup strategy level. On the other hand, if the demanded ransom is low, i.e., $r < \frac{L_j}{b_j^{high}}$, an organization does not care about the data loss due to ransomware attack even when that organization is under attack. In other words, that organization behaves like an organization that is not under ransomware attack, i.e., similar to Lemma 2.

3.3   Attacker's Best Response

Building on the characterization of the organizations' best responses, we now characterize the attacker's best-response strategies. Notice that the lemmas presented in the previous section show that an organization's best response does not depend on the identity of the organization, only on its group. Since we are primarily interested in studying equilibria, in which everyone plays a best response, we can make the following assumptions:

- All organizations within a group j play the same backup strategy, which is denoted by $\hat{b}_j$.
- $\frac{L_1}{b_1} \leq \frac{L_2}{b_2}$.

The second assumption is without loss of generality since we could easily re-number the groups.

In Lemma 4, we calculate the attacker's best-response demanded ransom given the attacker's effort and the organizations' backup strategies.

Lemma 4. *If the attacker's effort (a_1, a_2) is fixed, then its best-response ransom demand r^* is*

- $\frac{L_1}{b_1}$ *if* $|G_1| \, V_1(a_1, a_2)\frac{L_1}{b_1} > |G_2| \, V_2(a_1, a_2)\left(\frac{L_2}{b_2} - \frac{L_1}{b_1}\right)$
- $\frac{L_2}{b_2}$ *if* $|G_1| \, V_1(a_1, a_2)\frac{L_1}{b_1} < |G_2| \, V_2(a_1, a_2)\left(\frac{L_2}{b_2} - \frac{L_1}{b_1}\right)$
- *both $\frac{L_1}{b_1}$ and $\frac{L_2}{b_2}$ otherwise.*

Proof of Lemma 4 is provided in Appendix A.3.

Lemma 5 shows how the attacker divides its best-response attack effort between the two groups of organizations. Here, we assume that $a_1 + a_2 = a_{sum}$, where a_{sum} is a constant. Note that it is possible that the attacker decides not to attack either of the groups of organizations. The reason is that the benefit for the attacker from a ransomware attack may be lower than the cost of the attack. Hence, a rational attacker will abstain from attacking either of the groups.

Lemma 5. *The attacker's best-response attack effort (a_1^*, a_2^*) is as follows:*

- $a_1^* = 0$ *and* $a_2^* = a_{sum}$ *if* $|G_1| \cdot 1_{\left\{r \leq \frac{L_1}{b_1}\right\}} < |G_2| \cdot 1_{\left\{r \leq \frac{L_2}{b_2}\right\}}$ *and* $\frac{a_{sum}}{D + a_{sum}}|G_2| \cdot r \cdot 1_{\left\{r \leq \frac{L_2}{b_2}\right\}} > C_A \cdot a_{sum} + C_D$,
- $a_1^* = a_{sum}$ *and* $a_2^* = 0$ *if* $|G_1| \cdot 1_{\left\{r \leq \frac{L_1}{b_1}\right\}} > |G_2| \cdot 1_{\left\{r \leq \frac{L_2}{b_2}\right\}}$ *and* $\frac{a_{sum}}{D + a_{sum}}|G_1| \cdot r \cdot 1_{\left\{r \leq \frac{L_2}{b_2}\right\}} > C_A \cdot a_{sum} + C_D$,
- *any a_1^* between 0 and a_{sum} and $a_2^* = a_{sum} - a_1^*$ if* $|G_1| \cdot 1_{\left\{r \leq \frac{L_1}{b_1}\right\}} = |G_2| \cdot 1_{\left\{r \leq \frac{L_2}{b_2}\right\}}$ *and* $\frac{a_{sum}}{D + a_{sum}}|G_2| \cdot r \cdot 1_{\left\{r \leq \frac{L_2}{b_2}\right\}} > C_A \cdot a_{sum} + C_D$.
- $a_1^* = a_2^* = 0$ *otherwise.*

Proof of Lemma 5 is provided in Appendix A.4.

3.4 Equilibria

Proposition 1 provides the necessary and sufficient conditions for the attacker's strategy to abstain from attack, i.e., $a_1^* = a_2^* = 0$, and $\hat{b}_1^* = \sqrt{\beta\frac{F_1}{C_B}}$ and $\hat{b}_2^* = \sqrt{\beta\frac{F_2}{C_B}}$ is Nash equilibrium.

Proposition 1. *The attacker choosing not to attack and the organizations choosing backup efforts $\sqrt{\beta \frac{F_1}{C_B}}$ and $\sqrt{\beta \frac{F_2}{C_B}}$ is an equilibrium if and only if each of the following conditions are satisfied:*

$$-\frac{L_2 \cdot a_{sum} \cdot |G_2| \cdot 1_{\left\{r \le \frac{L_2}{b_2}\right\}}}{L_2(D+a_{sum})(C_A \cdot a_{sum}+C_D)} < \sqrt{\beta \frac{F_2}{C_B}} \ \text{and} \ |G_1| \cdot 1_{\left\{r \le \frac{L_1}{b_1}\right\}} \le |G_2| \cdot 1_{\left\{r \le \frac{L_2}{b_2}\right\}}$$

$$-\frac{L_2 \cdot a_{sum} \cdot |G_1| \cdot 1_{\left\{r \le \frac{L_1}{b_1}\right\}}}{L_2(D+a_{sum})(C_A \cdot a_{sum}+C_D)} < \sqrt{\beta \frac{F_2}{C_B}} \ \text{and} \ |G_1| \cdot 1_{\left\{r \le \frac{L_1}{b_1}\right\}} > |G_2| \cdot 1_{\left\{r \le \frac{L_2}{b_2}\right\}}$$

Proof of Proposition 1 is provided in Appendix A.5.

4   Numerical Illustrations

In this section, we present numerical results on our model. We first compare equilibria to social optima, and we study the effect of changing the values of key parameters (Sect. 4.1). We then investigate interdependence between multiple groups of organizations, which is caused by the strategic nature of attacks, and we again study the effect of changing key parameters (Sect. 4.2).

For any combination of parameter values, our game has at most one equilibrium, which we will plot in the figures below. However, for some combinations, the game does not have an equilibrium. In these cases, we used iterative best responses:

1. starting from an initial strategy profile,
2. we changed the attacker's strategy to a best response,
3. we changed the organization's strategy to a best response,
4. and then we repeated from Step 2.

We found that regardless of the initial strategy profile, the iterative best-response dynamics end up oscillating between two strategy profiles. Since these strategy profiles were very close, we plotted their averages in place of the equilibria in the figures below.

4.1   Equilibria and Social Optima

For clarity of presentation, we consider a single organization type in this subsection. The parameter values used in this study are as follows: $|G| = 100$, $W = 100$, $\beta = 0.9$, $F = 5$, $L = 5$, $T = 10$, $C_B = 1$, $D = 10$, $C_A = 10$, and $C_D = 10$ (unless stated otherwise).

Figure 1 shows the expected payoffs of an individual organization and the attacker for various values of the unit cost C_B of backup effort. In practice, the unit cost of backup effort may change, for example, due to technological improvements (decreasing the cost) or growth in the amount of data to be backed up (increasing the cost). When this cost is very low ($C_B < 0.5$), organizations can perform frequent and sophisticated backups, which means that the amount of

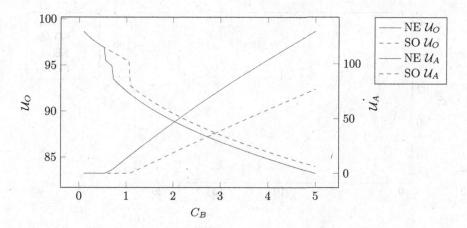

Fig. 1. The expected payoff of the attacker (red) and an individual organization (blue) in equilibrium (solid line —) and in social optimum (dashed line - - -) for various backup cost values C_B. (Color figure online)

data that may be compromised—and hence, the ransom that they are willing to pay—is very low. As a result, the attacker is deterred from deploying ransomware ($\mathcal{U}_A = 0$) since its income from ransoms would not cover its expenses. For higher costs ($0.5 \leq C_B < 1$), the organizations' equilibrium payoff is much lower since they choose to save on backups, which incentivizes the attacker to deploy ransomware and extort payments from them. In this case, the social optimum for the organizations is to maintain backup efforts and, hence, deter the attacker. For even higher costs ($C_B \geq 1$), deterrence is not socially optimal. However, the equilibrium payoffs are still lower since organizations shirk on backup efforts, which leads to more intense attacks and higher ransom demands.

Figure 2 shows the expected payoff of an individual organization and the attacker for various values of the unit cost C_A of attack effort. In practice, the unit cost of attack effort can change, e.g., due to the development of novel attacks and exploits (lowering the cost) or the deployment of more effective defenses (increasing the cost). Figure 2 shows phenomena that are similar to the ones exhibited in Fig. 1. When the attacker is at a technological advantage (i.e., when C_A is low), deterrence is not a realistic option for organizations. However, they can improve their payoffs—compared to the equilibrium—by coordinating and investing more in backups, thereby achieving social optimum. For higher attack costs ($10 < C_A \leq 15$), this coordination can result in significantly higher payoffs since deterrence becomes a viable option. For very high attack costs ($C_A > 15$), compromising an organization costs more than what the attacker could hope to collect with ransoms; hence, coordination is no longer necessary to deter the attacker.

Figure 3 shows how the organizations' backup efforts b and the attacker's payoff are effected by the behavioral discount factor β. With low values of β, organizations underappreciate future consequences; hence, they shirk on backup

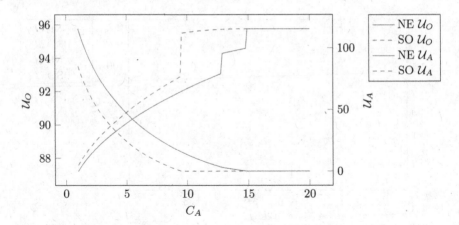

Fig. 2. The expected payoff of the attacker (red) and an individual organization (blue) in equilibrium (solid line —) and in social optimum (dashed line - - -) for various attack cost values C_A. (Color figure online)

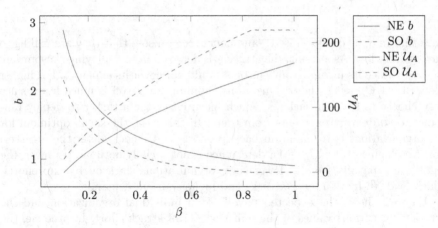

Fig. 3. The attacker's expected payoff (red) and the organizations' backup strategy (green) in equilibrium (solid line —) and in social optimum (dashed line - - -) for various discounting factor values β. (Color figure online)

efforts (as evidenced by low values of b). With high values of β, organizations care more about future losses, so they invest more in backup efforts (resulting in high values of b). We see that in all cases, there is a significant difference between the equilibrium and the social optimum. This implies that regardless of the organizations' appreciation of future consequences, coordination is necessary. In other words, low backup efforts cannot be attributed only to behavioral factors.

4.2   Interdependence

Now, we study the interdependence between two groups of organizations. We instantiate the parameters of both organizations (and the attacker) with the same values as in the previous subsection. Note that more numerical illustrations are available online in the extended version of the paper [20].

Figure 4 shows the payoffs of individual organizations from the two groups as well as the attacker, for various values of the costs L_1 and L_2 of permanent data loss. As expected, we see from the attacker's payoff (Fig. 4(c)) that as loss costs increase, organizations become more willing to pay higher ransoms, so the attacker's payoff increases. On the other hand, we observe a more interesting phenomenon in the organizations' payoffs. As the loss cost (e.g., L_1) of one group (e.g., group 1) increases, the payoff of organizations in that group (e.g., Fig. 4(a)) decreases. However, we also see an *increase* in the payoff (e.g., Fig. 4(b)) of organizations in the *other* group (e.g., group 2). The reason for this increase is in the strategic nature of attacks: as organizations in one group

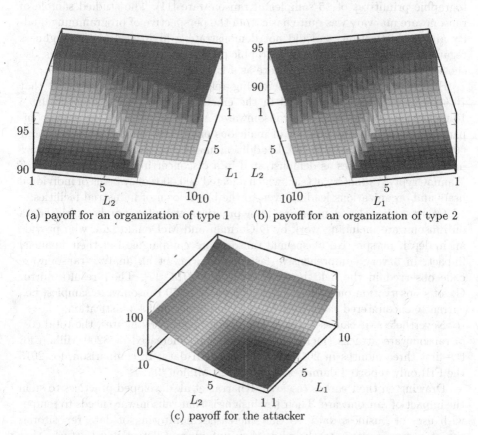

(a) payoff for an organization of type 1   (b) payoff for an organization of type 2

(c) payoff for the attacker

Fig. 4. The expected payoff of individual organizations of (a) type 1 and (b) type 2 as well as (c) the attacker in equilibrium for various data loss costs L_1 and L_2.

become more attractive targets, attackers are more inclined to focus their efforts on this group, which results in lower intensity attacks against the other group. This substitution effect, which can be viewed as a negative externality between groups of organizations, is strong when the attacker's efforts are focused (e.g., when ransomware is deployed using spear-phishing campaigns).

5 Related Work

Ransomware: Early work by Luo and Liao, in 2007 and 2009, respectively, represented first exploratory analyses of the ransomware phenomenon [22,23]. They focus on increased awareness (in particular, by employees) as a major means to diminish the effectiveness of ransomware attacks, which is a key recommendation regarding ransomware mirrored in the 2017 Verizon DBIR report ten years later: "stress the importance of software updates to anyone who'll listen [33]."

In 2010, Gazet investigated the quality of code, functionalities and cryptographic primitives of 15 samples of ransomware [11]. The studied sample of ransomware malware was quite basic from the perspective of programming quality and sophistication, and did not demonstrate a high level of thoroughness regarding the application of cryptographic primitives. However, the analysis also showed the ability to mass propagate as a key feature.

Highlighting ransomware's increasing relevance, Proofpoint reported that 70% of all malware encountered in the emails of its customer base during a 10-month interval in 2016 was ransomware. At the same time, the same company reported that the number of malicious email attachments grew by about 600% in comparison to 2015 [26]. In addition, many modern forms of ransomware have worm capabilities as demonstrated in a disconcerting fashion by the 2017 WannaCrypt/WannaCry attack, which affected 100,000s of systems of individual users and organizations leading even to the breakdown of industrial facilities.

Other studies also focus on providing practical examples of and empirical data on ransomware including work by O'Gorman and McDonald [25], who provide an in-depth perspective of specific ransomware campaigns and their financial impact. In a very comprehensive fashion, Kharraz et al. analyze ransomware code observed in the field between 2006 and 2014 [17]. Their results mirror Gazet's observation on a much broader pool of 1,359 ransomware samples, i.e., currently encountered ransomware lacks complexity and sophistication.

Nevertheless, it causes major harm. To cite just a few figures, the total cost of ransomware attacks (including paid ransoms) increased to $209 Million for the first three months in 2016 according to FBI data. In comparison, for 2015 the FBI only reported damages of about $24 Million [9].

Drawing on their earlier research, Kharraz et al. developed practices to stem the impact of ransomware. Their key insight is that ransomware needs to temper with user or business data, so that increased monitoring of data repositories can stop ransomware attacks while they unfold, and detect novel attacks that bypassed even sophisticated preventative measures [16]. Scaife et al. also present an early-warning detection approach regarding suspicious file activity [27].

Extending this line of work to a new context, Andronio et al. study ransomware in the mobile context and develop an automated approach for detection while paying attention to multiple typical behaviors of ransomware including the display of ransom notes on the device's screen [2]. Likewise, Yang et al. focus on mobile malware detection, and specifically ransomware identification [34].

Ransomware attacks appear to be predominantly motivated by financial motives, which supports the usage of an economic framework for their analysis. However, other related types of attacks such as malicious device bricking (see, for example, the BrickerBot attack focusing on IoT devices [29]) may be based on a purely destructive agenda, with less clearly identifiable motives.

While knowledge about the technical details and financial impact of ransomware is growing, we are unaware of any research which focuses on the strategic economic aspects of the interactions between cybercriminals that distribute ransomware and businesses or consumers who are affected by these actions.

Economics of Security: Game-theoretic models to better understand security scenarios have gained increased relevance due to the heightened professionalism of cybercriminals. Of central interest are models that capture interdependencies or externalities arising from actions by defenders or attackers [19].

A limited number of research studies focus on the modeling of the attack side. For example, Schechter and Smith capture different attacker behaviors [28]. In particular, they consider the cases of serial attacks where attackers aim to compromise one victim after another, and the case of a parallel attack, where attackers can automate their attacks to focus on multiple defenders at one point in time. We follow the latter approach, which has high relevance for self-propagating ransomware such as WannaCrypt/WannaCry.

Another relevant aspect of our work are incentives to invest in backup technologies, which have found only very limited consideration in the literature. Grossklags et al. investigate how a group of defenders individually decide on how to split security investments between preventative technologies and recovery technologies (called self-insurance) [12]. In their model, preventative investments are subject to interdependencies drawing on canonical models from the literature on public goods [31], while recovery investments are effective independent from others' choices. Fultz and Grossklags [10] introduce strategically acting attackers in this framework, who respond to preventative investments by all defenders. In our model, backup investments are also (partially) effective irrespective of others' investment choices. However, in the context of ransomware, pervasive investments in backup technologies can have a deterrence and/or displacement effect on attackers [5], which we capture with our work.

While we draw on these established research directions, to the best of our knowledge, our work is the first game-theoretic approach focused on ransomware.

6   Concluding Remarks

In this paper, we have developed a game-theoretic model, which is focused on key aspects of the adversarial interaction between organizations and ransomware attackers. In particular, we place significant emphasis on the modeling of security investment decisions for mitigation, i.e., level of backup effort, as well as the strategic decision to pay a ransom or not.

These factors are interrelated and also influence attacker behavior. For example, in the context of kidnappings by terrorists it has been verified based on incident data that negotiating with kidnappers and making concessions encourages substantially more kidnappings in the future [6]. We would expect a similar effect in the context of ransomware, where independently acting organizations who are standing with the "back against the wall" have to make decisions about ransom payments to get operations going again, or to swallow the bitter pill of rebuilding from scratch and not giving in to cybercriminals. Indeed, our analysis shows that there is a sizable gap between the decentralized decision-making at equilibrium and the socially optimal outcome. This raises the question whether organizations paying ransoms should be penalized? However, this (in turn) poses a moral dilemma, for example, when patient welfare at hospitals or critical infrastructure such at power plants are affected *now*.

An alternative pathway is to (finally) pay significantly more attention to backup efforts as a key dimension of overall security investments. The relative absence of economic research focused on optimal mitigation and recovery strategies is one key example of this omission. A laudable step forward is the recently released factsheet document by the U.S. Department of Health & Human Service on ransomware and the Health Insurance Portability and Accountability Act (HIPAA) [30]. It not only states that the encryption of health data by ransomware should be considered a security breach under HIPAA (even though no data is exfiltrated[3]), but also that having a data backup plan is a required security effort for all HIPAA covered organizations.

An interesting question for future research is the role of cyberinsurance in the context of ransomware, i.e., specifically policies including cyber-extortion. How would these policies have to be designed to achieve desirable outcomes? As discussed above, in the case of kidnappings one would worry about incentivizing future kidnappings by making concessions via kidnapping insurance [8]; however, the design space in the context of ransomware is significantly more complex, but also offers more constructive directions.

Acknowledgments. We thank the anonymous reviewers for their comments. The research activities of Jens Grossklags are supported by the German Institute for Trust and Safety on the Internet (DIVSI). Aron Laszka's work was supported in part by the National Science Foundation (CNS-1238959) and the Air Force Research Laboratory (FA 8750-14-2-0180).

---

[3] The reasoning is as follows: "When electronic protected health information (ePHI) is encrypted as the result of a ransomware attack, a breach has occurred because the ePHI encrypted by the ransomware was acquired (i.e., unauthorized individuals have taken possession or control of the information), and thus is a "disclosure" not permitted under the HIPAA Privacy Rule [30]".

A Proofs

A.1 Proof of Lemma 1

From Eq. (3), we have that the best-response strategy p_i^* of organization i is

$$p_i^* \in \underset{p \in \{0,1\}}{\operatorname{argmax}} \left[W_j - C_B \cdot b_i - \beta \left(\frac{F_j + (1 - p) \cdot L_j}{b_i} + T_j + p \cdot r \right) \right] \tag{9}$$

$$= \underset{p \in \{0,1\}}{\operatorname{argmax}} \, p \cdot \left(\frac{L_j}{b_i} - r \right). \tag{10}$$

Clearly, $p_i^* = 1$ is a best response if and only if $\frac{L_j}{b_i} - r \geq 0$, and $p_i^* = 0$ is a best response if and only if $\frac{L_j}{b_i} - r \leq 0$. $\qquad\square$

A.2 Proof of Lemma 2

From Eq. (2), we have that the best-response strategy b_i^* of organization i is

$$b_i^* \in \underset{b_i \in \mathbb{R}_+}{\operatorname{argmax}} \left[W_j - C_B \cdot b_i - \beta \frac{F_j}{b_i} \right]. \tag{11}$$

To find the maximizing b_i^*, we take the first derivative of the payoff, and set it equal to 0:

$$-C_B + \beta \frac{F_j}{b_i^{*2}} = 0 \tag{12}$$

$$b_i^* = \pm \sqrt{\beta \frac{F_j}{C_B}}, \tag{13}$$

Since $b_i \in \mathbb{R}_+$, the only local optima is $b_i^* = \sqrt{\beta \frac{F_j}{C_B}}$. Further, the payoff is a concave function of b_i as the second derivative is negative, which means that this b_i^* is the global optimum and, hence, a unique best response. $\qquad\square$

A.3 Proof of Lemma 4

The best-response ransom demand r^* is

$$r^* \in \underset{r \in \mathbb{R}_+}{\operatorname{argmax}} \left[\sum_j \sum_{i \in G_j} V_j(a_1, a_2) \cdot r \cdot p_i^*(r) \right] - C_A \cdot (a_1 + a_2) - C_D \tag{14}$$

$$= \underset{r \in \mathbb{R}_+}{\operatorname{argmax}} \sum_j \sum_{i \in G_j} V_j(a_1, a_2) \cdot r \cdot \mathbb{1}_{\left\{ r \leq \frac{L_j}{b_j} \right\}} \tag{15}$$

$$= \underset{r \in \mathbb{R}_+}{\operatorname{argmax}} \sum_j |G_j| \cdot V_j(a_1, a_2) \cdot r \cdot \mathbb{1}_{\left\{ r \leq \frac{L_j}{b_j} \right\}}. \tag{16}$$

Clearly, the optimum is attained at either $\frac{L_1}{\hat{b}_1}$ or $\frac{L_2}{\hat{b}_2}$. Since we assumed that $\frac{L_1}{\hat{b}_1} \leq \frac{L_2}{\hat{b}_2}$, we have that $r = \frac{L_1}{\hat{b}_1}$ is a best response if and only if

$$(|G_1| V_1(a_1, a_2) + |G_2| V_2(a_1, a_2)) \frac{L_1}{\hat{b}_1} \geq |G_2| V_2(a_1, a_2) \frac{L_2}{\hat{b}_2} \tag{17}$$

$$|G_1| V_1(a_1, a_2) \frac{L_1}{\hat{b}_1} \geq |G_2| V_2(a_1, a_2) \left(\frac{L_2}{\hat{b}_2} - \frac{L_1}{\hat{b}_1} \right). \tag{18}$$

Further, an analogous condition holds for $r = \frac{L_2}{\hat{b}_2}$ being a best response, which concludes our proof. $\qquad\square$

A.4 Proof of Lemma 5

Recall that the attacker's expected payoff is

$$\mathrm{E}\left[\mathcal{U}_A\right] = \left(\sum_j \sum_{i \in G_j} V_j(a_1, a_2) \cdot p_i \cdot r \right) - C_A \cdot (a_1 + a_2) - C_D \cdot 1_{\{a_1 > 0 \text{ or } a_2 > 0\}}. \tag{19}$$

Consider that $a_1 + a_2 = a_{\mathrm{sum}}$ and r are given, and $a_{\mathrm{sum}} > 0$. Under these conditions, the attacker's best strategy is

$$a_1^* \in \operatorname*{argmax}_{a_1 \geq 0} \left(\sum_j \sum_{i \in G_j} V_j(a_1, a_2) \cdot p_i^*(r) \cdot r \right) - C_A \cdot (a_1 + a_2) - C_D \tag{20}$$

$$= \operatorname*{argmax}_{a_1 \geq 0} \frac{a_1}{D + a_{\mathrm{sum}}} |G_1| \cdot 1_{\left\{ r \leq \frac{L_1}{\hat{b}_1} \right\}} + \frac{a_{\mathrm{sum}} - a_1}{D + a_{\mathrm{sum}}} |G_2| \cdot 1_{\left\{ r \leq \frac{L_2}{\hat{b}_2} \right\}}, \tag{21}$$

giving the non-negative payoff. The best strategy can be calculated readily. $\qquad\square$

A.5 Proof of Proposition 1

Lemma 5 shows the attacker's best-response attack effort for fixed effort level, i.e., a_{sum}. In this Lemma, for example, $a_1^* = 0$ and $a_2^* = a_{sum}$ is the attacker's best-response effort if $|G_1| \cdot 1_{\left\{ r \leq \frac{L_1}{\hat{b}_1} \right\}} < |G_2| \cdot 1_{\left\{ r \leq \frac{L_2}{\hat{b}_2} \right\}}$ and the resulting attacker's payoff is non-negative. According to Lemma 4, the attacker's best-response ransom demand is either $\frac{L_1}{\hat{b}_1}$ or $\frac{L_2}{\hat{b}_2}$ and without loss of generality, we have assumed that $\frac{L_1}{\hat{b}_1} \leq \frac{L_2}{\hat{b}_2}$.

For this case, the attacker's payoff is equal to:

$$\mathrm{E}\left[\mathcal{U}_A\right] = \frac{a_{sum}}{D + a_{sum}} |G_2| \cdot r \cdot 1_{\left\{ r \leq \frac{L_2}{\hat{b}_2} \right\}} - C_A \cdot a_{sum} - C_D. \tag{22}$$

If the above equation is negative, i.e.,

$$r < \frac{(D + a_{sum}) (C_A \cdot a_{sum} + C_D)}{a_{sum} \cdot |G_2| \cdot 1_{\left\{ r \leq \frac{L_2}{\hat{b}_2} \right\}}},$$

the attacker's best-response effort is $a_1^* = a_2^* = 0$. To satisfy the above condition, we replace r with $\frac{L_2}{b_2}$, which gives

$$\frac{L_2 \cdot a_{sum} \cdot |G_2| \cdot 1_{\left\{ r \leq \frac{L_2}{b_2} \right\}}}{L_2 \left(D + a_{sum} \right) \left(C_A \cdot a_{sum} + C_D \right)} < \hat{b}_2^*.$$

Further, the defender's best-response backup strategy when there is no attack, i.e., $a_1^* = a_2^* = 0$ is calculated based on Lemma 2. By inserting the value of $\hat{b}_2^*$ from Lemma 2, we can readily have the following:

$$\frac{L_2 \cdot a_{sum} \cdot |G_2| \cdot 1_{\left\{ r \leq \frac{L_2}{b_2} \right\}}}{L_2 \left(D + a_{sum} \right) \left(C_A \cdot a_{sum} + C_D \right)} < \sqrt{\beta \frac{F_2}{C_B}}.$$

Another condition can be calculated similarly. □

References

1. Acquisti, A., Grossklags, J.: What can behavioral economics teach us about privacy? In: Digital Privacy: Theory, Technologies, and Practices, pp. 363–379. Auerbach Publications (2007)
2. Andronio, N., Zanero, S., Maggi, F.: HELDROID: Dissecting and detecting mobile ransomware. In: Bos, H., Monrose, F., Blanc, G. (eds.) RAID 2015. LNCS, vol. 9404, pp. 382–404. Springer, Cham (2015). doi:10.1007/978-3-319-26362-5_18
3. Backblaze: Backup awareness survey, our 10th year, industry report. https://www.backblaze.com/blog/backup-awareness-survey/
4. Baddeley, M.: Information security: Lessons from behavioural economics. In: Workshop on the Economics of Information Security (WEIS) (2011)
5. Becker, G.: Crime and punishment: an economic approach. J. Polit. Econ. **76**(2), 169–217 (1968)
6. Brandt, P., George, J., Sandler, T.: Why concessions should not be made to terrorist kidnappers. Eur. J. Polit. Econ. **44**, 41–52 (2016)
7. Bruskin Research: Nearly one in four computer users have lost content to blackouts, viruses and hackers according to new national survey, survey conducted for Iomega Corporation (2001)
8. Fink, A., Pingle, M.: Kidnap insurance and its impact on kidnapping outcomes. Public Choice **160**(3), 481–499 (2014)
9. Finkle, J.: Ransomware: Extortionist hackers borrow customer-service tactics (2016). http://www.reuters.com/article/us-usa-cyber-ransomware-idUSKCN0X917X
10. Fultz, N., Grossklags, J.: Blue versus Red: towards a model of distributed security attacks. In: Dingledine, R., Golle, P. (eds.) FC 2009. LNCS, vol. 5628, pp. 167–183. Springer, Heidelberg (2009). doi:10.1007/978-3-642-03549-4_10
11. Gazet, A.: Comparative analysis of various ransomware virii. J. Comput. Virol. **6**(1), 77–90 (2010)
12. Grossklags, J., Christin, N., Chuang, J.: Secure or insure?: A game-theoretic analysis of information security games. In: Proceedings of the 17th International World Wide Web Conference, pp. 209–218 (2008)

13. Grossklags, J., Barradale, N.J.: Social status and the demand for security and privacy. In: De Cristofaro, E., Murdoch, S.J. (eds.) PETS 2014. LNCS, vol. 8555, pp. 83–101. Springer, Cham (2014). doi:10.1007/978-3-319-08506-7_5

14. IBM: IBM study: Businesses more likely to pay ransomware than consumers, industry report (2016). http://www-03.ibm.com/press/us/en/pressrelease/51230.wss

15. Kabooza: Global backup survey: About backup habits, risk factors, worries and data loss of home PCs, January 2009. http://www.kabooza.com/globalsurvey.html

16. Kharraz, A., Arshad, S., Mulliner, C., Robertson, W., Kirda, E.: UNVEIL: A large-scale, automated approach to detecting ransomware. In: Proceedings of the 25th USENIX Security Symposium (USENIX Security), pp. 757–772 (2016)

17. Kharraz, A., Robertson, W., Balzarotti, D., Bilge, L., Kirda, E.: Cutting the Gordian Knot: A look under the hood of ransomware attacks. In: Almgren, M., Gulisano, V., Maggi, F. (eds.) DIMVA 2015. LNCS, vol. 9148, pp. 3–24. Springer, Cham (2015). doi:10.1007/978-3-319-20550-2_1

18. KnowBe4: The 2017 endpoint protection ransomware effectiveness report, industry report (2017). https://www.knowbe4.com/hubfs/Endpoint%20Protection%20Ransomware%20Effectiveness%20Report.pdf

19. Laszka, A., Felegyhazi, M., Buttyan, L.: A survey of interdependent information security games. ACM Comput. Surv. **47**(2), 23:1–23:38 (2014)

20. Laszka, A., Farhang, S., Grossklags, J.: On the economics of ransomware. CoRR abs/1707.06247 (2017). http://arxiv.org/abs/1707.06247

21. Liao, K., Zhao, Z., Doupé, A., Ahn, G.J.: Behind closed doors: Measurement and analysis of CryptoLocker ransoms in Bitcoin. In: Proceedings of the 2016 APWG Symposium on Electronic Crime Research (eCrime) (2016)

22. Luo, X., Liao, Q.: Awareness education as the key to ransomware prevention. Inf. Syst. Secur. **16**(4), 195–202 (2007)

23. Luo, X., Liao, Q.: Ransomware: A new cyber hijacking threat to enterprises. In: Gupta, J., Sharma, S. (eds.) Handbook of Research on Information Security and Assurance, pp. 1–6. IGI Global (2009)

24. O'Donoghue, T., Rabin, M.: Doing it now or later. Am. Econ. Rev. **89**(1), 103–124 (1999)

25. O'Gorman, G., McDonald, G.: Ransomware: A growing menace. Symantec Security Response (2012)

26. Proofpoint: Threat summary: Q4 2016 & year in review, industry report. https://www.proofpoint.com/sites/default/files/proofpoint_q4_threat_report-final-cm.pdf

27. Scaife, N., Carter, H., Traynor, P., Butler, K.: Cryptolock (and drop it): Stopping ransomware attacks on user data. In: Proceedings of the IEEE International Conference on Distributed Computing Systems (ICDCS), pp. 303–312 (2016)

28. Schechter, S.E., Smith, M.D.: How much security is enough to stop a thief? In: Wright, R.N. (ed.) FC 2003. LNCS, vol. 2742, pp. 122–137. Springer, Heidelberg (2003). doi:10.1007/978-3-540-45126-6_9

29. Simon, R.: Mirai, BrickerBot, Hajime attack a common IoT weakness (2017). https://securingtomorrow.mcafee.com/mcafee-labs/mirai-brickerbot-hajime-attack-common-iot-weakness/

30. U.S. Department of Health & Human Service: Fact sheet: Ransomware and HIPAA (2016). https://www.hhs.gov/sites/default/files/RansomwareFactSheet.pdf

31. Varian, H.: System reliability and free riding. In: Camp, L., Lewis, S. (eds.) Economics of Information Security (Advances in Information Security), vol. 12, pp. 1–15. Kluwer Academic Publishers, Dordrecht (2004)

32. Venkat, S.: Lessons for telcos from the WannaCry ransomware attack, cerillion blog (2017). http://www.cerillion.com/Blog/May-2017/Lessons-for-Telcos-from-the-WannaCry-attack
33. Verizon: 2017 Data breach investigations report: Executive summary, industry report
34. Yang, T., Yang, Y., Qian, K., Lo, D.C.T., Qian, Y., Tao, L.: Automated detection and analysis for Android ransomware. In: Proceedings of the 1st IEEE International Conference on Big Data Security on Cloud (DataSec), pp. 1338–1343. IEEE (2015)
35. Young, A., Yung, M.: Cryptovirology: Extortion-based security threats and countermeasures. In: Proceedings of the 1996 IEEE Symposium on Security and Privacy, pp. 129–140 (1996)
36. Young, A., Yung, M.: Cryptovirology: The birth, neglect, and explosion of ransomware. Commun. ACM **60**(7), 24–26 (2017)

Deterrence of Cyber Attackers in a Three-Player Behavioral Game

Jinshu Cui[1,2(✉)], Heather Rosoff[2,3], and Richard S. John[1,2]

[1] Department of Psychology, University of Southern California,
Los Angeles, CA 90089, USA
jinshucu@usc.edu
[2] Center for Risk and Economic Analysis of Terrorism Events,
University of Southern California, Los Angeles, CA 90089, USA
[3] Sol Price School of Public Policy, University of Southern California,
Los Angeles, CA 90089, USA

Abstract. This study describes a three-player cyber security game involving an attacker, a defender, and a user. An attacker must choose to attack the defender or the user or to forego an attack altogether. Conversely, defender (e.g., system administrator) and user (e.g., individual system user) must choose between either a "standard" or "enhanced" security level. Deterrence is operationalized as a decision by an attacker to forego an attack. We conducted two behavioral experiments in which players were assigned to the cyber attacker role over multiple rounds of a security game and were incentivized based on their performance. The defender and user's decisions were based on a joint probability distribution over their two options known to the attacker. Coordination between the defender and user is manipulated via the joint probability distribution. Results indicate that attacker deterrence is influenced by coordination between defender and user.

Keywords: Deterrence · Cyber security · Expected utility

1 Introduction

Deterrence is defined as the use of threats (e.g., costs and losses) by a defender to convince an attacker to refrain from initiating an attack [1]. Literature on deterrence has focused on the use of rational choice and game theoretic models of decision making (game theory), as well as organizational theory and cognitive psychology [2].

Early literature describing deterrence in a cyber security context attempted to dissect motivations and strategies of attackers from a behavioral point of view, an approach that has become more and more common in cyber research [3–5]. Constructing a profile for cyber attackers and mapping their objectives is useful in predicting both target selection and means of attack [6,7] and provides valuable insight for strategizing security measures to promote deterrence.

© Springer International Publishing AG 2017
S. Rass et al. (Eds.): GameSec 2017, LNCS 10575, pp. 418–436, 2017.
DOI: 10.1007/978-3-319-68711-7_22

Much of the previous research on deterrence of cyber crime from a behavioral point of view is theory-driven [8,9], occasionally comparing cyber deterrence to concepts found in other fields such as criminology [10] and sociology [4]. These theories are often oriented toward building a network that is likely to effectively deter attackers, rather than mapping the relationship between defender strategies and attacker decision making.

Literature on deterrence of cyber crime using empirical approaches is fairly limited. One study by Chan and Yao [11] considered the perception of attackers by using actual attackers' self reports to examine the relationship between the likelihood deterrence by particular security measures and attacker characteristics such as hacking motivation and personal beliefs related to hacking. Guitton [12] took a retrospective approach by analyzing a database of previous cyber security attacks. Attribution was also explored in this study by analyzing media coverage pertaining to the attacks. Guitton [12] found that lack of attribution is correlated with number of attacks, and suggested that media coverage of cyber attacks may discourage attackers from taking action.

There also is a class of studies testing the effectiveness of different types of deterrence measures that can be employed by defenders, specifically different types of warning banners [13–15]. The experimental designs of these studies often involve decoy servers (also known as honeypots) to attract real life attackers to break into the system. The downside of this approach is that the information that can be collected about the attackers is limited, therefore little can be inferred about how attackers make decisions. While these approaches provide valuable data that can be compared against available theoretical frameworks, the next big step in developing a theory of deterrence of cyber crime involves collecting empirical data through controlled behavioral experiments of attacker decision making. This approach could offer better insight about the causal relationships between deterrence strategy and attacker behavior.

Previous research on deterrence suggests that increasing the certainty of punishment could increase deterrence, i.e., decrease crime rate [16–18]. In this study, we are interested in testing the effect of certainty on the deterrence of cyber attackers. We hypothesized that an increase in an attacker's perception of certainty of losses could deter the attacker more effectively.

There are three possible agents in a cyber environment: (1) an attacker, (2) a defender and (3) a user. Previous research on cyber security has focused on the relationship between attackers and defenders in general. In this study, we distinguish two types of players in a cyber security game who both engage in security measures against the attacker: the defender, who provides security for an entire information system, and the user, who only controls security for her own account. There is some degree of interaction between the two players. For example, a user can decide whether to close the firewall on her personal computer, or whether to open an email in the spam folder. Therefore, of particular interest is the attacker's decision when both the defender and the user are potential targets and have the option to interact and coordinate.

In this study, we describe a three-player cyber security game and collect empirical data through two behavioral experiments. In Experiment I, we focus on the effect of defender-user coordination on deterrence of cyber attackers. The attacker's expected values for attacking either the defender or user are constant, regardless of the coordination of the defender and user. However, an attacker's perception of the decision space may be influenced by the joint array of defender and user defenses. A rational expected value maximizing attacker should not employ a different strategy when expected values of attacking different targets are fixed. However, attacker preferences and behavior could be influenced by psychological effects such as certainty effects, risk aversion, etc. We hypothesized in Experiment I that an attacker's perception of the certainty of losses (i.e., certainty of a defense) could be influenced by defender-user coordination, which could thereby influence attacker deterrence. We tested a similar hypothesis in Experiment II with a modified game that includes less exogenous uncertainty. In addition, Experiment II tested the extent to which attacker deterrence is impacted by the magnitude of attacker expected values by varying the overall effectiveness of resources deployed by both the defender and the user.

The next section describes the three-player cyber security game utilized in our behavioral experiments. Sections 3 and 4 present methods and results of Experiments I and II, where humans played as cyber attackers in the three-player game. The last section discusses the findings from the two experiments and concludes the paper.

2 A Three-Player Cyber Security Game

2.1 Players

Attackers are agents that attempt to breach the security of their targets in order to obtain information. Attackers can be categorized according to many criteria. Among various types of attackers, the following entities present distinct threats: criminals, foreign governments, foreign military, non-state combatants, businesses, and terrorists [19]. Attackers may choose to target large networks, such as the database of a social media platform, or specific individuals, such as individual accounts on the social media platform.

System administrators (i.e., the defender) are assumed to be more knowledgeable than lay users. Therefore, we account for the possibility of an attacker discovering a vulnerability in the system or gaining an exploitation technique that can massively increase her chances of successfully breaching the defender's security, where the nature of gaining these capabilities is stochastic. For instance, an attacker can seek to identify a security bug in the system and to exploit that bug to attain her goal. An attacker can choose to immediately use that capability, or wait for a better time to use it, with the risk that either the system administrator discovers and eliminates that bug, or another attacker discovers it and uses it to her advantage.

Defenders are security professionals who are in charge of protecting the computer system that is also partly responsible for user security. For instance, the security professionals of a website are responsible for protecting all user data when an attacker tries to penetrate the website; however, such system security measures could also protect individual accounts associated with the website when an attacker targets individual users. Defenders should consider the level of security they impose based on several factors, including both cost and user satisfaction. Greater security is often associated with higher cost and more restrictions for users of the system, which can lead to lower user satisfaction. Because of defenders' expertise and access to resources, it is much more difficult for an attacker to break into the defender's security, involving a high level of risk and typically requiring a large investment of time and effort, a relatively low chance of success, potential legal consequences, but very high potential rewards related to attacker objectives, e.g., financial gains, obtaining sensitive information, signaling, etc. [19].

Users are individuals connected to a system, but who establish their personal security levels. The primary objective of individual users is efficiency and convenience in their access to a system rather than security. Therefore, user security measures are generally modest compared to those imposed by the system defender. There is substantial variability in how users apply their own security measures, accounting for aspects such as convenience or familiarity with cyber security protocols. However, in general, user security is expected to be much less sophisticated than the system security measures employed by the defender. It is typically easier to directly attack an individual user, but the reward is usually relatively low, unless the target is a high profile user.

2.2 Decision Spaces for Three Players

Attackers have the option to attack the system or individual users in the system. An attack towards the defender may have a large cost due to the sophistication of security measures, although a successful attack offers a larger reward. Success in discovering a capability to penetrate a system is stochastic and involves a significant allocation of time and energy. Without a capability, there is very little chance of success in attacking the system. Therefore, in the present game we assume that attackers must allocate effort to obtain a capability before attacking the system. As a result, attacking the system costs more and takes longer than attacking individual users.

Directly attacking a user is another option for the attacker. Such an attack requires fewer resources and a higher chance of success, but the reward is much less than a successful attack directed towards the system.

A third option for attackers is no attack. Eliminating attacks is characterized as deterrence and it may occur when costs and perceived expected losses outweigh perceived expected benefits. Choosing no attack incurs no cost or benefit. However, there is an opportunity cost for attackers who choose not to attack.

The defender's decision space consists of different levels of security she can impose on the system. A higher level of security offers greater protection from attack, both targeting the system and individual users, but is more onerous for users and is more expensive.

The user's decision space also consists of different security level settings, but the overall effectiveness of user defenses depends on both user and defender security settings. Another consequence of the user's dependency on the defender includes potential losses when an attacker successfully attacks the system. Users are always vulnerable to cyber attacks, whether they are personally targeted or the system they are depending on is attacked.

2.3   Uncertainties in Attacker's Choices

In the current study, we focus on attacker's choices in a three-player cyber security game. In particular, we are interested in deterrence of cyber attackers determined by strategy choices of the defender and the user. Attackers may choose to attempt to obtain a "capability", which is specifically relevant to the targeted system (defender) only. The probability of successfully obtaining a capability is assumed to be fixed. After obtaining a capability to penetrate a system, the success rate of compromising a system is related to defender's system security level, where it is harder to compromise the system when the defender imposes a high security level.

On the other hand, the probability of successfully attacking the user depends on both the defender and the user's security level. When the security level of the user is low, it is easy for the attacker to obtain personal information; however, if the defender has influenced the user to accept a high-security system, it is still hard for the attacker to break in. For instance, if the system does a good job of blocking suspicious attempts from an attacker (e.g., email filter) and the user trusts the filtering, the user is safe. Of course, the user can allow access to her account manually (e.g., open spam email), which can also increase the attacker's likelihood of success. In the first case, the defender and the user have the same high-security level, so their strategies are complementing (positively correlated). In the second case, the defender has a high-security level, whereas the user has a low-security level, so the two players' strategies are substituting (negatively correlated). In a third case, there is little or no relation between the defender and the user's strategies (independent). Attacking the system involves two costs, namely cost of obtaining a capability and the cost of compromising the system after obtaining the capability. The cost of attacking the user is lower than that of attacking the system.

There is no instant reward or penalty for an attacker attempting to obtain a capability to compromise a system; however, an attacker expects to receive a reward in the future when the capability is obtained and information is successfully obtained from the targeted system. The reward of compromising a system is much higher than compromising a user because the information obtained from a system contains all system users' information, which would be expected to include information for a large number of users. Likewise, the penalty is also higher for a failed system attack because the crime is more serious and likely to attract attention.

The next section presents two behavioral experiments conducted online. In Experiment I, we used the general three-player game described above and focused on attacker deterrence in terms of inhibiting attacks from players. We manipulated defender-user coordination and tested the influence of coordination on deterrence of cyber attacker players. In Experiment II, we used a modified game with less exogenous uncertainty for attackers' rewards and penalties. Experiment II also focuses on threat shifting, another form of deterrence, and whether threat shifting is influenced by the effectiveness of defenses employed by the defender and the user.

3 Experiment I

3.1 Method

In this section, we present a behavioral experiment conducted online. We implemented a version of the 3-way game motivated in Sect. 2 using a hypothetical Website attack scenario, where the attacker selects targets in repeated rounds with uncertainty about rewards and penalties contingent on the probabilistic defense strategies utilized by both the defender and user, determined by a joint probability distribution known to the attacker.

Design Overview. Players acted as attackers in a computer hacking game for 20 rounds. In each round, players chose one of three possible strategies: attacking the system, attacking individual users in the system, or no attack. The probability of success (i.e., system and individual users) depended on the defender and the user's implemented security levels, which are generated based on their joint probability distribution. The joint distribution of the defender and the user's implemented security levels is manipulated as either complementing, substituting, or nearly independent. Specifically, the probabilities of the defender and the user choosing a high or low-level security defense have a positive correlation, negative correlation or a small correlation (near zero). Each player is randomly assigned to one of the three conditions and is given a summary of the defense strategies used by the defender and the user over the last 100 (historical) attack attempts (see Table 1). In all three cases, the marginal probability of high security and low security for both the defender and the user is held constant (50%).

Scenario and Procedure. Players are asked to role-play an attacker targeting an e-commerce website for 20 rounds. The hypothetical target website hosts multiple customer databases and has over 2 million online customer users. The attacker's objective is to gain access to as much of the customers' personal information as possible in order to exploit customer financial information. In each round, the attacker must decide whether to attempt to (1) hack into one of the websites' customer databases and obtain information from approximately 800 customers, (2) hack into a subset of individual customer accounts by way of their personal computers and obtain information from approximately 200 customers, or (3) not carry out an attack at all. Attacking the database takes at least two rounds. Attackers must first expend resources to obtain the appropriate

Table 1. The defender and the user's security levels over 100 historical attack attempts. The joint probability distribution between defender and user is presented to the attacker to suggest a (a) positive correlation, (b) negative correlation, or (c) small correlation

User \ Defender	Standard	Enhanced
Standard	49	1
Enhanced	1	49

(a)

User \ Defender	Standard	Enhanced
Standard	1	49
Enhanced	49	1

(b)

User \ Defender	Standard	Enhanced
Standard	20	30
Enhanced	30	20

(c)

technical capability to carry out the attack (step 1) and if successful, execute the actual attack (step 2) in the subsequent round. The capability, once obtained, must be used in the subsequent round or it is lost.

The play of the defender (website administrator) and the users (customers) is simulated in a manner based only on the joint probability distributions provided to attacker players. The website administrator and customers execute either standard or enhanced security in each round. All customers are assumed to be using the same defense strategy. We manipulate the website administrator and customers' security strategies to be either complementing (positively correlated), substituting (negatively correlated), or nearly independent of each other (nearly uncorrelated).

Each attack alternative is associated with uncertainty about the attacker's probability of success, which depends on whether the website and customers follow standard security protocol or implement enhanced security. There are costs associated with attack execution, and rewards or penalties associated with each attack outcome. Attacker players' payoff in each round is calculated by the cost, reward and penalty correspondingly.

In each round, players are first presented with the decision framework, detailing attack alternatives and uncertain outcomes, as well as a summary of the joint probability distribution of defensive strategies used by the website administrator and customers over the last 100 attack attempts. Figure 1 shows the information provided to attacker players assigned to condition 1 (the defender and the user's defensive strategies are complementing). After reading and considering the information presented on the screen in each round, players make an attack decision.

Before beginning the first round of play, attacker players first watched a 5-minute video that describes the game and presents an example round. Players then play two practice rounds of the game and receive feedback. Players then play 20 rounds of the game. After attacker players select a target (or no attack) in each round, they are presented with the randomly generated website administrator and customers' security levels in that round, the attack outcome (success or not), and their net gain or loss from that round. The payoff for each round has a 25% chance of being selected after all 20 rounds for compensation; thus, on average, five rounds are selected for compensation for each player. At the end of 20 rounds, players are presented with a table indicating which rounds are randomly selected for compensation and their total compensation. In addition, attacker players received a $1 fixed show-up fee. Total compensation was capped at $10. The average attacker player payment was $3.80, including the show-up fee.

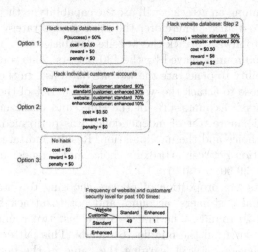

Fig. 1. Information display presented to players in condition 1

Respondents. One hundred and eighty players were recruited from TurkPrime, an affiliate of Amazon Mechanical Turk. Five players were removed for not completing the game. One hundred and seventy five players were included in the analysis. The median age ·of players was 31 years and 47% of the sample was female. The median gross annual income range was $30,000--$39,999.

3.2   Results

When an attacker has obtained a capability to penetrate the website database, the best choice is to execute the actual attack on the website database in the subsequent round, rather than attacking individual customers' accounts or no attack. In all three conditions of defender-user strategies, the expected value of attacking the website database is $4.50 once a capability is successfully obtained, whereas the expected value of attacking individual customers' accounts is only $0.50. Since an attacker capability expires if not used in the next round, players should always choose to attack the database website (step 2) once a capability is obtained (step 1). Failure to continue to step 2 is an indication that an attacker players did not understand game instructions. A total of 45 (26%) of all players chose not to continue to step 2 at least once following successfully obtaining a capability. Our analysis reports on the 130 players who always proceeded to step 2 once a capability was obtained (step 1).

The expected values of choosing the three alternatives are independent of the joint probability distributions of defender and user defense strategies. In particular, the expected values of attacking individual customers' accounts and no attack in each round are fixed at $0.50 and $0.00, whereas the expected value of attacking the website database per round is approximately $1.17. We calculated the expected value of attacking the website database by a simulation of 1000 trials, assuming an attacker will use the capability in the subsequent trial once it has been obtained. Therefore, the normative strategy for maximizing expected value is to always attack the website database.

Among the 2600 rounds played by the 130 players, 2096 rounds were played without the capability to penetrate the website database. In those no-capability rounds, players chose to attack the website database 51% of the time, to attack individual customers' accounts 43% of the time, and chose not to attack 6% of the time. A chi-square test of independence was performed to examine the relation between choice and the manipulation. Results indicated that there is a significant association between attacker's choice and defender-user relationship, $\chi^2(4, N = 2096) = 30.90$, $p < 0.01$.

Figure 2 depicts the proportion of players choosing to obtain a capability, to attack individual customers' accounts and not to attack by defender-user coordination over 20 rounds. The first two and last two rounds appear to be qualitatively different from the middle 16 rounds. This pattern was not unexpected, in that players are still learning the game in the beginning, and are playing in anticipation of the end during the last rounds. Notice that in the last two rounds the proportion of database attacks decreased, while attacks on customers' accounts and no attack increased. This may be due to avoidance of the 2-step database attack option near the end, and more conservative play (no attack) to avoid penalties near the end.

Therefore, we analyzed data from the middle 16 rounds, rounds 3 to 18. Results indicated that the proportion choosing the no attack option is significantly different between players in the substituting (negative correlation) condition compared to the nearly independent (small correlation) condition. Specifically, players were more likely to choose the no attack option when the correlation between the defender and the user's strategies was negative ($F(2, 127) = 2.360$, $p = 0.04$).

We then created 4 blocks of rounds, i.e., rounds 3 to 6, rounds 7 to 10, rounds 11 to 14, and rounds 15 to 18. Results indicated that the proportion of players choosing not to attack significantly increased over time ($F(1, 127) = 9.06$,

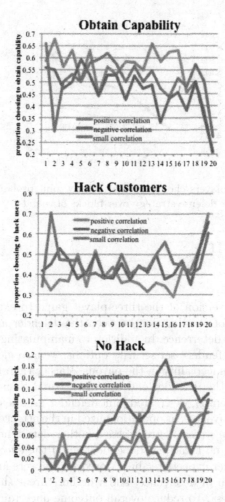

Fig. 2. Proportion of players choosing to obtain capability, to attack individual customers' accounts and not to attack by joint probability distribution of defender-user defense strategies over 20 rounds

$p < 0.01$). Again, the proportion of players choosing not to attack was significantly higher when the defender and user were substituting rather than nearly independent ($F(2, 127) = 2.447$, $p = 0.04$). There is no significant interaction effect between round block and the manipulated joint probability distribution of defender and user defenses ($F(2, 127) = 2.26$, $p = 0.11$). Figure 3 shows the proportion of players choosing no attack by the joint probability distribution characterizing the simulated defender and user's strategy over the four blocks of rounds.

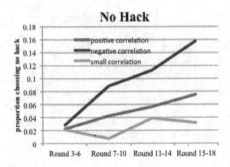

Fig. 3. Proportion of players choosing no attack by joint probability distribution of simulated defender-user defense strategy over blocks of rounds

4   Experiment II

4.1   Method

We used a modified version of the three-player game to extend our investigation of the influence of coordination between the defender and user's defensive strategies on attacker deterrence. In addition to manipulating the joint probability distributions of defender-user security options, we also modified the game to eliminate exogenous uncertainty in the payoffs. Rather than fixing the marginal defense strategies of the defender and user (i.e., equal chance of standard and enhanced in Experiment I), we created joint defender-user defense strategies that varied in overall effectiveness, as operationalized in the attacker players' expected value. As a result, we were able to assess whether deterrence (i.e., probability of attackers selecting no attack) is dependent upon the effectiveness of defense strategies, i.e., attacker expected value. Moreover, we were able to assess threat shifting, another form of deterrence, and whether threat shifting is influenced by defense effectiveness. To reduce overall outcome uncertainty, we removed all exogenous uncertainty so that the outcome of each attack decision is determined only by the joint implementation of the defender and the user's security levels. Figure 4 shows the information presented to attacker players in each round.

Specifically, players were randomly assigned to one of nine conditions of the defender and user's joint strategies. Conditions 1–3 are the same as the three conditions in Experiment I: (1) the defender and the user have matching defense levels nearly all of the time (complementing), (2) the defenders and the user have opposite defense levels nearly all of the time (substituting), or (3) the defender and the user's defense levels are randomly selected (independent), while the marginal probability of standard and enhanced security for both website administrator and customers is held at 50%. For the attacker's payoff space, the expected utilities of attacking the database (approximately $0.33) and individual accounts ($0.25) are constant across these three conditions. For conditions 4–6, either the defender or the user implements enhanced security level nearly all of the time (90%). For instance, in condition 4, the defender has enhanced security

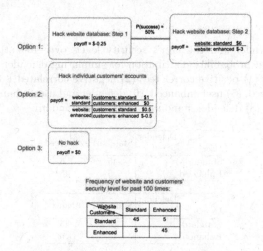

Fig. 4. Information presented to attacker players in Experiment II

90% of the time, while the user has enhanced security 50% of the time. For conditions 7–9, either the defender or the user implements a standard security level most of the time (90%). For instance, in condition 7, both the defender and the user have standard security 90% of the time. A full description of the nine conditions, elicited by the defender and the user's security levels over 100 previous attack attempts, is shown in Table 2.

Procedure. As in Experiment I, players assume the role of an attacker targeting an e-commerce website. In each round, players choose from one of three options: attacking the database, which takes at least two rounds; attacking individual customers' accounts in the system, which takes only one round; or not carrying out an attack.

Implementation of the defensive strategies of the defender and the user is simulated according to the joint probability distributions presented to attackers in Table 2. Each attack alternative is associated with different payoffs that depend on whether the website administrator and customers follow a standard security protocol or implement enhanced security.

Players first watch a video (about 5 min) that describes the game and presents an example round. Players then answer three questions that check whether they understand how the game is played. Players are allowed to proceed only if they answer all three questions correctly. Players who fail one or more questions do have a second chance to watch the video and answer the same three questions again. Failure to answer one of the three questions in the second pass disqualifies the player from going forward to play the game. Players who pass the attention-check then play two practice rounds of the game, followed by 30 actual rounds of the game. In each round, players are presented with the decision framework detailing attack alternatives and outcomes contingent on the joint defenses

Table 2. The defender and the user's security levels over 100 attack attempts in Experiment II. Joint probability distribution for simulating defender and user choices are manipulated as (1) positive correlated, (2) negative correlated, (3) not correlated, (4) defender enhanced, (5) user enhanced, (6) defender and user enhanced, (7) defender and user standard, (8) defender standard, or (9) user standard

User \ Defender	Standard	Enhanced
Standard	45	5
Enhanced	5	45

(1)

User \ Defender	Standard	Enhanced
Standard	5	45
Enhanced	45	5

(2)

User \ Defender	Standard	Enhanced
Standard	25	25
Enhanced	25	25

(3)

User \ Defender	Standard	Enhanced
Standard	5	45
Enhanced	5	45

(4)

User \ Defender	Standard	Enhanced
Standard	5	5
Enhanced	45	45

(5)

User \ Defender	Standard	Enhanced
Standard	5	5
Enhanced	5	85

(6)

User \ Defender	Standard	Enhanced
Standard	85	5
Enhanced	5	5

(7)

User \ Defender	Standard	Enhanced
Standard	45	5
Enhanced	45	5

(8)

User \ Defender	Standard	Enhanced
Standard	45	45
Enhanced	5	5

(9)

implemented by defender and user, as well as a summary of the joint strategy used by the website administrator and customers over the last 100 attempted attacks. Following each attacker choice, players see the generated strategies of the website administrator and customers in that round, as well as the attacker payoff. The payoff for each round has a 20% chance of being selected at the end of the 30 rounds for compensation; hence, players are compensated for six rounds on average. Following completion of 30 rounds, players are presented with a table indicating which rounds are selected for compensation and their total compensation. Players also received a $1 fixed show-up fee. Total compensation is capped at $10. The average payment was approximately $3.00.

Respondents. Four hundred and ninety-seven players were recruited from TurkPrime. The median age of players was 34 years, and 48% of the sample was female. The median gross annual income range was $30,000–$39,999.

4.2 Results

We first analyzed whether choosing not to attack (deterrence) is affected by coordination between the defender and the user (i.e., keeping the marginal defense constant in conditions 1 to 3). We found a significant effect of defender-user coordination for the last 8 rounds. Consistent with Experiment I, the proportion of no-attack rounds is significantly higher for the negative correlation (substituting) condition compared with the no correlation (independent) condition for rounds 23 to 30 ($F(2, 160) = 2.41$, $p = 0.047$). We expected the influence of coordination would be greatest at the end of the sequence of trials, after players experience several rounds of play. Figure 5 displays the proportion of players choosing not to attack by defender-user coordination (conditions 1 to 3) over 30 rounds. The solid line in the figure marks round 23.

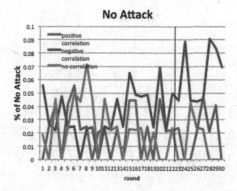

Fig. 5. Proportion of players choosing not to attack by defender-user relationship over 30 rounds

We also investigated whether deterrence (selection of no attack) is affected by the marginal probability of the defender and the user's defensive strategies. Results indicated that deterrence (no attack) decreased significantly as defense effectiveness was reduced (from conditions 4–6 to conditions 1–3 to conditions 7–9) ($F(2, 494) = 10.41$, $p < 0.001$). Specifically, the probability of attackers choosing not to attack was significantly higher when at least one player has enhanced security in place nearly all of the time (conditions 4 to 6), compared to strategies in which at least one player has standard security in place nearly all of the time (conditions 7 to 9) ($p < 0.001$). Figure 6 shows the proportion of players choosing no attack across conditions 4 to 9 over 30 rounds.

We also examined threat shifting, i.e., attackers switching from one target to the other (e.g., from database to individual customers' accounts). Results clearly indicated that the probability of attacking the database or individual accounts is influenced by the marginal effectiveness of defenses implemented by the defender and the user ($F(3, 221) = 8.86$, $p < 0.001$). Specifically, the probability of attacking the database decreased significantly when the defender has enhanced security in place nearly all of the time (condition 4) rather than standard security (condition 8) ($p = 0.02$); when the user has standard security in place nearly all of the time (condition 9) rather than enhanced security (condition 5) ($p < 0.001$); when the defender (condition 4) rather than user (condition 5) has enhanced security nearly all of the time ($p < 0.001$); and when the user (condition 9) rather than defender (condition 8) has standard security nearly all of the time ($p = 0.01$). The results suggest that threat was shifted from the database when the defender had more enhanced security in place or when the user had more standard security in place.

We found reciprocal results for attackers threat shifting away from attacking individual accounts when the user implemented enhanced security nearly all of the time, or when the database defenders implemented standard security nearly all of the time. Figure 7 shows the proportion of players choosing to attack the

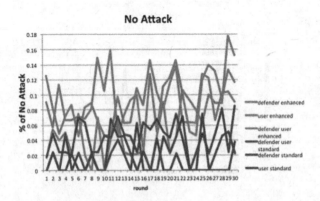

Fig. 6. Proportion of players choosing not to attack across conditions 4 to 9 over 30 rounds

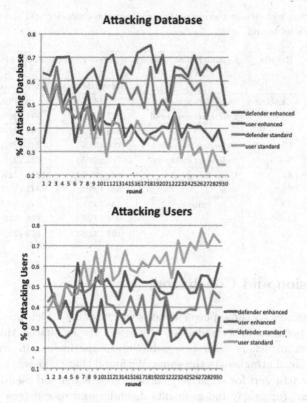

Fig. 7. Proportion of players choosing to attack database and individual accounts over 30 rounds

database and the individual accounts across conditions 4, 5, 8, and 9 over 30 rounds.

Finally, we analyzed the sequence of attackers' decision making and examined whether threat is shifted by the realized outcome from the previous round. Table 3 summarizes the number of rounds in which attackers chose each of the three alternatives across different outcomes from the previous round. Results indicated that after obtaining a capability in the previous round, attacker's choice in the next round is affected by the revealed defender's strategy in the previous round; respondents were more likely to forgo the second step of defender attack and switch to a user attack or be deterred after learning that the defender had enhanced security in the previous round ($\chi^2(2, N = 2978) = 182.02$, $p < 0.001$). Moreover, respondents were more likely to discontinue attacking individual users accounts and switch to the database after learning the user had enhanced security in the previous round ($\chi^2(2, N = 5687) = 204.38, p < 0.001$).

Table 3. Attacker's choice in the current round per the defender and the user's strategies in the previous round

Attacker choice (previous round)	Results	Defender/user strategy	Attacker choice (current round)		
			Attack database	Attack user	No attack
Attack database	Capability	Defender: standard	1305 (89%)	148 (10%)	10 (0.7%)
		Defender: enhanced	1052 (69%)	394 (26%)	69 (5%)
	No cap		2056 (71%)	764 (27%)	63 (2%)
	$6	Defender: standard	786 (68%)	342 (29%)	35 (3%)
	$-3	Defender: enhanced	777 (69%)	321 (29%)	20 (2%)
Attack user	$1	User: standard	441 (25%)	1240 (72%)	50 (3%)
	$0.5		330 (25%)	943 (72%)	22 (3%)
	$0	User: enhanced	448 (40%)	649 (58%)	19 (2%)
	$-0.5		691 (45%)	786 (51%)	68 (4%)
No attack	$0		191 (33%)	154 (26%)	239 (41%)

5   Discussion and Conclusion

This study describes a three-player cyber security game which manipulates the relationships between cyber defender and user. We evaluated the deterrence effect on cyber attackers with two behavioral experiments in which players assumed the role of attackers in the game. We found that there was greater deterrence on cyber attackers for negatively correlated defender and user defenses than for independent (or nearly independent) defender and user defenses. This suggests that fixed resources (in this case website defender and individual customers) were perceived to be more effective when defense strategies are coordinated in a manner in which one defender was highly likely to present an enhanced defense, while the other defender was highly likely to be in a standard defense. For negatively correlated defenses, the attacker is nearly certain to face an enhanced defense from one of the targets, while there is only a 75% chance of facing one or more enhanced defenses for independent (uncoordinated) defender user strategies. Therefore, negatively correlated defenses could increase attackers' perceptions of certainty of loss and increase the deterrence effect, which is consistent with the deterrence literature. The effect was stronger under exogenous outcome uncertainty in Experiment I.

We also found that allocation of greater resources (greater marginal probability of enhanced security for defender and/or user) could inhibit attackers from choosing to attack. In addition, we found that threat could be shifted from the first target to the second when the first target has enhanced security more often, or when the second target has standard security more often. These findings again confirm that certainty of punishment is positively related to deterrence.

Our results also suggest that attackers' target choices are autocorrelated with implemented defense by defender and user. Respondents tended to switch from one target to the other after learning the first target was protected previously. This implies that attackers expect both the database defender and individual user to engage in consistent security levels over time. Attackers expect defenders to exhibit a pattern of behavior similar to the "hot hand" effect. This finding is consistent with the literature, which suggests that positive recency is a psychological default [20–22].

As expected, an attacker was more likely to attack the system than to attack individual users. This is because the expected value of attacking the system is higher. However, there were a non-negligible number of players choosing to attack users repeatedly. Players who exhibit loss aversion [23] may choose to attack users since there is no penalty associated with a user attack (Experiment I). Another explanation is that some players may prefer the simpler, one-step attack on users, since it takes an uncertain number of rounds (minimum of 2 rounds, average of about 3 rounds) to attack the system. A third possibility is that the attacker is risk averse, and prefers the option of attacking the user, with lower expected value and lower variance, compared to attacking the system, with higher expected value and greater variance. Further research is needed to disentangle these three possibilities.

In the current experiments, players were provided with full information in the form of the joint probability distribution of defender and user strategy in every round. We anticipate that attackers may behave differently when the distribution of defender and user strategies is learned through experience. It is not clear that attackers recognize the inherent uncertainty in the defender and the user's strategies when full information is not provided. One possible hypothesis is that attackers will become overconfident in their uncertainty predictions of defender and user strategy selection, relying on recency and the "law of small numbers" [24] in the absence of full information. Further research is warranted to address the extent to which attacker choices are strongly influenced by recency.

There are some other elements in the current three-way game that warrant further exploration. For instance, the probability of obtaining a "capability" is fixed in both versions of the current game while in reality it might depend on the defender's security level. In both versions of the current game, the attacker is aware of only the joint strategies of the other two players. In reality, an attacker may have other knowledge of the defender and the user, such as their cyber investments, quality of security taskforce, whether there is a cost to switch from one security level to another, or whether the defender and the user select security levels sequentially, with communication, or simultaneously. All of these parameters might impact the attacker's assessment of attack success probability, and thereby influence her choice of target or decision not to attack.

References

1. Huth, P.K.: Deterrence and international conflict: empirical findings and theoretical debates. Annu. Rev. Polit. Sci. **2**(1), 25–48 (1999)
2. Jentleson, B.W., Whytock, C.A.: Who "won" Libya? The force-diplomacy debate and its implications for theory and policy. Int. Secur. **30**(3), 47–86 (2006)
3. Pfleeger, S.L., Caputo, D.D.: Leveraging behavioral science to mitigate cyber security risk. Comput. Secur. **31**(4), 597–611 (2012)

4. Summers, T.C., Lyytinen, K.J., Lingham, T., Pierce, E.A.: How hackers think: a study of cybersecurity experts and their mental models. In: Third Annual International Conference on Engaged Management Scholarship, Atlanta, Georgia (2013)
5. Crossler, R.E., Johnston, A.C., Lowry, P.B., Hu, Q., Warkentin, M., Baskerville, R.: Future directions for behavioral information security research. Comput. Secur. **32**, 90–101 (2013)
6. Hoath, P., Mulhall, T.: Hacking: motivation and deterrence, Part 1. Comput. Fraud Secur. **1998**(4), 16–19 (1998)
7. Barber, R.: Hackers profiled-who are they and what are their motivations? Comput. Fraud Secur. **2001**(2), 14–17 (2001)
8. Wible, B.: A site where hackers are welcome: using hack-in contests to shape preferences and deter computer crime. Yale Law J. **112**(6), 1577–1623 (2003)
9. Hua, J., Bapna, S.: How can we deter cyber terrorism? Inf. Secur. J. Glob. Perspect. **21**(2), 102–114 (2012)
10. Sharma, R.: Peeping into a hacker's mind: can criminological theories explain hacking? (2007). https://ssrn.com/abstract=1000446 or http://dx.doi.org/10.2139/ssrn.1000446
11. Chan, S.H., Yao, L.J.: An empirical investigation of hacking behavior. Rev. Bus. Inf. Syst. **9**(4), 41–58 (2011)
12. Guitton, C.: Criminals and cyber attacks: the missing link between attribution and deterrence. Int. J. Cyber Criminol. **6**(2), 1030–1043 (2012)
13. Maimon, D., Alper, M., Sobesto, B., Cukier, M.: Restrictive deterrent effects of a warning banner in an attacked computer system. Criminology **52**(1), 33–59 (2014)
14. Jones, H.M.: The restrictive deterrent effect of warning messages on the behavior of computer system trespassers. Doctoral dissertation (2014)
15. Wilson, I.I., Henry, T.: Restrictive deterrence and the severity of hackers' attacks on compromised computer systems. Doctoral dissertation (2014)
16. Gibbs, J.: Crime, punishment and deterrence. Southwest. Soc. Sci. Q. **48**(4), 515–530 (1968)
17. Tittle, C.: Crime rates and legal sanctions. Soc. Forces **16**(4), 409–423 (1969)
18. Durlauf, S.N., Nagin, D.S.: The deterrent effect of imprisonment. In: Cook, P.J., Ludwig, J., McCrary, J. (eds.) Controlling Crime: Strategies and Tradeoffs, pp. 43–94. University of Chicago Press, Chicago (2010)
19. Parker, D.B.: Fighting Computer Crime: A New Framework for Protecting Information. Wiley, New York (1998)
20. Wilke, A., Barrett, H.C.: The hot hand phenomenon as a cognitive adaptation to clumped resources. Evol. Hum. Behav. **30**(3), 161–169 (2009)
21. Scheibehenne, B., Wilke, A., Todd, P.M.: Expectations of clumpy resources influence predictions of sequential events. Evol. Hum. Behav. **32**(5), 326–333 (2011)
22. Tyszka, T., Zielonka, P., Dacey, R., Sawicki, P.: Perception of randomness and predicting uncertain events. Think. Reason. **14**(1), 83–110 (2008)
23. Kahneman, D., Tversky, A.: Choices, values, and frames. Am. Psychol. **39**(4), 341–350 (1984)
24. Tversky, A., Kahneman, D.: Belief in the law of small numbers. Psychol. Bull. **76**(2), 105–110 (1971)

Information Leakage Games

Mário S. Alvim[1], Konstantinos Chatzikokolakis[2], Yusuke Kawamoto[3](✉),
and Catuscia Palamidessi[4]

[1] Universidade Federal de Minas Gerais, Belo Horizonte, Brazil
[2] CNRS and École Polytechnique, Palaiseau, France
[3] AIST, Tsukuba, Japan
yusuke.kawamoto.aist@gmail.com
[4] INRIA and École Polytechnique, Palaiseau, France

Abstract. We consider a game-theoretic setting to model the interplay
between attacker and defender in the context of information flow, and
to reason about their optimal strategies. In contrast with standard game
theory, in our games the utility of a mixed strategy is a convex function of the distribution on the defender's pure actions, rather than the
expected value of their utilities. Nevertheless, the important properties
of game theory, notably the existence of a Nash equilibrium, still hold for
our (zero-sum) leakage games, and we provide algorithms to compute the
corresponding optimal strategies. As typical in (simultaneous) game theory, the optimal strategy is usually mixed, i.e., probabilistic, for both the
attacker and the defender. From the point of view of information flow,
this was to be expected in the case of the defender, since it is well known
that randomization at the level of the system design may help to reduce
information leaks. Regarding the attacker, however, this seems the first
work (w.r.t. the literature in information flow) proving formally that in
certain cases the optimal attack strategy is necessarily probabilistic.

1 Introduction

A fundamental problem in computer security is the leakage of sensitive information due to correlation of *secret information* with *observable information* publicly
available, or in some way accessible, to the attacker. Correlation in fact allows
for the use of Bayesian inference to guessing the value of the secret. Typical
examples are *side channels attacks*, in which (observable) physical aspects of the
system, such as the execution time of a decryption algorithm, may be exploited
by the attacker to restrict the range of the possible (secret) encryption keys. The
branch of security that studies the amount of information leaked by a system
is called *Quantitative Information Flow* (QIF), and it has seen growing interest
over the past decade. See for instance [3, 4, 10, 15, 27], just to mention a few.

In general, it has been recognized that randomization can be very useful to
obfuscate the link between secrets and observables. Examples include various
anonymity protocols (for instance, the dining cryptographers [9] and Crowds
[23]), and the renown framework of differential privacy [11]. The *defender* (the
system designer, or the user) is, therefore, typically probabilistic. As for the

© Springer International Publishing AG 2017
S. Rass et al. (Eds.): GameSec 2017, LNCS 10575, pp. 437–457, 2017.
DOI: 10.1007/978-3-319-68711-7_23

attacker, most works in the literature consider only *passive attacks*, limited to observing the system's behavior. Notable exceptions are the works of Boreale and Pampaloni [4], and of Mardziel et al. [18], which consider *adaptive attackers* who interact with and influence the system. We note that, however, [4] does not consider probabilistic strategies for the attacker. As for [18], although their model allows them, none of their extensive case-studies needs probabilistic attack strategies to maximize leakage. This may seem surprising, since, as mentioned before, randomization is known to be useful (and, in general, crucial) for the defender to undermine the attack and protect the secret. Thus there seems to be an asymmetry between attacker and defender w.r.t. probabilistic strategies in QIF. Our thesis is that there is indeed an asymmetry, but this does not mean that the attacker has nothing to gain from randomization: when the defender can change his own strategy according to the attacker's actions, it becomes advantageous for the attacker to try to be *unpredictable* and, consequently, adopt a probabilistic strategy. For the defender, while randomization is useful for the same reason, it is also useful because *it reduces the information leakage*, and since information leakage constitutes the gain of the attacker, this reduction influences his strategy. This latter aspect introduces the asymmetry mentioned above.

In the present work, we consider scenarios in which both attacker and defender can make choices that influence the system during the attack. We aim, in particular, at analyzing the attacker's strategies that can maximize information leakage, and the defender's most appropriate strategies to counterattack and keep the system as secure as possible. As argued before, randomization can help both attacker and defender make their moves unpredictable. The most suitable framework for analyzing this kind of interplay is, naturally, game theory, where the use of randomization can be modeled by the notion of *mixed strategies*, and where the interplay between attacker and defender, and their struggle to achieve the best result for themselves, can be modeled in terms of *optimal strategies* and *Nash equilibrium*. It is important to note, however, that one of the two advantages that randomization has for the defender, namely the reduction of information leakage, has no counterpart in standard game theory. Indeed, we demonstrate that this property makes the utility of a mixed strategy be a convex function of the distribution of the defender. In contrast, in standard game theory the utility of a mixed strategy is the expectation of the utility of the pure strategies of each player, and therefore it is an affine function on each of the players' distributions. As a consequence, we need to consider a new kind of games, which we call *information leakage games*, where the utility of a mixed strategy is a function affine on the attacker's strategy, and convex on the defender's. Nevertheless, the fundamental results of game theory, notably the minimax theorem and the existence of Nash equilibria, still hold for our zero-sum leakage games. We also propose algorithms to compute the optimal strategy, namely, the strategies for the attacker and the defender that lead to a Nash equilibrium, where no player has anything to gain by unilaterally changing his own strategy.

For reasoning about information leakage, we employ the well-established information-theoretic framework, which is by far the most used in QIF. A central notion in this model is that of *vulnerability*, which intuitively measures how easily the secret can be discovered (and exploited) by the attacker. For the sake of generality, we adopt the notion of vulnerability as any convex and continuous function [2,4], which has been shown to subsume most previous measures of the QIF literature [2], including *Bayes vulnerability* (a.k.a. min-vulnerability [8,27]), *Shannon entropy* [25], *guessing entropy* [19], and *g-vulnerability* [3].

We note that vulnerability is an expectation measure over the secrets. In this paper we assume the utility to be such average measure, but, in some cases, it could be advantageous for the defender to adopt different strategies depending on the value of the secret. We leave this refinement for future work.

The main contributions of this paper are the following:

- We define a general framework of *information leakage games* to reason about the interplay between attacker and defender in QIF scenarios.
- We prove that, in our framework, the utility is a convex function of the mixed strategy of the defender. To the best of our knowledge, this is a novelty w.r.t. traditional game theory, where the utility of a mixed strategy is defined as expectation of the utilities of the pure strategies.
- We provide methods for finding the solution and the equilibria of leakage games by solving a convex optimization problem.
- We show examples in which Nash equilibria require a mixed strategy. This is, to the best of our knowledge, the first proof in QIF that in some cases the optimal strategy of the attacker must be probabilistic.
- As a case study, we consider the Crowds protocol in a MANET (Mobile Ad-hoc NETwork). We study the case in which the attacker can add a corrupted node as an attack, the defender can add an honest node as a countermeasure, and we compute the defender component of the Nash equilibrium.

Plan of the paper. In Sect. 2 we review the basic notions of game theory and QIF. In Sect. 3 we introduce some motivating examples. In Sect. 4 we discuss the difference of our leakage games from those of standard game theory. In Sect. 5 we prove the convexity of the utility of the defender. In Sect. 6 we present algorithms for computing the Nash equilibria and optimal strategies for leakage games. In Sect. 7 we apply our framework to a version of the Crowds protocol. In Sect. 8 we discuss related work. Section 9 concludes.

2 Preliminaries

In this section we review some basic notions from game theory and QIF.

We use the following notation. Given a set $\mathcal{I}$, we denote by $\mathbb{D}\mathcal{I}$ the *set of all probability distributions* over $\mathcal{I}$. Given $\mu \in \mathbb{D}\mathcal{I}$, its *support* $\mathsf{supp}(\mu)$ is the set of its elements with positive probabilities, i.e., $\mathsf{supp}(\mu) = \{i \in \mathcal{I} : \mu(i) > 0\}$. We write $i \leftarrow \mu$ to indicate that a value $i \in \mathcal{I}$ is sampled from a distribution μ on $\mathcal{I}$.

2.1 Two-Player, Simultaneous Games

We review basic definitions from *two-player games*, a model for reasoning about the behavior of strategic players. We refer to [22] for more details.

In a game, each player has at its disposal a set of *actions* that he can perform, and obtains some payoff (gain or loss) depending on the outcome of the actions chosen by both players. The payoff's value to each player is evaluated using a *utility function*. Each player is assumed to be *rational*, i.e., his choice is driven by the attempt to maximize his own utility. We also assume that the set of possible actions and the utility functions of both players are *common knowledge*.

In this paper we only consider *finite games*, namely the cases in which the set of actions available to each player is finite. Furthermore, we only consider simultaneous games, meaning that each player chooses actions without knowing the actions chosen by the other. Formally, such a game is defined as a tuple[1] $(\mathcal{D}, \mathcal{A}, u_d, u_a)$, where $\mathcal{D}$ is a nonempty set of *defender's actions*, $\mathcal{A}$ is a nonempty set of *attacker's actions*, $u_d : \mathcal{D} \times \mathcal{A} \rightarrow \mathbb{R}$ is the *defender's utility function*, and $u_a : \mathcal{D} \times \mathcal{A} \rightarrow \mathbb{R}$ is the *attacker's utility function*.

Each player may choose an action deterministically or probabilistically. A *pure strategy* of the defender (resp. attacker) is a deterministic choice of an action, i.e., an element $d \in \mathcal{D}$ (resp. $a \in \mathcal{A}$). A pair (d, a) is a *pure strategy profile*, and $u_d(d, a)$, $u_a(d, a)$ represent the defender's and the attacker's utilities.

A *mixed strategy* of the defender (resp. attacker) is a probabilistic choice of an action, defined as a probability distribution $\delta \in \mathbb{D}\mathcal{D}$ (resp. $\alpha \in \mathbb{D}\mathcal{A}$). A pair (δ, α) is called a *mixed strategy profile*. The defender's and the attacker's *expected utility functions* for mixed strategies are defined, respectively, as:

$$U_d(\delta, \alpha) \overset{\text{def}}{=} \underset{\substack{d \leftarrow \delta \\ a \leftarrow \alpha}}{\mathbb{E}} u_d(d, a) = \sum_{\substack{d \in \mathcal{D} \\ a \in \mathcal{A}}} \delta(d)\alpha(a)u_d(d, a)$$

$$U_a(\delta, \alpha) \overset{\text{def}}{=} \underset{\substack{d \leftarrow \delta \\ a \leftarrow \alpha}}{\mathbb{E}} u_a(d, a) = \sum_{\substack{d \in \mathcal{D} \\ a \in \mathcal{A}}} \delta(d)\alpha(a)u_a(d, a)$$

A defender's mixed strategy $\delta \in \mathbb{D}\mathcal{D}$ is a *best response* to an attacker's mixed strategy $\alpha \in \mathbb{D}\mathcal{A}$ if $U_d(\delta, \alpha) = \max_{\delta' \in \mathbb{D}\mathcal{D}} U_d(\delta', \alpha)$. Symmetrically, $\alpha \in \mathbb{D}\mathcal{A}$ is a *best response* to $\delta \in \mathbb{D}\mathcal{D}$ if $U_a(\delta, \alpha) = \max_{\alpha' \in \mathbb{D}\mathcal{A}} U_d(\delta, \alpha')$. A *mixed-strategy Nash equilibrium* is a profile (δ^*, α^*) such that δ^* is a best response to α^* and vice versa. Namely, no unilateral deviation by any single player provides better utility to that player. If δ^* and α^* are point distributions concentrated on some $d^* \in \mathcal{D}$ and $a^* \in \mathcal{A}$, respectively, then (δ^*, α^*) is a *pure-strategy Nash equilibrium*, and will be denoted by (d^*, a^*). While not all games have a pure strategy Nash equilibrium, every finite game has a mixed strategy Nash equilibrium.

2.2   Zero-Sum Games and Minimax Theorem

A game $(\mathcal{D}, \mathcal{A}, u_d, u_a)$ is *zero-sum* if for any $d \in \mathcal{D}$ and any $a \in \mathcal{A}$, $u_d(d, a) = -u_a(d, a)$, i.e., the defender's loss is equivalent to the attacker's gain. For brevity, in zero-sum games we denote by u the attacker's utility function u_a, and by U

---

[1] Following the convention of *security games*, we set the first player to be the defender.

the attacker's expected utility U_{a}.[2] Consequently, the goal of the defender is to minimize U, and the goal of the attacker is to maximize it.

In simultaneous zero-sum games the Nash equilibrium corresponds to the solution of the *minimax* problem (or equivalently, the *maximin* problem), namely, the profile (δ^*, α^*) such that $U(\delta^*, \alpha^*) = \min_\delta \max_\alpha U(\delta, \alpha)$. The von Neumann's minimax theorem ensures that such solution (which always exists) is stable:

Theorem 1 (von Neumann's minimax theorem). *Let $\mathcal{X} \subset \mathbb{R}^m$ and $\mathcal{Y} \subset \mathbb{R}^n$ be compact convex sets, and $U : \mathcal{X} \times \mathcal{Y} \to \mathbb{R}$ be a continuous function such that $U(x, y)$ is convex in $x \in \mathcal{X}$ and concave in $y \in \mathcal{Y}$. Then it is the case that $\min_{x \in \mathcal{X}} \max_{y \in \mathcal{Y}} U(x, y) = \max_{y \in \mathcal{Y}} \min_{x \in \mathcal{X}} U(x, y)$.*

A related property is that, under the conditions of Theorem 1, there exists a *saddle point* (x^*, y^*) s.t., for all $x \in \mathcal{X}$ and $y \in \mathcal{Y}$, $U(x^*, y) \leq U(x^*, y^*) \leq U(x, y^*)$.

2.3 Quantitative Information Flow

Finally, we briefly review the standard framework of quantitative information flow, which is used to measure the amount of information leakage in a system.

Secrets and vulnerability. A *secret* is some piece of sensitive information the defender wants to protect, such as a user's password, social security number, or current location. The attacker usually only has some partial knowledge about the value of a secret, represented as a probability distribution on secrets called a *prior*. We denote by $\mathcal{X}$ the set of possible secrets, and we typically use π to denote a prior belonging to the set $\mathbb{D}\mathcal{X}$ of probability distributions over $\mathcal{X}$.

The *vulnerability* of a secret is a measure of the utility of the attacker's knowledge about the secret. In this paper we consider a very general notion of vulnerability, following [2], and define a vulnerability $\mathbb{V}$ to be any continuous and convex function of type $\mathbb{D}\mathcal{X} \to \mathbb{R}$. It has been shown in [2] that these functions coincide with the set of g-vulnerabilities, and are, in a precise sense, the most general information measures w.r.t. a set of basic axioms.[3]

Channels, posterior vulnerability, and leakage. Systems can be modeled as information theoretic channels. A *channel* $C : \mathcal{X} \times \mathcal{Y} \to \mathbb{R}$ is a function in which $\mathcal{X}$ is a set of *input values*, $\mathcal{Y}$ is a set of *output values*, and $C(x, y)$ represents the conditional probability of the channel producing output $y \in \mathcal{Y}$ when input $x \in \mathcal{X}$ is provided. Every channel C satisfies $0 \leq C(x, y) \leq 1$ for all $x \in \mathcal{X}$ and $y \in \mathcal{Y}$, and $\sum_{y \in \mathcal{Y}} C(x, y) = 1$ for all $x \in \mathcal{X}$.

---

[2] Conventionally in game theory the utility u is set to be that of the first player, but we prefer to look at the utility from the point of view of the attacker to be in line with the definition of utility as *vulnerability*, as we will introduce in Sect. 2.3.

[3] More precisely, if posterior vulnerability is defined as the expectation of the vulnerability of posterior distributions, the measure respects the data-processing inequality and yields non-negative leakage iff vulnerability is convex.

A distribution $\pi \in \mathbb{D}\mathcal{X}$ and a channel C with inputs $\mathcal{X}$ and outputs $\mathcal{Y}$ induce a joint distribution $p(x, y) = \pi(x)C(x, y)$ on $\mathcal{X} \times \mathcal{Y}$, with marginal probabilities $p(x) = \sum_y p(x, y)$ and $p(y) = \sum_x p(x, y)$, and conditional probabilities $p(x|y) = p(x,y)/p(y)$ if $p(y) \neq 0$. For a given y (s.t. $p(y) \neq 0$), the conditional probabilities $p(x|y)$ for each $x \in \mathcal{X}$ form the *posterior distribution* $p_{X|y}$.

A channel C in which $\mathcal{X}$ is a set of secret values and $\mathcal{Y}$ is a set of observable values produced by a system can be used to model computations on secrets. Assuming the attacker has prior knowledge π about the secret value, knows how a channel C works, and can observe the channel's outputs, the effect of the channel is to update the attacker's knowledge from a prior π to a collection of posteriors $p_{X|y}$, each occurring with probability $p(y)$.

Given a vulnerability $\mathbb{V}$, a prior π, and a channel C, the *posterior vulnerability* $\mathbb{V}[\pi, C]$ is the vulnerability of the secret after the attacker has observed the output of C. Formally: $\mathbb{V}[\pi, C] \overset{\text{def}}{=} \sum_{y \in \mathcal{Y}} p(y)\mathbb{V}[p_{X|y}]$.

The *information leakage* of a channel C under a prior π is a comparison between the vulnerability of the secret before the system was run—called the *prior* vulnerability—and the posterior vulnerability of the secret. The leakage reflects by how much the observation of the system's outputs increases the utility of the attacker's knowledge about the secret. It can be defined either *additively* ($\mathbb{V}[\pi, C] - \mathbb{V}[\pi]$), or *multiplicatively* ($\mathbb{V}[\pi,C]/\mathbb{V}[\pi]$).

3   A Motivating Example

We present some simple examples to motivate our information leakage games.

3.1   The Two-Millionaires Problem

The "two-millionaires problem" was introduced by Yao in [33]. In the original formulation, there are two "millionaires", Alice and Don, who want to discover who is the richest among them, but neither wants to reveal to the other the amount of money that he or she has.

We consider a (conceptually) asymmetric variant of this problem, where Alice is the attacker and Don is the defender. Don wants to learn whether or not he is richer than Alice, but does not want Alice to learn anything about the amount x of money he has. To this purpose, Don sends x to a trusted server Jeeves, who in turn asks Alice, privately, what is her amount a of money. Jeeves then checks which among x and a is greater, and sends the result y back to Don.[4] However, Don is worried that Alice may intercept Jeeves' message containing the result of the comparison, and exploit it to learn more accurate information about x by tuning her answer a appropriately (since, given y, Alice can deduce whether a is an upper or lower bound on x). We assume that Alice may get to know Jeeves' reply, but not the messages from Don to Jeeves.

---

[4] The reason to involve Jeeves is that Alice may not want to reveal a to Don, either.

We will use the following information-flow terminology: the information that should remain secret (to the attacker) is called *high*, and what is visible to (and possibly controllable by) the attacker is called *low*. Hence, in the program run by Jeeves a is a *low input* and x is a *high input*. The result y of the comparison (since it may be intercepted by the attacker) is a *low output*. The problem is to avoid the *flow of information* from x to y (given a).

One way to mitigate this problem is to use randomization. Assume that Jeeves provides two different programs to ensure the service. Then, when Don sends his request to Jeeves, he can make a random choice d among the two programs 0 and 1, sending d to Jeeves along with the value x. Now if Alice intercepts the result y, it will be less useful to her since she does not know which of the two programs has been run. As Don of course knows which program was run, the result y will still be just as useful to him.[5]

In order to determine the best probabilistic strategy that Don should apply to select the program, we analyze the problem from a game-theoretic perspective. For simplicity, we assume that x and a both range in $\{0,1\}$. The two alternative programs that Jeeves can run are shown in Table 1.

Table 1. The two programs run by Jeeves.

Program 0	Program 1
High Input: $x \in \{0,1\}$	High Input: $x \in \{0,1\}$
Low Input: $a \in \{0,1\}$	Low Input: $a \in \{0,1\}$
Output: $y \in \{T, F\}$	Output: $y \in \{T, F\}$
return $x \leq a$	return $x \geq a$

The combined choices of Alice and Don determine how the system behaves. Let $\mathcal{D} = \{0,1\}$ represent Don's possible choices, i.e., the program to run, and $\mathcal{A} = \{0,1\}$ represent Alice's possible choices, i.e., the value of the low input a. We shall refer to the elements of $\mathcal{D}$ and $\mathcal{A}$ as *actions*. For each possible combination of actions d and a, we can construct a channel C_{da} with inputs $\mathcal{X} = \{0,1\}$ (the set of possible high input values) and outputs $\mathcal{Y} = \{T, F\}$ (the set of possible low output values), modeling the behavior of the system *from the point of view of the attacker*. Intuitively, each channel entry $C_{da}(x,y)$ is the probability that the program run by Jeeves (which is determined by d) produces output $y \in \mathcal{Y}$ given that the high input is $x \in \mathcal{X}$ and that the low input is a. The resulting four channel matrices are represented in Table 2. Note that channels C_{01} and C_{10} do not leak any information about the input x (output y is constant), whereas channels C_{00} and C_{11} completely reveal x (output y is in a bijection with x).

We want to investigate how the defender's and the attacker's strategies influence the leakage of the system. For that we can consider the (simpler) notion of posterior vulnerability, since, for a given prior, the value of leakage is in a one-to-one (monotonic) correspondence with the value of posterior vulnerability. For this example, we consider posterior Bayes vulnerability [8,27], defined as

---

[5] Note that d should not be revealed to the attacker: although d is not sensitive information in itself, knowing it would help the attacker figure out the value of x.

Table 2. The two-millionaires system, from the point of view of the attacker.

	$a = 0$			$a = 1$		
$d = 0$ $(x \leq a?)$	C_{00}	$y = T$	$y = F$	C_{01}	$y = T$	$y = F$
	$x = 0$	1	0	$x = 0$	1	0
	$x = 1$	0	1	$x = 1$	1	0
$d = 1$ $(x \geq a?)$	C_{10}	$y = T$	$y = F$	C_{11}	$y = T$	$y = F$
	$x = 0$	1	0	$x = 0$	0	1
	$x = 1$	1	0	$x = 1$	1	0

$\mathbb{V}[\pi, C] = \sum_y \max_x C(x, y)\pi(x)$. Intuitively, Bayes vulnerability measures the probability of the adversary guessing the secret correctly in one try, and it can be shown that $\mathbb{V}[\pi, C]$ coincides with the converse of the Bayes error.

For simplicity, we assume a uniform prior distribution π_u. It has been shown that, in this case, the posterior Bayes vulnerability of a channel C can be computed as the sum of the greatest elements of each column of C, divided by the high input-domain size [7]. Namely, $\mathbb{V}[\pi_u, C] = \sum_y \max_x C(x,y)/|\mathcal{X}|$. It is easy to see that we have $\mathbb{V}[\pi_u, C_{00}] = \mathbb{V}[\pi_u, C_{11}] = 1$ and $\mathbb{V}[\pi_u, C_{01}] = \mathbb{V}[\pi_u, C_{10}] = 1/2$. Thus we obtain the utility table shown in Table 3, which is similar to that of the well-known "matching-pennies" game.

As in standard game theory, there may not exist an optimal pure strategy profile. The defender as well as the attacker can then try to minimize/maximize the system's vulnerability by adopting a mixed strategy δ and α, respectively. A crucial task is *evaluating the vulnerability* of the system under such mixed strategies. This evaluation is naturally performed from the point of view of the attacker, who knows his own choice a, but *not the defender's choice d*. As a consequence, the attacker sees the system as the convex combination $C_{\delta a} = \sum_d \delta(d) C_{ad}$, i.e., a probabilistic choice between the channels representing the defender's actions. Hence, the overall vulnerability of the system will be given by the vulnerability of $C_{\delta a}$, averaged over all attacker's actions.

Table 3. Utility table for the two-millionaires game.

$\mathbb{V}$	$a = 0$	$a = 1$
$d = 0$	1	$1/2$
$d = 1$	$1/2$	1

We now define formally the ideas illustrated above.

Definition 1. *An information-leakage game is a tuple $(\mathcal{D}, \mathcal{A}, C)$ where $\mathcal{D}, \mathcal{A}$ are the sets of actions of the attacker and the defender, respectively, and $C = \{C_{da}\}_{da}$ is a family of channel matrices indexed on pairs of actions $d \in \mathcal{D}, a \in \mathcal{A}$. For a given vulnerability $\mathbb{V}$ and prior π, the utility of a pure strategy (d, a) is given by $\mathbb{V}[\pi, C_{da}]$. The utility $\mathbb{V}(\delta, \alpha)$ of a mixed strategy (δ, α) is defined as:*

$$\mathbb{V}(\delta, \alpha) \overset{\text{def}}{=} \underset{a \leftarrow \alpha}{\mathbb{E}} \, \mathbb{V}[\pi, C_{\delta a}] = \sum_a \alpha(a) \mathbb{V}[\pi, C_{\delta a}] \quad \text{where} \quad C_{\delta a} \overset{\text{def}}{=} \sum_d \delta(d) C_{ad}$$

In our example, δ is represented by a single number p: the probability that the defender chooses $d = 0$ (i.e., Program 0). From the point of view of the attacker,

Table 4. The two-millionaires mixed strategy of the defender, from the point of view of the attacker, where p is the probability the defender picks action $d = 0$.

Utility table for $a = 0$

C_{p0}	$y = T$	$y = F$
$x = 0$	1	0
$x = 1$	$1 - p$	p

Utility table for $a = 1$

C_{p1}	$y = T$	$y = F$
$x = 0$	p	$1 - p$
$x = 1$	1	0

once he has chosen a, the system will look like a channel $C_{pa} = p\, C_{0a} + (1-p)\, C_{1a}$. For instance, in the case $a = 0$, if x is 0 Jeeves will send T with probability 1, but, if x is 1, Jeeves will send F with probability p and T with probability $1 - p$. Similarly for $a = 1$. Table 4 summarizes the various channels modelling the attacker's point of view. It is easy to see that $\mathbb{V}[\pi_u, C_{p0}] = (1+p)/2$ and $\mathbb{V}[\pi_u, C_{p1}] = (2-p)/2$. In this case $\mathbb{V}[\pi_u, C_{pa}]$ coincides with the expected utility with respect to p, i.e., $\mathbb{V}[\pi_u, C_{pa}] = p\,\mathbb{V}[\pi_u, C_{0a}] + (1 - p)\,\mathbb{V}[\pi_u, C_{1a}]$.

Assume now that the attacker choses $a = 0$ with probability q and $a = 1$ with probability $1 - q$. The utility is obtained as expectation with respect to the strategy of the attacker, hence the total utility is: $\mathbb{V}(p, q) = q\,(1+p)/2 + (1-p)\,(2-p)/2$, which is affine in both p and q. By applying standard game-theoretic techniques, we derive that the optimal strategy is $(p^*, q^*) = (1/2, 1/2)$.

In the above example, things work just like in standard game theory. However, in the next section we will show an example that fully exposes the difference of our games with respect to those of standard game theory.

3.2 Binary Sum

The previous example is an instance of a general scenario in which a user, Don, delegates to a server, Jeeves, a certain computation that requires also some input from other users. Here we will consider another instance, in which the function to be computed is the binary sum $\oplus$. We assume Jeeves provides the programs in Table 5. The resulting channel matrices are represented in Table 6.

Table 5. The two programs for $\oplus$ and its complement.

Program 0

```
High Input: x ∈ {0,1}
Low Input: a ∈ {0,1}
Output: y ∈ {0,1}
return x ⊕ a
```

Program 1

```
High Input: x ∈ {0,1}
Low Input: a ∈ {0,1}
Output: y ∈ {0,1}
return x ⊕ a ⊕ 1
```

We consider again Bayes posterior vulnerability as utility. It is easy to see that we have $\mathbb{V}[\pi_u, C_{00}] = \mathbb{V}[\pi_u, C_{11}] = \mathbb{V}[\pi_u, C_{01}] = \mathbb{V}[\pi_u, C_{10}] = 1$. Thus for the pure strategies we obtain the utility table shown in Table 7. This means that all pure strategies have the same utility 1 and therefore they are all equivalent. In standard game theory this would mean that also the mixed strategies have the same utility 1, since they are defined as expectation. In our case, however,

Table 6. The binary-sum system, from the point of view of the attacker.

	$a = 0$			$a = 1$		
$d = 0$ $(x \oplus a)$	C_{00}	$y = 0$	$y = 1$	C_{01}	$y = 0$	$y = 1$
	$x = 0$	1	0	$x = 0$	0	1
	$x = 1$	0	1	$x = 1$	1	0

	C_{10}	$y = 0$	$y = 1$	C_{11}	$y = 0$	$y = 1$
$d = 1$ $(x \oplus a \oplus 1)$	$x = 0$	0	1	$x = 0$	1	0
	$x = 1$	1	0	$x = 1$	0	1

the utility of a mixed strategy of the defender is convex on the distribution, so it may be convenient for the defender to adopt a mixed strategy. Let $p, 1 - p$ be the probabilities of the defender choosing Program 0 and Program 1, respectively. From the point of view of the attacker, for each of his choices of a, the system will appear as the probabilistic channel C_{pa} represented in Table 8.

Table 7. Utility table for the binary-sum game.

$\mathbb{V}$	$a = 0$	$a = 1$
$d = 0$	1	1
$d = 1$	1	1

Table 8. The binary-sum mixed strategy of the defender, from the point of view of the attacker, where p is the probability the defender picks action $d = 0$.

	$a = 0$			$a = 1$		
	C_{p0}	$y = T$	$y = F$	C_{p1}	$y = T$	$y = F$
	$x = 0$	p	$1 - p$	$x = 0$	$1 - p$	p
	$x = 1$	$1 - p$	p	$x = 1$	p	$1 - p$

It is easy to see that $\mathbb{V}[\pi_u, C_{p0}] = \mathbb{V}[\pi_u, C_{p1}] = 1 - p$ if $p \leq 1/2$, and $\mathbb{V}[\pi_u, C_{p0}] = \mathbb{V}[\pi_u, C_{p1}] = p$ if $p \geq 1/2$. On the other hand, with respect to a mixed strategy of the attacker the utility is still defined as expectation. Since in this case the utility is the same for $a = 0$ and $a = 1$, it remains the same for any strategy of the attacker. Formally, $\mathbb{V}(p, q) = q \mathbb{V}[\pi_u, C_{p0}] + (1 - q) \mathbb{V}[\pi_u, C_{p1}] = \mathbb{V}[\pi_u, C_{p0}]$, which does not depend on q and it is minimum for $p = 1/2$. We conclude that the point of equilibrium is $(p^*, q^*) = (1/2, q^*)$ for any value of q^*.

4   Leakage Games Vs. Standard Game Theory Models

In this section we explain the differences between our information leakage games and standard approaches to game theory. We discuss: (1) why the use of vulnerability as a utility function makes our games non-standard w.r.t. von

Neumann-Morgenstern's treatment of utility, (2) why the use of concave utility functions to model risk-averse players does not capture the behavior of the
attacker in our games, and (3) how our games differ from traditional convex-
concave games.

4.1   The von Neumann-Morgenstern's Treatment of Utility

In their treatment of utility, von Neumann and Morgenstern [29] demonstrated
that the utility of a mixed strategy equals the expected utility of the corresponding pure strategies when a set of axioms is satisfied for player's preferences over
probability distributions (a.k.a. *lotteries*) on payoffs. Since in our leakage games
the utility of a mixed strategy is *not* the expected utility of the corresponding
pure strategies, it is relevant to identify how exactly our framework fails to meet
von Neumann and Morgenstern (vNM) axioms.

Let us first introduce some notation. Given two mixed strategies σ, σ' for a
player, we write $\sigma \preceq \sigma'$ (or $\sigma' \succeq \sigma$) when the player prefers σ' over σ, and $\sigma \sim \sigma'$
when the player is indifferent between σ and σ'. Then, the vNM axioms can be
formulated as follows [24]. For every mixed strategies σ, σ' and σ'':

A1 *Completeness*: it is either the case that $\sigma \preceq \sigma'$, $\sigma \succeq \sigma'$, or $\sigma \sim \sigma'$.
A2 *Transitivity*: if $\sigma \preceq \sigma'$ and $\sigma' \preceq \sigma''$, then $\sigma \preceq \sigma''$.
A3 *Continuity*: if $\sigma \preceq \sigma' \preceq \sigma''$, then there exist $p \in [0,1]$ s.t. $p\sigma + (1-p)\sigma'' \sim$
σ'.
A4 *Independence*: if $\sigma \preceq \sigma'$ then for any σ'' and $p \in [0,1]$ we have $p\sigma + (1-p)\sigma'' \preceq p\sigma' + (1-p)\sigma''$.

For any fixed prior π on secrets, the utility function $u(C) = \mathbb{V}[\pi, C]$ is a total
function on $\mathcal{C}$ ranging over the reals, and therefore it satisfies axioms A1, A2 and
A3 above. However, $u(C)$ does not satisfy A4, as the next example illustrates.

Example 1. Consider the following three channel matrices from input set $\mathcal{X} = \{0,1\}$ to output set $\mathcal{Y} = \{0,1\}$, where ϵ is a small positive constant:

C_1	$y=0$	$y=1$
$x=0$	$1-\epsilon$	ϵ
$x=1$	ϵ	$1-\epsilon$

C_2	$y=0$	$y=1$
$x=0$	1	0
$x=1$	0	1

C_3	$y=0$	$y=1$
$x=0$	0	1
$x=1$	1	0

If we focus on Bayes vulnerability, it is clear that an attacker would prefer C_2
over C_1, i.e., $C_1 \preceq C_2$. However, for the probability $p = 1/2$ we would have:

$pC_1 + (1-p)C_3$	$y=0$	$y=1$
$x=0$	$(1-\epsilon)/2$	$(1+\epsilon)/2$
$x=1$	$(1+\epsilon)/2$	$(1-\epsilon)/2$

and

$pC_2 + (1-p)C_3$	$y=0$	$y=1$
$x=0$	$1/2$	$1/2$
$x=1$	$1/2$	$1/2$

Since channel $pC_1 + (1-p)C_3$ clearly reveals no less information about the secret
than channel $pC_2 + (1-p)C_3$, we have that $pC_1 + (1-p)C_3 \succeq pC_2 + (1-p)C_3$,
and the axiom of independence is not satisfied.

It is actually quite natural that vulnerability does not satisfy independence: a convex combination of two "leaky" channels (i.e., high-utility outcomes) can produce a "non-leaky" channel (i.e., a low-utility outcome). As a consequence, the traditional game-theoretic approach to the utility of mixed strategies does not apply to our information leakage games. However the existence of Nash equilibria is still granted, as we will see in Sect. 5, Corollary 1.

4.2   Risk Functions

At a first glance, it may seem that our information leakage games could be expressed with some clever use of the concept of *risk-averse players* (in our case, the attacker), which is also based on convex utility functions (cf. [22]). There is, however, a crucial difference: in the models of risk-averse players, the utility function is convex *on the payoff of an outcome of the game*, but the utility of a mixed strategy is still *the expectation of the utilities of the pure strategies*, i.e., it is linear on the distributions. On the other hand, the utility of mixed strategies in our information leakage games is *convex on the distribution of the defender*. This difference arises precisely because in our games utility is defined as the vulnerability of the channel perceived by the attacker, and, as we discussed, this creates an extra layer of uncertainty for the attacker.

4.3   Convex-Concave Games

Another well-known model from standard game-theory is that of convex-concave games, in which each of two players can choose among a continuous set of actions yielding convex utility for one player, and concave for the other. In this kind of game the Nash equilibria are given by *pure strategies* for each player.

A natural question would be why not represent our systems as convex-concave games in which the pure actions of players are the mixed strategies of our leakage games. Namely, the real values p and q that uniquely determine the defender's and the attacker's mixed strategies, respectively, in the two-millionaires game of Sect. 3, could be taken to be the choices of pure strategies in a convex-concave game in which the set of actions for each player is the real interval $[0, 1]$.

This mapping from our games to convex-concave games, however, would not be natural. One reason is that utility is still defined as expectation in the standard convex-concave games, in contrast to our games. Consider two strategies p_1 and p_2 with utilities u_1 and u_2, respectively. If we mix them using the coefficient $q \in [0, 1]$, the resulting strategy $q \, p_1 + (1 - q) \, p_2$ will have utility $u = q \, u_1 + (1 - q) \, u_2$ in the standard convex-concave game, while in our case the utility would in general be strictly smaller than u. The second reason is that a pure action corresponding to a mixed strategy may not always be realizable. To illustrate this point, consider again the two-millionaires game, and the defender's mixed strategy consisting in choosing Program 0 with probability p and Program 1 with probability $1 - p$. The requirement that the defender has a pure action corresponding to p implies the existence of a program (on Jeeves' side) that makes internally a probabilistic choice with bias p and, depending on the outcome, executes Program 0 or Program 1. However, it is not granted that Jeeves disposes of such a program. Furthermore, Don would not know what choice has actually been made, and thus the program would not achieve the same functionality, i.e., let Don know who is the richest. (Note that Jeeves should not communicate to Don the result of the choice, because of the risk that Alice intercepts it.) This latter consideration underlines a key practical aspect of leakage games, namely, the defender's advantage over the attacker due to his knowledge of the result of his own random choice (in a mixed strategy). This advantage would be lost in a convex-concave representation of the game since the random choice would be "frozen" in its representation as a pure action.

5   Convexity of Vulnerability w.r.t. Channel Composition

In this section we show that posterior vulnerability is a convex function of the strategy of the defender. In other words, given a set of channels, and a probability distribution over them, the vulnerability of the composition of these channels according to the distribution is smaller than or equal to the composition of their vulnerabilities. As a consequence, we derive the existence of the Nash equilibria.

In order to state this result formally, we introduce the following notation: given a channel matrix C and a scalar a, aC is the matrix obtained by multiplying every element of C by a. Given two *compatible* channel matrices C_1 and C_2, namely matrices with the same indices of rows and columns[6], $C_1 + C_2$ is obtained by adding the cells of C_1 and C_2 with same indices. Note that if μ is a probability distribution on $\mathcal{I}$, then $\sum_{i \in \mathcal{I}} \mu(i) C_i$ is a channel matrix.

Theorem 2 (Convexity of vulnerability w.r.t. channel composition).
Let $\{C_i\}_{i \in \mathcal{I}}$ be a family of compatible channels, and μ be a distribution on $\mathcal{I}$. Then, for every prior distribution π, and every vulnerability $\mathbb{V}$, the corresponding posterior vulnerability is convex w.r.t. to channel composition. Namely, for any probability distribution μ on $\mathcal{I}$, we have $\mathbb{V}[\pi, \sum_i \mu(i) C_i] \leq \sum_i \mu(i) \mathbb{V}[\pi, C_i]$.

Proof. Define $p(y) = \sum_x \pi(x) \sum_i \mu(i) C_i(x, y)$. Then:

$$
\begin{aligned}
\mathbb{V}[\pi, \textstyle\sum_i \mu(i) C_i] &= \sum_y p(y) \, \mathbb{V}\left[\frac{\pi(\cdot) \sum_i \mu(i) C_i(\cdot, y)}{p(y)}\right] && \text{(by def. of posterior } \mathbb{V}) \\
&= \sum_y p(y) \, \mathbb{V}\left[\sum_i \mu(i) \frac{\pi(\cdot) C_i(\cdot, y)}{p(y)}\right] \\
&\leq \sum_y p(y) \sum_i \mu(i) \, \mathbb{V}\left[\frac{\pi(\cdot) C_i(\cdot, y)}{p(y)}\right] && (*) \\
&= \sum_i \mu(i) \sum_y p(y) \, \mathbb{V}\left[\frac{\pi(\cdot) C_i(\cdot, y)}{p(y)}\right] \\
&= \sum_i \mu(i) \, \mathbb{V}[\pi, C_i] && \text{(by def. of posterior } \mathbb{V})
\end{aligned}
$$

where $(*)$ follows from the convexity of $\mathbb{V}$ w.r.t. the prior (cf. Sect. 2.3).   □

---

[6] Note that two channel matrices with different column indices can always be made compatible by adding appropriate columns with 0-valued cells in each of them.

The existence of Nash equilibria immediately follows from the above theorem:

Corollary 1. *For any (zero-sum) information-leakage game there exist a Nash equilibrium, which in general is given by a mixed strategy.*

Proof. Given a mixed strategy (δ, α), the utility $\mathbb{V}(\delta, \alpha)$ given in Definition 1 is affine (hence concave) on α. Furthermore, by Theorem 2, $\mathbb{V}(\delta, \alpha)$ is convex on δ. Hence we can apply the von Neumann's minimax theorem (Sect. 2.2), which ensures the existence of a saddle point, i.e., a Nash equilibrium.   □

6   Computing Equilibria of Information Leakage Games

Our goal is to solve information leakage games, in which the success of an attack a and a defence d is measured by a vulnerability measure $\mathbb{V}$. The attack/defence combination is a pure strategy profile (d, a) in this game, and is associated with a channel C_{da} modeling the behavior of the system. The attacker clearly knows his own choice a, whereas the defender's choice is assumed to be hidden. Hence the utilty of a mixed strategy profile (δ, α) will be given by Definition 1, that is:

$$\mathbb{V}(\delta, \alpha) = \sum_a \alpha(a) \, \mathbb{V}\left[\pi, \sum_d \delta(d) \, C_{da}\right]$$

Note that $\mathbb{V}(\delta, \alpha)$ is convex on δ and affine on α, hence Theorem 1 guarantees the existence of an equilibrium (i.e. a saddle-point) (δ^*, α^*) which is a solution of both the minimax and the maximin problems. The goal in this section is to compute a) a δ^* that is part of an equilibrium, which is important in order to optimize the defence, and b) the utility $\mathbb{V}(\delta^*, \alpha^*)$, which is important to provide an upper bound on the effectiveness of an attack when δ^* is applied.

This is a convex-concave optimization problem for which various methods have been proposed in the literature. If $\mathbb{V}$ is twice differentiable (and satisfies a few extra conditions) then the Newton method can be applied [6]; however, many such measures, most notably Bayes-vulnerability, our main vulnerability measure of interest, are not differentiable. For non-differentiable functions, [21] proposes a subgradient method that iterates on both δ, α at each step. We have applied this method and it does indeed converge to $\mathbb{V}(\delta^*, \alpha^*)$, with one important caveat: the solution δ that it produces is not necessarily an equilibrium (note that $\mathbb{V}(\delta, \alpha) = \mathbb{V}(\delta^*, \alpha^*)$ does not guarantee that (δ, α) is a saddle point). Producing an optimal δ^* is of vital importance in our case.

The method we propose is based on the idea of solving the minimax problem $\hat{\delta} = \operatorname{argmin}_\delta \max_\alpha \mathbb{V}(\delta, \alpha)$, since its solution is guaranteed to be part of an equilibrium.[7] To solve this problem, we exploit the fact that $\mathbb{V}(\delta, \alpha)$ is affine on α (not just concave). For a fixed δ, maximizing $\sum_a \alpha(a) \, \mathbb{V}\left[\pi, \sum_d \delta(d) \, C_{da}\right]$ simply involves picking the a with the highest $\mathbb{V}\left[\pi, \sum_d \delta(d) \, C_{da}\right]$ and assigning probability 1 to it. Hence, our minimax problem is equivalent to $\hat{\delta} = \operatorname{argmin}_\delta f(\delta)$

---

[7] Note that this is true only for δ, the α-solution of the minimax problem is not necessarily part of an equilibrium; we need to solve the maximin problem for this.

where $f(\delta) = \max_a \mathbb{V}[\pi, \sum_d \delta(d) C_{da}]$; that is, we have to minimize the max of finitely many convex functions, with δ being the only variables.

For this problem we can employ the *projected subgradient* method, given by:

$$\delta^{(k+1)} = P(\delta^{(k)} - \alpha_k g^{(k)})$$

where $g^{(k)}$ is any subgradient of f on $\delta^{(k)}$ [5]. Note that the subgradient of a finite max is simply a subgradient of any branch that gives the max at that point. $P(x)$ is the projection of x on the domain of f; in our case the domain is the probability simplex, for which there exist efficient algorithms for computing the projection [30]. Finally α_k is a step-size, for which various choices guarantee convergence [5]. In our experiments we found $\alpha_k = 0.1/\sqrt{k}$ to perform well.

As the starting point $\delta^{(1)}$ we take the uniform distribution; moreover the solution can be approximated to within an arbitrary $\epsilon > 0$ by using the stopping criterion of [5, Sect . 3.4]. Note that the obtained $\hat{\delta}$ approximates the equilibrium strategy δ^*, while $f(\hat{\delta})$ approximates $\mathbb{V}(\delta^*, \alpha^*)$. Hence we achieve both desired goals, as formally stated in the following result.

Proposition 1. *If $\mathbb{V}$ is Lipschitz then the subgradient method discussed in this section converges to a δ^* that is part of an equilibrium of the game. Moreover, let $\hat{\delta}$ be the solution computed within a given $\epsilon > 0$, and let (δ^*, α^*) be an equilibrium. Then it holds that:*

$$\mathbb{V}(\hat{\delta}, \alpha) - \epsilon \le \mathbb{V}(\hat{\delta}, \alpha^*) \le \mathbb{V}(\delta, \alpha^*) + \epsilon \qquad \forall \delta, \alpha$$

which also implies that $f(\hat{\delta}) - \mathbb{V}(\delta^*, \alpha^*) \le \epsilon$.

Proof (Sketch). $\operatorname{argmin}_\delta f(\delta)$ is equivalent to the minimax problem whose δ-solution is guaranteed to be part of an equilibrium. Convergence is ensured by the subgradient method under the Lipschitz condition, and given that $\|\delta^{(1)} - \delta^*\|$ is bounded by the distance between the uniform and a point distribution. $\square$

Finally, of particular interest is the Bayes-vulnerability measure [8,27], given by $\mathbb{V}[\pi, C] = \sum_y \max_x \pi(x) C(x, y)$, since it is widely used to provide an upper bound to all other measures of information leakage [3]. For this measure, $\mathbb{V}$ is Lipschitz and the subgradient vector $g^{(k)}$ is given by $g_d^{(k)} = \sum_y \pi(x_y^*) C_{da^*}(x_y^*, y)$ where a^*, x_y^* are the ones giving the max in the branches of $f(\delta^{(k)})$. Note also that, since f is piecewise linear, the convex optimization problem can be transformed into a linear one using a standard technique, and then solved by linear programming. However, due to the large number of max branches of $\mathbb{V}$, this conversion can be a problem with a huge number of constraints. In our experiments we found that the subgradient method described above is significantly more efficient than linear programming.

Note also that, although the subgradient method is general, it might be impractical in applications where the number of attacker or defender actions is very large. Application-specific methods could offer better scalability in such cases, we leave the development of such methods as future work.

7 Case Study

In this section, we apply our game-theoretic analysis to the case of anonymous communication on a mobile ad-hoc network (MANET). In such a network, nodes can move in space and communicate with other nearby nodes. We assume that nodes can also access some global (wide area) network, but such connections are neither anonymous nor trusted. Consider, for instance, smartphone users who can access the cellular network, but do not trust the network provider. The goal is to send a message on the global network without revealing the sender's identity to the provider. For that, users can form a MANET using some short-range communication method (e.g., bluetooth), and take advantage of the local network to achieve anonymity on the global one.

Crowds [23] is a protocol for anonymous communication that can be employed on a MANET for this purpose. Note that, although more advanced systems for anonymous communication exist (e.g. Onion Routing), the simplicity of Crowds makes it particularly appeling for MANETs. The protocol works as follows: the *initiator* (i.e., the node who wants to send the message) selects some other node connected to him (with uniform probability) and forwards the request to him. A *forwarder*, upon receiving the message, performs a probabilistic choice: with probability p_f he keeps forwarding the message (again, by selecting uniformly a user among the ones connected to him), while with probability $1 - p_f$ he delivers the message on the global network. Replies, if any, can be routed back to the initiator following the same path in reverse order.

Anonymity comes from the fact that the *detected* node (the last in the path) is most likely not the initiator. Even if the attacker knows the network topology, he can infer that the initiator is most likely a node close to the detected one, but if there are enough nodes we can achieve some reasonable anonymity guarantees. However, the attacker can gain an important advantage by deploying a node himself and participating to the MANET. When a node forwards a message to this *corrupted* node, this action is observed by the attacker and increases the probability of that node being the initiator. Nevertheless, the node can still claim that he was only forwarding the request for someone else, hence we still provide some level of anonymity. By modeling

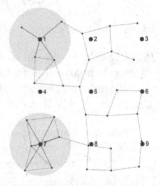

Fig. 1. A MANET with 30 users in a 1km × 1km area.

the system as a channel, and computing its posterior Bayes vulnerability [27], we get the probability that the attacker guesses correctly the identity of the initiator, after performing his observation.

In this section we study a scenario of 30 nodes deployed in an area of 1 km × 1 km, in the locations illustrated in Fig. 1. Each node can communicate with others up to a distance of 250 m, forming the network topology shown in the graph. To compromise the anonymity of the system, the attacker plans to deploy a corrupted node in the network; the question is which is the *optimal location* for such a node. The answer is far from trivial: on the one side being connected to many nodes is beneficial, but at the same time these nodes need to be "vulnerable", being close to a highly connected clique might not be optimal. At the same time, the administrator of the network is suspecting that the attacker is about to deploy a corrupted node. Since this action cannot be avoided (the network is ad-hoc), a countermeasure is to deploy a *deliverer* node at a location that is most vulnerable. Such a node directly delivers all messages forwarded to it on the global network; since it never generates messages its own anonymity is not an issue, it only improves the anonymity of the other nodes. Moreover, since it never communicates in the local network its operation is invisible to the attacker. But again, the optimal location for the new deliverer node is not obvious, and most importantly, the choice depends on the choice of the attacker.

To answer these questions, we model the system as a game where the actions of attacker and defender are the locations of newly deployed corrupted and honest nodes, respectively. We assume that the possible locations for new nodes are the nine ones shown in Fig. 1. For each pure strategy profile (d, a), we construct the corresponding network and use the PRISM model checker to construct the corresponding channel C_{da}, using a model similar to the one of [26]. Note that the computation considers the specific network topology of Fig. 1, which reflects the positions of each node at the time when the attack takes place; the corresponding channels need to be recomputed if the network changes in the future. As leakage measure we use the posterior Bayes vulnerability (with uniform prior π), which is the attacker's probability of correctly guessing the initiator given his observation in the protocol. According to Definition 1, for a mixed strategy profile (δ, α) the utility is $\mathbb{V}(\delta, \alpha) = \mathbb{E}_{a \leftarrow \alpha} \mathbb{V}[\pi, C_{\delta a}]$.

The utilities (posterior Bayes vulnerability %) for each pure profile are displayed in Table 9. Note that the attacker and defender actions substantialy affect the effectiveness of the attack, with the probability of a correct guess ranging between 5.46% and 9.5%. Based on the results of Sect. 6, we can then compute the best strategy for the defender, which turns out to be (probabilities expressed as %):

Table 9. Utility for each pure strategy profile.

		Attacker's action								
		1	2	3	4	5	6	7	8	9
Defender's action	1	7.38	6.88	6.45	6.23	7.92	6.45	9.32	7.11	6.45
	2	9.47	6.12	6.39	6.29	7.93	6.45	9.32	7.11	6.45
	3	9.50	6.84	5.46	6.29	7.94	6.45	9.32	7.11	6.45
	4	9.44	6.92	6.45	5.60	7.73	6.45	9.03	7.11	6.45
	5	9.48	6.91	6.45	6.09	6.90	6.13	9.32	6.92	6.44
	6	9.50	6.92	6.45	6.29	7.61	5.67	9.32	7.11	6.24
	7	9.50	6.92	6.45	5.97	7.94	6.45	7.84	7.10	6.45
	8	9.50	6.92	6.45	6.29	7.75	6.45	9.32	6.24	6.45
	9	9.50	6.92	6.45	6.29	7.92	6.24	9.32	7.11	5.68

$$\delta^* = (34.59, 3.48, 3.00, 10.52, 3.32, 2.99, 35.93, 3.19, 2.99)$$

This strategy is part of an equilibrium and guarantees that for any choice of the attacker the vulnerability is at most 8.76%, and is substantially better that the best pure strategy (location 1) which leads to a worst vulnerability of 9.32%. As expected, δ^* selects the most vulnerable locations (1 and 7) with the highest probability. Still, the other locations are selected with non-negligible probability, which is important for maximizing the attacker's uncertainty about the defense.

8   Related Work

There is an extensive literature on game theory models for security and privacy in computer systems, including network security, vehicular networks, cryptography, anonymity, location privacy, and intrusion detection. See [17] for a survey.

In many studies, security games have been used to model and analyze utilities between interacting agents, especially an attacker and a defender. In particular, Korzhyk et al. [16] present a theoretical analysis of security games and investigate the relation between Stackelberg and simultaneous games under various forms of uncertainty. In application to network security, Venkitasubramaniam [28] investigates anonymous wireless networking, which they formalize as a zero-sum game between the network designer and the attacker. The task of the attacker is to choose a subset of nodes to monitor so that anonymity of routes is minimum whereas the task of the designer is to maximize anonymity by choosing nodes to evade flow detection by generating independent transmission schedules.

Khouzani et al. [14] present a framework for analyzing a trade-off between usability and security. They analyze guessing attacks and derive the optimal policies for secret picking as Nash/Stackelberg equilibria. Khouzani and Malacaria [13] investigate properties of leakage when perfect secrecy is not achievable due to the limit on the allowable size of the conflating sets, and show the existence of universally optimal strategies for a wide class of entropy measures, and for g-entropies (the dual of g-vulnerabilities). In particular, they show that designing a channel with minimum leakage is equivalent to Nash equilibria in a corresponding two-player zero-sum games of incomplete information for a range of entropy measures.

Concerning costs of security, Yang et al. [32] propose a framework to analyze user behavior in anonymity networks. Utility is modeled as a combination of weighted cost and anonymity utility. They also consider incentives and their impact on users' cooperation.

Some security games have considered leakage of information about the defender's choices. For example, Alon et al. [1] present two-player zero-sum games where a defender chooses probabilities of secrets while an attacker chooses and learns some of the defender's secrets. Then they show how the leakage on the defender's secrets influences the defender's optimal strategy. Xu et al. [31] present zero-sum security games where the attacker acquires partial knowledge on the security resources the defender is protecting, and show the defender's optimal strategy under such attacker's knowledge. More recently, Farhang et al. [12] present two-player games where utilities are defined taking account of information leakage, although the defender's goal is different from our setting. They consider a model where the attacker incrementally and stealthily obtains partial information on a secret, while the defender periodically changes the secret after some time to prevent a complete compromise of the system. In particular, the defender is not attempting to minimize the leak of a certain secret, but only to make it useless (for the attacker). Hence their model of defender and utility is totally different from ours. To the best of our knowledge there have been no works exploring games with utilities defined as information-leakage measures.

Finally, in game theory Matsui [20] uses the term "information leakage game" with a meaning different than ours, namely, as a game in which (part of) the strategy of one player may be leaked in advance to the other player, and the latter may revise his strategy based on this knowledge.

9 Conclusion and Future Work

In this paper we introduced the notion of information leakage games, in which a defender and an attacker have opposing goals in optimizing the amount of information leakage in a system. In contrast to standard game theory models, in our games the utility of a mixed strategy is a convex function of the distribution of the defender's actions, rather than the expected value of the utilities of the pure strategies in the support. Nevertheless, the important properties of game theory, notably the existence of a Nash equilibrium, still hold for our zero-sum leakage games, and we provided algorithms to compute the corresponding optimal strategies for the attacker and the defender.

As future research, we would like to extend leakage games to scenarios with repeated observations, i.e., when the attacker can repeatedly observe the outcomes of the system in successive runs, under the assumption that both the attacker and the defender may change the channel at each run. Furthermore, we would like to consider the possibility to adapt the defender's strategy to the secret value, as we believe that in some cases this would provide significant advantage to the defender. We would also like to consider the cost of attack and of defense, which would lead to non-zero-sum games.

Acknowledgments. The authors are thankful to Arman Khouzani and Pedro O. S. Vaz de Melo for valuable discussions. This work was supported by JSPS and Inria under the project LOGIS of the Japan-France AYAME Program, and by the project Epistemic Interactive Concurrency (EPIC) from the STIC AmSud Program. Mário S. Alvim was supported by CNPq, CAPES, and FAPEMIG. Yusuke Kawamoto was supported by JSPS KAKENHI Grant Number JP17K12667.

References

1. Alon, N., Emek, Y., Feldman, M., Tennenholtz, M.: Adversarial leakage in games. SIAM J. Discret. Math. **27**(1), 363–385 (2013)
2. Alvim, M.S., Chatzikokolakis, K., McIver, A., Morgan, C., Palamidessi, C., Smith, G.: Axioms for information leakage. In: Proceedings of CSF, pp. 77–92 (2016)
3. Alvim, M.S., Chatzikokolakis, K., Palamidessi, C., Smith, G.: Measuring information leakage using generalized gain functions. In: CSF, pp. 265–279 (2012)
4. Boreale, M., Pampaloni, F.: Quantitative information flow under generic leakage functions and adaptive adversaries. Log. Meth. Comput. Sci. **11**(4:5), 1–31 (2015)
5. Boyd, S., Mutapcic, A.: Subgradient methods. Lecture notes of EE364b. Stanford University, Winter Quarter 2007 (2006)
6. Boyd, S., Vandenberghe, L.: Convex Optimization. Cambridge University Press, New York (2004)

7. Braun, C., Chatzikokolakis, K., Palamidessi, C.: Quantitative notions of leakage for one-try attacks. In: Proceedings of MFPS. ENTCS, vol. 249, pp. 75–91. Elsevier (2009)
8. Chatzikokolakis, K., Palamidessi, C., Panangaden, P.: On the Bayes risk in information-hiding protocols. J. Comput. Secur. **16**(5), 531–571 (2008)
9. Chaum, D.: The dining cryptographers·problem: unconditional sender and recipient untraceability. J. Cryptol. **1**, 65–75 (1988)
10. Clark, D., Hunt, S., Malacaria, P.: A static analysis for quantifying information flow in a simple imperative language. J. Comput. Secur. **15**, 321–371 (2007)
11. Dwork, C., McSherry, F., Nissim, K., Smith, A.: Calibrating noise to sensitivity in private data analysis. In: Halevi, S., Rabin, T. (eds.) TCC 2006. LNCS, vol. 3876, pp. 265–284. Springer, Heidelberg (2006). doi:10.1007/11681878_14
12. Farhang, S., Grossklags, J.: FlipLeakage: a game-theoretic approach to protect against stealthy attackers in the presence of information leakage. In: Zhu, Q., Alpcan, T., Panaousis, E., Tambe, M., Casey, W. (eds.) GameSec 2016. LNCS, vol. 9996, pp. 195–214. Springer, Cham (2016). doi:10.1007/978-3-319-47413-7_12
13. Khouzani, M., Malacaria, P.: Relative perfect secrecy: universally optimal strategies and channel design. In: Proceedings of CSF, pp. 61–76. IEEE (2016)
14. Khouzani, M.H.R., Mardziel, P., Cid, C., Srivatsa, M.: Picking vs. guessing secrets: a game-theoretic analysis. In: Proceedings of CSF, pp. 243–257 (2015)
15. Köpf, B., Basin, D.A.: An information-theoretic model for adaptive side-channel attacks. In: Proceedings of CCS, pp. 286–296. ACM (2007)
16. Korzhyk, D., Yin, Z., Kiekintveld, C., Conitzer, V., Tambe, M.: Stackelberg vs. nash in security games: an extended investigation of interchangeability, equivalence, and uniqueness. J. Artif. Intell. Res. **41**, 297–327 (2011)
17. Manshaei, M.H., Zhu, Q., Alpcan, T., Bacşar, T., Hubaux, J.-P.: Game theory meets network security and privacy. ACM Comput. Surv. **45**(3), 25:1–25:39 (2013)
18. Mardziel, P., Alvim, M.S., Hicks, M.W., Clarkson, M.R.: Quantifying information flow for dynamic secrets. In: Proceedings of S&P, pp. 540–555 (2014)
19. Massey, J.L.: Guessing and entropy. In: Proceedings of ISIT, p. 204. IEEE (1994)
20. Matsui, A.: Information leakage forces cooperation. Games Econ. Behav. **1**(1), 94–115 (1989)
21. Nedić, A., Ozdaglar, A.: Subgradient methods for saddle-point problems. J. Optim. Theor. Appl. **142**(1), 205–228 (2009)
22. Osborne, M.J., Rubinstein, A.: A Course in Game Theory. MIT Press, Cambridge (1994)
23. Reiter, M.K., Rubin, A.D.: Crowds: anonymity for web transactions. ACM Trans. Inf. Syst. Secur. **1**(1), 66–92 (1998)
24. Rubinstein, A.: Lecture Notes in Microeconomic Theory, 2nd edn. Princeton University Press, Princeton (2012)
25. Shannon, C.E.: A mathematical theory of communication. Bell Syst. Tech. J. **27**(379–423), 625–656 (1948)
26. Shmatikov, V.: Probabilistic analysis of anonymity. In: CSFW, pp. 119–128 (2002)
27. Smith, G.: On the foundations of quantitative information flow. In: de Alfaro, L. (ed.) FoSSaCS 2009. LNCS, vol. 5504, pp. 288–302. Springer, Heidelberg (2009). doi:10.1007/978-3-642-00596-1_21
28. Venkitasubramaniam, P., Tong, L.: A game-theoretic approach to anonymous networking. IEEE/ACM Trans. Netw. **20**(3), 892–905 (2012)
29. Von Neumann, J., Morgenstern, O.: Theory of Games and Economic Behavior. Princeton University Press, Princeton (2007)

30. Wang, W., Carreira-Perpinán, M.A.: Projection onto the probability simplex: an efficient algorithm with a simple proof, and an application. arXiv preprint arXiv:1309.1541 (2013)
31. Xu, H., Jiang, A.X., Sinha, A., Rabinovich, Z., Dughmi, S., Tambe, M.: Security games with information leakage: modeling and computation. In: Proceedings of IJCAI, pp. 674–680 (2015)
32. Yang, M., Sassone, V., Hamadou, S.: A game-theoretic analysis of cooperation in anonymity networks. In: Degano, P., Guttman, J.D. (eds.) POST 2012. LNCS, vol. 7215, pp. 269–289. Springer, Heidelberg (2012). doi:10.1007/978-3-642-28641-4_15
33. Yao, A.C.: Protocols for secure computations. In: IEEE 54th Annual Symposium on Foundations of Computer Science, pp. 160–164 (1982)

Optimal Patrol Planning for Green Security Games with Black-Box Attackers

Haifeng Xu[1]([✉]), Benjamin Ford[1], Fei Fang[2], Bistra Dilkina[3],
Andrew Plumptre[4], Milind Tambe[1], Margaret Driciru[5], Fred Wanyama[5],
Aggrey Rwetsiba[5], Mustapha Nsubaga[6], and Joshua Mabonga[6]

[1] University of Southern California, Los Angeles, USA
{haifengx,benjamif,tambe}@usc.edu
[2] Carnegie Mellon University, Pittsburgh, USA
feifang@cmu.edu
[3] Georgia Institute of Technology, Atlanta, GA, USA
bdilkina@cc.gatech.edu
[4] Wildlife Conservation Society, New York City, NY, USA
aplumptre@wcs.org
[5] Uganda Wildlife Authority, Kampala, Uganda
{margaret.driciru,fred.wanyama,aggrey.rwetsiba}@ugandawildlife.org
[6] Wildlife Conservation Society, Kampala, Uganda
{mnsubuga,jmabonga}@wcs.org

Abstract. Motivated by the problem of protecting endangered animals, there has been a surge of interests in optimizing patrol planning for conservation area protection. Previous efforts in these domains have mostly focused on optimizing patrol routes against a specific boundedly rational poacher behavior model that describes poachers' choices of areas to attack. However, these planning algorithms do not apply to other poaching prediction models, particularly, those complex machine learning models which are recently shown to provide better prediction than traditional bounded-rationality-based models. Moreover, previous patrol planning algorithms do not handle the important concern whereby poachers infer the patrol routes by partially monitoring the rangers' movements. In this paper, we propose OPERA, a general patrol planning framework that: (1) generates optimal implementable patrolling routes against a black-box attacker which can represent a wide range of poaching prediction models; (2) incorporates entropy maximization to ensure that the generated routes are more unpredictable and robust to poachers' partial monitoring. Our experiments on a real-world dataset from Uganda's Queen Elizabeth Protected Area (QEPA) show that OPERA results in better defender utility, more efficient coverage of the area and more unpredictability than benchmark algorithms and the past routes used by rangers at QEPA.

Haifeng Xu and Benjamin Ford are both first authors of this paper.

S. Rass et al. (Eds.): GameSec 2017, LNCS 10575, pp. 458–477, 2017.
DOI: 10.1007/978-3-319-68711-7_24

1   Introduction

Worldwide, wildlife conservation agencies have established protected areas to protect threatened species from dire levels of poaching. Unfortunately, even in many protected areas, species' populations are still in decline [1,3]. These areas are protected by park rangers who conduct patrols to protect wildlife and deter poaching. Given that these areas are vast, however, agencies do not have sufficient resources to ensure rangers can adequately protect the entire park.

At many protected areas, rangers collect observational data while on patrol, and these observations on animals and illegal human activities (e.g., poaching, trespassing) are commonly recorded into a park-wide database (e.g., SMART, Cybertracker). Once enough patrols have been conducted, a patrol manager will analyze the data and generate a new patrolling strategy to execute. However, given the vast area and limited financial budgets of conservation agencies, improving the efficiency of ranger patrols is an important goal in this domain.

Following the success of automated planning tools used in domains such as fare enforcement and seaport protection [15,20], novel planning tools have also been proposed and applied to ranger patrol planning. Work in [11] developed a new game-theoretic model that optimized against its proposed poacher behavior model to generate randomized patrol strategies. However, they did not account for spatial constraints (i.e., are two areas adjacent?) in their planning and thus cannot guarantee the implementability of their proposed strategies. Moreover, the planning in [11] is specific to one poacher behavior model and cannot be applied to different predictive models. Critchlow et al. [1] demonstrated the potential for automated planning tools in the real world via a successful field test. However, the planning process in [1] is deterministic and thus is predictable to poachers.

In this paper, we present OPERA (Optimal patrol Planning with Enhanced RAndomness), a general patrol planning framework with the following key features. First, OPERA optimally generates patrols against a black-box poaching prediction model. Unlike other approaches in this domain that can only optimize against their specified prediction model [7,11], OPERA is capable of optimizing against a wide range of prediction models. Second, OPERA optimizes directly over the space of feasible patrol routes and guarantees implementability of any generated patrol strategy. Lastly, OPERA incorporates entropy maximization in its optimization process to ensure that the generated strategies are sufficiently randomized and robust to partial information leakage – i.e., a frequently observed phenomenon in practice whereby poachers try to infer the patroller's patrolling route by monitoring part of the patroller's movements [10,13,18].

We evaluate OPERA on a real-world data set from Uganda's Queen Elizabeth Protected Area (QEPA). Our experiments show that, compared to benchmark heuristic planning algorithms, OPERA results in significantly better defender utility and more efficient coverage of the area. Moreover, the experiments also show that the new entropy maximization procedure results in patrol routes that are much more unpredictable than those routes generated by classical techniques. This effectively mitigates the issue of partial information leakage. Finally, we integrate OPERA with a predictive model of a bagging ensemble of decision trees to generate patrolling routes for QEPA, and compare these routes with the past routes used by rangers at QEPA. The experiments show that OPERA is able to detect all the attacks that are found by past ranger patrolling and also predicted by the predictive model. Moreover, OPERA results in better attack detection and more efficient coverage of the area than the past ranger routes.

2   Related Work

Prior work in planning wildlife patrols has also generated patrol strategies based on a predictive model [11]. In [11], poacher behavior was modeled via a two-layered graphical model and a randomized patrolling strategy was planned in a Stackelberg Security Game (SSG) framework. Similarly, SSGs have been applied to the problem of interdicting rhino poachers [7]. In [7], optimal interdiction strategies were generated for rangers by solving an SSG. However, patrol strategies generated in [11] were not guaranteed to be implementable in the form of patrol routes that satisfy spatial constraints, while [7] optimized over a very small set of patrols specified a priori. In contrast, our scalable approach optimizes over the space of all feasible patrol routes and is guaranteed to generate executable routes. Additionally, both the patrol strategy generation approaches in [7,11] were constrained to each of their own adversary behavior models, while our black-box approach can generate a patrol strategy for a wide range of adversary frameworks and corresponding behavior models.

Green Security Games [5,19] have been introduced to model the interaction in domains such as wildlife protection, fishery protection, and forest protection. In [5], a multi-stage game is used to model the repeated interactions in these domains. In the case where the defender's strategy in one stage can affect the attackers' behavior in future stages, look-ahead planning algorithms were proposed [5] to compute a sequence of defender strategies against attackers that follow a specific behavior model. OPERA can also handle multi-stage planning to generate a sequence of strategies to use, but OPERA additionally introduces a novel and scalable approach to handle black-box attackers.

Other work in this domain has resulted in the successful field testing of planned patrols [1,4]. Critchlow et al. [1] generates patrol strategy by reorganizing historical patrol effort values such that areas of highest predicted activity would receive the highest patrol effort. However, such reorganization leads to a deterministic patrol strategy that can be easily exploited by poachers. Fang et al. [4] introduced a patrol planning tool that incorporated spatial constraints to plan detailed patrol routes. However, it relied on a specific type of attacker behavior model [12] while OPERA can optimize against any black-box attacker model that can be approximated by a piece-wise linear function of the patrolling effort.

3   Green Security Games with Black-Box Attackers

In this section, we provide an overview of Green Security Games (GSGs) and how they can work with a black-box attacker model.

3.1   Green Security Games

GSGs are security games that specifically focus on the unique challenges present in conservation domains (e.g., protecting wildlife, fisheries); GSGs focus on protecting threatened natural resources, with limited defender capacity, from repeated outside attacks. Like most of the previous work in GSGs [5,19], we consider a discrete setting where the conservation area is divided into N discrete grid cells, each treated as a *target*. Let $[N]$ denote the set of all cells, among which one cell is designated as the *patrol post*. Any patrol route must originate from and return to the patrol post. W.l.o.g., we will treat cell 1 as the patrol post throughout the paper. There is one patroller resource (e.g., ranger team) who patrols $[N]$ cells each day. Due to real-world spatial constraints, from one cell the patroller can only move to neighboring cells. We assume that traversing each cell requires one unit of time, and we let T denote the upper bound of the total units of time that the patroller can patrol each day. As a result, the patroller can traverse at most T cells each day. These spatial constraints can be captured by a time-unrolled graph G (e.g., Fig. 1). Any node $v_{t,i}$ in G denotes the cell i at time t. The edges in G, only connecting two consecutive time steps, indicate feasible spatial traversals from one cell to another within a unit of time. Recall that cell 1 is the patrol post, so a feasible patrol route will be a path in G starting from $v_{1,1}$ and ending at $v_{T,1}$ (e.g., the dotted path in Fig. 1).

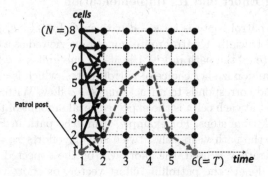

Fig. 1. An example of the time-unrolled graph

3.2   Black-Box Attackers

Unlike previous security game models which explicitly assume an attacker behavior model (rational or boundedly rational), we assume that for each cell i there is a black-box function g_i, which takes as input certain measure of the defender's patrolling effort l_i^1, l_i^0 at the current and previous period respectively, and outputs a prediction of attacks at cell i. Note that the dependence of the prediction on static features (e.g., animal density, distance & terrain features) is integrated into the function form of g_i and thus will not be an explicit input into g_i. This is motivated by the recent surge of research efforts in using machine learning models to predict attacks in conservation areas [8,9,19]. Each of these models can be treated as a black-box function of the attacker's behavior, though we note that the function g_i can also capture perfectly rational or other models of boundedly rational attackers. In this paper, we assume that the function g_i outputs the predicted number of *detected attacks* at cell i (i.e., attacks that happen and are also detected by the patroller) since the real-world data and corresponding machine learning model we use fit this task. However, we remark that our framework and algorithms are also applicable to other forms of g_i.

We wish to optimize the design of patrol routes against such black-box attackers. Of course, one cannot hope to develop a general, efficient algorithm that works for an arbitrary function g_i. We thus make the following assumption: g_i depends *discretely* on the *patrolling effort* at cell i. More specifically, we assume that there are $m+1$ levels of patrolling effort at any time period, ranging increasingly from 0 to m, and $g_i : \{0, 1, ..., m\}^2 \to \mathbb{R}$ takes level l_i^1, l_i^0 as input. We remark that this discretization is a natural choice since it can be viewed as a *piecewise constant* approximation for a continuous attacker behavior model. It's worth noting that some machine learning models in this domain indeed use discrete patrolling levels as input features [9]. The output of g_i can be any number (e.g., 0–1 for classifiers, a real number for regression).

3.3 Patrolling Effort and Its Implementation

Recall that each patrol route is an $s - d$ path in G for $s = v_{1,1}$ and $d = v_{T,1}$. Equivalently, and crucially, a patrol route can also be viewed as a one unit *integer flow* from s to d (e.g., the path in Fig. 1 is also a one-unit $s - d$ flow). We allow the defender to randomize her choice of patrol routes, which is called a *defender mixed strategy* and corresponds to a fractional $s - d$ flow. With any period, the patrolling effort x_i at each cell i is the expected total amount of time units spent at cell i during T time steps. For example, using the path in Fig. 1, the effort x_2 equals 2 since the path visits cell 2 twice and the efforts $x_5 = 1, x_7 = 0$, etc. When a mixed strategy is used, the effort will be the expected time units. Let $\mathbf{x} = (x_1, ..., x_N)$ denote the patrolling effort vector, or effort vector for short. One important property of the patrolling effort quantity is that it is additive across different time steps. Such additivity allows us to characterize feasible effort vectors using a flow-based formulation. An effort vector is *implementable* if there exists a mixed strategy that results in the effort vector.

Lemma 1. *The effort vector* $(x_1, ..., x_N) \in \mathbb{R}_+^N$ *within any period is implementable if and only if there exists a flow f in G such that*

$$x_i = \sum_{t=1}^{T} \left[\sum_{e \in \sigma^+(v_{t,i})} f(e) \right], \qquad for \ i = 1, ..., N. \tag{1}$$
$$f \ is \ a \ feasible \ 1\text{-}unit \ s - d \ flow \ in \ G$$

where $\sigma^+(v_{t,i})$ *is the set of all edges entering node* $v_{t,i}$.

Proof. This simply follows from the observation that any mixed strategy is a 1-unit fractional $s - d$ flow and the definition that x_i is the aggregated effort at cell i from all the T time steps.   □

Let $\alpha_0 < \alpha_1 ... < \alpha_m < \alpha_{m+1}$ be $m + 2$ threshold constants which determine the patrol level of any patrolling effort. By default, we always set $\alpha_0 = 0$ and $\alpha_{m+1} = +\infty$. The patrolling effort x_i has level $l \in \{0, 1, ..., m\}$ if $x_i \in [\alpha_l, \alpha_{l+1})$. In most applications, these thresholds are usually given together with the function $g_i(l_i^1, l_i^0)$.

4 Patrolling Route Design to Maximize Attack Detection

Recall that the black-box function g_i in our setting predicts the number of detected attacks at cell i. In this section, we look to design the optimal (possibly randomized) patrol route so that the induced patrol effort maximizes the *total number of detected attacks* $\sum_{i \in [N]} g_i(l_i^1, l_i^0)$. For simplicity, we first restrict our attention to the planning of the patrol routes at only *current* period without looking into the future periods. We illustrate at the end of this section that how our techniques can be generalized to the planning with look-ahead.

When designing the current patrol routes, the patrolling level l_i^0 at the previous period has already happened, thus is fixed. Therefore, only the input l_i^1 for g_i is under our control. To that end, for notational convenience, we will simply view g_i as a function of l_i^1. In fact, we further omit the superscript "1", and use l_i as the variable of function g_i for simplicity. We start by proving the NP-hardness of the problem. The underlying intuition is that patrolling a cell at different levels will result in different "values" (i.e., the number of detected attacks). Given a fixed budget of patrolling resources, the designer needs to determine which cell has what patrolling level so that it maximizes the total "value". This turns out to encode a Knapsack problem.

Theorem 1. *It is NP-hard to compute the optimal patrolling strategy.*

The proof of Theorem 1 precisely tracks the intuition above, and is omitted here due to space limit.[1] Because of the NP-hardness, the problem of patrolling optimization to maximize attack detection (with inputs: a time-unrolled graph G with $N \times T$ nodes, $\{g_i(j)\}_{i \in [N], j \in [m]}$, $\{\alpha_l\}_{l \in [m]}$) is unlikely to have an efficient polynomial time algorithm. Next, we propose a novel mixed integer linear

---

[1] All missing proofs in this paper can be found in an online appendix.

program (MILP) to solve the problem. We start by observing that the following abstractly described Mathematical Program (MP), with integer variables $\{l_i\}_{i=1}^{N}$ and real variables $\{x_i\}_{i=1}^{N}$, encodes the problem.

$$
\begin{aligned}
&\text{maximize } \sum_{i=1}^{N} g_i(l_i) \\
&\text{subject to } \alpha_{l_i} \leq x_i \leq \alpha_{l_i+1}, &&\text{for } i \in [N]. \\
&\qquad\quad l_i \in \{0,1,...,m\}, &&\text{for } i = 1,...,N. \\
&\qquad\quad (x_1,...,x_N) \text{ is an implementable effort vector}
\end{aligned}
\tag{2}
$$

We remark that in MP (2), the constraint $\alpha_{l_i} \leq x_i < \alpha_{l_i+1}$ is relaxed to $\alpha_{l_i} \leq x_i \leq \alpha_{l_i+1}$. This is without loss of generality since, in practice, if $x_i = \alpha_{l_i+1}$ for some cell i, we can decrease x_i by a negligible amount of effort and put it anywhere feasible. This will not violate the feasibility constraint but makes x_i strictly less than α_{l_i+1}.

Though MP (2) has complicated terms like $g_i(l_i)$ and α_{l_i} which are *non-linear* in the variable l_i, we show that it can nevertheless be reformulated as a compactly represented MILP. The main challenge here is to eliminate these non-linear terms. To do so, we introduce m new *binary* variables $\{z_i^j\}_{j=1}^{m}$, for each i, to encode the integer variable l_i and linearize the objective and constraints of MP (2) using the new variables. By properly constraining the new variables $\{z_i^j\}_{i,j}$, we obtain the following novel MILP (3), which we show is equivalent to MP (2). MILP (3) has binary variables $\{z_i^j\}_{i \in [N], j \in \{1,...,m\}}$ (thus mN binary variables), continuous effort value variables $\{x_i\}_{i \in [N]}$ and flow variables $\{f(e)\}_{e \in E}$. Note however, $g_i(j)$'s are constants given by the black-box attacker model. By conventions, $\sigma^+(v)$ ($\sigma^-(v)$) denotes the set of edges that enter into (exit from) any node v.

$$
\begin{aligned}
&\text{maximize } \sum_{i=1}^{N} \left(g_i(0) + \sum_{j=1}^{m} z_i^j \cdot [g_i(j) - g_i(j-1)] \right) \\
&\text{subject to } x_i \geq \sum_{j=1}^{m} z_i^j \cdot [\alpha_j - \alpha_{j-1}], &&\text{for } i = 1,...,N. \\
&\qquad\quad x_i \leq \alpha_1 + \sum_{j=1}^{m} z_i^j \cdot [\alpha_{j+1} - \alpha_j], &&\text{for } i = 1,...,N. \\
&\qquad\quad z_i^1 \geq z_i^2 ... \geq z_i^m, &&\text{for } i = 1,...,N. \\
&\qquad\quad z_i^j \in \{0,1\}, &&\text{for } i = 1,...,N, j = 1,...,m. \\
&\qquad\quad x_i = \sum_{t=1}^{T} \left[\sum_{e \in \sigma^+(v_{t,i})} f(e) \right], &&\text{for } i = 1,...,N. \\
&\qquad\quad \sum_{e \in \sigma^+(v_{t,i})} f(e) = \sum_{e \in \sigma^-(v_{t,i})} f(e), &&\text{for } i = 1,...,N; t = 2,...,T-1. \\
&\qquad\quad \sum_{e \in \sigma^+(v_{T,1})} f(e) = \sum_{e \in \sigma^-(v_{1,1})} f(e) = 1 \\
&\qquad\quad 0 \leq x_i \leq 1, \quad 0 \leq f(e) \leq 1, &&\text{for } i = 1,...,N; e \in E.
\end{aligned}
\tag{3}
$$

Theorem 2. *MILP (3) is equivalent to the Mathematical Program (2).*

Proof. By Lemma 1 and noticing that variable $f(e)$ for all $e \in E$ represents a one-unit flow on graph G, it is easy to verify that the last four sets of constraints in MILP (3) are precisely a mathematical formulation for the constraint "$(x_1, ..., x_N)$ is an implementable effort vector". We therefore only prove that the first four sets of constraints in MILP (3) encode the first two constraints of MP (2). Moreover, the objective functions in MILP (3) and MP (2) are equivalent.

We start by examining the constraints. Since $z_i^1 \geq z_i^2 ... \geq z_i^m$ and $z_i^j \in \{0, 1\}$, we know that any feasible $\{z_i^j\}_{j=1}^m$ corresponds to a $l_i \in \{0, 1, ..., m\}$ such that $z_i^j = 1$ for all $j \leq l_i$ and $z_i^j = 0$ for all $j > l_i$ ($l_i = 0$ means $z_i^j = 0$ for all j). Conversely, given any $l_i \in \{0, 1, ..., m\}$, we can define $z_i^j = 1$ for all $j \leq l_i$ and $z_i^j = 0$ for all $j > l_i$ as a feasible choice of $\{z_i^j\}_{j=1}^m$. That is, there is a one-to-one mapping from feasible $\{z_i^j\}_{j=1}^m$ to feasible l_i for any cell i. Utilizing this one-to-one mapping, we have

$$\sum_{j=1}^m z_i^l \cdot [\alpha_j - \alpha_{j-1}] = \sum_{j=1}^{l_i} [\alpha_j - \alpha_{j-1}] = \alpha_{l_i}.$$

$$\sum_{j=1}^m z_i^j \cdot [\alpha_{j+1} - \alpha_j] + \alpha_1 = \alpha_1 + \sum_{j=1}^{l_i} [\alpha_{j+i} - \alpha_j] = \alpha_{l_i+1}.$$

This shows that any feasible $\{z_i^j\}_{j=1}^m$ encodes an $l_i \in \{0, 1, ..., m\}$ and the first two constraints in MILP (3) are equivalent to $x_i \geq \alpha_{l_i}$ and $x_i \leq \alpha_{l_i+1}$, respectively. The argument for the objective function is similar. In particular, $g_i(0) + \sum_{j=1}^m z_i^j \cdot [g_i(j) - g_i(j-1)] = g_i(0) + \sum_{j=1}^{l_i} [g_i(j) - g_i(j-1)] = g_i(l_i)$. This proves that MILP (3) is equivalent to MP (2), as desired.   □

Generalizations. The techniques above can be easily generalized to handle more general tasks and models. First, it works for any defender objective that is linear in g_i's, not necessarily the particular one in MP (2). For example, if the attacks at different cells have different impacts, we can generalize the objective to be a weighted sum of g_i's. Second, the assumption that g_i depends discretely on the patrol level is equivalent to assume that g_i is a piece-wise *constant* function of x_i. This can be further generalized. Particularly, when g_i is a piece-wise *linear* function of x_i, we can still use a similar MILP to solve the problem with a slightly more involved formulation of the objective (e.g., [17]). Finally, when g_i is a continuous function in x_i, its piece-wise linear approximation usually serves as a good estimation of g_i.

Generalization to Route Design with Look-Ahead

We now illustrate how the previous techniques can be generalized to patrol design with look-ahead. The designer will plan for multiple periods and need to take into account the effect of current patrolling on the next period's prediction. For simplicity, we focus on planning for two periods: the current period 1 and the next period 2. The approach presented here can be generalized to planning for any small number of periods. Moreover, in the real-world domain we focus on, there is usually no need for a long-term patrol plan because patrolling resources and environments are dynamic – new plans will need to be frequently designed. The optimal planning for two periods can be formulated as the following mathematical program (MP). Note that here we bring back the omitted superscripts for l_i's to indicate different time periods. Moreover, we use g^1, g^2 to denote the prediction function at period 1, 2 respectively.

$$
\begin{aligned}
\text{maximize } & \sum_{i=1}^{N} g_i^2(l_i^2, l_i^1) + \sum_{i=1}^{N} g_i^1(l_i^1, l_i^0) \\
\text{subject to } & \alpha_{l_i^2} \leq x_i^2 \leq \alpha_{l_i^2+1}, && \text{for } i = 1, ..., N. \\
& \alpha_{l_i^1} \leq x_i^1 \leq \alpha_{l_i^1+1}, && \text{for } i = 1, ..., N. \\
& l_i^2 \in \{0, 1, ..., m\}, && \text{for } i = 1, ..., N. \\
& l_i^1 \in \{0, 1, ..., m\}, && \text{for } i = 1, ..., N. \\
& \mathbf{x}^2, \mathbf{x}^1 \text{ are both implementable effort vectors.}
\end{aligned}
\qquad (4)
$$

MP (4) can also be reformulated as an MILP by employing the techniques above. More precisely, we can introduce binary variables $\{z_i^j\}_{j=1}^m$ and $\{t_i^j\}_{j=1}^m$ to encode l_i^2 and l_i^1 respectively. The additional challenge is to represent $g_i^2(l_i^2, l_i^1)$ as a linear function. To do so, we can equivalently view g_i^2 as a function of $l_i = (m+1)l_i^2 + l_i^1 \in \{0, 1, ..., m^2 + 2m\}$, and introduce $m^2 + 2m$ additional binary variables $c_i^1, ..., c_i^{m^2+2m}$ for each i, such that $1 \geq c_i^1 \geq ... \geq c_i^{m^2+2m} \geq 0$ and $\sum_{j=1}^{m^2+2m} c_i^j = l_i = (m+1)\sum_{j=1}^m z_i^j + \sum_{j=1}^m t_i^j$. This guarantees that $c_i^j = 1$ for all $j \leq l_i$ and $c_i^j = 0$ otherwise. Thus $g_i^2(l_i^2, l_i^1) = g_i^2(l_i) = g_i^2(0) + \sum_{j=1}^{m^2+2m} c_i^j \cdot [g_i^2(j) - g_i^2(j-1)]$. So the objective and all the constraints are linear in these variables. Note that this approach introduces $N(m^2 + 4m)$ binary variables.

5   Increasing Unpredictability via Entropy Maximization

The algorithms in Sect. 4 output only a flow $\{f_e\}_{e \in E}$ together with the corresponding effort vector. To implement this effort vector in the real world, one needs to decompose the flow to an executable mixed strategy, i.e., a distribution over deterministic patrolling routes. The classical approach is to use a standard flow decomposition algorithm. Unfortunately, these algorithms often output a decomposition with very small number of route choices. For example, in one real-world patrol post we tested, the resulted optimal mixed strategy essentially randomizes over only two patrol routes, as depicted in Fig. 2. Despite its optimality, such a mixed strategy is problematic due to its lack of randomness and unpredictability. First, since there are only two routes, the poacher can quickly learn these patrolling routes. Then, knowing these two routes, a poacher can easily figure out where the patroller will be at any time during the day by simply looking at whether their initial move is to the northeast or southwest since this initial move uniquely indicates which route the patroller takes.

To overcome this issue, we seek to compute a mixed strategy that implements the (same) optimal effort vector but is the "most random" in the sense that it has the maximum possible (Shannon) entropy. The underlying intuition is that the increased randomness will make patrolling more unpredictable even when poachers can observe part of the patroller's movement. A thorough experimental justification of the max-entropy approach is done in Sect. 8.

We start by formulating the problem of computing the mixed strategy that implements the effort vector while maximizing entropy. Let set $\mathcal{P}$ denote the set

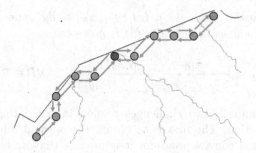

Fig. 2. Visualization of two patrol routes.

of all $s-d$ paths. Our goal is to implement $\{x_i\}_{i\in[N]}$ as a distribution over $\mathcal{P}$ with the maximum entropy, a task which we term the *max-entropy decomposition* of $\{x_i\}_{i\in[N]}$. We will view any $P \in \mathcal{P}$ as a set of nodes in G that specifies the ranger's position at each time step (these nodes uniquely determine the $s - d$ path). Let $P_i = \{v_{t,i} : \exists t,$ s.t. $v_{t,i} \in P\}$ denote those nodes corresponding to cell i, so path P patrols cell i with $|P_i|$ units of effort, where $|P_i|$ is the cardinality of P_i. The max-entropy decomposition can be formulated as the following program with variable θ_P representing the probability of picking path P.

$$
\begin{aligned}
\text{maximize } & -\textstyle\sum_{P\in\mathcal{P}} \theta_P \log \theta_P \\
\text{subject to } & \textstyle\sum_{P\in\mathcal{P}} |P_i|\theta_P = x_i, \text{ for } i \in E. \\
& \textstyle\sum_{P\in\mathcal{P}} \theta_P = 1 \\
& \theta_P \geq 0, \qquad\qquad \text{for } P \in \mathcal{P}.
\end{aligned}
\tag{5}
$$

Observe that program (5) is a convex program (CP) since entropy is a concave function. However, the major challenge of solving CP (5) is that the size of $\mathcal{P}$ (i.e., the total number of $s-d$ paths in G) is exponential in T and therefore so is the total number of variables in CP (5). Indeed, though $T = 12$ in our real-world setting, this results in about 10^{10} variables in CP (5). Such a convex program cannot be efficiently solved by any state-of-the-art optimization software.

To overcome this challenge, we instead examine the *Lagrangian dual* of CP (5) and utilize a well-known characterization of the optimal solution to CP (5) in terms of the optimal solution of its Lagrangian dual – an unconstrained convex program with variables $\{y_i\}_{i\in[N]}$, as follows:

$$
\text{Dual of CP(5)}: \quad \min \quad H(\mathbf{y}) = \sum_{i=1}^{N} x_i \cdot y_i + \ln\left[\sum_{P\in\mathcal{P}} \exp\left(-\sum_{i=1}^{N} |P_i| y_i\right)\right] \tag{6}
$$

Note that $H(\mathbf{y})$ is a convex function. The following well-known lemma characterizes the optimal solution of CP (5) in terms of the optimal solution of CP (6).

Lemma 2. (Adapted from [16]). *Let $\{y_i^*\}_{i\in[N]}$ be the optimal solution to CP (6). Then the optimal solution to CP (5) is given by:*

$$\theta_P^* = \frac{\exp(-\sum_{i=1}^N |P_i|y_i^*)}{\sum_{P'\in\mathcal{P}} \exp(-\sum_{i=1}^N |P_i'|y_i^*)}, \qquad \forall P \in \mathcal{P} \tag{7}$$

Despite of Lemma 2, two challenges remain in computing the maximum entropy decomposition. The first is to obtain the optimal solution to CP (6). Though CP (6) is a convex program, it is unclear that its objective function can be even evaluated efficiently since $\sum_{P\in\mathcal{P}} \exp(-\sum_{i=1}^N |P_i|y_i)$ is a summation of exponentially many terms. To overcome this challenge, we design an efficient dynamic program (DP) to compute the term $\sum_{P\in\mathcal{P}} \exp(-\sum_{i=1}^N |P_i|y_i)$ for any given input $\{y_i\}_{i\in[N]}$. The second challenge is that we cannot *explicitly* output the optimal solution $\{\theta_P^*\}_{P\in\mathcal{P}}$ since it takes exponential time to even write down these many variables. We therefore instead develop a *sampling algorithm* and prove that it samples a path $P \in \mathcal{P}$ with the desired probability θ_P^* in $\mathrm{poly}(N,T)$ time. Next, we elaborate our algorithms while omitting formal proofs due to space limit (formal proofs can be found in the online appendix).

For notational convenience, let $C(\mathbf{y})$ denote the term $\sum_{P\in\mathcal{P}} \exp(-\sum_{i=1}^N |P_i|y_i)$. To compute $C(\mathbf{y})$, we utilize the natural chronological order along the temporal dimension for nodes in graph G and build a dynamic programming table $DP(t,i)$, for $t \in [T]$ and $i \in [N]$, such that $DP(t,i) = \sum_{P\in\mathcal{P}(t,i)} \exp(-\sum_{i=1}^N |P_i|y_i)$ where $\mathcal{P}(t,i)$ is the set of paths from s to the node $v_{t,i}$. We initialize $DP(1,1) = y_1$ (recall $s = v_{1,1}$) and $DP(1,i) = 0$ for all $i > 1$, and then use the following update rule to return $DP(T,1)$ (recall $d = v_{T,1}$):

$$DP(t,i) = \sum_{e:e=(v_{t-1,i'},v_{t,i})} DP(t-1,i') \cdot \exp(-y_i).$$

Correctness of the algorithm follows a textbook argument. Utilizing this DP, one can efficiently evaluate the objective value of CP (6), and solve the unconstrained optimization problem via any black-box optimization tool (e.g., `fmincon` in MAT-LAB).

Next, we take $\{y_i^*\}_{i\in[N]}$ as input and efficiently samples an $s-d$ path $P \in \mathcal{P}$ with probability θ_P^*, as defined in Eq. (7). The algorithm starts from the node d ($=v_{T,1}$) and at any time t and location *loc*, samples its predecessor node $v_{t-1,i}$ with probability $p_e = \frac{\exp(-y_{loc}^*)\cdot DP(t-1,i)}{DP(t,loc)}$ where $e = (v_{t-1,i}, v_{t,loc})$. Full details are in Algorithm 1. The following theorem summarizes its correctness.

Theorem 3. *Algorithm 1 correctly samples P with probability θ_P^* for any $P \in \mathcal{P}$ and runs in $\mathrm{poly}(N,T)$ time, where $\{\theta_P^*\}_{P\in\mathcal{P}}$ is the optimal solution to CP (5).*

Algorithm 1. Max-Entropy Implementation of the Effort Vector

Input: : Effort values at each cell $\{x_i\}_{i \in [N]}$.

Output: : a random path $P \in \mathcal{P}$ which implements $\{x_i\}$ and maximizes entropy.

1: Compute the optimal solution $\{y_i^*\}_{i \in [N]}$ to CP (6) by utilizing the DP.
2: Build the DP table $DP(t, i)$ with $\{y_i^*\}_{i \in [N]}$;
3: Initialize: $P = \{v_{T,1}\}$, Define $loc = 1$;
4: **for** $t = T$ to 2 **do**
5:     Sample an edge $e = (v_{t-1,i}, v_{t,loc})$ for all such edges that exist, with probability

$$p_e = \frac{\exp(-y_{loc}^*) \cdot DP(t-1, i)}{DP(t, loc)};$$

6:     Let $e = (v_{t-1,i^*}, v_{t,loc})$ be the sampled edge above, and add v_{t-1,i^*} to P.
7:     Update $loc = i^*$.
8: **end for**
9: **return** P.

6   Real-World Dataset

Our analysis focuses on a real-world wildlife crime dataset from Uganda's Queen Elizabeth Protected Area (QEPA). QEPA spans approximately $2{,}520\,\text{km}^2$ and is patrolled by wildlife park rangers. While on patrol, they collect data on animal sightings and signs of illegal human activity (e.g., poaching, trespassing). In addition to this observational data, the dataset contains terrain information (e.g., slope, vegetation), distance data (e.g., nearest patrol post), animal density, and the kilometers walked by rangers in an area (i.e., effort).

We divide QEPA into $1\,\text{km}^2$ grid cells and compute several features based on the dataset's contents (e.g., observations, terrain, effort). Additionally, we group the observations and effort values (i.e., the values that change over time) into a series of month-long time steps. Finally, we compute two effort features, previous effort and current effort, that represent the amount of patrolling effort expended by rangers in the previous time step and current time step, respectively. Because effort is a continuous value (0 to ∞), we discretize the effort values into m effort groups (e.g., $m = 2$: `high` and `low`).

7   Predictive Model Analysis

In this section, we analyze an example attack prediction model that can predict poaching activity for the real-world dataset described in Sect. 6 using an ensemble of decision trees [6]. This model can provide the black-box attack function $g_i(l_i)$ for OPERA and will be used to evaluate OPERA in Sect. 8.

The goal of the analysis is two-fold. First, we analyze the performance of the prediction model to verify that it is a realistic model that can provide $g_i(l_i)$.

Second, we analyze how the prediction model's predictions change as a function of ranger effort, which can provide intuition on why planning patrols using OPERA can help increase the efficiency of patrols.

Note that although the hybrid model proposed in [6] is currently the best performing predictive model in this domain, conducting such an analysis on this model may be confounded by its complexity. For instance, the hybrid model's reaction to a change in effort may be due to the underlying bagging ensemble's reaction or it may be due to a reaction in the Markov Random Field that boosts predictions under specific conditions. For scientific rigor, we instead focus on the analysis of a single model's reactivity – the bagging ensemble (which outperforms the previously most accurate model in [9]).

7.1   Ensemble Model

Bagging (Bootstrap aggregation technique) is an ensemble method (in this case applied to decision trees) where each tree is trained on a bootstrapped subset of the whole dataset. The subsets are generated by randomly choosing, with replacement, M observations where M is the dataset size. Once all trees in the ensemble are trained, the ensemble's predictions are generated by averaging the predictions from each tree. We trained a bagging ensemble using the *fitcensemble* function in MATLAB 2017a. For this model, the best training period consists of 5 years of data (based on repeated analyses for different train/test splits). Described in Sect. 6, the 11 input features consist of terrain and geospatial features, and two patrol effort features (one for previous time step's effort and one for current effort). Each data point's label corresponds to whether an attack was detected at that cell. For the training set, a label will be 1 if at any point in the training period an attack was detected (0 otherwise). For the test set, a label will be 1 if an attack was detected during the current time step.

We present results for a bagging ensemble on a three-month time scale where the ensemble is trained on 5 years of data (effort values are in three-month chunks) and is used to predict detections for a test period of three months. The test set corresponds to September through November 2016, and the training set contains data for 2,129 patrolled cells from September 2012 to August 2016.

In Table 1, we present prediction performance results as verification that subsequent analyses are done on a realistic model. We also present baseline results from common boosting models – AdaBoost and RUSBoost [14]. Additionally, we present a baseline, TrainBaseline, where if an attack was detected at a cell in the training data, the baseline will predict a detected attack for the test data (for cells that were not patrolled in the training data, and thus there is no training sample for that cell, a uniform random binary prediction is made). Due to the large class imbalance present in the dataset (many more negative labels than positives), we compute the area under a Precision-Recall curve (PR-AUC[2]) instead

---

[2] Because TrainBaseline makes binary predictions and thus does not have continuous prediction values, PR-AUC is not computed for TrainBaseline.

of the standard ROC curve (which is not as informative for such a dataset) [2]. We also present F1, Precision, and Recall scores.

Table 1. Model performances

Model	F1	Precision	Recall	PR-AUC
TrainBaseline	0.4	0.25	0.96	-
RUSBoost	0.21	0.12	0.96	0.28
AdaBoost	0.49	0.35	0.82	0.50
Bagging	**0.65**	**0.52**	**0.86**	**0.79**

As can be seen, the Bagging model outperforms all other models in terms of F1, Precision, and PR-AUC. While Bagging does not always score the highest in recall, its precision score greatly outperforms the other models' precision. In practical terms, this means that the Bagging model will predict far less false positives and can thus better ensure that the patrol generation algorithm is not needlessly sending rangers to areas where they won't detect attacks.

7.2   Effort Function Analysis

The goal of the patrol generation algorithm is to allocate effort such that rangers' detections of attack (poaching activity) are maximized. For the following analysis, we examine how the bagging ensemble's predictions change as a function of ranger effort. For example, if we increase effort in an area over a period of three months, will rangers detect an attack in that area in any of the three months?

For this analysis, we present the changes in (1) the model's detected attack predictions $g_i(l_i)$ and (2) the model's detected attack prediction probabilities when the effort in the current time step is changed. Both values are outputted by MATLAB's *predict* function for our learned ensemble. We refer to effort group 0 as low and group 1 as high; an increase in allocated effort, for example, would result in l_i changing from low to high. Results for changes in predictions and prediction probabilities are shown in Tables 2 and 3 respectively.

Table 2. Prediction changes as function of current effort

Effort change	Neg to Pos	Pos to Neg	No change (Pos)	No change (Neg)
Low to high	119	30	172	1693
High to low	2	110	122	274

In Table 2, for each type of change in effort (low to high or high to low), there are three possible outcomes for a prediction change: a negative prediction (no detection) can change to a positive prediction (detected attack), referred to as **Neg to Pos**, positive can change to negative (**Pos to Neg**), and there can be no change in the prediction (for either the positive or negative prediction cases). Given these outcomes, we make the following observations. First, there are a substantial number of cells whose corresponding detection predictions do not change as a result of changes in effort. In the case of the unchanged positive predictions, these are predicted to be high-risk cells where rangers will find poaching activity even if they allocate relatively low effort values to it. For unchanged negative predictions, these correspond to low-risk cells that are essentially predicted to not be attacked at all. Second, while there are substantially more instances of predicted detections increasing as a result of increasing effort, there are still some instances of predicted detections decreasing as a result of increasing effort. However, because there is not a rational explanation for this trend, these rare instances are likely due to noise in the model. Finally, we make the same observation regarding the case where detections mostly decrease as a result of decreasing effort while detections increase at only two cells.

Table 3. Prediction probability changes as function of current effort

Effort change	Inc	Mean Inc	Dec	Mean Dec	No change(Pos)	No change(Neg)
Low to high	1546	0.16	423	0.09	4	41
High to low	142	0.09	358	0.22	0	8

As for the prediction probability changes in Table 3, we examine changes in the prediction probability with increases and decreases referred to as **Inc** and **Dec** respectively (i.e., any increase or decrease), the mean changes in prediction probability for the increase and decrease cases (referred to as **Mean Inc** and **Mean Dec** respectively), and also in the instances where there was no change in the probability for both the positive (i.e., probability ≥ 0.50) and negative (i.e., probability < 0.50) cases. First, when effort is increased, many more cells are predicted to have a substantial increased prediction probability (mean change of 16%). While there are a non-trivial number of cells with a decrease in their prediction probability, the mean decrease is approximately half that of the mean increase, with the difference being statistically significant ($\alpha < 0.01$), and is thus interpreted as noise. Second, when effort is decreased, there are many more cells with a decrease in prediction probability than increase. Additionally, the mean decrease in prediction probability is more than twice that of the mean increase (22% vs 9%) and is also statistically significant ($\alpha < 0.01$). Finally, as with the prediction changes in Table 2, a few cells are low-risk and increasing effort will not result in a corresponding increase in predicted detection probability. While changes in predicted probability do not necessarily correspond to changes in actual predictions (0/1), the shifts in probability do provide a concrete indication of the actual impacts that coverage has on the model's predictions.

8   Experimental Evaluations

8.1   Evaluation of the Patrol Optimization Algorithm

We start by experimentally testing OPERA using the aforementioned real-world data and bagging ensemble predictive model. Particularly, the inputs to all the tested algorithms are specified as follows: graph G is constructed according to the real-world terrain in QEPA; the function g_i's, together with the corresponding $\{\alpha_i\}_{i=1}^m$, are precisely the predictive model described in Sect. 7 for the test period September through November 2016; $T = 12$ as suggested by domain experts. Since we are not aware of any previous patrol generation algorithm that deals with attackers described by a black-box machine learning model[3], we instead compare our patrol optimization algorithms with the following two heuristic planning algorithms. Note that we will also compare OPERA with its preliminary version without entropy maximization, i.e., the Optimal patrol Planning by flow Decomposition (OPD).

GREED: a heuristic patrol planning algorithm that, at any cell i, greedily picks the next cell j that satisfies: 1. it is feasible to go from i to j; 2. patrolling cell j at high is more *beneficial*, (results in more predicted attacks than patrolling j at low). If there are multiple such cells, pick one uniformly at random; if there are no such cells, then pick any neighbor cell uniformly at random. To guarantee that the patrol path starts and ends at the patrol post, this procedure continues until time $\lceil T/2 \rceil$ and then the patroller returns via its outgoing route.

RAND: a heuristic patrol planning algorithm that is similar to GREED except that at any cell i, it chooses a neighbor cell j to go uniformly at random without considering the prediction model.

There are 39 patrol posts at QEPA. We test the algorithms on the real data/model at patrol post 11, 19 and 24, which are the three posts that have the most attacks in the three months of our testing. In our data, all posts have less than 100 cells/targets reachable from the post by a route of maximum duration $T = 12$ (equivalently, a 12-cells long route) and all the algorithms scale very well to this size (the MILP takes at most 2 s in any tested instance). We thus focus on comparing these algorithms' ability in detecting attacks under multiple criteria, as follows:

– **#Detection**: total number of detected attacks under the prediction model. Since the prediction model we adopt is a 0–1 classification algorithm, in this case **#Detection** also equals the number of cells at which the corresponding patrolling levels result in detected attacks. However, here we exclude those cells for which high or low patrol effort results in the same prediction because patrolling levels at these cells do not make a difference to the criterion.
– **#Cover**: total number of cells that are patrolled with high. Note that due to limited resources, not every cell – in fact, only a small fraction of the cells – can be covered with high.

---

[3] Most previous algorithms either require knowledge of the patroller's and poacher's payoffs [5,7] which are not available in our setting or generates patrolling strategies that are not guaranteed to be implementable [11].

– **#Routes**: the number of different patrol routes in 90-day route samples (corresponding to a 3-month patrolling period).
– **Entropy**: The entropy of the empirical distribution of the 90 samples.

Note that the last two criteria are used particularly to test the unpredictability of these algorithms in an environment with partial observations by the attacker. A higher value of **#Routes** means that the patroller has more choices of patrol routes, thus less explorable by the poacher. **Entropy** is a natural measure to quantify uncertainty. The experimental results for patrol post 11 and 19 are shown in Tables 4 and 5, respectively. The results for post 24 are similar to that for post 19, thus are omitted here to avoid repetition. For the **#Detection** criterion, a/b means that out of the b cells for which low or high makes a prediction difference, a of them are "hit" correctly – i.e., patrolled at the right level that results in predicted attack detection – by the patrolling algorithm. For example, in Table 4, the "15/19" comes from the follows: there are 19 cells at the post for which patrol level high or low makes a difference in attack detection; The patrol levels by OPERA result in positive attack detections in 15 out of these 19 cells. For the **#Cover** criterion, a/b means that out of b cells in total, a cells are patrolled with high. From the analysis in Sect. 7 we know that compared to the low patrol level, the high patrol level is more likely to, though not always, result in attack detections. Therefore, a larger **#Cover** value will be preferred in our comparisons.

Table 4. Comparisons of different criteria for patrol post 11

	#Detection	#Cover	#Routes	Entropy
OPERA	15/19	20/47	61	4.0
OPD	15/19	20/47	10	2.0
GREED	5/19	4/47	84	4.4
RAND	4/19	6/47	89	4.5

Table 5. Comparisons of different criteria for patrol post 19

	#Detection	#Cover	#Routes	Entropy
OPERA	6/6	24/72	22	2.6
OPD	6/6	24/72	6	1.3
GREED	2/6	2/72	1	0
RAND	2/6	6/72	90	4.5

As we can see from both tables, OPERA and OPD[4] result in significantly more detected attacks and cells with high coverage than the GREED and RAND heuristics. GREED results in slightly more detected attacks than RAND, but RAND covers more cells with high. This is because GREED biases towards cells that need more patrolling, thus easily gets concentrated on these cells. For the #Routes and Entropy criteria, RAND is the most unpredictable (as expected). GREED is unstable. Particularly, at patrol post 19, GREED always chooses the same path. This is because it reaches a cell for which high is better and gets stuck at the same cell always due to its greedy choice. This is a critical drawback of GREED. In fact, the same phenomenon is also observed at post 24. Clearly, OPERA exhibits more unpredictability than OPD, and is more stable than GREED. This shows that among these tested algorithms, OPERA provides the best balance among unpredictability, stability and the ability in detecting attacks and covering more cells.

8.2 Comparisons with the Past Patrol Routes

We now compare the patrol routes generated by OPERA with the past patrol routes used by rangers at QEPA. We still adopt the measures in the above Sect. 8.1. Since there is no ground truth to compare with (for the past patrolling, we do not know what happened at those cells that are not patrolled), as an approximation we will treat the bagging ensemble predictive model described in Sect. 7 as the ground truth. This is a reasonable choice since [6] recently shows that this model outperforms all previous poaching prediction models and provides relatively accurate predictions on the QEPA dataset.

Table 6. Comparisons of different criteria at different patrol posts

Criteria	Post 11		Post 19		Post 24	
	OPERA	Past	OPERA	Past	OPERA	Past
#Detections	15/19	4/19	6/6	5/6	4/4	3/4
#Cover	20/47	6/47	24/72	11/72	20/59	14/59
#Routes	61	4	22	33	34	5
#Entropy	4.0	1.2	2.6	3	2.8	1.4

The results are jointly presented in Table 6. As we can see, the patrol routes generated by OPERA clearly outperform past patrolling in terms of the #Detections and #Cover criteria. Particularly, the routes we generate can detect attacks on most (if not all) cells by properly choosing their patrolling levels and also result in more cells covered with high. In terms of unpredictability,

---

[4] Note: they always have the same #Detection and #Cover since they are both optimal.

the past patrolling does not have a stable performance. Particularly, it follows only a few routes at post 11 and 24 with low unpredictability but takes many different routes at post 19 with high unpredictability. This is a consequence of various factors at different posts, like patroller's preferences, location of the patrol post (e.g., inside or at the boundary of the area), terrain features, etc. On the other hand, OPERA always comes with good unpredictability guarantee. This shows the advantage of OPERA over the past patrolling.

9   Conclusion

In this paper, we presented a general patrol planning framework OPERA. It can optimize against a wide range of prediction models and generate implementable patrol strategies. In addition, OPERA maximizes the randomness of generated strategies and increases robustness against partial information leakage (i.e., poachers may infer the patroller's patrolling route by monitoring part of the patroller's movements). Experimental results on a real-world data set from Uganda's Queen Elizabeth Protected Area (QEPA) show that OPERA results in better defender strategies than heuristic planning algorithms and the past real patrol routes used by rangers at QEPA in terms of defender utility, coverage of the area and unpredictability.

Acknowledgement. Part of this research is supported by NSF grant CCF-1522054. Fei Fang is partially supported by the Harvard Center for Research on Computation and Society fellowship.

References

1. Critchlow, R., Plumptre, A.J., Alidria, B., Nsubuga, M., Driciru, M., Rwetsiba, A., Wanyama, F., Beale, C.M.: Improving law-enforcement effectiveness and efficiency in protected areas using ranger-collected monitoring data. Conserv. Lett. (2016). https://doi.org/10.1111/conl.12288. ISSN 1755-263X
2. Davis, J., Goadrich, M.: The relationship between precision-recall and ROC curves. In: Proceedings of the 23rd International Conference on Machine Learning. ICML (2006)
3. Di Marco, M., Boitani, L., Mallon, D., Hoffmann, M., Iacucci, A., Meijaard, E., Visconti, P., Schipper, J., Rondinini, C.: A retrospective evaluation of the global decline of carnivores and ungulates. Conserv. Biol. **28**(4), 1109–1118 (2014)
4. Fang, F., Nguyen, T.H., Pickles, R., Lam, W.Y., Clements, G.R., An, B., Singh, A., Tambe, M., Lemieux, A.: Deploying PAWS: field optimization of the protection assistant for wildlife security. In: Twenty-Eighth IAAI Conference (2016)
5. Fang, F., Stone, P., Tambe, M.: When security games go green: designing defender strategies to prevent poaching and illegal fishing. In: Twenty-Fourth International Joint Conference on Artificial Intelligence (2015)

6. Gholami, S., Ford, B., Fang, F., Plumptre, A., Tambe, M., Driciru, M., Wanyama, F., Rwetsiba, A., Nsubaga, M., Mabonga, J.: Taking it for a test drive: a hybrid spatio-temporal model for wildlife poaching prediction evaluated through a controlled field test. In: Proceedings of the European Conference on Machine Learning & Principles and Practice of Knowledge Discovery in Databases, ECML PKDD 2017 (2017)
7. Haas, T.C., Ferreira, S.M.: Optimal patrol routes: interdicting and pursuing rhino poachers. Police Pract. Res. 1–22 (2017). Routledge
8. Haghtalab, N., Fang, F., Nguyen, T.H., Sinha, A., Procaccia, A.D., Tambe, M.: Three strategies to success: learning adversary models in security games. In: Proceedings of the Twenty-Fifth International Joint Conference on Artificial Intelligence, pp. 308–314. AAAI Press (2016)
9. Kar, D., Ford, B., Gholami, S., Fang, F., Plumptre, A., Tambe, M., Driciru, M., Wanyama, F., Rwetsiba, A., Nsubaga, M., Mabonga, J.: Cloudy with a chance of poaching: adversary behavior modeling and forecasting with real-world poaching data. In: Proceedings of the 16th Conference on Autonomous Agents and MultiAgent Systems, AAMAS 2017, pp. 159–167 (2017)
10. Moreto, W.: To conserve and protect: Examining law enforcement ranger culture and operations in Queen Elizabeth National Park, Uganda. Ph.D. thesis, Rutgers University-Graduate School-Newark (2013)
11. Nguyen, T.H., Sinha, A., Gholami, S., Plumptre, A., Joppa, L., Tambe, M., Driciru, M., Wanyama, F., Rwetsiba, A., Critchlow, R., et al.: Capture: a new predictive anti-poaching tool for wildlife protection. In: Proceedings of the 2016 International Conference on Autonomous Agents & Multiagent Systems, pp. 767–775. International Foundation for Autonomous Agents and Multiagent Systems (2016)
12. Nguyen, T.H., Yang, R., Azaria, A., Kraus, S., Tambe, M.: Analyzing the effectiveness of adversary modeling in security games. In: AAAI (2013)
13. Nyirenda, V.R., Chomba, C.: Field foot patrol effectiveness in Kafue national park, Zambia. J. Ecol. Nat. Environ. 4(6), 163–172 (2012)
14. Seiffert, C., Khoshgoftaar, T.M., Van Hulse, J., Napolitano, A.: Rusboost: a hybrid approach to alleviating class imbalance. IEEE Trans. Syst. Man Cybern. Part A Syst. Hum. 40(1), 185–197 (2010)
15. Shieh, E., An, B., Yang, R., Tambe, M., Baldwin, C., DiRenzo, J., Maule, B., Meyer, G.: Protect: a deployed game theoretic system to protect the ports of the united states. In: Proceedings of the 11th International Conference on Autonomous Agents and Multiagent Systems, vol. 1, pp. 13–20. International Foundation for Autonomous Agents and Multiagent Systems (2012)
16. Singh, M., Vishnoi, N.K.: Entropy, optimization and counting. In: STOC, pp. 50–59. ACM (2014)
17. Wolsey, L.A.: Integer programming. Wiley-Interscience, New York (1998)
18. Xu, H., Tambe, M., Dughmi, S., Noronha, V.L.: The curse of correlation in security games and principle of max-entropy. CoRR abs/1703.03912 (2017)
19. Yang, R., Ford, B., Tambe, M., Lemieux, A.: Adaptive resource allocation for wildlife protection against illegal poachers. In: Proceedings of the 2014 International Conference on Autonomous Agents and Multi-agent Systems, pp. 453–460. International Foundation for Autonomous Agents and Multiagent Systems (2014)
20. Yin, Z., Jiang, A.X., Tambe, M., Kiekintveld, C., Leyton-Brown, K., Sandholm, T., Sullivan, J.P.: Trusts: scheduling randomized patrols for fare inspection in transit systems using game theory. AI Mag. 33(4), 59 (2012)

Short Papers

Security Games with Probabilistic Constraints on the Agent's Strategy

Corine M. Laan[1,2,3]([⊠]), Ana Isabel Barros[2,4], Richard J. Boucherie[1], and Herman Monsuur[3]

[1] Stochastic Operations Research, University of Twente, Enschede, Netherlands
c.m.laan@utwente.nl
[2] TNO - Defense, Safety and Security, Den Hague, Netherlands
[3] Netherlands Defense Academy, Den Helder, Netherlands
[4] Institute for Advanced Study, Amsterdam, Netherlands

Abstract. This paper considers a special case of security games dealing with the protection of a large area divided in multiple cells for a given planning period. An intruder decides on which cell to attack and an agent selects a patrol route visiting multiple cells from a finite set of patrol routes such that some given operational conditions on the agent's mobility are met. For example, the agent might be required to patrol some cells more often than others. In order to determine strategies for the agent that deal with these conditions and remain unpredictable for the intruder, this problem is modeled as a two-player zero-sum game with probabilistic constraints such that the operational conditions are met with high probability. We also introduce a variant of the basic constrained security game in which the payoff matrices change over time, to allow for the payoff that may change during the planning period.

Keywords: Game theory · Probability constraints · Defense applications

1 Introduction

This paper considers a special case of a security game dealing with the protection of a large area for a given time period where the agent's strategy set is restricted. The area consists of several cells containing assets to be protected. An intruder decides on which cell to attack, while the agent needs to select a patrol route that visits multiple cells. The agent's strategy is constrained by existing governmental guidelines that require that some cells should be patrolled more often than others. This problem can be modeled as a two-player zero-sum game with probabilistic constraints.

In the literature there are several models considering patrolling games (e.g., [1,5,8]). Also, many models consider constraints on the agent's or intruder's strategy set. For example in [2,6,15], the authors require constraints on the agent's strategy because only a limited number of resources is available, and

© Springer International Publishing AG 2017
S. Rass et al. (Eds.): GameSec 2017, LNCS 10575, pp. 481–493, 2017.
DOI: 10.1007/978-3-319-68711-7_25

in [17] the authors consider constraints on both the agent's and the intruder's strategy set.

Often, linear constraints are considered in constrained games. For instance, in [3] a two-person zero-sum game with linear constraints is introduced. More recently, [10] described a bimatrix game with linear constraints on the strategy of both players. In [14], the author considers nonlinear ratio type constraints. Our security game models situations where operational conditions have to be met with high probability, which results in nonlinear probabilistic constraints.

An example application of our model lies in countering illegal or unreported and unregulated fishing. These illicit activities endanger the economy of the fishery sector, fish stocks and the marine environment and require the monitoring of large areas with scarce resources subject to national regulations. To support the development of patrols against illegal fishing, in [7] a decision support system is developed. This system models the interaction between different types of illegal fishers and the patrolling forces as a repeated game. More recently, [4] introduced a game theoretical approach wherein a generalization of Stackelberg games is used to derive sequential agent strategies that learn from adversary behavior. However, these papers do not consider constraints to the patroller's strategy.

The main contribution of this paper is that we introduce a new model to cope with the conditions on the agent's random strategy that have to be met with high probability. Because of the random nature of the strategies, it cannot be guaranteed that the conditions are always met. By introducing probabilistic constraints, we assure that the conditions are met with high probability. In practice the payoff matrices may change over time, in the fishery case, due to weather conditions, seasonal fluctuations or other circumstances. Therefore, we introduce an extension of the model to deal with multiple payoff matrices.

This paper is organized as follows. In Sect. 2, we introduce the new security game model with constraints on the agent's strategy. In Sect. 3, we present an extension of the model in which multiple payoff matrices are considered. Finally, in Sect. 4 we give examples of the model and present computational results.

2   Model with Constant Payoff

This section describes the model assuming that the gain an intruder obtains by successfully visiting a cell is constant over the planning period. We first provide a general description of a constrained security game over multiple cells in Sect. 2.1. For each cell, there is a condition on the minimal number of visits per time period for that cell. We discuss the probability that these conditions are met for each cell separately in Sect. 2.2, which gives a lower bound for the game value. In the application of countering illegal fishing, governmental guidelines require that some cells should be patrolled more than others because some regions are more vulnerable. The conditions on the number of visits have to be met for all cells simultaneously. These simultaneous conditions are discussed in Sect. 2.3.

2.1   Constrained Game

We consider a security game with constraints on the strategy sets (see [11], Chap. 3.7). Let $C = \{1, ..., N_C\}$ be the set of cells that can be attacked by an intruder and let $R = \{1, ..., N_R\}$ be the set of routes that can be chosen by the agent. The matrix A indicates which cells are visited by each route, such that a_{ij} equals 1 if route i includes cell j and 0 otherwise. Let M be the payoff matrix, such that m_{ij} is the payoff for the intruder if the agent chooses route i and the intruder attacks cell j, $i = 1, ..., N_R$, $j = 1, ..., N_C$:

$$m_{ij} = ((1 - d_j)a_{ij} + (1 - a_{ij}))\, g_j, \quad i = 1, ..., N_R, \ j = 1, ..., N_C, \tag{1}$$

where g_j is the intruder's gain if the intruder successfully attacks cell j and d_j is the probability that the intruder is caught if the agent's chosen route i includes cell j. The game is repeated N_D times (e.g. days), our planning period. We assume that only one intruder is present in the area. If that intruder is caught, then another will replace him. The overall aim from an intruders perspective is to maximize the total payoff over the time period.

Remark 1. Note that the model described in this section assumes that each intruder attacks one cell each day. By changing the payoff matrix and the actions of the agent and the intruder, the model can be extended to other situations. □

The intruder attempts to maximize the payoff by choosing which cell to attack, so the action set of the intruder is given by C. The agent tries to catch the intruder by selecting a route, so the action set of the agent is given by R. The agent minimizes the payoff by deciding on the probability p_i, $i = 1, ..., N_R$, that route i is chosen, while the intruder maximizes the payoff by selecting the probability q_j, $j = 1, ..., N_C$, that cell j is attacked. The strategy of the agent is constrained by the conditions $f(p) \geq 0$, determined by the minimum number of times each cell is visited by the agent. In Sects. 2.2 and 2.3, we will elaborate on these conditions. The value of the game, V, equals the expected payoff per day. Optimal strategies can be found by solving the following mathematical program:

$$
\begin{aligned}
V = \min_{p} \max_{q} \quad & p^T M q \\
\text{s.t.} \quad & f(p) \geq 0, \\
& \sum_{i=1}^{N_R} p_i = 1, \ \sum_{j=1}^{N_C} q_j = 1, \\
& p, q \geq 0.
\end{aligned}
\tag{2}
$$

Taking the dual of the inner linear program $\max_q \{p^T M q | \sum_{j=1}^{N_C} q_j = 1, q \geq 0\}$, the minmax formulation (2) can be rewritten to obtain the value of the game and optimal strategies for the agent:

$$V = \min_{p,z} \quad z$$

$$\text{s.t.} \quad e^T z \geq p^T M,$$

$$f(p) \geq 0, \tag{3}$$

$$\sum_{i=1}^{N_R} p_i = 1, \ p \geq 0,$$

where e is the row vector with only ones. Note that there only exists a value for this game if the set $\{p | f(p) \geq 0, \sum_{i=1}^{N_R} p_i = 1, p \geq 0\}$ is not empty.

Remark 2. For clearness of presentation, we model the game as a zero-sum game. Note that a similar model applies if we consider a bimatrix game in which the agent and the intruder have different payoff matrices. In bimatrix games, the game value is calculated using quadratic programming (see for example [12], Chap. 13.2) instead of linear programming, but the probabilistic constraints can be implemented similarly. In addition, in the same manner, conditions on the intruder's strategy set can be added.   □

2.2   Conditions on the Number of Visits to a Cell

In this subsection, we consider conditions on the number of visits for each cell separately to obtain a lower bound for V. Let N_D be the number of days in the planning period. The strategy of the agent is constrained by the minimum number of visits v_j to each cell j, $j = 1, ..., N_C$, over the entire period N_D, that must be realized with at least probability $1 - \epsilon$. Given any strategy p, the probability that cell j is visited by the agent is $a_j p$, where a_j is the row vector of the j-th column of A.

Let X_j, $j = 1, ..., N_C$, be the random variable that records the number of visits to cell j during the planning period. The probability that cell j is visited equals $a_j p$. As there are N_D successive days, X_j is binomially distributed with parameters N_D and $a_j p$. The constraint on the number of visits then reads $P(X_j \geq v_j) \geq (1 - \epsilon)$, i.e.,

$$\sum_{k=v_j}^{N_D} \frac{N_D!}{k!(N_D - k)!} (a_j p)^k (1 - a_j p)^{N_D - k} \geq 1 - \epsilon,$$

which can be implemented in (3) by choosing $f(p) = (f_1(p), f_2(p), ... f_{N_C}(p))$ with $f_j(p) = P(X_j \geq v_j) - (1 - \epsilon)$.

For large N_D, the binomial distribution becomes intractable for implementation. Therefore, we use the following approximation. For large N_D, the binomially distributed X_j can be approximated by the normally distributed $\tilde{X}_j$ with mean $N_D a_j p$ and variance $N_D a_j p(1 - a_j p)$ (see [13], Chap. 1.8):

$$P(X_j \geq v_j) = 1 - P(X_j < v_j) \approx 1 - P(\tilde{X}_j \leq v_j),$$

yielding

$$f_j(p) = \epsilon - \Phi\left(\frac{v_j - N_D a_j p}{\sqrt{N_D a_j p(1 - a_j p)}}\right), \tag{4}$$

where $\Phi(x)$ is the cumulative distribution function for the standard normal distribution.

Considering the conditions for each cell separately gives a relaxation of the original conditions, where the minimum number of visits has to be obtained for all cells simultaneously. If we replace $f(p)$ in (3) by the constraints in (4), we obtain the following lower bound for the game value V:

$$V_L = \min_{p,z} \quad z$$

$$\text{s.t.} \quad e^T z \geq p^T M,$$

$$\Phi\left(\frac{v_j - N_D a_j p}{\sqrt{N_D a_j p(1 - a_j p)}}\right) \leq \epsilon, \quad j = 1, ..., N_C, \tag{5}$$

$$\sum_{i=1}^{N_R} p_i = 1, \quad p \geq 0.$$

In order to linearize these constraints, we can determine for each cell j all possible values of $a_j p$ such that $\epsilon - P(\tilde{X}_j \leq v_j) \geq 0$ using the table of the standard normal distribution. The constraints in (5) can be replaced by the linear constraint $p^T A \geq \tilde{b}$, where $\tilde{b}_j$ is determined by the minimum probability for each cell such that the conditions are met with probability $1 - \epsilon$.

Visits to cells are correlated via the routes. Therefore, we are interested in the joint probability:

$$P(X_1 \geq v_1, X_2 \geq v_2, ..., X_{N_C} \geq v_{N_C}),$$

that we will discuss in the next section.

2.3   Conditions on All Cells Simultaneously

In this section, we discuss the condition on the minimum number of visits for all cells simultaneously. Let Y_i, $i = 1, ..., N_R$, be the random variable that specifies the number of times that route i is selected. $Y = (Y_1, Y_2, ..., Y_{N_R})$ is multinomially distributed with parameters N_D and p:

$$P(Y_1 = v_1, Y_2 = v_2, ..., Y_{N_R} = v_{N_R}) = N_D! \prod_{i=1}^{N_R} \frac{p_i^{v_i}}{v_i!}.$$

For large N_D, Y_i, $i = 1, ..., N_R$ can be approximated by the multivariate normally distributed $\tilde{Y}_i$ with expectation $N_D p_i$, variance $N_D p_i (1 - p_i)$ and covariance $Cov(\tilde{Y}_i, \tilde{Y}_{i'}) = -N_D p_i p_{i'}$, $i' = 1, ..., N_R$ (see [13], Chap. 1.8).

The number of times cell j is visited, X_j, can then be expressed as $X_j = \sum_{i=1}^{N_R} a_{ij} Y_i$ and using the approximation $\tilde{Y}$ for Y, X_j can be approximated by a normally distributed $\tilde{X}_j$ with expectation, variance and covariance (see [13], Chap. 1.4), $j = 1, .., N_C$:

$$E(\tilde{X}_j) = N_D a_j p, \qquad Var(\tilde{X}_j) = N_D a_j p (1 - a_j p),$$

$$Cov(\tilde{X}_j, \tilde{X}_{j'}) = \sum_{i=1}^{N_R} \sum_{i'=1}^{N_R} a_{ij} a_{i'j'} Cov(\tilde{Y}_i, \tilde{Y}_{i'}).$$

The probability that the conditions are met for all cells is:

$$P(X_1 \geq v_1, X_2 \geq v_2, ..., X_{N_C} \geq v_{N_C}) \approx P(\tilde{X}_1 \geq v_1, \tilde{X}_2 \geq v_2, ..., \tilde{X}_{N_C} \geq v_{N_C})$$

$$= \frac{1}{\sqrt{|\Sigma|(2\pi)^{N_C}}} \int_{v_1}^{\infty} \int_{v_2}^{\infty} ... \int_{v_{N_C}}^{\infty} e^{-\frac{1}{2}(v-\mu)' \Sigma^{-1} (v-\mu)} dv_{N_C} ... dv_1, \quad (6)$$

where Σ is the covariance matrix and μ is a vector with all expected values. This can be implemented in (3) by choosing $f(p)$ as

$$f(p) = P(\tilde{X}_1 \geq v_1, \tilde{X}_2 \geq v_2, ..., \tilde{X}_{N_C} \geq v_{N_C}) - (1 - \epsilon). \qquad (7)$$

The constraint described above is not linear and cumbersome to implement in a mathematical program. To simplify the model, we use a lower bound for the probability that the conditions are met and implement this lower bound.

A lower bound for the probability that the conditions for all cells are met is:

$$P(\tilde{X}_1 \geq v_1, ..., \tilde{X}_{N_C} \geq v_{N_C}) \geq 1 - \sum_{j=1}^{N_C} P(\tilde{X}_j < v_j). \qquad (8)$$

This lower bound can be used to simplify the mathematical program as follows:

$$f(p) = \epsilon - \sum_{j=1}^{N_C} \Phi\left(\frac{v_j - N_D a_j p}{\sqrt{N_D a_j p (1 - a_j p)}}\right).$$

Replacing $f(p)$ in (3) by a lower bound in the condition, results in an upper bound for the game value V:

$$V_U = \min_{p,z} \quad z$$

$$\text{s.t.} \quad e^T z \geq p^T M,$$

$$\sum_{j=1}^{N_C} \Phi\left(\frac{v_j - N_D a_j p}{\sqrt{N_D a_j p (1 - a_j p)}}\right) \geq \epsilon, \qquad (9)$$

$$\sum_{i=1}^{N_R} p_i = 1, \ p \geq 0,$$

Combining this upper bound and the lower bound obtained in Sect. 2.2, we readily obtain the following result:

Lemma 1. *For V_L given in* (5) *and V_U given in* (9) *we have $V_L \leq V \leq V_U$* □

In Sect. 4, we investigate the impact of this approximation modeling approach on the game value.

Remark 3. We may linearize this program by approximating the normal distribution for each cell j by a piecewise linear function as described in [16], Chap. 9.2. However, we use in the result section the mathematical program stated in (9) since this model is still solvable for realistic instances.   □

3 Generalization: Multiple Payoff Matrices

The previous section considers games with constant payoff. This section considers a generalization to situations where payoff can change over time due to, e.g., weather conditions or seasonal fluctuations resulting in multiple payoff matrices.

3.1 Constrained Game

Consider the game with multiple payoff matrices $M^{(k)}$, $k = 1, ..., N_M$, of size $N_R \times N_C$. Let $\mu^{(k)}$ be the probability that the payoff matrix is $M^{(k)}$, with $\sum_{k=1}^{N_M} \mu^{(k)} = 1$. Moreover let $q^{(k)}$ and $p^{(k)}$ be strategies of the agent and the intruder when the payoff matrix is $M^{(k)}$. The value of the game is the expected payoff per day and can be found by solving the following optimization problem:

$$
\begin{aligned}
V = \min_{p} \max_{q} \quad & \sum_{k=1}^{N_M} \mu^{(k)} (p^{(k)})^T M^{(k)} q^{(k)} \\
\text{s.t.} \quad & f(p) \geq 0, \\
& \sum_{i=1}^{N_R} p_i^{(k)} = 1, \quad \sum_{i=1}^{N_C} q_i^{(k)} = 1, \quad k = 1, ..., N_M, \\
& p, q \geq 0,
\end{aligned}
\tag{10}
$$

where $p^T = (p^{(1)}, ..., p^{(N_M)})$ and $q^T = (q^{(1)}, ..., q^{(N_M)})$. In the next section, we discuss the constraint $f(p) \geq 0$ if multiple payoff matrices are considered.

3.2 Conditions for Games with Multiple Payoff Matrices

The conditions on the minimal number of visits for all cells during the planning period can be constructed following the same reasoning as in Sect. 2. Now, the number of visits for cell j is the sum of the number of visits for cell j for each payoff matrix. Let $X_j^{(k)}$, $j = 1, ..., N_C$, $k = 1, ..N_M$, be the random variable describing the number of visits to cell j when the payoff matrix is M_k and let

$\tilde{X}_j^{(k)}$ be the approximation of $X_j^{(k)}$. $N_D^{(k)}$ is the number of periods that the payoff matrix is $M^{(k)}$. We are interested in the following probability:

$$P(\tilde{X}_1^{(1)} + ... + \tilde{X}_1^{(N_M)} \geq v_1, ..., \tilde{X}_{N_C}^{(1)} + ... + \tilde{X}_{N_C}^{(N_M)} \geq v_{N_C}),$$

with $E(\tilde{X}_j^{(k)})$, $Var(\tilde{X}_j^{(k)})$, and $Cov(\tilde{X}_j^{(k)})$ calculated as in Sect. 2.3 with $N_D^{(k)}$ and $p^{(k)}$. Since $\tilde{X}_j^{(k)}$ and $\tilde{X}_{j'}^{(k)}$ are independent if $j \neq j'$, we have:

$$E(\tilde{X}_j) = \sum_{k=1}^{N_M} N_D^{(k)} a_j p^{(k)}, \qquad Var(\tilde{X}_j) = \sum_{k=1}^{N_M} N_D^{(k)} a_j p^{(k)} (1 - a_j p^{(k)}),$$

$$Cov(\tilde{X}_j, \tilde{X}_{j'}) = \sum_{k=1}^{N_M} \sum_{k'=1}^{N_M} Cov(X_j^{(k)}, X_{j'}^{(k')}).$$

To make sure that the conditions are met with high probability we define,

$$f(p) = P(\tilde{X}_1 \geq v_1, ..., \tilde{X}_{N_C} \geq v_{N_C}) - (1 - \epsilon),$$

where $P(\tilde{X}_1 \geq v_1, ..., \tilde{X}_{N_C} \geq v_{N_C})$ equals (6). Similarly as in Sect. 2.3, a lower bound for this probability is given in (8). Taking the dual of the inner LP of (10) and using this lower bound, optimal strategies for the agent and the intruder can be found by solving:

$$V_U = \min_{p,z} \sum_{k=1}^{N_M} z^{(k)}$$

$$\text{s.t.} \quad e^T z^{(k)} \geq \mu^{(k)} (p^{(k)})^T M^{(k)}, \quad k = 1, ..., N_M,$$

$$\sum_{j=1}^{N_C} \Phi \left(\frac{v_j - \sum_{k=1}^{N_M} N_D^{(k)} a_j p^{(k)}}{\sqrt{\sum_{k=1}^{N_M} N_D^{(k)} a_j p^{(k)} (1 - a_j p^{(k)})}} \right) \geq \epsilon, \qquad (11)$$

$$\sum_{i=1}^{N_R} p_i^{(k)} = 1, \quad k = 1, ..., N_M,$$

$$p \geq 0,$$

where $z = (z^{(1)}, ..., z^{(N_M)})$. In the next section, we will illustrate this model.

4 Results

In this section, we give computational results and examples to illustrate our models. In Sect. 4.1, we investigate the approximation error introduced in Sect. 2.3. Thereafter, we give two examples to illustrate our model in Sect. 4.2.

4.1   Computational Results

This section investigates the error introduced by using the lower bound in (8). Solving (3) with $f(p)$ given in (7) numerically is computationally intractable for networks with more than two or three routes and cells. Therefore, we have compared the relative difference between the lower and upper bounds of V, see Lemma 1. We have randomly generated 100 payoff matrices, conditions and routes for different network sizes. Table 1 shows the average relative difference between the upper and lower bound with 95%-confidence interval between brackets. The last columns gives the average running time in seconds for (9). The results are implemented in Matlab version R2016b [9] on an Intel(R) Core(TM) i7 CPU, 2.4 GHz, 8 GB of RAM. As the results in Table 1 show, (9) gives a good approximation of the game value V and can be solved in reasonable time. The size of more realistic examples, as encountered in the patrolling against illegal fishing context, is comparable to the size of these randomly generated instances.

Table 1. Average relative difference of upper bound V_U and lower bound V_L ($\epsilon = 0.05$).

# Cells	# Routes	Error	Running time
10	5	0.8% (± 1.0%)	0.217 s
20	15	1.9% (± 1.8%)	0.347 s
30	25	2.2% (± 1.4%)	0.819 s

4.2   Illustrative Examples

This section presents some examples to illustrate the models described in this paper. The results in this section are obtained by implementation of (9) and (11). Consider an area with 12 cells and 9 routes. The routes are chosen such that the cells are evenly spread over all routes, see Table 2. Suppose $N_M = 2$ and the payoff matrices are constructed using (1), where $d_j = 0.9$, $j = 1, ..., N_C$ and $g^{(k)}$ is the intruder's gain. Figure 1 depicts payoff matrices $M^{(1)}$, $M^{(2)}$ and two example routes, Routes 1 and 8. The white cells have a gain of 1, the light gray cells have a gain of 2 and the dark gray cell have a gain of 3.

Constant Payoff Matrix. Consider the games with payoff matrices $M^{(1)}$ and $M^{(2)}$ separately. Suppose that the planning period for both payoff matrices is $N_D = 100$. Table 3 shows the game values for different conditions. For example, a condition of 0.1 means that the minimum number of visits equals 10. The second and the third column give the game value of both games for the conditions specified in the first column. The first row shows the value of the game without conditions on the number of visits to the cells, the second row considers the game in which all nodes must be visited at least 10 times, and the third row

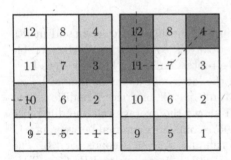

Fig. 1. Payoff matrices $M^{(1)}$, Route 1 (left) and $M^{(2)}$, Route 8 (right).

Table 2. Possible routes.

Routes	Cells visited by route
1	1, 5, 9, 10
2	2, 3, 8, 12
3	3, 7, 6, 10
4	4, 7, 6, 9
5	1, 2, 3, 4
6	3, 4, 7, 12
7	2, 5, 6, 9
8	4, 7, 11, 12
9	2, 5, 10, 11

considers the game in which Nodes 1-4 must be visited at least 30 times and the other nodes at least 10 times.

Table 3 indicates that the game value increases if more conditions are imposed on the agent's strategy. However, the increase of the game value depends on the payoff matrix. For example, the extra condition on Nodes 1–4 does not increase the game value for payoff matrix $M^{(1)}$, since the intruder's gain for these nodes is high and the agent is already patrolling these cells more often, as the results below indicate.

Table 3. Expected payoff per day for different conditions ($\epsilon = 0.05$).

Conditions (fraction)	Payoff $M^{(1)}$	Payoff $M^{(2)}$	Average	Combined
None	1.10	1.58	1.34	1.34
All nodes: 0.1	1.23	1.64	1.44	1.34
Nodes 1–4: 0.3, Nodes 5–12: 0.1	1.23	2.14	1.69	1.35

Figure 2 displays the agent's strategy for the different payoff matrices without conditions. The color of each cell is determined by the gain of the intruder and the number within each cell shows the fraction of the time period that the cell

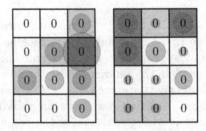

Fig. 2. Agent's strategy for the game without conditions.

should be visited. The agent's strategy is shown by the circles in each cell. The probability that a cell is visited is proportional to the radius of the circle in that specific cell. For example in Fig. 2, the probability that cell 3 is visited equals 1 for $M^{(1)}$ and 0.24 for $M^{(2)}$. Figure 3 displays the agent's strategy when conditions as given in Table 3 are considered. For all cases, it is clear that cells with a high gain for the intruder are visited more often.

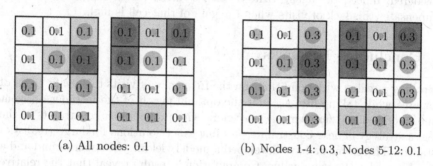

(a) All nodes: 0.1   (b) Nodes 1-4: 0.3, Nodes 5-12: 0.1

Fig. 3. Agent's strategy for different conditions.

Multiple Payoff Matrices. The previous example considers the game with a constant payoff matrix such that for each game the conditions on the minimum number of visits have to be met. Now, we consider multiple payoff matrices simultaneously. Suppose that the total planning period $N_D = 200$ and both payoff matrices $M^{(1)}$ and $M^{(2)}$ have equal probability, so $\mu^{(1)} = \mu^{(2)} = 0.5$. Again, routes and conditions are given in Tables 2 and 3. A condition of 0.1 means that the total number of visits is 20, but it is, for example, allowed that there are only 5 visits when the payoff matrix is $M^{(1)}$ and 15 when the payoff matrix is $M^{(2)}$. This is the benefit of playing the game repeatedly and considering multiple payoff matrices simultaneously. In the last column of Table 3 the value of the game in which the conditions are combined for multiple payoff matrices is shown. If there are no conditions on the number of visits to the cells, the game value is just the average of both games with constant payoff, which is shown

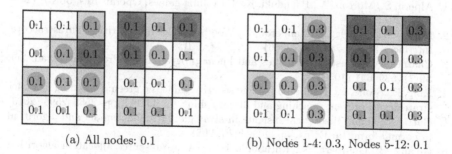

(a) All nodes: 0.1   (b) Nodes 1-4: 0.3, Nodes 5-12: 0.1

Fig. 4. Agent's strategy if multiple payoff matrices are considered simultaneously.

in the second last column of Table 3. However, when conditions are considered, the value of the combined game is lower than the average of both games with constant payoff, because the agent has more flexibility in meeting the conditions.

Figure 4 shows the agent's strategy for the combined game with conditions given in Table 3. Comparing the results with those in Fig. 3 reveals that the agent has more flexibility in meeting the constraints when multiple payoff matrices are considered. Indeed the agent visits a cell less often when the gain is low and compensates this lack of visits when the gain of that cell is high.

5   Concluding Remarks

Patrolling a region with conditions on the frequency of visits to specific parts of that area while taking into account the optimal payoff of the intruder or agent can be modeled as a zero-sum security game with probabilistic constraints on the agent's strategy. These constraints prohibit exact solutions for large (realistic) instances. Therefore, we have developed a model yielding an upper bound and a lower bound for the game value. Computational results reveal that the relative difference between the upper and lower bound for the instances considered is less than 2.5% and that instances of realistic size can be solved within seconds.

In practice, the agent's strategy is constrained by existing guidelines. Numerical examples show that as the number of conditions increases, the agent's loss will increase. However, if multiple payoff matrices are considered, the agent has more flexibility in meeting the conditions and the loss of the agent is reduced.

In this paper, we have assumed that only one intruder is present in the area, that the payoff of intruders is known and that the agent decides on a strategy in advance. For future research, it would be interesting to investigate the case where not all payoff matrices are known in advance and multiple intruders attack simultaneously. Also, considering a more dynamic strategy of the agent, for example by taking into account extra information about the payoff and cells that already have been visited, should be pursued.

References

1. Alpern, S., Morton, A., Papadaki, K.: Patrolling games. Operat. Res. **59**(5), 1246–1257 (2011)
2. Brown, G., Carlyle, M., Salmeron, J., Wood, K.: Defending critical infrastructure. Interfaces **36**(6), 530–544 (2006)
3. Charnes, A.: Constrained games and linear programming. Proc. Nat. Acad. Sci. **39**(7), 639–641 (1953)
4. Fang, F., Stone, P., Tambe, M.: When security games go green: Designing defender strategies to prevent poaching and illegal fishing. In: IJCAI, pp. 2589–2595 (2015)
5. Gatti, N.: Game theoretical insights in strategic patrolling: model and algorithm in normal-form. In: ECAI, pp. 403–407 (2008)
6. Golany, B., Goldberg, N., Rothblum, U.G.: A two-resource allocation algorithm with an application to large-scale zero-sum defensive games. Comput. Oper. Res. **78**, 218–229 (2017)

7. Haskell, W., Kar, D., Fang, F., Tambe, M., Cheung, S., Denicola, E.: Robust protection of fisheries with compass. In: Twenty-Sixth IAAI Conference (2014)
8. Lin, K.Y., Atkinson, M.P., Chung, T.H., Glazebrook, K.D.: A graph patrol problem with random attack times. Oper. Res. **61**(3), 694–710 (2013)
9. MATLAB. version 9.1 (R2016b). The MathWorks Inc., Natick, Massachusetts (2016)
10. Meng, F., Zhan, J.: Two methods for solving constrained bi-matrix games. Open Cybern. Syst. J. **8**, 1038–1041 (2014)
11. Owen, G.: Game Theory, 3rd edn. Academic Press, London (1995)
12. Peters, H.: Game Theory: A Multi-leveled Approach, 1st edn. Springer, Heidelberg (2008)
13. Ross, S.: Stochastic Processes. John Wiley & Sons Inc, New York (1996)
14. Semple, J.: Constrained games for evaluating organizational performance. Eur. J. Oper. Res. **96**(1), 103–112 (1997)
15. Washburn, A., Lee, E.L.T.C.: Allocation of clearance assets in IED warfare. Nav. Res. Logistics (NRL) **58**(3), 180–187 (2011)
16. Winston, W.L.: Operation Research, Applications and Algorithms. Brooks/Cole, Belmont (2004)
17. Wood, K.R.: Bilevel network interdiction models: formulations and solutions. In: Wiley Encyclopedia of Operations Research and Management Science (2011)

On the Cost of Game Playing: How to Control the Expenses in Mixed Strategies

Stefan Rass[1]([⊠]), Sandra König[2], and Stefan Schauer[2]

[1] Institute of Applied Informatics, System Security Group,
Universitaet Klagenfurt, Klagenfurt, Austria
stefan.rass@aau.at
[2] Center for Digital Safety and Security,
Austrian Institute of Technology, Vienna, Austria
{sandra.koenig,stefan.schauer}@ait.ac.at

Abstract. Game theory typically assumes rational behavior of the players when looking for optimal solutions. Still in case of a mixed equilibrium, it allows players to choose any strategy from the mix in each repetition of the game as long as the optimal frequencies are met in the long run. Which strategy is chosen in a specific round may not be purely random but also depend on what strategy has just been played.

In many cases, playing a particular strategy is tied to cost or efforts. For instance, adding a new defensive strategy (e.g., applying a new virus scanner) requires some investment (implementation cost), but playing the strategy may incur some efforts as well (playing cost: a virus scan takes time and consumes resources, so too frequent scanning appears undesirable). If a security system successfully repels an attack, the attacker is most likely coming back using a different attack vector. Thus we here study repeated games in order to respond to changing attacks.

The effort to play a strategy may be quite dependent on what has been played before, and the switch from the last strategy to the new one, in the next instance (repetition) of the game, may come at what we call a *switching cost*. These can create an incentive to *not* play a certain mixed strategy. In cases when there equilibrium is unique, a player may have an incentive to nonetheless deviate from it to save costs, and thus gain more (only in a different way). So, the strategy plan should depend on the equilibrium and the (switching) cost for playing it.

The matter is essentially more complex than only asking for how to play a mixed strategy most efficiently; instead, we need to incorporate the switching costs into the game as an additional goal to be optimized. Those costs are indeed dependent on the equilibrium of the game itself. Thus, the usual dependency of the equilibrium on the payoffs is herein augmented by the converse dependency of the payoffs on the equilibrium. To handle this circular dependency, we will apply a generalized game-theoretic model that allows payoffs to be random variables (rather than real numbers). We show how to solve this new form of game and illustrate the method with an example.

Keywords: Game theory · Security · Security economics · Equilibrium

© Springer International Publishing AG 2017
S. Rass et al. (Eds.): GameSec 2017, LNCS 10575, pp. 494–505, 2017.
DOI: 10.1007/978-3-319-68711-7_26

1 Introduction and Motivation

Playing a repeated game usually assumes that strategies can be changed without efforts or costs between instances of the same game. While this is certainly true for many classical games (like board games, the battle of sexes, etc.), playing games in security is different, since the players will enforce themselves to repeat their efforts and change their strategies over time.

Suppose a game has a Nash equilibrium in pure strategies. Then both players can straightforwardly implement their individually optimal action. The implementation cost arises only once, and since there is no change needed (the equilibrium is pure), there is no cost to switch to a new strategy. Things are thus trivial in this case, since the defense is static.

However, what happens if the game has only equilibria in mixed strategies (such as can be expected for the example to follow in Sect. 1.2)? In that case, a player is forced to change his/her behavior over repetitions of the game, where switching from one pure-strategy played in the last round to the new strategy in the next round of the game is tied to some cost. For instance, a security officer would certainly not like to reconfigure a file server or firewall every day in order to play a mixed strategy, just because the equilibrium prescribes it.

1.1 Efficiently Implementing the Equilibrium Is Not Enough

The problem is essentially *not* as simple as preferring "more pure" strategies (in the sense of entailing less frequent action changes) over more uniformly mixed ones. That would be easy by computing multiple Nash equilibria and going for the one with smallest Shannon entropy (zero entropy would correspond to a pure strategy). From a security perspective, the opponent (in this case player 2) could attempt to forecast player 1's moves, based on the hypothesis that this player seeks to minimize the costs for its next choice. That is, if a player would implement an equilibrium mixed strategy by switching such that the costs for playing the equilibrium strategy is minimized, then the opponent can gain quite some idea what the next move of the defender will be. The problem is most obvious for pure strategies, where the defense is static and the adversary can aim at a non-moving target. However, also pseudorandomly chosen defense actions can be an advantage for the attacker, say, if two defensive actions are chosen alternatingly to play an equilibrium $(0.5, 0.5)$ over two defensive strategies. In both cases, the attacker, knowing the current action a_t at time t, can form a hypothesis $\Pr(a_{t+1}|a_t)$ about the next action a_{t+1} of the defender. This hypothesis can, for example, be based on considering which moves a_{t+1} are easier/cheaper to make from a_t than others. If the current action has been a_t, and action a_1 is much cheaper to do next than another action a_2, then the attacker may predict the defender's random next move A_{t+1} to be a_1 rather than a_2. Note that the defender, in any case, will follow an equilibrium distribution, so the unconditional likelihood to play a_1 or a_2 is as the equilibrium prescribes, but the *conditional* likelihoods $\Pr(A_{t+1} = a_1|a_t)$ and $\Pr(A_{t+1} = a_2|a_t)$ for the next action A_{t+1} can be quite different.

Here, we seek to prevent predictions of the defender's moves based on costs incurred by the current state of the system under protection. That is, the defender plays a repeated game, in which the costs to take actions depend on what has happened in the past instances of the (same) game. Ideally, actions should be taken stochastically independent though still w.r.t. cost minimization. Technically, if a_{n+1} depends on a_n, then the game is not really static any more. Neither does it call for a treatment like a dynamic game, as we will show below. Our main goal is thus *not* on finding the "cheapest to play" equilibria among those that exist, but rather to adapt the equilibrium so that it can be played in the usual way while at the same time coming in as cheap as possible.

1.2   A Short Example

For a more detailed example, consider a game-theoretically optimized physical surveillance setting: there is an enterprise building with rooms $R_1, \ldots, R_n$ to be visited by a security guard repeatedly at random (cf. [7] for a similar scenario). In a game-theoretic model, let a visit to the i-th room R_i corresponds to the i-th pure strategy in the action set of player 1 (the security guard). It may be good to check one spot R_j, and soon after this check the room R_j again, simply because the adversary may not expect this (since the adversary may have been hiding during the first visit, and could be caught upon the unexpected revisit of the security guard). Also, player 1 cannot leave any room unvisited at all, and neither would it be effective to check only one fixed location, say R_1, all the time; as would correspond to a pure strategy in the surveillance game, which cannot be an equilibrium. Since all locations of the premise under surveillance need to be checked anyway, an equilibrium would expectedly exist only in mixed strategies. However, it would be inefficient to send the security guard through a sequence of rooms whose distance is maximal. That is, we would surely prefer checking nearby rooms at once, leaving far away rooms for later visitations. Also, rooms may have different importance levels for the attacker (e.g., induced by different security clearances in the enterprise), and the frequency at which rooms are to be checked may differ. This takes us away from an optimal strategy being a humble shortest round trip route (like in a traveling salesperson instance), since the security guard has to visit all the rooms with prescribed (equilibrium/optimal) frequencies. The attacker, in turn, seeks entry to rooms with highest importance (e.g., damage potential).

The game play is repeated, since the defender (player 1) needs to repeatedly check all rooms at random. Moreover, the visitations should not all be in the same sequence all the time (for this would let the attacker hide easily), so the defender needs to change the strategy in each round of the game, but the cost of playing strategy "visit room R_i" depends on the distance between the current and the next room to search. Thus, playing a mixed strategy equilibrium induces costs not when the strategy is played but mostly when the strategy is *changed*. If the attacker knows that room R_i has been checked last, then rooms in the proximity of R_i are much more likely candidates to be visited next than far remote locations. This hypothesis is a *conditional* probability over defense actions, and can give the attacker an advantage, although the *unconditional* probability to check the rooms is optimal.

Taking costs into account when playing mixed strategies turns out to be slightly tricky: since the current action has been randomly sampled from the equilibrium distribution, the cost for the next randomly chosen action is itself a random variable, whose distribution depends on the equilibrium. Conversely, if cost matters for optimization (e.g., if the overall round trip time for the security guard should also be kept low for efficiency), a mixed equilibrium should also depend on the cost for switching the strategies. This circular dependency of equilibria and costs is studied hereafter.

1.3   Related Work

It turns out that the last point induces theoretical as well as practical difficulties, at least in the standard instances of game theory: the switch from one strategy to another in a repeated game is, in the *usual* game theoretic setting, neglected (assumed with zero cost). Including this in a game theoretic model makes the game sequential or stochastic, so that the next payoff (structure) depends on the past game's instance. This in turn complicates matters of modeling and solving the games, and recent work of [12] approached the issue using Kullback-Leibler (KL) divergence as a measure of "mixedness" of strategies to favor pure strategies over randomized ones. More specifically, the KL divergence between two mixed strategies is taken as a measure for the switching cost when the system is defended with a new strategy as time goes by. We may bear in mind this idea when we approach the problem subsequently in our distribution-valued framework, but will allow for arbitrary costs to be used.

The issue of predictability of players moves, which could be based on known costs, has been examined in [2], where predictive adjustments of strategy choices can even lead players away from an equilibrium. This is an independent theoretical reason to avoid such information in a general gameplay. Deviations from the predicted optimal behavior by game theory are well known and have been frequently reported (and empirically verified) [1]. Parts of these deviations may root in complex mental processes, and one possibility is a player's tendency to stick with what has worked well in the past. That is, there is an element of "lazyness" or inertia in how players behave, which we can capture explicitly by considering costs. In zero sum games, the saddle point value is provably independent of the chosen equilibrium. This suggests that players are "indifferent" between each of the pure strategies – an assumption that has been questioned in past literature [3]. Dropping this indifference hypothesis means taking some pure strategies as more attractive than others; though not uniformly so, but instead dependent on what has been played recently. This leads to a cost-based model again. Somewhat looser related is also the question (and work) about equilibria and welfare optimality [5]; precisely asking for which equilibrium should be chosen to the good of the community. In a purely non-cooperative setting like security, the same question of welfare optimality can be imposed by a player on itself, which is yet another form of playing cost-optimally.

In security in particular, some attack vectors aim at keeping the defender busy in order to take away defensive resources from the actual location of the real attack. A denial-of-service (DoS) attack can be seen as a simple version of that, although the incentive of a DoS can be pure distraction only. In assigning costs to certain randomized strategies, we can penalize playing a strategy too often, which corresponds to the aforementioned decoy measures by the attacker. The penalty then refers to a switch from one strategy a to the same strategy a in the next round and diversionary maneuvers in security are one example where pure strategies can be more expensive to play than mixed strategies.

While the problem of accounting for switching costs in playing the equilibrium yields to quite technical optimization problems and sequential games in the standard setting, the approach discussed in this work follows entirely classical routes, but uses the framework of distribution-valued games [9] as a purely technical vehicle to this end.

1.4   Structure of the Article and Our Contribution

In the following Sect. 2, we will introduce some preliminary concepts and notations, which are required to formally describe the costs of switching strategies in Sect. 3. The application of the theoretical framework developed in Sect. 3 is then conceptually explained using a short numerical example to conclude Sect. 3. Finally, Sect. 4 summarizes the findings and highlights some open issues and discussions.

The main contribution of this work is extending the game theoretic model to a more diverse understanding of payoffs: a game rewards players for what strategies they have chosen, but in real life, a change of behavior also affects the payoffs. This dimension is typically not considered in game theoretic models, and studied explicitly here (similar to dynamic systems, whose trajectories depend on the current state, but also the speed of motion by the first order derivative).

2   Preliminaries

Vectors are printed in bold-face, and random variables and sets appear as uppercase letters in normal font. If a random variable X has distribution F, we write $X \sim F$. Distributions on finite and ordered sets are described by a vector $\boldsymbol{x}$ of probabilities to take each of the elements. The notation $d \stackrel{\boldsymbol{x}}{\leftarrow} PS$ means a random draw of an element d from the set PS, with distribution $\boldsymbol{x}$. That is, if d is the i-th element in the (canonic) ordering of PS, then $\Pr(d) = x_i$, being the i-th coordinate of the vector $\boldsymbol{x}$ describing the (categorical) distribution over PS. Hereafter, we will treat $\boldsymbol{x}$ as a probability vector (having all nonnegative elements that sum up to 1). It will be synonymously called a distribution, when the respective support PS is clear from the context.

For vectors, the relation $\boldsymbol{x} \leq \boldsymbol{y}$ is understood as $x_i \leq y_i$ for all $i = 1, 2, \ldots, n$, when $\boldsymbol{x} = (x_1, \ldots, x_n)$ and $\boldsymbol{y} = (y_1, \ldots, y_n)$. The strict $<$-relation is defined alike. The complement relation is $\boldsymbol{x} \leq_1 \boldsymbol{y}$, meaning that there is a coordinate i for which $x_i \leq_1 y_i$, irrespectively of the values of the other coordinates.

2.1   Game Setup

Without loss of generality, let a game be played between two players (the generalization to $n > 2$ players will be obvious), called 1 and 2, with associated pure strategy sets PS_1 and PS_2. Let us write $S(PS)$ to mean the simplex over a (strategy) set PS containing all probability distributions supported on PS. For a security game, we adopt the defenders perspective, acting as player 1 in the game, and assume the defender's moves to be tied to some cost upon a switch from strategy $i \in PS_1$ to strategy $j \in PS_1$. Assuming a zero-sum competition for a worst-case security analysis, the attacker (player 2) has the same payoff structure as the defender, only with opposite signs.

Let the defender be *minimizing* two objectives: one is expected damage (i.e., the security goal is risk minimization), the other is cost minimization for achieving the security goal (e.g., the cost for reconfigurations, malware scans, system reinstallments, etc.). The first objective is modeled by some function $u_1^{(1)}$. The second objective is denoted as $u_1^{(2)}$, and depends on the actions of both players. The treatment here is thus a two-objective game (in which the attacker maximizes the same objectives, as we have a zero-sum competition).

In the general case of a multi-objective game (MOG), we can allow for several security goals (e.g., availability, intrusion detection rates, or similar), besides the cost for playing a mixed strategy (as only one among the other goals to be optimized). In a MOG, each player i may have $d_i \geq 2$ goals described by d_i payoff functions $u_i^{(k)}$ for $k = 1, 2, \ldots, d_i$, defined over $S(PS_i) \times S(PS_{-i})$, where PS_{-i} is the joint strategy space of players other than i. For our two-player (zero-sum) security MOG, the defender (and hence also the attacker) have vector-valued payoff functions $\boldsymbol{u}_1, -\boldsymbol{u}_1 : S(PS_1) \times S(PS_2) \to \mathbb{R}$.

Let a two-player zero-sum MOG $\Gamma = (\{1,2\}, \{PS_1, PS_2\}, \{\boldsymbol{u}_1, -\boldsymbol{u}_1\})$ be given. A *Pareto-Nash equilibrium* (for a minimizing player 1) is a strategy profile $(\boldsymbol{x}^*, \boldsymbol{y}^*) \in S(PS_1) \times S(PS_2)$ for which

$$\boldsymbol{u}_1(\boldsymbol{x}, \boldsymbol{y}^*) \geq_1 \boldsymbol{u}_1(\boldsymbol{x}^*, \boldsymbol{y}^*) \geq_1 \boldsymbol{u}_1(\boldsymbol{x}^*, \boldsymbol{y}) \quad \forall \boldsymbol{x} \in S(PS_1), \boldsymbol{y} \in S(PS_2). \quad (1)$$

Since $\geq_1$ coincides with $\geq$ in the 1-dimensional case, (1) degenerates into a normal equilibrium for single-objective games, thus it appears as the natural solution concept for MOGs. The existence and computation of Pareto-Nash equilibria have been studied in [4], who gave a simple conversion of a MOG into a scalar game, which we will use hereafter. In a nutshell, the method is the following: for each player i, choose a vector $\boldsymbol{0} < \boldsymbol{\alpha}_i \in \mathbb{R}^{d_i}$ with $\|\boldsymbol{\alpha}_i\|_1 = 1$ and scalarize the payoffs for the i-th player as $\boldsymbol{\alpha}_i^T \cdot \boldsymbol{u}_i$. Then, it can be shown (see [4]) that a Nash equilibrium in the so-scalarized game is a Pareto-Nash equilibrium in the original game. We will apply this procedure below.

2.2   Stochastic Payoffs and Orders

Picking up on our introductory explanation, the cost to play the j-th strategy in the next repetition of the game depends on what we did recently. To ease derivations in the following, let us consider the cost to play strategy j as the random variable $S_{\to j}$, whose distribution depends on the equilibrium being played (and conversely, the cost $S_{\to j}$ will also have an impact on the equilibrium).

Minimizing the cost can be done either by minimizing some scalar quantity derived from $S_{\to j}$ (typically its average), or by minimizing the entire random variable. The latter is done using (total) stochastic orders [11]. For our purposes, it will be enough to bear in mind that such a suitable total order exists, denoted as $\preceq$ and studied in [8], and that game theory, to the extent relevant here, together will all necessary equilibrium concepts can equivalently be formulated in stochastic orders (just like as in $\leq$-order in $\mathbb{R}$) [9]. Specifically, we will let our payoffs be from the set $\mathcal{F}$ of distributions that are absolutely continuous w.r.t. the counting or Lebesgue measure, and have a compact support within $[1, \infty)$. This set can be totally ordered (as proven in [8]), by an efficiently decidable relation $\preceq$. It can be shown that real valued payoffs can be represented by distributions from $\mathcal{F}$ in an $\leq$-order-preserving manner [6], and that any two distributions in $\mathcal{F}$ can be represented by real numbers so that the $\preceq$-ordering on $\mathcal{F}$ corresponds to the $\leq$-ordering on $\mathbb{R}$ [9].

In the following, we will use distributions to model the game payoffs and costs, as a mere technical vehicle to derive the results. Thus, the technicalities and theory behind the stochastic order will be of no further interest. The only important fact used in the following is the payoffs for player 1 in our MOG being defined as the vector-valued mapping $\boldsymbol{u}_1 : S(PS_1) \times S(PS_2) \to \mathcal{F}^d$, where we will take $d = 2$ for simplicity. This corresponds to two goals, one of which is the primary security goal to be optimized, and the other goal (second coordinate function in $\boldsymbol{u}_1$) being the (random) cost for playing mixed strategies. Now, let us become specific on the latter.

3   Costs for Playing Mixed Strategies

Let us assume n and m strategies to be available to the defender and the attacker, respectively. Further, let S_{ij} be the cost incurred by switching to the next (pure) i-th strategy *from* the current (pure) j-th strategy for the *defender*. For mixed strategies, the last choice of both players is a random variable, so that the cost for switching to strategy j in the next round is itself a random variable (since we move to action j from a random starting point). Note that these costs are purely due to the moves of player 1, and the analogous costs of player 2 in switching its own behavior is irrelevant for player 1 here. Finally, let $F_{S_{ij}}$ be the respective distribution function of the costs of playing the i-th strategy when the j-th strategy has just been played.

The random variables S_{ij} for $i, j = 1, \ldots, n$ constitute the switching matrix $\boldsymbol{S} = (F_{S_{ij}}) \in \mathcal{F}^{n \times n}$, with $F_{S_{ij}}(x)$ being the respective distribution function telling the probability to pay a cost $S_{ij} \leq x$ for any given cost x. Consider any stage of the game, where the attacker is playing a random move according to a mixed strategy $\boldsymbol{x} \in S(PS_1)$, where $PS_1 = \{1, \ldots, n\}$. Let us write $S_{\to i}$ for the random cost arising from a switch to the (random) strategy $d_i \xleftarrow{\boldsymbol{x}} PS_1$, from any currently played random strategy $d_j \in PS_1$.

Using the law of total probability, and letting d_j be the current action, the distribution of $S_{\to i}$ is given by

$$\Pr(S_{\to i} \le c) = \sum_{j=1}^{n} \Pr(S_{\to i} \le c|d_j) \cdot \Pr(d_j).$$

Now, observe that $\Pr(S_{\to i} \le c|d_j) = F_{S_{ij}}(c)$, since this is the distribution of the cost when switching to d_i from strategy d_j. Moreover, $\Pr(d_j) = x_j$, where $\boldsymbol{x} = (x_1, \ldots, x_n) \in S(PS_1)$ is the current mixed strategy. Consequently, the vector of switching costs for all strategies can be written as

$$\begin{pmatrix} F_{S_{\to 1}} \\ \vdots \\ F_{S_{\to n}} \end{pmatrix} = \left(\sum_{j=1}^{n} F_{S_{ij}} \cdot x_j \right)_{i=1}^{n} = \boldsymbol{S} \cdot \boldsymbol{x}.$$

This payoff is independent of the attacker's behavior $\boldsymbol{y}$ (at most it can be indirectly be influenced by the attacker acting towards forcing the defender to frequent strategy changes), and the overall cost for player 1 is then determined by the randomized strategy $\boldsymbol{x}$ in the usual way (by using the law of total probability once more), thus giving a cost of

$$u_{switch}(\boldsymbol{x}, \boldsymbol{y}) = u_1^{(2)}(\boldsymbol{x}, \boldsymbol{y}) = \boldsymbol{x}^T \boldsymbol{S} \boldsymbol{x}.$$

This costs add another security goal to the defender's losses. Collecting the primary goal $u_1^{(1)}$ of player 1 with assigned priority $0 < \alpha < 1$ as a term $\alpha \cdot u_1^{(1)}$, the payoff for the defender, now accounting for the cost of switching strategies, becomes

$$u_1(\boldsymbol{x}, \boldsymbol{y}) = (1 - \alpha)u_{switch}(\boldsymbol{x}, \boldsymbol{y}) + \alpha u_1^{(1)}(\boldsymbol{x}, \boldsymbol{y}), \tag{2}$$

where $u_1^{(1)}(\boldsymbol{x}, \boldsymbol{y}) = \boldsymbol{x}^T \boldsymbol{A} \boldsymbol{y}$ has the usual form for matrix games, with the payoff matrix $\boldsymbol{A} \in \mathcal{F}^{n \times m}$ of all loss distributions. Unfortunately, fictitious play loses its guarantee of convergence here, since an inspection of the classical convergence proof in [10] reveals that this relies on the linearity of the payoff functional. However, even despite the strategies are still finitely many, our setting nonetheless does not admit the usual bilinear payoff functional, since the strategy of the defender must be chosen as a best reply not only to the attacker's behavior, but also to itself. This "self-dependence" induces the quadratic term u_{switch}, which in turn renders fictitious play seemingly inapplicable here (unless its convergence is reproved under the new setting, which is a separate question of research that we leave unanswered in this work).

However, a standard analysis can be done, only bearing in mind that the random payoffs are stochastically ordered, so that minima and maxima are w.r.t. $\preceq$: The problem is finding the saddle point $\min_{\boldsymbol{x} \in S(PS_1)} \max_{\boldsymbol{y} \in S(PS_2)} u_1(\boldsymbol{x}, \boldsymbol{y}) = \max_{\boldsymbol{y} \in S(PS_2)} \min_{\boldsymbol{x} \in S(PS_1)} u_1(\boldsymbol{x}, \boldsymbol{y})$ by virtue of Nash's theorem (in the usual version), since u_1 is continuous and the game is finite [6]. Since the switching efforts are relevant for the defender (only), suppose its chosen (mixed) strategy is x. Then, the attacker aims to act towards

$$\max_{y \in S(PS_2)} (x^T A y + x^T S x),$$

in which $x^T S x$ is a constant term (the distribution of the random switching cost). Letting $e_i \in \mathbb{R}^m$ be the vector of all zeros except the i-th element being 1, then $\text{argmax}_{y \in S(PS_2)}(x^T A y) = \text{argmax}_i(x^T A e_i)$. Substituting $v := \max_i(x^T A e_i + x^T S x)$, the resulting problem becomes $\min v$ subject to $v \succeq x^T A e_i + x^T S x$ for $i = 1, 2, \ldots, m$. Thus, we arrive almost at the well-known form of the optimization problem to solve zero-sum games; only with the exception that it is nonlinear now:

$$\left. \begin{array}{rl} & v \to \min \\ \text{subject to} & v \succeq \alpha \cdot x^T S x + (1 - \alpha) x^T A e_i \quad \text{for } i = 1, \ldots, m; \\ & \sum_{j=1}^n x_j = 1; \\ & x_i \geq 0, \quad \text{for } i = 1, \ldots, n. \end{array} \right\} \quad (3)$$

For a practical solution, there is no need to deal with the technicalities of the stochastic orders or distributions, since the $\preceq$-order chosen here can be decided equivalently by representing the distributions by some real-valued representatives (as outlined in Sect. 2.2 and rigorously dealt with in [9, Theorem 15]). In "converting" all distributions (v, and the matrices A, S) to real values, we can replace the $\succeq$ relation by the humble $\geq$ relation and approach (3) as any standard constrained nonlinear optimization problem.

In its real-valued form, the problem may not be convex, since the matrix S, in its numeric representation, may not be positive (semi-)definite. Therefore, to ease optimization matters, it may be advisable to solve the Lagrangian dual program instead, which is in any case concave.

Remark 1. The cost for *playing* a strategy can, as well, be put into the diagonal of S, which then means the cost of switching from i to itself. Typically this value can be regarded as zero, but putting a random cost here makes a pure strategy come at exactly this cost, since $e_i^T A e_i = F_{ii}$ in the quadratic cost term. Thus, the diagonal entries cover the special cases of costs arising from *playing* a strategy, while the off-diagonal elements are costs of *changing* a strategy.

Summarizing our findings, we can state the following result:

Theorem 1. *Let $A = (F_{ij})_{i,j=1}^{n,m}$ model a finite and repeated two-player game (where player 1 has n and player 2 has m strategies), and let $S = (F_{S_{ij}})_{i,j=1}^{n,m}$ be a matrix of cost (distributions), where $S_{ij} \sim F_{S_{ij}}$ is the cost (distribution) of player 1 when switching to strategy i from the previously played strategy j.*

Then, for any $0 < \alpha < 1$, the optimum of (3) is a Pareto-Nash equilibrium in the MOG in which player 1 has two goals, modeled by A and S, opposing player 2 who engages in a zero-sum competition per goal of player 1.

A few remarks appear in order regarding Theorem 1:

- It remains valid with the only change of $\succeq$ into $\geq$, when the payoff matrices are defined over $\mathbb{R}$ instead of probability distribution space $\mathcal{F}$. This change amounts to defining all distributions as degenerate (Dirac masses), but casts things back to the more familiar setting of classical games.
- The cases $\alpha = 0$ and $\alpha = 1$ are permitted, though (3) degenerates to either a standard matrix game (linear program), or to triviality when nothing but switching cost counts to that all pure strategies are equally optimal (since no switching means no cost).
- It straightforwardly extends to MOGs with d goals so that the switching cost becomes the $(d+1)$-th goal. In that case, α will be a vector of nonzero weights assigned to all goals, only one of them being the switching cost (according to the method of [4]).

We shall illustrate the technique using a standard setup with payoffs and costs being real values. In the framework of distribution-valued games [6], this simply corresponds to using degenerate distributions.

3.1 Example 1

Let A and S be matrices of payoffs and switching costs and let player 1 be minimizing.

To compare the results of to the usual setting (change of strategy goes for free) to the one with switching costs, let us start with the simpler case first. The payoff and cost matrices are given by

$$A = \begin{pmatrix} 4 & 8 & 6 \\ 9 & 3 & 8 \\ 7 & 6 & 3 \end{pmatrix}, \qquad \text{and} \qquad S = \begin{pmatrix} 0 & 1 & 1 \\ 2 & 0 & 1 \\ 1 & 3 & 0 \end{pmatrix}$$

Analyzing the game in the usual way, e.g., by solving (3) with $\alpha = 0$ to "deactivate" the switching cost matrix S, yields $v_0 = val(A) \approx 6.0889$ at an equilibrium strategy $x_0^* \approx (0.511, 0.311, 0.178)$.

Instantiating and solving (3) with $\alpha = 0.5$, corresponding to equal priority on game payoffs (losses) and switching cost, gives $v \approx 3.36$ at the new equilibrium $x^* \approx (0.2, 0, 0.8)$. Doubling v to cancel out the equal weights $\alpha = 1/2$ for both goals, gives the total sum cost in the game, coming to $2 \cdot v = u_1(x^*, y^*) + u_{switch}(x^*, y^*) \approx 6.72$. Naturally, this loss is higher than in the conventional game without switching, so let us look how costly both equilibria are in playing them:

$$(x_0^*)^T S x_0^* \approx 0.88 \qquad \text{and} \qquad (x^*)^T S x^* \approx 0.32.$$

Not surprisingly, x^* is cheaper to play than x_0^*. However, the costs of 0.88 have to be added to the loss $v_0 = 6.0889$, making the total payoff $v_0 + 0.88 \approx 6.9689$. So, a disregard of the switching costs comes at additional cost and is suboptimal compared to the "more informed" approach. This makes the account worthwhile.

On the contrary, the necessary tradeoff to optimize multiple goals can be quite big, as the next example shows.

3.2   Example 2

Consider slightly different payoff structure (taking the same switching cost matrix S):

$$A = \begin{pmatrix} 4 & 9 & 7 \\ 8 & 3 & 6 \\ 6 & 8 & 3 \end{pmatrix}$$

Disregarding switching costs, we get $v_0 = val(A) \approx 6.08889$ attained at $x_0^* \approx (0.422, 0.467, 0.111)$ and coming in at cost $(x_0^*)^T S x_0^* \approx 0.892346$. Taking the switching cost into account gives $v \approx 3.69388$, playable at cost $(x^*)^T S x^* \approx 0.816326$ (cheaper, as expected), but being more expensive in total: we have $2 \cdot v \approx 7.38776$, as opposed to $v_0 + 0.986173 \approx 6.98123$. This is a case where the additional cost from the deviation from x_0^* is higher than the cost reduction from playing the cheaper equilibrium x^*. This is intrinsically attributable to the Pareto-Nash equilibrium tradeoff, and becomes effective due to the low cost values in S. If those costs would increase, say, if we redo the calculation with $2 \cdot S$, then the cost-optimized equilibrium becomes profitable again. This means that the magnitude of the costs modeled by S should bear a meaningful relation to the magnitude of the other goals in the MOG, when switching cost is an issue.

4   Conclusion

In this article, we extended classical games with mixed strategy equilibria by adding costs to changing from one strategy to another. We propose a framework which integrates these costs into the general payoff matrix of the game and thus making them an additional goal of the players to optimize. This introduces a circular dependency between the equilibrium and the game's payoffs, which can be derived straightforwardly using distribution-valued payoffs.

Our main objective in this work was studying the role of cost for playing a mixed strategy in a game, which is usually out of scope in a normal game-theoretic analysis. The practical relevance of this is motivated by security games, where mixed strategies correspond to (frequent) changes of configurations, passwords, or similar. Practically, people may evade efforts tied to the prescribed actions, which leads to unwanted deviations from the optimal (equilibrium) behavior (often observed as "workarounds" to bypass cumbersome security precautions even without any adversarial intention). The point made here is that for an accurate game-theoretic model, these aspects need to be part of the game.

Playing mixed strategy equilibria is typically also a matter of overcoming some inertia, which is the preference to repeat what has been successful in the past. The incentive to stay with a strategy or to change it in the next round may be independent of the game dynamics (payoff structure), but in general calls for more than a cost-optimal way of playing a given mixed strategy. It is indeed not difficult to incorporate costs in the game, and to make the equilibrium playable at minimal cost.

As an independent consideration, the costs incurred by switching strategies in repeated games may contribute to explanations as to why individuals tend to deviate from what game theory predicts in their real behavior. In this context, keeping the payoffs stochastic may help modeling occasionally irrational behavior of individuals, which is quite human but difficult to bring into game theory. Investigating the degree to which switching costs could explain irrationality and be useful to model bounded rationality, could be a natural next step upon this work.

Acknowledgment. This work is partially supported by the European Commission's Project No. 608090, HyRiM (Hybrid Risk Management for Utility Networks) under the 7th Framework Programme (FP7-SEC-2013-1).

References

1. Barraclough, D.J., Conroy, M.L., Lee, D.: Prefrontal cortex and decision making in a mixed-strategy game. Nature **7**, 404–410 (2004)
2. Bloomfield, R.: Learning a mixed strategy equilibrium in the laboratory. J. Econ. Behav. Organ. **25**(3), 411–436 (1994)
3. Chiappori, P.-A., Levitt, S., Groseclose, T.: Testing mixed-strategy equilibria when players are heterogeneous: The case of penalty kicks in soccer. Am. Econ. Rev. **92**(4), 1138–1151 (2002)
4. Lozovanu, D., Solomon, D., Zelikovsky, A.: Multiobjective games and determining pareto-nash equilibria. Buletinul Academiei de Stiinte a Republicii Moldova Matematica **3**(49), 115–122 (2005)
5. Maskin, E.: Nash equilibrium and welfare optimality. Rev. Econ. Stud. **66**(1), 23 (1999)
6. Rass, S.: On game-theoretic risk management (part one) - towards a theory of games with payoffs that are probability-distributions. ArXiv e-prints (2015)
7. Rass, S., Alshawish, A., Abid, M.A., Schauer, S., Zhu, Q., De Meer, H.: Physical intrusion games: optimizing surveillance by simulation and game theory. IEEE Access **5**, 8394–8407 (2017)
8. Rass, S., König, S., Schauer, S.: Decisions with uncertain consequences-a total ordering on loss-distributions. PLoS ONE **11**(12), e0168583 (2016). Journal Article
9. Rass, S., König, S., Schauer, S.: Defending against advanced persistent threats using game-theory. PLoS ONE **12**(1), e0168675 (2017). Journal Article
10. Robinson, J.: An iterative method for solving a game. Ann. Math. **54**, 296–301 (1951)
11. Shaked, M., George Shanthikumar, J.: Stochastic Orders. Springer, New York (2006)
12. Zhu, Q., Başar, T.: Game-theoretic approach to feedback-driven multi-stage moving target defense. In: Das, S.K., Nita-Rotaru, C., Kantarcioglu, M. (eds.) GameSec 2013. LNCS, vol. 8252, pp. 246–263. Springer, Cham (2013). doi:10. 1007/978-3-319-02786-9_15

Dynamics of Strategic Protection Against Virus Propagation in Heterogeneous Complex Networks

Yezekael Hayel[1($\boxtimes$)] and Quanyan Zhu[2]

[1] LIA/CERI, University of Avignon, Avignon, France
`yezekael.hayel@univ-avignon.fr`
[2] Tandon School of Engineering, NYU, New York, USA
`quanyan.zhu@nyu.edu`

Abstract. With an increasing number of wide-spreading cyber-attacks on networks such as the recent WannaCry and Petya Ransomware, protection against malware and virus spreading in large scale networks is essential to provide security to network systems. In this paper, we consider a network protection game in which heterogeneous agents decide their individual protection levels against virus propagation over complex networks. Each agent has his own private type which characterizes his recovery rate, transmission capabilities, and perceived cost. We propose an evolutionary Poisson game framework to model the heterogeneous interactions of the agents over a complex network and analyze the equilibrium strategies for decentralized protection. We show the structural results of the equilibrium strategies and their connections with replicator dynamics. Numerical results are used to corroborate the analytical results.

1 Introduction

Complex networks are useful frameworks to model large-scale network systems and understand emerging macroscopic behaviors arising from local interactions in which nodes communicate, share information, and make interdependent decisions [1–3]. One key application of complex network is to study how computer viruses and diseases propagate over large networks [4–7]. One motivating example is the propagation of a disease in public transportation in which individuals interact randomly in groups of different sizes through the sharing of the infrastructures such as buses, cars, and subways. The size of the interacting population varies in the population, and each is characterized by his type, which indicates the recovery rate or the capability to recover when infected. The spreading of the virus through transportation networks is essentially different from the classical epidemic models which describe the virus spreading over a homogeneous population through a pairwise interaction [8,9]. To capture the epidemic spreading over interactions of different sizes and heterogeneity of the nodes, we develop an evolutionary Poisson game framework in which nodes are not limited

© Springer International Publishing AG 2017
S. Rass et al. (Eds.): GameSec 2017, LNCS 10575, pp. 506–518, 2017.
DOI: 10.1007/978-3-319-68711-7_27

to pairwise interactions, and the size of local interaction is modeled by a Poisson random variable. In addition, individuals are characterized by their private types to model the heterogeneous types of the individuals. The framework is illustrated in Fig. 1. Each node has a type, and the heterogeneous interactions are depicted in groups.

The evolutionary Poisson game framework allows us to study the virus protection game problem in which heterogeneous individuals decide to choose whether or not implement protection schemes. A protected individual cannot be infected and therefore protects the others by taking this decision. However, an individual who chooses not to protect himself can lead to his infection by other nodes and the spreading of the virus over the network. The local decisions of the individuals will contribute to the macroscopic spreading of viruses. The analysis of the evolutionary game framework provides means to understand the emerging behaviors of the complex networks under strategic protection decisions. We show that the Nash equilibrium of the game can be fully characterized and the strategic protection decision has a particular threshold structure. We also show that there is an equivalence between the Nash equilibrium and the rest point of the replicator dynamics. The interaction structure considered in this short paper is complex. In fact, every local interaction involves a random number of selfish individuals as illustrated on Fig. 1.

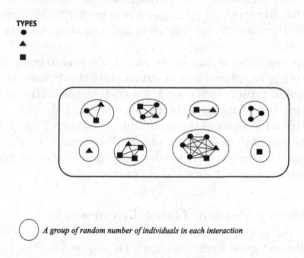

A group of random number of individuals in each interaction

Fig. 1. Heterogeneous framework of virus propagation in a complex network.

With an increasing number of wide-spreading cyber-attacks on networks such as WannaCry and Petya Ransomware [10], studying optimal protection strategies against virus spreading is an important research topic. Compartmental models of virus propagation have been well-studied in the literature. In [11], the recovery rate of each node is determined optimally to minimize the global infection rate, considering a centralized optimization problem. Recently, game-theoretic solutions are proposed to tackle the problem of controlling epidemic processes in a distributed way. For example in [12], a non-cooperative epidemic containment game has been proposed, and the goal is to provide the information on how many users are infected to help the decision of the nodes. The authors have combined the epidemic propagation of the worm with a game-theoretic description of the user behavior into a nonlinear dynamical system. One of their results shows that the learning rate of users has an impact on the proportion of nodes infected at equilibrium. In [13], the authors consider a problem of controlling an epidemic outbreak by distributing resources throughout the network of contacts: antidote allocation and vaccination in each node. The goal of their paper is to determine, in a centralized manner, the cost-optimal distribution of vaccines and antidotes to maximize the exponential decay rate of the infection disease, given a total fixed budget. In [2], the authors have considered an optimal control problem of interdependent epidemics spreading over complex networks. The designed control strategy globally optimizes the trade-off between the control cost and the severity of epidemics in the network. It has been shown that the control can lead to switching between three types of equilibria of the interdependent epidemics network. In our work, we consider a large-scale complex network and analyze the decentralized protection game through the use of the proposed Poisson evolutionary games, which is an extension of the classical evolutionary game framework [16,17]. This work aims to determine the structure of the equilibrium strategy for decentralized virus protection explicitly. Moreover, the main result is obtained by considering that costs and contamination parameters are heterogeneous and type-dependent.

The paper is organized as follows. In Sect. 2, the evolutionary Poisson game (EPG) framework is introduced and we explain how this framework is a suitable model for complex networks. Section 3 describes the decentralized protection game considering heterogeneity into the population of decision makers. This induces a non-trivial analysis of the equilibrium structure. Finally, we illustrate in Sect. 3.4, the equilibrium structure obtain theoretically, by showing numerically convergence of replicator dynamics equations in this complex networks context. We give some conclusions and perspectives in Sect. 4.

2   Evolutionary Poisson Game Framework

Evolutionary Poisson game framework has been proposed in [16,17]. This framework is an extension of the standard evolutionary game framework to multiple individuals in interaction, particularly a random number, instead of pairwise interactions involving only two individuals. A similar concept of evolutionary stable strategies and replicator dynamics can be proposed in such context. We focus in this paper on describing the dynamic of the evolution of the protection strategy into a heterogeneous population through the use of the well-known replicator dynamics. The replicator dynamics, introduced in [18], serve to highlight the role of selection from a dynamic perspective. It is formalized by a system of ordinary differential equations, and it establishes that the evolution of the size of the populations (the proportion of individuals playing a given pure strategy) depends on the utility obtained during pairwise interactions. We then adapt this framework to a random number of individuals in each local interaction.

2.1 Standard Evolutionary Games

The standard evolutionary game framework is described as follows. Consider an infinitely large population of players, where each player repeatedly meets a randomly selected individual within the population. This *pairwise* interaction framework is central in evolutionary game models, and it leads to a matrix game representation [19]. Each individual has a finite pure action space $A = \{a_0, \ldots, a_{K-1}\}$, with the cardinality of set A is $|A| = K$. Let $\Delta(A) = \{p \in [0,1]^K \mid \sum_{a_i \in A} p_i = 1\}$ be the set of mixed actions, that are probability measures over the action space. Then p_i is the probability to play action a_i. We define by $U(p, q)$ the expected payoff of an individual playing p against an opponent using q, where $p, q \in \Delta(A)$. It is important to note that a mixed strategy p can also be interpreted as a distribution of pure strategies among all the individuals. This is also called the *strategy profile* of the population. The population profile at time t is given by the vector $p(t) \in \Delta(A)$, and then for each $a_i \in A$, $p_i(t)$ denotes the proportion of individuals that plays pure strategy (action) a_i at time t. The replicator dynamics describe how the distribution of pure actions (the strategy profile) evolves in time depending on the interactions between individuals. We denote by e_i the unit vector with the element equal to 1 at $(i + 1)^{th}$ position, i.e., $e_0 = (1, 0, \ldots, 0)$ and $e_{K-1} = (0, \ldots, 0, 1)$. The replicator dynamics for any action (pure strategy) $a_i \in A$ are expressed by the following equations:

$$\forall a_i \in A, \quad \dot{p}_i(t) = p_i(t)[U(p(t), p(t)) - U(e_i, p(t))], \tag{1}$$

where $U(e_i, p(t))$ is the immediate utility of an individual playing i against the population profile $p(t)$ and $U(p(t), p(t))) = \sum_{i=0}^{K-1} p_i(t)U(e_i, p(t))$ is the average immediate utility of the population. A strategy will sustain if its utility is lower than the utility averaged over all the strategies used in the whole population. Recall that we deal with utility function, and not fitness, then each individual looks for minimizing its utility. The *folk theorem of evolutionary games* allows to establish a strict connection between the stable points of the replicator dynamics and the Nash equilibria of the game [20].

2.2 Type-Dependent Poisson Games

Poisson game is a game theoretic framework that captures two important features of an interacting complex systems: the number of decision makers in interaction and their types. Myerson introduced in [14] the framework of Poisson game by considering a random number of players in interaction with different types. This type of game has been recently used to study decentralized resource allocation in complex network settings [21,22]. The number of players involved in a local interaction follows a Poisson process with rate λ, typically $\lambda \gg 1$. We denote by $\Omega := \{1, \ldots, T\}$ is the set of types of players, and each one belongs to one type $\tau \in \Omega$. The probability of a player being of type τ is given by $r(\tau)$, and then the number of players of type τ in each local interaction is a Poisson random variable with parameter $\lambda r(\tau)$. In standard Poisson Games defined by Myerson in [14], the utility of an individual depends on the number of players that choose the same action and not on their type. Extension of the framework to type-dependent utilities as been proposed in our previous paper [16,17]. Then, we define the utility $u_\tau(a, \mathbf{x})$ of a player of type τ playing action $a \in \mathcal{A}$ against $\mathbf{x} = (x_1, \ldots, x_T)$ where for each type τ, x_τ is a vector of size K, where $x_\tau(b)$ is the number of players of type τ who choose action $b \in \mathcal{A}$. Note that $\mathbf{x}$ can be also described as a type/action matrix of size $T \times K$. The expected utility of a player of type τ who plays action $a \in \mathcal{A}$ while the rest of the players are expected to play using strategy $\sigma \in \Delta(\mathcal{A}) \times \ldots \times \Delta(\mathcal{A})$ (we have that $\sigma := (\sigma_1, \ldots, \sigma_T)$ with for all type $\tau \in \Omega$, $\sigma_\tau \in \Delta(\mathcal{A})$) is given by:

$$U_\tau(a, \sigma) = \sum_{\mathbf{x} \in Z(\mathcal{A}) \times \ldots \times Z(\mathcal{A})} \mathbb{P}(\mathbf{X} = \mathbf{x}|\sigma) u_\tau(a, \mathbf{x}), \qquad (2)$$

where $x \in (Z(\mathcal{A}))^T$ and $Z(\mathcal{A})$ denotes the set of elements $w \in \mathbb{N}^K$ such that $w(c)$ is a non-negative integer for all $c \in \mathcal{A}$ ($w(c)$ represents the number of players choosing action c). The relation between σ and x is given through the following decomposition property of the Poisson distribution:

$$\mathbb{P}(\mathbf{X} = \mathbf{x}|\sigma) = \prod_{\tau \in \Omega} \prod_{b \in \mathcal{A}} \mathbb{P}(X_\tau(b) = x_\tau(b)|\sigma),$$

$$= \prod_{\tau \in \Omega} \prod_{b \in \mathcal{A}} e^{-\lambda r(\tau)\sigma_\tau(b)} \frac{(\lambda r(\tau)\sigma_\tau(b))^{x_\tau(b)}}{x_\tau(b)!}.$$

where players play according to the mixed strategy σ. Here, for any type $\tau \in \Omega$, $\sigma_{\tau,b}$ is the probability that a player of type τ chooses the pure action $b \in \mathcal{A}$. Thus, the vector σ_τ can be seen as the strategy profile for type-τ individuals. Therefore, the expected utility of type τ player choosing pure action a can be seen as an infinite number of successive interactions with different individuals where the number of individuals at each interaction is a random Poisson variable. Based on the expected utility description, we define a Nash equilibrium as follows:

Definition 1. *The strategy vector $\sigma^* \in \Delta(\mathcal{A}) \times \ldots \times \Delta(\mathcal{A})$ is a pure Nash equilibrium if*

$$\forall \tau \in \Omega, \quad \sigma_\tau^* \in B_\tau(\overset{*}{\sigma}),$$

with

$$B_\tau(\sigma) = \{b \in \mathcal{A} : b \in \arg\min_{a \in \mathcal{A}} U_\tau(a, \sigma)\}.$$

We can also extend this framework to the mixed-strategy Nash equilibrium by considering the set of best-response mixed strategies $\Delta(B_\tau(\sigma))$. The replicator dynamics describe the evolution of the proportion of individuals of each type

τ that choose a pure action at any given time. We denote by $\sigma(t)$ the vector $\sigma(t) := (\sigma_1(t), \ldots, \sigma_T(t))$. The replicator dynamics in this heterogeneous context is then composed of T coupled differential equations, where for any type $\tau \in \Omega$ and action a, this evolution is described as:

$$\dot{\sigma}_{\tau,a}(t) = \sigma_{\tau,a}(t) \left(U_\tau(\sigma_\tau(t), \sigma(t)) - U_\tau(a, \sigma(t)) \right), \tag{3}$$

where $U_\tau(\sigma_\tau(t), \sigma(t))$ is the immediate average utility of the type τ individuals. Indeed, individuals compare their utility with the average inside their class because some types may have different actions and associated utilities. Hence it can be described as follows:

$$U_\tau(\sigma_\tau(t), \sigma(t)) = \sum_{b \in \mathcal{A}} \sigma_{\tau,b}(t) U_\tau(b, \sigma(t)).$$

3   Decentralized Protection Game

3.1   Game Model

We describe in this section the mathematical model of our decentralized protection game in a heterogeneous complex network. Each individual, represented as a node, has a type. We denote by $\Omega := \{1, \ldots, T\}$ is the set of types of players, and each one belongs to one type $\tau \in \Omega$. All individuals have the same discrete action space $\mathcal{A} = \{a_0, a_1\}$ in which action a_0 means that the individual decides not to protect itself and the action a_1 he decides to be protected. To be protected depends on the context of the application.

The utility (cost) perceived by each individual depends on his type τ, his action and the vector $\mathbf{x} = (x_1, \ldots, x_T)$ in which x_τ denotes the number of individuals of type τ that are not protected (choose action a_0). The cost of protection is not negligible and then, any individual that decides to be protected perceived the following cost:

$$u_\tau(a_1, \mathbf{x}) = \bar{C}.$$

On the other hand, any individual that decides not to be protected perceived a cost if and only if the virus becomes a pandemic, i.e. the virus propagates to all individuals. We consider a Susceptible-Infected-Susceptible (SIS) propagation model. Each individual becomes infected following a Poisson process with rate β. Also, an infected individual of type τ recovers following a Poisson process with rate δ_τ. It is assumed that the infection and curing processes are independent. An important value is the ratio $\rho_\tau := \beta / \delta_\tau$, which is called the effective spreading rate for type τ individuals. Our model is an inhomogeneous SIS epidemic model, and the analysis of the long term behavior of the epidemic is analyzed in [15]. Particularly, the condition on which the virus becomes pandemic under certain topology conditions is analytically described in this paper. Considering that all interactions occur into a complete graph with $N := \sum_{\tau=1}^{T} x_\tau$ nodes between the individuals of different types, the virus propagates to all nodes if and only if the following condition holds:

$$\sum_{\tau=1}^{T} \frac{x_\tau}{1+\rho_\tau} \leq \sum_{\tau=1}^{T} x_\tau - 1.$$

This condition can be simplified and then the cost for an individual that decides not to be protected is defined as:

$$u_\tau(a_0, \mathbf{x}) = \begin{cases} C_\tau & \text{if} \quad \sum_{\tau=1}^{T} \frac{x_\tau \rho_\tau}{1+\rho_\tau} \geq 1, \\ 0 & \text{otherwise.} \end{cases}$$

In this problem, the cost of an unprotected individual depends on its type τ and the action of all the other individuals.

3.2 Equilibrium Analysis

Each individual decides whether to be protected (i.e., action a_0), and pay the price for it, or to have a risky behavior (i.e., action a_1) and not to be protected. In the latter, it will incur a cost depending on the decision of the other individuals. We then look for a stable strategic situation in which no individuals have the interest to deviate from his decision. A decision can be a mixed strategy, in other words, a probability distribution over the action set $\mathcal{A}$. Note that our framework incorporates heterogeneity in the individuals which is described through the type parameter. Then, the strategy of each individual depends on his type, and we denote by $\sigma_\tau \in [0, 1]$, the probability that a type-τ individual chooses action a_0. As we have still only 2 actions, we limit our equilibrium analysis to this scalar for each type τ. Thus, for each type τ, the strategy σ_τ is defined in the simplex $\Delta(\mathcal{A})$ and, we consider symmetric equilibrium within each type, which means that, at equilibrium, all individuals of the same type play the same strategy.

The infection cost C_τ perceived by each individual depends on his type τ. Then, the expected utility of a type-τ individual playing action a_0 against a population strategy given by the vector σ, is given by:

$$U_\tau(a_0, \sigma) = \sum_{\mathbf{x} \in \mathbb{N}^T} \mathbb{P}(\mathbf{X} = \mathbf{x}|\sigma) u_\tau(a_0, \mathbf{x}),$$

$$= C_\tau \left(\sum_{\mathbf{x}:\sum_{\tau=1}^{T} \frac{x_\tau \rho_\tau}{1+\rho_\tau} \geq 1} \mathbb{P}(\mathbf{X} = \mathbf{x}|\sigma) \right) := C_\tau \mathcal{F}(\sigma),$$

with $\sigma = (\sigma_1, \ldots, \sigma_T)$ the strategy vector for all types of individuals. The function $\mathcal{F}(\cdot)$ is defined by:

$$\mathcal{F}(\sigma) = \sum_{\mathbf{x}: \sum_{\tau=1}^{T} \frac{x_\tau \rho_\tau}{1 + \rho_\tau} \geq 1} \mathbb{P}(\mathbf{X} = \mathbf{x}|\sigma),$$

$$= 1 - \sum_{\mathbf{x}: \sum_{\tau=1}^{T} \frac{x_\tau \rho_\tau}{1 + \rho_\tau} < 1} \mathbb{P}(\mathbf{X} = \mathbf{x}|\sigma),$$

$$= 1 - \sum_{\mathbf{x}: \sum_{\tau=1}^{T} \frac{x_\tau \rho_\tau}{1 + \rho_\tau} < 1} \prod_{\tau \in \Omega} \prod_{b \in \mathcal{A}} e^{-\lambda r(\tau) \sigma_\tau(b)} \frac{(\lambda r(\tau) \sigma_\tau(b))^{x_\tau(b)}}{x_\tau(b)!}.$$

The expected utility of a type-τ individual playing action a_1 is:

$$U_\tau(a_1, \sigma) = \sum_{\mathbf{x} \in \mathbb{N}^T} \mathbb{P}(\mathbf{X} = \mathbf{x}|\sigma) u_\tau(a_1, \mathbf{x}),$$

$$= \sum_{\mathbf{x} \in \mathbb{N}^T} \mathbb{P}(\mathbf{X} = \mathbf{x}|\sigma) \bar{C} = \bar{C}.$$

We have the following result on the equilibrium, considering without loss of generality that we order the types as $C_1 < C_2 < \ldots < C_T$. We consider the following assumption:

Assumption 1. *The following equation:*

$$\mathcal{F}(1, \ldots, 1, \sigma_t, 0, \ldots, 0) = \frac{\bar{C}}{C_t}, \tag{4}$$

has a unique solution in the interval $[0, 1]$. This solution is denoted by σ_t^*.

The function $\mathcal{F}(\cdot)$ has a complex structure, and proving some interesting properties analytically is intractable. Besides, this assumption comes from intuitive reasoning and leads to the following proposition on the structure of the equilibrium.

Proposition 1. *There exists a Nash equilibrium σ^* for our heterogeneous EPG, given in the following items:*

- *If $\bar{C} \geq C_T$, then the unique equilibrium is $\sigma^* = (1, \ldots, 1)$,*
- *Otherwise, under Assumption 1, $\sigma^* = (1, \ldots, 1, \sigma_t^*, 0, \ldots, 0)$, with σ_t^* the unique solution of Eq. 4) is an equilibrium.*

Proof. The proof has two parts.

First, we consider the case where $\bar{C} \geq C_T$. In this extreme case, the strategy profile $\sigma^* = (1, \ldots, 1)$ is the unique equilibrium. In fact, for any type τ individual we have:

$$U_\tau(a_1, \overset{*}{\sigma}) = \bar{C} > U_\tau(a_0, \overset{*}{\sigma}) = C_\tau.$$

Then, the strategy profile $\sigma^* = (1, \ldots, 1)$ is an equilibrium in this case as no individual of any type has an interest to change his action. More, it is unique as the action a_0 is the best action for any type of individuals.

Second, when $\bar{C} < C_T$, we assume that there exists a type i such that the following equation:

$$\mathcal{F}(1,\ldots,1,\sigma_i^*,0,\ldots,0) = \frac{\bar{C}}{C_i},$$

has a unique solution $\sigma_i^* \in [0,1]$. Considering the strategy profile $\sigma^* = (1,\ldots,1,\sigma_i^*,0,\ldots,0)$ we have that for all types $j \leq i-1$:

$$U_j(a_0,\overset{*}{\sigma}) = C_j\mathcal{F}(\overset{*}{\sigma}) < C_i\mathcal{F}(\overset{*}{\sigma}) = \bar{C}.$$

Then, no type-j individual has an interest to deviate from action a_0. Type-i individuals are at equilibrium as $U_i(a_0,\sigma^*) = \bar{C}$. Finally, for any individual of type $l > i$, we have that:

$$U_l(a_0,\overset{*}{\sigma}) = C_l\mathcal{F}(\overset{*}{\sigma}) > C_i\mathcal{F}(\overset{*}{\sigma}) = \bar{C},$$

and then all type-l individuals invest, i.e., choose action a_1. Thus, given the existence of the solution σ_i^*, the strategy profile σ^* is an equilibrium.

3.3   Replicator Dynamics

The previous section gives a characterization of the Nash equilibrium as a threshold-type. By the way, this characterization is performed through the analysis of a nontrivial Eq. (4). Another way to study the equilibrium is to consider the rest points of the replicator dynamics. In fact, based on Evolutionary game theory, Nash equilibrium of large population games can be expressed as rest points of the replicator dynamics equations. In our heterogeneous framework, for each type τ population, a replicator dynamic equation describes the evolution of the proportion of type τ individuals that play action a_0 at each instant. Then, based on the replicator dynamics Eq. (3), we have the following differential equations modeling the proportion of protected individuals for each type evolution in time:

$$\forall \tau \in \Omega, \quad \dot{\sigma}_\tau(t) = \sigma_\tau(t)(1 - \sigma_\tau(t))[\bar{C} - C_\tau\mathcal{F}(\sigma_1(t),\ldots,\sigma_T(t))].$$

In order to find the equilibrium, we can study the rest points of a non-linear system of T ordinal differential equations. Indeed, following the folk theorem of evolutionary game theory [20].

Theorem 1. *The folk theorem of evolutionary game theory [20] states that:*

   i. *any Nash equilibrium profile is a rest point of the replicator equation;*
   ii. *if a Nash equilibrium profile is strict then it's asymptotically stable;*
   iii. *if a rest point is the limit of an interior orbit for $t \to \infty$, then it is a Nash equilibrium profile;*
   iv. *any stable rest point of the replicator dynamics is a Nash equilibrium profile.*

We illustrate this theorem with several numerical tests in the following section.

3.4 · Numerical Illustrations

Section 3.3 has described the connection between the rest point of the replicator dynamics and the equilibrium point of the evolutionary Poisson game. We illustrate the replicator dynamics of protection strategies by showing the convergence of the coupled dynamical systems in different scenarios. In Fig. 2, the equilibrium obtained is $\sigma^* = (1, 0.22)$. The parameters are: $\delta_1 = 10$, $\delta_2 = 5$, $\beta = 1$, $\bar{C} = 2$, $C_1 = 5$, $C_2 = 8$, $r(1) = 0.2$ and $\lambda = 20$. Type-1 individuals have the lowest infection cost, and then no individuals of this type protect itself, whereas 78% of type-2 individuals adopt protection. Initial points are $\sigma_1(0) = 0.1$ and $\sigma_2(0) = 0.8$. We denote also that type-1 dynamic achieves a rest point (i.e. $\dot{\sigma}_1(t) = 0$) which is not stable due to the instability of the type-2 dynamic. Hence, this illustrates on simulations the interaction between types of individuals.

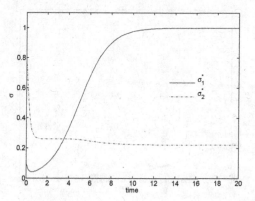

Fig. 2. Convergence of the coupled replicator dynamics in an heterogeneous network with 2 types of individual.

Numerical experiments with more complex network structures are proposed in Figs. 3 and 4. The number of types is extended to 4 types. We observe in Fig. 3 the convergence of the coupled replicator dynamics with 4 types to the equilibrium $\sigma^* = (1, 1, 0.66, 0)$. The parameters are: $\delta_1 = 20$, $\delta_2 = 15$, $\delta_3 = 10$, $\delta_4 = 5$, $\beta = 1$, $\bar{C} = 2$, $C_1 = 5$, $C_2 = 8$, $C_3 = 10$, $C_4 = 15$, $r = (0.2, 0.2, 0.3, 0.3)$ and $\lambda = 20$. The equilibrium is in accordance with Proposition 1. Initial conditions of the dynamical system are chosen to be $\sigma_1(0) = 0.1$, $\sigma_2(0) = 0.8$, $\sigma_3(0) = 0.4$ and $\sigma_4(0) = 0.6$.

Finally, we illustrate the impact of the protection cost on the convergence of the replicator dynamics. On Fig. 4, we observe the convergence to the same threshold type of equilibrium, i.e. $\sigma^* = (1, 0.92, 0, 0)$, by changing the protection cost to the value $\bar{C} = 0.005$. Initial conditions are chosen as follows: $\sigma_1(0) = \sigma_2(0) = \sigma_3(0) = \sigma_4(0) = 0.5$. In this case, the protection cost is so low that type 3 and type 4 individuals are all protected against the epidemics.

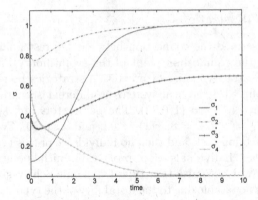

Fig. 3. Convergence of the coupled replicator dynamics with 4 types.

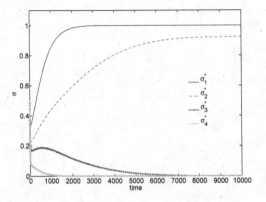

Fig. 4. Convergence of the coupled replicator dynamics with 4 types and the following parameters: $\delta_1 = 20$, $\delta_2 = 15$, $\delta_3 = 10$, $\delta_4 = 5$, $\beta = 1$, $\bar{C} = 0.005$, $C_1 = 5$, $C_2 = 8$, $C_3 = 10$, $C_4 = 15$, $r = (0.2, 0.2, 0.3, 0.3)$ and $\lambda = 20$.

4   Conclusions and Perspectives

This paper has proposed evolutionary Poisson games to study decentralized protection strategies inside the heterogeneous population of interacting agents like networks or connected devices. We have determined explicitly the structure of the decision equilibrium protection strategies of individuals interacting in a heterogeneous complex network. The threshold type structure of the equilibrium provides useful perspectives for controlling and optimizing such complex systems. For example, it can be used to determine an appropriate vaccination campaign at a lower cost under the strategic behavior of individuals. With a growing number of interconnected devices in future communication networks, the proposed framework can be used to model heterogeneous complex networks and their protection mechanisms. One future work would aim to leverage machine learning techniques to develop data-driven and self-adaptive algorithms to changing configurations of the systems.

References

1. Barrat, A., Barthelemy, M., Vespignani, A.: Dynamical Processes on Complex Networks. Cambridge university Press, Cambridge (2008)
2. Chen, J., Zhu, Q.: Optimal control of interdependent epidemics in complex networks. In: SIAM Conference on Network Science, Boston, Massachusetts, 15–16 July 2016
3. Chen, J., Zhu, Q.: Interdependent network formation games with an application to critical infrastructures. In: Proceedings of 2016 American Control Conference, Boston, USA, 6-8 July 2016
4. Gubar, E., Zhu, Q.: Optimal control of multi-strain epidemic processes in complex networks. In: Proceedings of the 7th EAI International Conference on Game Theory for Networks (GameNets), Knoxville, TN, 9-10 MAY 2017
5. Taynitskiy, V., Gubar, E., Zhu, Q.: Optimal impulsive control of epidemic spreading of heterogeneous malware. In: Proceedings of 20th IFAC World Congress, Toulouse, France, 9-14 July 2017
6. Taynitskiy, V., Gubar, E., Zhu, Q.: Optimal security policy for protection against heterogeneous malware. In: International conference on NETwork Games, COntrol and Optimization (NETGCOOP), Avignon, France (2016)
7. Gubar, E., Zhu, Q.: Optimal control of Influenza epidemic model with virus mutations. In: Proceedings of European Control Conference (ECC), Zurich, Switzerland (2013)
8. Hethcote, H.W.: Qualitative analyses of communicable disease models. Math. Biosci. **28**(3–4), 335–356 (1976)
9. Pastor-Satorras, R., Vespignani, A.: Epidemic spreading in scale-free networks. Phys. Rev. Lett. **86**(14), 3200 (2001)
10. O'Gorman, G., McDonald, G.: Ransomware: A growing menace. Symantec Corporation (2012)
11. Gourdin, E., Omic, J., Van Mieghem, P.: Optimization of network protection against virus spread. In: Proceedings of the 8th International Workshop on the Design of Reliable Communication Networks (2001)
12. Theodorakopoulos, G., Le Boudec, J.Y., Baras, J.: Selfish response to epidemic propagation. IEEE Trans. Autom. Control **58**(2), 363–376 (2013)
13. Preciado, V.M., Zargham, M., Enyioha, C., Jadbabaie, A., Pappas, G.: Optimal resource allocation for network protection against spreading processes. IEEE Trans. Control Netw. Syst. **1**(1), 99–108 (2014)
14. Myerson, R.: Large Poisson games. J. Econ. Theor. **94**(1), 7–45 (2000)
15. Van Mieghem, P., Omic, J.: In-homogeneous virus spread in networks (2013). arXiv:1306.2588
16. Hayel, Y., Zhu, Q.: Epidemic protection over heterogeneous networks using evolutionary Poisson games. IEEE Trans. Inf. Forensics Secur. **12**(8), 1786–1800 (2017)
17. Hayel, Y., Zhu, Q.: Evolutionary Poisson games for controlling large population behaviors. In: Proceedings of 54th IEEE Conference on Decision and Control (CDC), Osaka, pp. 300–305 (2015)
18. Taylor, P., Jonker, L.: Evolutionarily stable strategies and game dynamics. Math. Biosci. **40**, 145–156 (1978)

19. Hofbauer, J., Sigmund, K.: Evolutionary Games and Population Dynamics. Cambridge University Press, Cambridge (1998)
20. Hofbauer, J., Sigmund, K.: Evolutionary game dynamics. Am. Math. Soc. **40**, 479–519 (2003)
21. Filio-Rodriguez, O., Primak, S., Kontorovich, V., Shami, A.: A game theory interpretation for multiple access. In: Cognitive Radio Networks with Random Number of Secondary Users in CoRR (2013)
22. Simhon, E., Starobinski, D.: Advance reservation games and the price of conservatism. ACM SIGMETRICS Perform. Eval. Rev. **42**(3), 33 (2014)

Three Layer Game Theoretic Decision Framework for Cyber-Investment and Cyber-Insurance

Deepak K. Tosh[1][(✉)], Iman Vakilinia[2], Sachin Shetty[3], Shamik Sengupta[2], Charles A. Kamhoua[4], Laurent Njilla[5], and Kevin Kwiat[5]

[1] Department of Computer Science, Norfolk State University, Norfolk, VA, USA
dktosh@nsu.edu
[2] Department of Computer Science and Engineering,
University of Nevada, Reno, NV, USA
{ivakilinia,ssengupta}@unr.edu
[3] Virginia Modeling Analysis and Simulation Center,
Old Dominion University, Norfolk, VA, USA
sshetty@odu.edu
[4] Network Security Branch, Army Research Laboratory, Adelphi, MD, USA
charles.a.kamhoua.civ@mail.mil
[5] Cyber Assurance Branch, Air Force Research Laboratory, Rome, NY, USA
{laurent.njilla,kevin.kwiat}@us.af.mil

Abstract. Cyber-threat landscape has become highly complex, due to which isolated attempts to understand, detect, and resolve cybersecurity issues are not feasible in making a time constrained decisions. Introduction of cyber-threat information (CTI) sharing has potential to handle this issue to some extent, where knowledge about security incidents is gathered, exchanged across organizations for deriving useful information regarding the threat actors and vulnerabilities. Although, sharing security information could allow organizations to make informed decision, it may not completely eliminate the risks. Therefore, organizations are also inclined toward considering cyber-insurance for transferring risks to the insurers. Also, in networked environment, adversaries may exploit the information sharing to successfully breach the participating organizations. In this paper, we consider these players, i.e. organizations, adversary, and insure, to model a three layer game, where players play sequentially to find out their optimal strategies. Organizations determine their optimal self-defense investment to make while participating in CTI sharing and cyber-insurance. The adversary looks for an optimal attack rate while the insurer targets to maximize its profit by offering suitable coverage level to the organizations. Using backward induction approach, we conduct subgame perfect equilibrium analysis to find optimal strategies for the involved players. We observe that when cyber-insurance is not

Approved for Public Release; Distribution Unlimited: 88ABW-2017-3219, 05 Jul 2017. This work was supported by the Office of the Assistant Secretary of Defense for Research and Engineering (OASD (R&E)) agreement FA8750-15-2-0120, Department of Homeland Security Grant 2015-ST-061-CIRC01 and National Science Foundation (NSF), Award #1528167.

© Springer International Publishing AG 2017
S. Rass et al. (Eds.): GameSec 2017, LNCS 10575, pp. 519–532, 2017.
DOI: 10.1007/978-3-319-68711-7_28

considered, attacker prefers to increase its rate of attack. This motivates the organizations to consider cyber-insurance option for transferring the risks on their critical assets.

Keywords: Cyber-insurance · Cyber-threat information sharing · Game theory · CYBEX

1   Introduction

Growing utilization of cyberspace invites malicious adversaries to exploit unpatched vulnerabilities of Internet users/organizations for various profitable reasons. The cyber attacks are becoming sophisticated and complex day by day, where the adversaries target the victims and persist until their objectives are pursued. Therefore, the attackers always try to stay one step ahead of victims and use advanced tactics, techniques, and procedures (TTPs) to achieve their goals. While it is becoming difficult for the cyberspace users to detect and prevent cyber-malicious activities using the traditional signature-based security measures, independent efforts to address such issues are turning out to be ineffective in practice. Due to this, security researchers, and policy makers are emphasizing on enforcing mutual collaborative efforts from private organizations and government institutions for collecting, sharing and analyzing threat information. This could help in deriving proactive cyber-intelligence [6] to efficiently identify structured information regarding novel attack campaigns, compromised resources, TTPs, actors behind the scene etc. and take defensive actions in a timely manner [4].

The significance of cybersecurity information sharing has lead governments and regulators to mandate/encourage such sharing. In U.S., the Cybersecurity Information Sharing Act (CISA) [1] bill motivates for collaborative sharing among private and public sector organizations by providing liability protections to the sharing parties. EU has also launched several cross-sector and intra-sector initiatives to enhance the EU Member States' capability for preparedness, cooperation, information exchange, coordination, and response to cyber threats. Furthermore, ITU-T has approved a CYBersecurity information EXchange (CYBEX) [16] framework that facilitates organization and sharing cyber-threat information, such as knowledge of threats, vulnerabilities, incidents, risks, and mitigations and their associated remedies.

On the other hand, the inescapable fact is that it is impossible to achieve perfect/near-perfect cybersecurity protection. Therefore, organizations also rely on cyber-insurance to transfer the cyber-related risks on their critical assets. Cyber-insurance is a risk management technique via which cyber-risks are transferred to an insurance company, in return for a periodic fee, i.e., the insurance premium. Cyber-insurance can indemnify various costs impacted from cyber-attacks causing data destruction, extortion, theft, hacking, and denial of service attacks. Although, insurance could indirectly improve security of an organization, it also demands investment on self-defense mechanisms. The possibility of correlated risks from other entities caused due to networked environment also hints to participate in collaborative CTI sharing. Having noticed the usefulness of CTI sharing and cyber-insurance, it is important to model the mutual inter-action between organizations and insurer in presence of an adversary, where organizations opt for both risk mitigation approaches to maintain socially efficient security level.

In this work, we consider organizations, adversaries, and insurers as three category of players in the game with periodic strategic interactions. The organization aims to find its optimal self-defense investment while participating in information sharing and undertaking cyber-insurance. However, adversarial attacks negatively affect the organization, which also costs the adversary. Therefore, it requires an optimal and balanced attack rate that will cause maximum disruption to the organizations. At the same time, organizations consider cyber-insurance to protect their critical assets by paying a premium to the insurers, who provide a certain level of coverage on the event of a cyber-breach. Thus, the insurer aims to offer an optimal coverage level to the organizations so that both insured and insurer maximize their payoffs. We consider a sequential interaction of three players, starting with the organization who decides its self-defense investment, followed by the adversary who chooses the attack rate. Then, the insurer plays its strategy to decide what coverage level it must offer an organization by observing its self-defense investment and attacker's attack rate. The game repeats further with the same interactions. We rigorously analyze the model when insurance is not undertaken by the defending organizations and show that this situation is not beneficial for them as the attacker's optimal strategy is to increase its attack rate.

This paper is organized as follows: Sect. 2 presents brief overview on prior research works. The system model of the 3-layer game is described in Sect. 3. The utility models and game formulation is presented in Sect. 4. In the Sect. 5, we analyze the sequential interaction game to derive Stackleberg equilibrium. The numerical results are presented in Sect. 6. Finally, Sect. 7 concludes the paper.

2   Related Works

This topic has gained significant attention and is being investigated by government, policy makers, economists, non-profit organizations, industries, cyber-security and network professionals with researches in this particular area still emerging [8,9,22]. Considering the need of cybersecurity information sharing, Gordon et al. [11] analyzed the economic (dis)advantages of this activity and derived its relationship with accounting aspects of an organization. Using game theoretic models, [18,19] prove that information exchange activity improves the social welfare as well as security level of the firms. Incentivization schemes for inducing sharing nature is discussed in [13,17,20]. Authors of [7] have proposed a game theoretic model to determine the IT security investment levels and compare it with the outcome of a decision theoretic approach that considers various components, such as vulnerability, payoff from investment etc. Authors of [10] applied functional dependency network analysis to model the attack propagation for a set of correlated organizations and analyze the sharing behaviors.

On the other hand, cyber-insurance market is emerging [15] due to the high occurrence of targeted cyber breaches over the years. However, the components such as interdependent security, correlated risks, and information asymmetries [3,5] make it challenging to model appropriate policies for the organizations. Nash equilibrium analysis and social optima concepts are applied to model security games in [12] that consider above three components into account and decide how investment can be used for both public good (protection) and a private good (insurance). Full insurance and partial insurance coverage models are proposed in [14] and study the impact of cooperation on self-defense investments. Another quantitative framework is proposed in [23] that applies optimization technique to provide suggestions to the network users and operators on investments toward cybersecurity insurance by minimizing the overall cyber risks. Sequential interaction of three player groups, when organization consider both strategies for risk management, is not undertaken in prior research. This paper particularly models this problem and solve for Stackleberg equilibria for the formulated game.

3   System Model

In this section, we define the players of the game and the interactions among the organizations and insurance vendors in presence of adversaries in the cybersecurity information sharing framework. These players: organizations, insurer, and adversary, interplay in the sharing system to achieve their objectives as briefed in the following.

- *Organizations* operate their business essentials with the use of network systems and hence they are vulnerable to data breaches, and service disruptions. So, minimizing the losses from the adversary's cyber attacks is primary focus of the organizations.
- *Insurers* are the entities or companies, who assess the cyber-risks of the organizations and formulate potential insurance policies to cover financial damages (fully or partially) at the occurrence of successful cyber attacks at a cost of periodic subscription fee in the form of premiums.
- *Adversaries* are the malice users, who always try to take advantage of the communication network and system loopholes in the organizations to perform data breaches. Furthermore, attackers look deep into the shared cyber-threat information among the organizations to exploit them with the gained knowledge.

3.1   System Overview

As described above, our system involves three categories of players, where organizations are considered to be interconnected with each other and runs similar applications for their operations. Thus, the attacker has opportunity to attack individual organizations directly or indirectly via exploiting some other organization. So, the loss for an organization on a successful attack event can have two different components, namely direct and indirect loss. To avoid such losses, organizations typically invest in self-defense. However, this strategy may not completely alleviate the chances of getting compromised. Therefore, the organizations participate in a threat information sharing framework, where they exchange vulnerability related information, such as patches, and fixes with each other, to foster their proactive defense abilities.

In addition to the strategies of self-defense investment and CTI sharing, organizations also prefer to transfer some of the risks to third parties via insuring their critical assets. Thus, insurance companies come into the picture, who directly interact with organizations to evaluate their risk factors and offer coverage for assets that are candidates of cyber-exploitation. Typical categories of cyber-insurance coverage includes critical data breach, business interruption, destruction of data/software, denial of service, ransomware etc. Since cyber-insurance has become a critical component of cybersecurity risk management, we consider insurance provider as a player in our game model and analyze the impact of insurance coverage on the investment decisions of organizations and attack rate of adversaries.

3.2   Threat Model

In our proposed game model, it is considered that the organizations are vulnerable to cyber-breaches and the adversary is rational in nature. With assumption that the adversary can observe the strategies undertaken by the organizations, it might alter its attacking strategy to maximize its profit. This observation can be partial or full depending on how it conducts the reconnaissance phase prior to the attack. In our model, we assume the attacker has complete information about the defending organization's strategies. The two important parameters for the defenders are self-defense investment as well as amount of threat-information sharing among each other, which have (in)direct impacts on the rational adversary's overall utility. If the investment toward self-defense is observed to be higher, then the attacker's probability of successfully compromising the organization may be low. While at the same time, we believe that if the threat related information sharing is improved in the system, it helps the participants to enhance their security practice based on the community knowledge. Thus, observing this information, the adversary's attack strategy will be different and its payoff will be impacted.

3.3   Model Description

In our model, we consider a set $\mathcal{N}$ of n risk-averse and rational organizations that operate in a networked environment, who also participate in the cybersecurity information sharing process. The organizations may or may not cooperate in disclosing the investment information with each other because of competitive advantages, however they voluntarily share CTI information to proactively defend future cyber attacks. Considering the market existence for cyber-insurance and self-defense, organizations look forward to invest in both strategies to reduce their risks on cyber-exploitation.

We consider that each organization i has a fixed portion T of total assets that it tries to protect. For risk reduction, the organization $i \in \mathcal{N}$ chooses to invest in self-defense, which is represented as $s_i \in [0, 1]$ and each self-defense investment maps to a particular risk probability $p(s_i)$. It is intuitive to state that the probability function is decreasing with respect to s_i, hence it is assumed that $p(s_i)$ is continuous and twice differentiable, i.e. $p'(s_i) < 0$ and $p''(s_i) > 0$. Also, the probability diminishes to zero, when the investment reaches to very high quantity, i.e. $\lim_{s_i \to \infty} p(s_i) = 0$. Similarly, we assume that every organization i involved in CTI information exchange shares $l_i \in [0, 1]$ amount of cybersecurity information voluntarily with the others. This knowledge to the adversary may help to use against other organizations in performing data breach or service disruptions. So, we define another risk probability $q(\{l_i : i \in \mathcal{N}\})$ with respect to sharing of threat information. It represents the probability of getting breached when the system participants share $L = \{l_1, l_2, \cdots l_n\}$ amount of information individually with each other. We consider that this sharing risk function depends on every organization's CTI sharing strategy because others' information brings new insights for an organization, while sharing own information may have adversarial impact. Hence, the characteristics of sharing based risk probability function $q(L)$ has following properties. For an organization i, $q(L)$ increases with increase in l_i, however it decreases when l_j increases for any $j \neq i$. Thus, $\frac{\partial q}{\partial l_i} > 0$ and $\frac{\partial^2 q}{\partial l_i^2} < 0$, but for any $j \neq i$, $\frac{\partial q}{\partial l_j} < 0$ and $\frac{\partial^2 q}{\partial l_j^2} > 0$. Although this risk probability infers that the more information an organization shares, the probability of cyber incident increases, it is also noted that this risk goes down if other organizations also collaboratively exchange their threat knowledge.

Although organizations are vulnerable to direct attacks from the adversaries depending on their self-defense investment and information sharing, there exists the possibility of indirect risks because of other organizations in the same network. The probability of direct attack to an organization i can be represented as: $\mathbf{P}_{dir}^i = 1 - (1 - p(s_i))(1 - q(L))$, where the second term defines the probability of not getting direct attack from investing s_i amount in self-defense and sharing $L = \{l_i, l_{-i}\}$ amount of threat information. Considering the dissemination of threat information is perfect in nature, the probability of indirect risk ($\mathbf{P}_{ind}^i$) for an organization i depends on its own self-defense investment as well as sharing strategy of every other firm. This indirect risk can be interpreted as, $\mathbf{P}_{ind}^i(s_{-i}, L, n) = 1 - \prod_{j \neq i}^n (1 - p(s_j))(1 - q(L))$, where, s_{-i} is the vector of self-investment values of all organizations except i. Hence, the total risk probability of organization i can be expressed as combination of both direct and indirect risk, $\mathbf{P}_i = \mathbf{P}_{dir}^i + (1 - \mathbf{P}_{dir}^i)\mathbf{P}_{ind}^i$.

$$\mathbf{P}_i(S, L, n) = 1 - \prod_{k=1}^n (1 - p(s_k))(1 - q(L)) \tag{1}$$

We can notice that with increase in information sharing (l_i), $q(L)$ increases and the second component of Eq. 1 decreases, thereby $\mathbf{P}_i(S, L, n)$ increases. Whereas, if the self-defense investment increases, $p(s_i)$ decreases and hence, the overall risk probability decreases. This characteristics is expected for any candidate probability function that is used to model cyber risk. As we mentioned earlier that cyber risks may not be completely eliminated, organizations also prefer to transfer their risk to insurance vendors. We consider a coverage function $G(X) : \mathbb{R} \to \mathbb{R}$ that maps the loss X to an organization due to cyber attack event to the appropriate coverage level to offer. The firm pays a premium of amount $M(\alpha)$ periodically depending on the coverage level offered by the insurer. For simplicity, we consider a linear function for the coverage: $G(X) = \alpha X$, where $\alpha \in [0, 1]$ is the coverage level for the organizations and we denote the insurance policy as $\{\alpha, M\}$. In the next section, we formulate the individual objectives for the three group of players.

In this paper, the interactions of the players are considered to be sequential and the state of organizations' security can be either *compromised* or *protected* depending on the attacker's strategy. As simultaneous interactions rarely occur in reality, periodic strategic [21] interaction is focused in this paper. We initiate the game from the organization's side, where it decides the amount of self-defense investment to make, then attacker plays its strategy of choosing the attack rate to maximize its impact, and then the insurer plays its strategy of coverage by assessing organizations' security state. While attackers look for periodic optimal attack strategies, the organizations seek to find the right self-defense investment to make so that net payoff can be maximized.

4   Game Formulation

In the periodic interaction game, we assume that z is the number of times an organization is at compromised state, i.e. $(1 - z)$ times on average it is protected by its self-defense and CTI sharing. Considering the adversary attacks the organizations at a rate θ_a and organizations' defense rate is ψ, the mean value of z [21] can be:

$$\mathbb{E}(z) = \begin{cases} 1 - \frac{\psi}{2\theta_a} & \text{if } \theta_a \geq \psi \\ \frac{\theta_a}{2\psi} & \text{Otherwise} \end{cases} \qquad (2)$$

4.1   Organization's Payoff Model

Given the organization i invests s_i towards self-defense and shares l_i amount of information with CYBEX, while other organizations share l_{-i} amount, i's net payoff model involves two components. First it gets rewarded from the security investment by a factor $I(s_i)$ and from others' shared information $B(l_{-i})$. Second, the loss protection by taking cyber-insurance coverage level α. However, the costs involved are: (1) insurance premium (M), which is a function of coverage level, and (2) cost of defending at rate ψ. By combining all the components, the net payoff of organization i can be:

$$U_i(s_i, \theta_a, \{\alpha, M\}) = (1-z)I(s_i)B(l_{-i}) + z\alpha \mathbf{P}_i T - \psi C_d(s_i) - M(\alpha) \quad (3)$$

where, the first component is the estimated benefit out of self-investment and information sharing, when the organization is at protected state. The second component is the gain out of insurance coverage on the asset of value T. We can observe here that with higher coverage level, the expected gain $(z\alpha \mathbf{P}_i T)$ of an organization improves because it does not lose the total value of assets upon any cyber incident, rather a fraction of T depending on what coverage the organization has opted for. The third component represents the cost of defending, where, $C_d(.)$ is the cost of each defense which depends on the self-defense investment. Finally, the last component is the premium cost that is dependent on the coverage level undertaken.

4.2 Adversary Payoff Model

The strategy of the adversary is to suitably vary its attack rate (θ_a) to cause maximum disruption to the organization by driving it to the compromised state. Thus, the attacker's benefit lies in average number of times the organization i is at compromised state, i.e. z. However, it involves a cost C_a to perform attack. Furthermore, the payoff is considered to decline as the information exchange activity and self-defense investment of organization i are increased, which forms the third component of Eq. 4.

$$U_a(s_i, \theta_a, \{\alpha, M\}) = z - C_a\theta_a - (1-l_i)^b I(s_i) \quad (4)$$

where, $b \geq 1$ is an exponent to define the relevancy of information for the adversary.

4.3 Insurance Vendor's Payoff Model

The insurer's utility typically depends on the periodically collected premium amount $(M(\alpha))$. Whereas, the insurer must have to pay the coverage amount when an organization is affected due to cyber-breach, which constitutes the cost component of the insurance vendor and expressed in the following.

$$U_{Ins}(s_i, \theta_a, \{\alpha, M\}) = nM(\alpha) - \alpha \mathbf{P}_i zT \quad (5)$$

5 Equilibrium Analysis

Considering the three group of players interacting sequentially in the system, we study the subgame perfect equilibrium under two different cases: (1) no cyber-insurance coverage is considered, and (2) in presence of insurance. In both cases, the organization first decides its self-defense investment at the starting of the game interaction, after which the adversary observes the actions taken by the organization and finds it optimal attack rate by maximizing its objective. If the organization have considered the cyber insurance for its risk transferring process, then the insurance vendor observes the prior actions played by the organization and the adversary to decide the optimal coverage level to recommend the organization along with its revised premium for the next interaction period in the game. Now, depending on the rate of attack by adversary and defense rate of the organization, we derive the subgame perfect equilibrium of our periodic interaction game.

5.1   No Cyber-Insurance Scenario

As the periodic interaction game, consisting of three group of players, modeled in the previous scenario mimics the characteristics of a sequential game, finding subgame perfect equilibrium will give more insights on the players' stable strategy. In the first scenario, we plan to analyze the game with consideration that the organizations are skipping cyber-insurance to avoid premium cost. Hence, $\alpha = 0$ in the organization i's payoff model as presented in Eq. 3. Thus, the organization i and attacker's payoff will be simplified as:

$$U_i(s_i, \theta_a) = (1 - z)I(s_i)B(l_{-i}) - \psi C_d(s_i) \tag{6}$$

$$U_a(s_i, \theta_a) = z - C_a\theta_a - (1 - l_i)^b I(s_i) \tag{7}$$

Theorem 1. *Given the adversary's attack rate is higher than organization's rate of defense and linear investment model, $I(s_i) = ks_i$, the Subgame Perfect Nash Equilibrium (SPNE) strategy for the organization-adversary game without the insurer is (s_i^*, θ_a^*).*

$$(s_i^*, \theta_a^*) = \begin{cases} \left(\dfrac{kB(l_{-i})C_d'^{-1}\sqrt{\widehat{K}}}{\sqrt{8C_a}}, \psi \right) & \text{if } \psi(\psi^2 + 1) \geq 2C_a \\[3ex] \left(\dfrac{kB(l_{-i})C_d'^{-1}\sqrt{\psi + \widehat{K}}}{\sqrt{8C_a}}, \dfrac{\sqrt{2C_a}}{\sqrt{\psi + \widehat{K}}} \right) & \text{Otherwise} \end{cases}$$

where, $\widehat{K} = k^2(1 - l_i)^b B(l_{-i})C_d'^{-1}$.

Proof. Considering, $\theta_a \geq \psi$, the mean value of z is $1 - \frac{\psi}{2\theta_a}$ and replacing it in Eq. 6, the net payoff of organization i is $U_i(s_i, \theta_a) = \frac{\psi}{2\theta_a}ks_iB(l_{-i}) - \psi C_d(s_i) - M(\alpha)$. Since, organization plays first it would try to find its optimal s_i by maximizing Eq. 6. Thus, differentiating $U_i(s_i, \theta_a)$ w.r.t. s_i, we have, $\frac{\partial U_i}{\partial s_i} = \frac{\psi k B(l_{-i})}{2\theta_a} - \psi C_d'(s_i)$. Since cost of defense is always positive and an increasing function, it is easy to see that $\frac{\partial^2 U_i}{\partial s_i^2} < 0$. Thus, there exists an optimal self-defense investment s_i^*, which can be represented as $\frac{kB(l_{-i})C_d'^{-1}}{2\theta_a}$, where $C_d'^{-1}$ is the inverse first order differential of defense cost function. Now, the adversary's payoff becomes $1 - \frac{\psi + k^2(1-l_i)^b B(l_{-i})C_d'^{-1}}{2\theta_a} - C_a\theta_a$, which is maximized at $\theta_a^* = \max\left\{\psi, \sqrt{\frac{2C_a}{\psi+\widehat{K}}}\right\}$. Now, if $\theta_a^* = \psi$, then attacker's payoff is $U_a = 0.5 - C_a\psi - \frac{\widehat{K}}{2\psi}$, which will be maximum only when $\psi = \sqrt{\frac{2C_a}{\widehat{K}}}$. Thus, $\psi^2 \geq \frac{2C_a}{\psi+\widehat{K}}$, due to which $\theta_a^* = \psi$. In other cases, adversary's payoff will be less than the above best-response. Therefore, the subgame perfect Nash equilibrium in this case will be $\left(\frac{k^2(B(l_{-i})C_d'^{-1})^{3/2}(1-l_i)^{b/2}}{\sqrt{8C_a}}, \psi\right)$. If $\theta_a^* = \sqrt{\frac{2C_a}{\psi+\widehat{K}}}$, the adversary payoff will be $U_a = 1 - \frac{(\psi+\widehat{K})^{1.5}}{\sqrt{8C_a}} - \frac{\sqrt{8C_a}}{\sqrt{\psi+\widehat{K}}}$. This scenario happens only if $\psi < \sqrt{\frac{2C_a}{\psi+\widehat{K}}}$ and the attacker will achieve maximum at a particular value of defense rate $\psi = \frac{8}{3}C_a - \widehat{K}$. Therefore, the subgame perfect equilibrium in this case will be $\left(\frac{kB(l_{-i})C_d'^{-1}\sqrt{\psi+\widehat{K}}}{\sqrt{8C_a}}, \sqrt{\frac{2C_a}{\psi+\widehat{K}}}\right)$.

5.2   Undertaking Cyber-Insurance

In this scenario, we consider the presence of insurer in the game and organizations undertake insurance coverage to reduce risk of cyber-loss. The insurer tries to find out the optimal insurance level to offer the organizations. In order to find the subgame perfect equilibrium of the three layer game between organizations, adversary, and insurer, we use backward induction approach. The interactions among players occur in the following manner: (i) first the organizations decide their self-defense investment (s_i), (ii) by observing this action of organizations, the adversary tunes its attack rate (θ_a) for the considered stage, (iii) finally, by observing the strategies of organization's self-defense investment and the adversary's attack rate, the insurer decides what coverage level to offer the organizations. In our model, the organizations are considered to be homogeneous and share similar characteristics. The backward induction procedure is given below.

1. Organization i finds investment $s_i(\theta_a, \alpha) = \text{argmax}_{s_i} U_i(s_i, \theta_a, \{\alpha, M\})$ for any θ_a, α.
2. After replacing $s_i(\theta_a, \alpha)$ in the attacker's payoff, it determines its attack rate, $\theta_a(\alpha) = \text{argmax}_{\theta_a} U_a(s_i(\theta_a, \alpha), \theta_a, \{\alpha, M\})$, for any α.
3. Now, the insurer finds coverage $\alpha^* = \text{argmax}_\alpha U_{Ins}(s_i(\theta_a(\alpha), \alpha), \theta_a(\alpha), \{\alpha, M\})$.
4. Find the attacker's optimal attack rate $\theta_a^* = \theta_a(\alpha^*)$ and organization i's optimal investment $s_i^* = s_i(\theta_a^*, \alpha^*)$, which form the SPNE $(s_i^*, \theta_a^*, \alpha^*)$ for our proposed three layer game.

Proposition 1. *Assuming a 2-organization and an opportunistic attack scenario, where the cost of defense is linear and rate of attack is higher than defense rate, the optimal security investment function for organization 1 is the following.*

$$s_1(\theta_a, \alpha) = \sqrt{\frac{(2\theta_a - \psi)(1 - p(s_2))(1 - q(L))tvT\alpha}{\psi(2\theta_a - kl_2)}} - \epsilon \tag{8}$$

Proof. In an opportunistic attack scenario, the probability of breach as a function of self-defense investment can be modeled [2] as $p(s_i) = \frac{tv}{s+\epsilon}$, where $\epsilon \ll 1$, and $v \in (0,1)$ is organization's intrinsic vulnerability, and t is affinity factor to receive a particular type of attack. Also, when $\theta_a > \psi$, $z = 1 - \frac{\psi}{2\theta_a}$. Thus, organization 1's payoff can be rewritten as, $U_1(s_1) = \frac{\psi}{2\theta_a}ks_1l_2 - \left(1 - \frac{\psi}{2\theta_a}\right)T\alpha(1 - \prod_{i=1}^{2}(1 - p(s_i))(1 - q(L))) - \psi C_d(s_1) - M(\alpha)$. The first order differential of above expression, by replacing $C_d(s_1) = s_1$, will be $U_1'(s_1) = \frac{\psi}{2\theta_a}kB(l_{-i}) + \left(1 - \frac{\psi}{2\theta_a}\right)T\alpha p'(s_1)(1 - q(L))(1 - p(s_2)) - \psi$, where $p'(s_1) = \frac{-vt}{(s_1+\epsilon)^2}$. Thus the optimal self-defense investment function $s_1(\theta_a, \alpha)$ can be obtained by solving $U_1'(s_1) = 0$, or $\psi(1 - \frac{kl_2}{2\theta_a}) = (1 - \frac{\psi}{2\theta_a})(1 - p(s_2))(1 - q(L))\alpha\frac{vt}{(s_1+\epsilon)^2}$, which gives rise the expression presented in Eq. 8. Using similar analysis, we can also derive the optimal self-defense investment function, $s_2(\theta_a, \alpha)$ for organization 2.

Observations: From the Eq. 8, we can observe that optimal self-defense investment is a function of attacker's as well as insurer's strategy. As the attacker's attack rate increases, the self-defense investment also needs to be updated accordingly in order to reduce the impact. Similarly, as the insurance level is higher, it demands to have more self-defense investment in order to keep harden the security, which can be inferred from Eq. 8 that $s_i \propto \sqrt{\alpha}$.

Finding optimal attack rate and coverage level: Now after deriving the optimal self-defense investments for both organizations, the attacker will tune its corresponding attack rates for maximizing the impact. So, to find the optimal rate of attack, $s_i(\theta_a, \alpha), i = 1, 2$ can be replaced in attacker's payoff model before solving for θ_a^*. Considering the case of $\theta_a > \psi$, the payoff of attacker becomes, $U_a(\theta_a, \alpha) = (1 - \frac{\psi}{2\theta_a}) - C_a\theta_a - (1 - l_i)^b ks_i(\theta_a, \alpha)$. Thus, to find attack rate function $\theta_a(\alpha)$, the first order differential of U_a must be equated to zero and solve, which leads to $U_a'(\theta_a, \alpha) = \frac{\psi}{2\theta_a^2} - C_a - (1 - l_i)^b \frac{\partial s_i}{\partial \theta_a} = 0$. The optimal attack rate must satisfy the condition $\frac{\psi}{2\theta_a^2} = C_a + (1 - l_i)^b \frac{\psi - kl_2}{\sqrt{2\theta_a - \psi}(2\theta_a - kl_2)^{1.5}C_0}$, where $C_0 = \sqrt{\frac{(1-p(s_2))(1-q(L))tvT\alpha}{\psi}}$. After finding $\theta_a(\alpha)$, the insurer's payoff needs to be maximized to find optimal coverage level, i.e. $\underset{\alpha}{\operatorname{argmax}} \{2M(\alpha) - \alpha T \sum_{i=1}^{2}((1 - \frac{\psi}{2\theta_a(\alpha)}))\mathbf{P}_i(s_i((\theta_a(\alpha), \alpha)))\}$. Finding a close form expression is difficult in this situation, which therefore needs to be solved numerically and we plan to extend this in our future work.

6   Numerical Results and Discussion

In this section, we evaluate our game theoretic model numerically to verify the existence of subgame perfect Nash equilibrium that we found earlier theoretically. To conduct the experiments, we consider two connected organizations and one adversary in the system, where sharing amount from each is fixed to 0.5. It is also assumed that the information dissemination is perfect and shared CTI information are not corrupted by the adversary. Information relevancy coefficient b is fixed as 1. Since our model parameters are normalized between 0 and 1, a quadratic function is used to model the organization's cost of defense, $C_d(.)$, thus, $C'_d > 0$. The cost of attack (C_a) is kept constant throughout the experiments as 0.1. We have varied the defense rate parameter of the organization to observe the strategic differences and the impacts on overall payoffs of both the organization as well as adversary. This scenario does not involve insurer.

In Fig. 1(a), we present the strategy landscape of both players for different possible defense rates of the organization. It is observed that adversary prefers to increase its attack rate at equilibrium when it observes the rate of defense is increasing beyond a certain threshold, which is 0.2 in our simulation instance. However, below this threshold the adversary's optimal attack rate decreases because of strategy reversal as derived in Theorem 1. For the considered parameter set, the optimal equilibrium strategy of both players change at $\psi = 0.2$, because the corresponding condition $\psi(\psi^2+1) \geq 2C_a$ is satisfied. We can observe this trend in Fig. 1(b), which represents the payoff variation of both players under the same circumstances. The adversary's net payoff decreases beyond this threshold because the optimal attack rate increases which involves cost C_a for every attack. On the other hand, the organization's payoff increases slowly as its defense rate is increased. This is because the optimal investment in this phase has not changed. Before the strategy reversal point $\psi = 0.2$, the payoff of organization was decreasing due to their low investment. The take-away is that under no-insurance situation, attacker will prefer to raise its attack rate irrespective of organization's self-defense investment strategy. Thus, it motivates the organizations to consider the cyber-insurance option to improve their security standards.

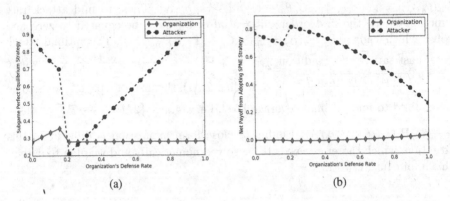

(a)   (b)

Fig. 1. (a) Subgame perfect equilibrium strategy w.r.t. organization's defense rate, (b) SPNE Payoff w.r.t. organization's defense rate

7   Conclusions and Future Research

The initiative to share cyber-threat information for addressing critical cyberse-curity issues is important but it may not completely eradicate the possibilities of uncertain losses. Cyber-insurance is an alternative to transfer such risks to insur-ers. In this paper, we model a three layer game among organizations, adversary, and insurer to study the best decision strategy of self-defense investment for organizations, optimal attack rate for adversary and optimal coverage level for insurers. Modeling the interactions as sequential form, we used backward induc-tion to solve for subgame perfect equilibrium the involved players. Numerical results show that attacker's prefer to increase their attack rate when the orga-nization's defense rate is increased. In future, we plan to extend this research to find insights when the insurer plays before of the attacker and analyze the model for deriving optimal coverage level from the insurer's perspective.

References

1. Cybersecurity information sharing act (cisa). https://www.congress.gov/bill/114th-congress/senate-bill/754
2. Huang, C.D., Behara, R.S.: Economics of information security investment in the case of concurrent heterogeneous attacks with budget constraints. Int. J. Prod. Econ. **141**(1), 255–268 (2013)
3. Anderson, R., Moore, T.: The economics of information security. Science **314**(5799), 610–613 (2006)
4. Barnum, S.: Standardizing cyber threat intelligence information with the struc-tured threat information expression (stix). MITRE Corporation 11 (2012)
5. Böhme, R., Schwartz, G., et al.: Modeling cyber-insurance: towards a unifying framework. In: WEIS (2010)
6. Burger, E.W., Goodman, M.D., Kampanakis, P., Zhu, K.A.: Taxonomy model for cyber threat intelligence information exchange technologies. In: Proceedings of the 2014 ACM Workshop on Information Sharing & Collaborative Security, pp. 51–60. ACM (2014)
7. Cavusoglu, H., Raghunathan, S., Yue, W.T.: Decision-theoretic and game-theoretic approaches to it security investment. J. Manage. Inf. Syst. **25**(2), 281–304 (2008)
8. Dandurand, L., Serrano, O.S.: Towards improved cyber security information shar-ing. In: 5th International Conference on Cyber Conflict, pp. 1–16. IEEE (2013)
9. de Fuentes, J.M., González-Manzano, L., Tapiador, J., Peris-Lopez, P.: Pracis: privacy-preserving and aggregatable cybersecurity information sharing. Comp. Secur. **69**, 127–141 (2016)
10. Garrido-Pelaz, R., González-Manzano, L., Pastrana, S.: Shall we collaborate?: a model to analyse the benefits of information sharing. In: Proceedings of the 2016 ACM on Workshop on Information Sharing and Collaborative Security, pp. 15–24. ACM (2016)
11. Gordon, L.A., Loeb, M.P., Lucyshyn, W.: Sharing information on computer sys-tems security: an economic analysis. J. Account. Public Policy **22**(6), 461–485 (2003)
12. Grossklags, J., Christin, N., Chuang, J.: Secure or insure?: a game-theoretic analy-sis of information security games. In: Proceedings of the 17th international confer-ence on World Wide Web, pp. 209–218. ACM (2008)

13. Khouzani, M.H.R., Pham, V., Cid, C.: Strategic discovery and sharing of vulnerabilities in competitive environments. In: Poovendran, R., Saad, W. (eds.) GameSec 2014. LNCS, vol. 8840, pp. 59–78. Springer, Cham (2014). doi:10.1007/978-3-319-12601-2_4

14. Pal, R., Golubchik, L.: Analyzing self-defense investments in internet security under cyber-insurance coverage. In: 2010 IEEE 30th International Conference on Distributed Computing Systems (ICDCS), pp. 339–347. IEEE (2010)

15. Pal, R., Golubchik, L., Psounis, K., Hui, P.: Will cyber-insurance improve network security? a market analysis. In: INFOCOM, 2014 Proceedings IEEE, pp. 235–243. IEEE (2014)

16. Rutkowski, A., et al.: Cybex: the cybersecurity information exchange framework (x. 1500). ACM SIGCOMM Comput. Comm. Rev. **40**(5), 59–64 (2010)

17. Tosh, D.K., Sengupta, S., Kamhoua, C.A., Kwiat, K.A., Martin, A.: An evolutionary game-theoretic framework for cyber-threat information sharing. In: IEEE International Conference on Communications, ICC, pp. 7341–7346 (2015)

18. Tosh, D.K., Sengupta, S., Mukhopadhyay, S., Kamhoua, C., Kwiat, K.: Game theoretic modeling to enforce security information sharing among firms. In: IEEE 2nd International Conference on Cyber Security and Cloud Computing (CSCloud), pp. 7–12 (2015)

19. Vakilinia, I., Sengupta, S.: A coalitional game theory approach for cybersecurity information sharing. In: Military Communications Conference, (MILCOM). IEEE (2017)

20. Vakilinia, I., Tosh, D.K., Sengupta, S.: 3-way game model for privacy-preserving cybersecurity information exchange framework. In: Military Communications Conference, (MILCOM). IEEE (2017)

21. Van Dijk, M., Juels, A., Oprea, A., Rivest, R.L.: Flipit: the game of "stealthy takeover". J. Cryptol. **26**(4), 655–713 (2013)

22. Wang, T., Kannan, K.N., Ulmer, J.R.: The association between the disclosure and the realization of information security risk factors. Inf. Syst. Res. **24**(2), 201–218 (2013)

23. Young, D., Lopez, J., Rice, M., Ramsey, B., McTasney, R.: A framework for incorporating insurance in critical infrastructure cyber risk strategies. Int. J. Crit. Infrastruct. Prot. **14**, 43–57 (2016)

Author Index

Printed in the United States
By Bookmasters